Sheet Metal Drafting Using SolidWorks®

James M. Kirkpatrick
Eastfield College

Prentice Hall
Boston Columbus Indianapolis New York San Francisco Upper Saddle River
Amsterdam Cape Town Dubai London Madrid Milan Munich Paris Montreal Toronto Delhi
Mexico City São Paulo Sydney Hong Kong Seoul Singapore Taipei Tokyo

Editorial Director: Vernon R. Anthony
Acquisitions Editor: Sara Eilert
Editorial Assistant: Doug Greive
Director of Marketing: David Gesell
Senior Marketing Manager: Harper Coles
Senior Marketing Coordinator: Alicia Wozniak
Marketing Assistant: Crystal Gonzalez
Project Manager: Maren L. Miller
Senior Managing Editor: JoEllen Gohr
Associate Managing Editor: Alexandrina Benedicto Wolf
Senior Operations Supervisor: Pat Tonneman
Operations Specialist: Deidra Skahill
Art Director: Jayne Conte
Cover Designer: Suzanne Behnke
Cover Image: Fotolia: © Eky Chan
AV Project Manager: Janet Portisch
Full-Service Project Management: Niraj Bhatt/Aptara®, Inc.
Composition: Aptara®, Inc.
Printer/Binder: Edwards Brothers
Cover Printer: Lehigh-Phoenix Color/Hagerstown
Text Font: Bookman

Credits and acknowledgments borrowed from other sources and reproduced, with permission, in this textbook appear on the appropriate page within the text. Unless otherwise stated, all artwork has been provided by the author.

Library of Congress Cataloging-in-Publication Data

Kirkpatrick, James M.
Sheet metal drafting using SolidWorks/James M. Kirkpatrick.
p. cm.
ISBN-13: 978-0-13-267583-3
ISBN-10: 0-13-267583-8
1. Sheet-metal work—Patternmaking. 2. Metal-work—Drawings. I. Title.
TS250.K57 2012
671.8′23—dc22

2011010235

10 9 8 7 6 5 4 3 2 1

Prentice Hall
is an imprint of

PEARSON

www.pearsonhighered.com

ISBN 10: 0-13-267583-8
ISBN 13: 978-0-13-267583-3

Preface

Sheet Metal Drafting Using SolidWorks® teaches basic drafting techniques and sheet metal drafting using SolidWorks. The text begins with simple solid models and proceeds to complex sheet metal assemblies with drawings, exploded views, parts lists, revisions, and other items related to the field. The outline of the book is shown in the table of contents.

The strength of this text is that it presents material in such a way that users can learn SolidWorks Sheet Metal and basic drafting techniques by actually using the program in real-world situations. A wide variety of tutorials, exercises, and review questions enables users to reinforce the information learned in each chapter. Menus in the margin show you how to locate the command you are using. Numerous illustrations in the text capture the essence of this powerful program and its importance in the manufacturing industries.

Each chapter has objectives at the beginning, and review questions including multiple-choice, true-false, matching, and general questions at the end. The text was written for the student version of SolidWorks, so online students and students who work at home can get the program free from SolidWorks at institutions that have such an agreement. Student access codes must be obtained from a class instructor.

The library of forming tools may or may not be available for some students, so two of the projects use a workaround to make a model similar to those that can be made more easily with forming tools. Those two projects are Project 6-5, a louver, and Project 7-2b, a support for a raking tool.

The text is intended for high school technology students, freshman and sophomore college technology students, and industrial users who need to learn SolidWorks Sheet Metal. Students need no prerequisite for this course, and the text is being class tested in 2011. A solutions manual with answers to review questions, semester tests, and sample syllabi is also available.

Hallmark Features

Progresses from Basic Commands to Complex Modeling Projects

- Builds confidence and basic skills before moving on to more complex assignments.
- Ensures students have mastered the fundamental features and commands of the SolidWorks program before they apply it to more complex problems.
- Guides readers step-by-step through each new SolidWorks command.
- Encourages students to learn commands and features on their own.

Provides over 90 Exercises

- Gives students the opportunity to work with a variety of real-world situations.

Includes over 500 Figures

- Helps students by allowing them to compare their work and progress with the many figures available.
- Shows many drawings to scale so students can assess and check their understanding of chapter material.

Download Instructor Resources from the Instructor Resource Center

To access supplementary materials online, instructors need to request an instructor access code. Go to www.pearsonhighered.com/irc to register for an instructor access code. Within 48 hours of registering, you will receive a confirming e-mail including an instructor access code. Once you have received your code, locate your text in the online catalog and click on the Instructor Resources button on the left side of the catalog product page. Select a supplement, and a login page will appear. Once you have logged in, you can access instructor material for all Prentice Hall textbooks. If you have any difficulties accessing the site or downloading a supplement, please contact Customer Service at http://247pearsoned.custhelp.com/.

SolidWorks® Learning License

Students may get trial access to the SolidWorks® software student version. Instructors should ask their Pearson sales rep and bookstore to package this book with ISBN-10: 0-13-275865-2 at no additional cost. To contact your Pearson Rep, please visit www.pearsonhighered.com/educator.

Acknowledgments

The author is grateful to SolidWorks® Corporation for providing such a powerful and interesting program for our use.

Contents

chapterone

1 Setting Up the First Models, Drawing, and Modifying Them Using Drawing Constructions

CHAPTER OBJECTIVES

After completing this chapter, you will be able to:

- Start a New drawing and Open an existing one.
- Describe the SolidWorks screen and begin working with it.
- Correctly make startup options.
- Make 2D sketches using drawing constructions and dimension them.
- Make simple extruded models.
- Use dynamic viewing commands.
- Correctly answer questions regarding SolidWorks commands used in this chapter.

Introduction

This book presents basic commands and settings in SolidWorks together with some sheet metal projects. It then proceeds to develop your skills in sheet metal parts, assemblies, exploded views, and parts lists. This chapter begins with starting a new drawing.

Starting a New Drawing

Step 1. Start SolidWorks.

Step 2. Click **New...** (Figure 1-1).

The **New SolidWorks Document** dialog box appears.

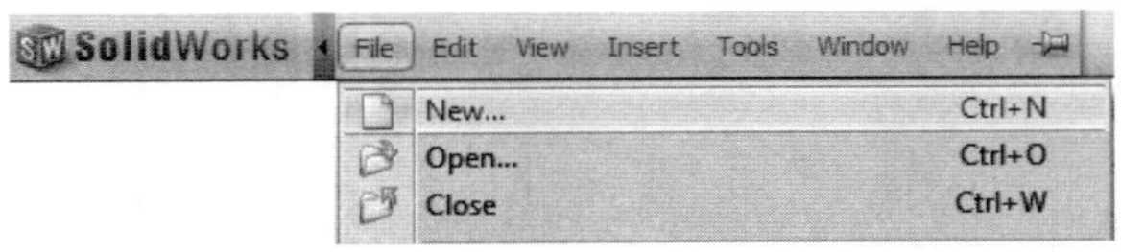

Figure 1-1
Start a New Drawing

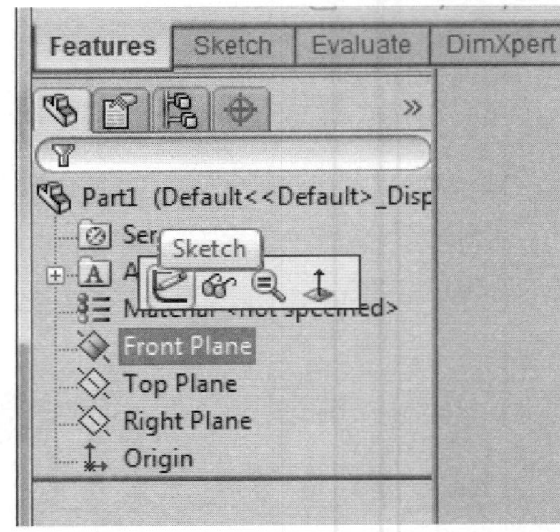

Figure 1-2
Sketch Mode

Step 3. Click **Part** and click **OK.**

The SolidWorks screen appears.

Step 4. Click the **Right Plane** and click **Sketch** so that the Sketch mode is active (Figure 1-2).

The Solidworks Screen for Sketching

Figure 1-3 shows the SolidWorks screen in the Sketch mode. Now, let's examine the SolidWorks screen.

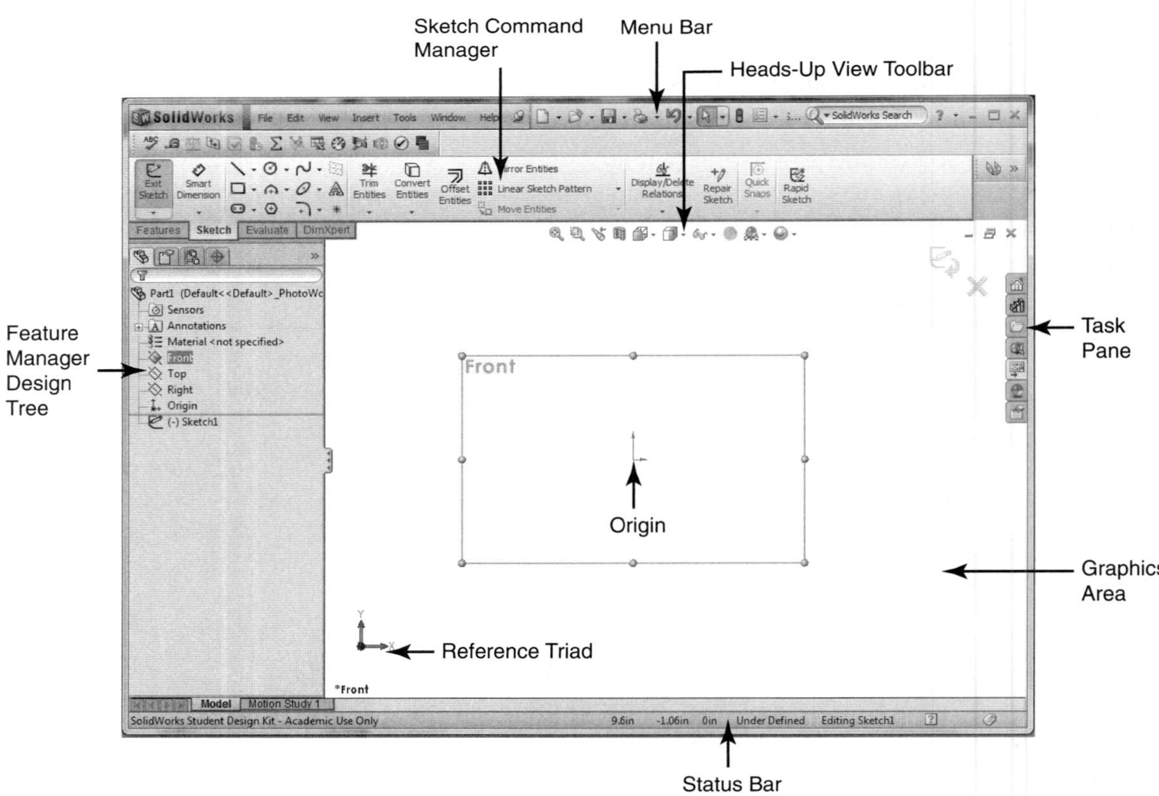

Figure 1-3
SolidWorks Screen in Sketch Mode

Feature Manager

The **Feature Manager** (Figure 1-4) is located on the upper left of the graphics area. It gives an overview of the currently open part, drawing, or assembly document. The **Feature Manager** consists of four tabs:

Feature Manager Design Tree: This tab allows you to show, hide, and edit features.

Property Manager: This tab opens when objects are selected and commands are activated. It allows you to type values in the required boxes to complete commands.

Configuration Manager: This tab allows you to make, select, and view two or more configurations of parts and assemblies.

DimXpert Manager: This tab allows you to list the tolerances that have been defined for parts tools in the DimXpert.

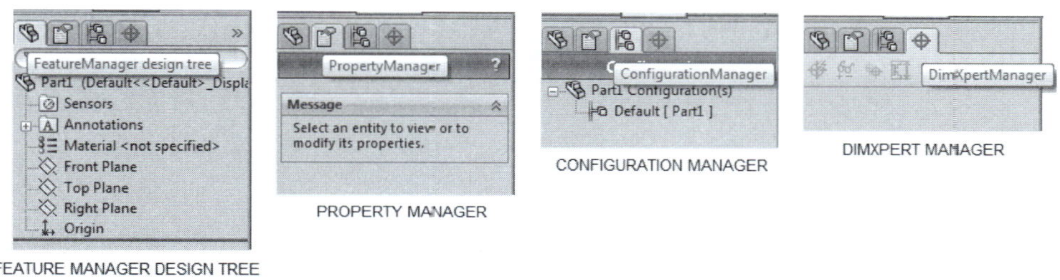

Figure 1-4
Feature Manager

Menu Bar

The menu bar contains a set of commands that are often used as shown in Figure 1-5.

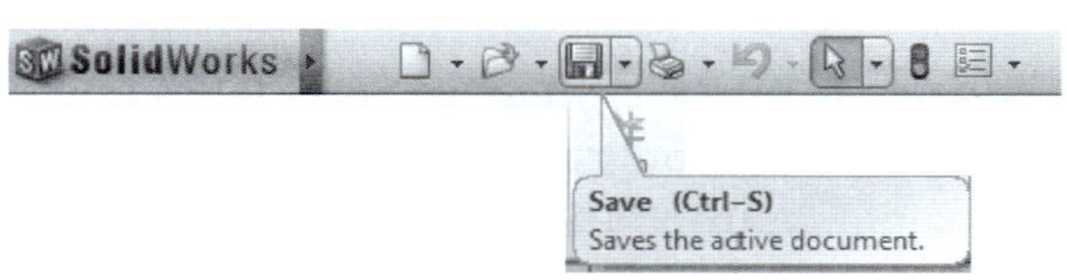

Figure 1-5
Menu Bar

Menu Bar Pull-Down Menus

These menus are displayed when you move your cursor over the SolidWorks logo. They contain commands and settings that are used in all modes: parts, drawings, and assemblies. These menus are shown in Figure 1-6.

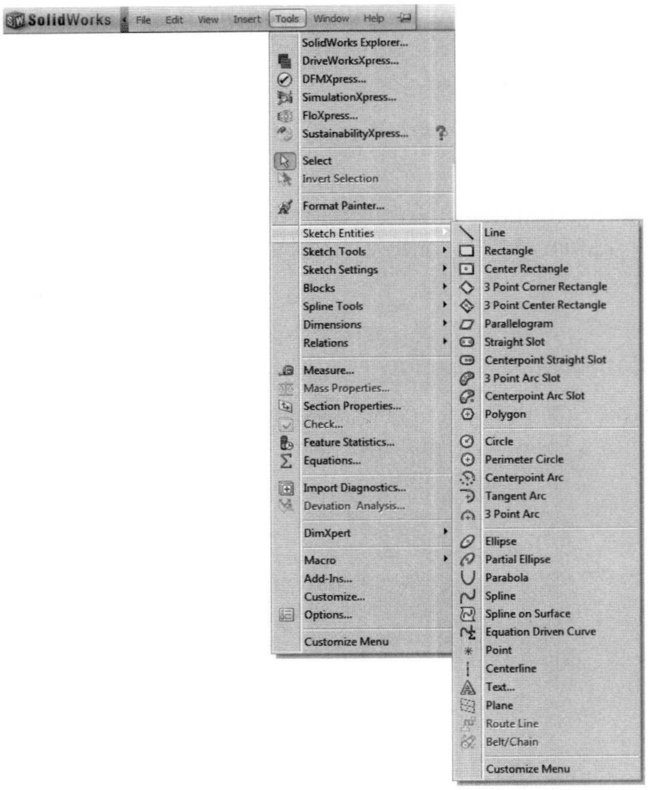

Figure 1-6
Menu Bar Menus with the Tools Menu Displayed

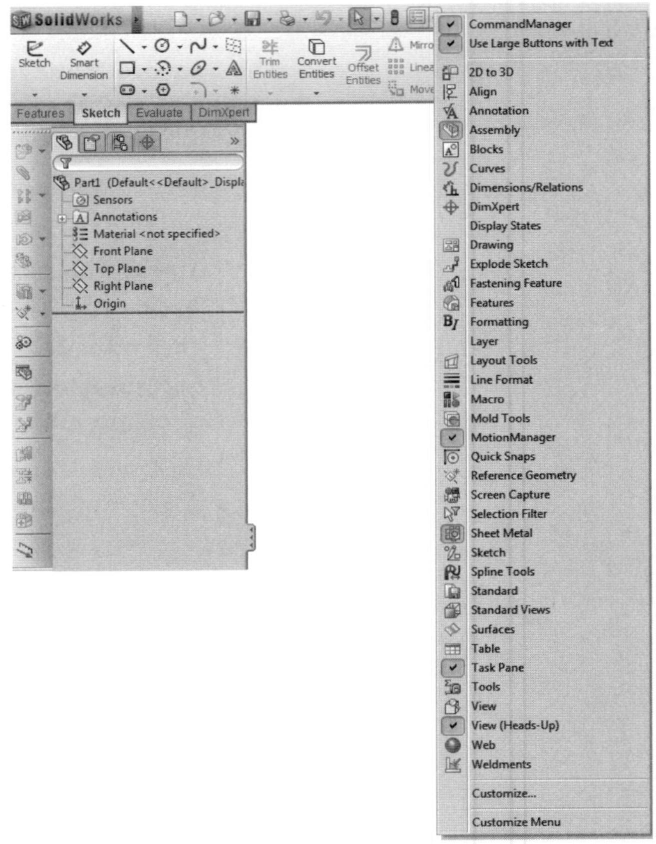

Figure 1-7
Right-Click: Any Toolbar and Click: Command Manager

Sketch Command Manager

The **Sketch Command Manager** gives you another way to access commands other than by using the **Sketch** toolbar. To turn on the **Sketch Command Manager,** right-click any toolbar and then click on **Command Manager** from the list of toolbars as shown in Figure 1-7.

Heads-Up View Toolbar

The **Heads-Up View** toolbar (Figure 1-8) allows you to view models in any of the standard orthographic views including isometric. It also allows you to pan, zoom, rotate, and change the way models are displayed in terms of their transparency.

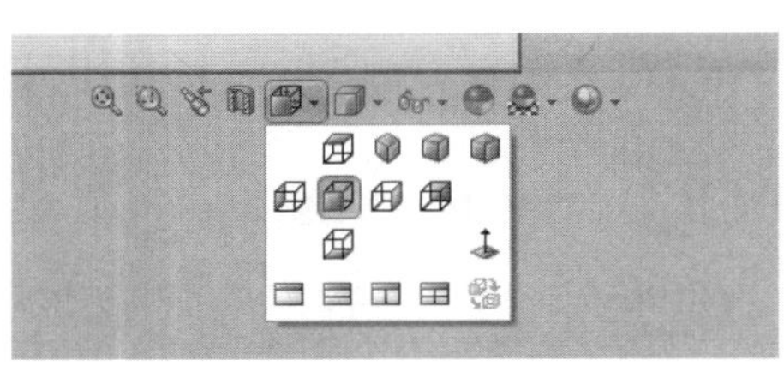

Figure 1-8
Heads-Up View Toolbar

Task Pane

When you start a SolidWorks session, the **Task** pane is displayed. The **Task** pane has the following icons:

SolidWorks Resources: The **SolidWorks Resources** icon displays the following options: **Getting Started, Community, Online Resources,** and **Tip of the Day.**

Design Library: The **Design Library** icon has parts, assemblies, and other items. You will find the Design Library to be a very useful tool in your work.

File Explorer: The **File Explorer** icon gives you the same features as Windows Explorer.

SolidWorks Search: The **SolidWorks Search** icon is a search tool that allows you to find items that are difficult to find. It is similar to all the other search tools you have used on your computer.

View Palette: The **View Palette** icon allows you to insert views of parts by clicking and dragging them into the drawing. It also allows you to browse to find the part and its views.

Appearances/Scenes: The **Appearances/Scenes** icon enables you to display models in a photorealistic manner using the library of **Appearances and Scenes**.

Custom Properties: The **Custom Properties** icon allows you to insert custom and configuration properties directly into SolidWorks files.

Graphics Area

The graphics area is where you make parts, assemblies, and drawings. It is the area where all the action takes place. Here you will also assign materials to parts, make scenes, and animate your parts and assemblies.

Status Bar

The status bar tells you the status of parts, drawings, and assemblies. It also tells you where your cursor is and whether the part has enough dimensions and relations to be fully defined. "Fully defined" means that the sketch is ready to be made into a solid model. Sometimes your sketch can be made into a solid even though it is not fully defined.

Origin

The origin is the location of 0,0,0 for parts, drawings, and assemblies. This is a very important reference point that you will use in almost every sketch, part, and assembly. You will need to consider the origin when planning all of your SolidWorks documents. The model origin is blue, and the sketch origin is red.

Reference Triad

The **Reference Triad** helps you view models and appears in the graphics area. It is for reference only.

Now that you have examined the screen in Sketch mode, you are ready to set up your first part and make your first model.

Project 1-1: **Make a Solid Model of Figure 1-9**

If you have exited SolidWorks, open it again and start a new part as you did at the beginning of this chapter.

Step 1. Click **Options** on the menu bar, select the **Document Properties** tab, and then click **Units** in the list to the left.

Step 2. Make sure the **Units** are correctly set as shown in Figure 1-10 [notice that **IPS** (inch, pound, second) is checked], then click **OK.**

MIKE KIRKPATRICK

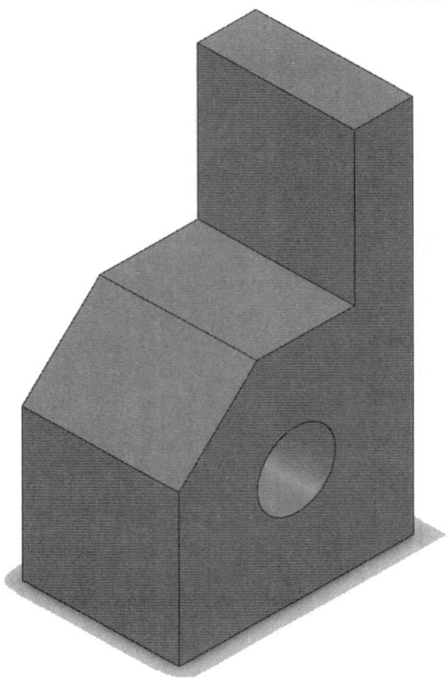

Figure 1-9
Project 1-1

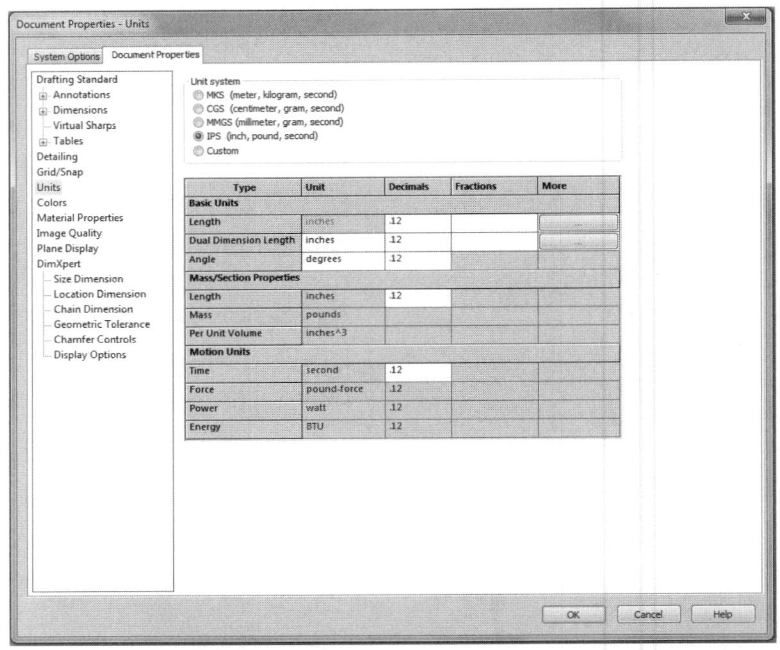

Figure 1-10
Set Units to Inch, Pound, Second

Step 3. Use the **Circle** command to draw a circle at the origin.

1. Click the **Circle** icon in the ribbon at the top of the screen.
2. Click the red origin in the center of the graphics area.
3. Move your mouse so the readout shows about **.25** inch and click.

Step 4. Use the **Smart Dimension** command to dimension the circle as shown in Figure 1-11.

1. Click the **Smart Dimension** icon in the ribbon at the top of the screen.
2. Click any point on the circle and click a point outside the circle.

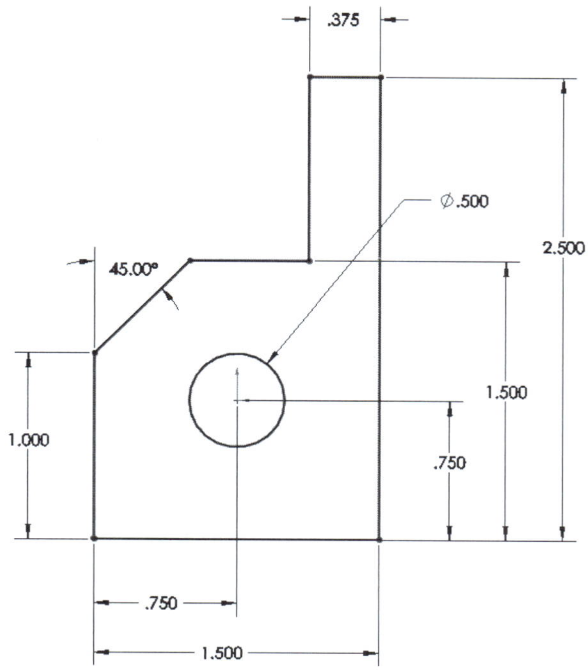

Figure 1-11
The Complete Sketch on the Right Plane

3 When the **Modify text** box appears, type **.5** and press **<Enter>.**

4 Press the **<Shift>** and **<Z>** keys to zoom in on the drawing, or press the **<Z>** key to zoom out.

Step 5. Use the **Line** command to draw lines in the approximate location shown in Figure 1-11. (The exact location and size are not important because you will dimension the lines in Step 6.)

1 Click the **Line** icon in the ribbon at the top of the screen.

2 Click the left end of the horizontal line at the bottom of the figure, move your mouse directly to the right, and click the right end of the line.

3 Continue drawing lines until your drawing looks approximately like Figure 1-11.

Step 6. Use the **Smart Dimension** command to dimension the lines as shown in Figure 1-11.

1 Click the **Smart Dimension** icon in the ribbon at the top of the screen.

2 Click the lower left corner of the figure, then click the origin and move the dimension (0.750) to the location shown.

3 When the **Modify text** box appears, type **.75** and press **<Enter>.**

4 To dimension the bottom line, click any point on the line and move the dimension 1.500 to the location shown.

5 Continue dimensioning lines until your drawing looks like Figure 1-11. (To dimension the angle, click the vertical line, then click the angled line.)

6 Press the **<Shift>** and **<Z>** keys to zoom in on the drawing, or press the **<Z>** key to zoom out.

The sketch is now complete.

Step 7. Click **Exit Sketch,** then **Features,** and then **Extrude Boss/Base.**

Step 8. Make the settings as shown in Figure 1-12 (be sure that 1.00 is the setting for D1), then click **OK** (the green check mark at the top of the **Boss-Extrude** box).

The model is now complete.

Step 9. Select the **Isometric** view from the **Heads-Up View** toolbar as shown in Figure 1-13.

MIKE KIRKPATRICK

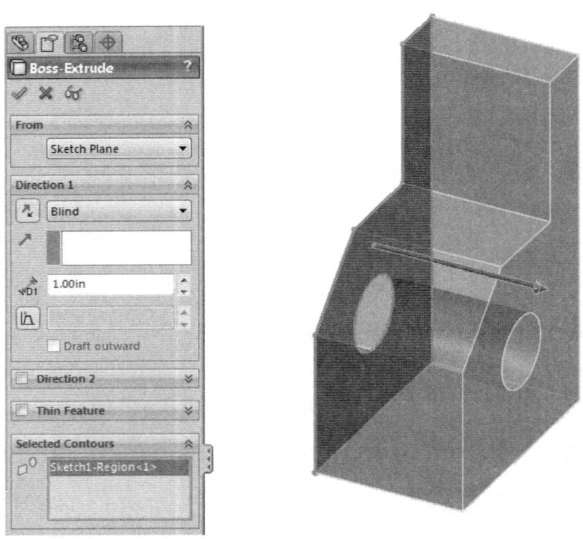

Figure 1-12
The Extruded Model

Figure 1-13
Project 1-1 Complete

Step 10. Put your name on the drawing (Figure 1-14) as described next:

1 Click **Print,** then **Header/Footer...,** then **Custom Header...,** and then **Font....**

2 Click **Century Gothic,** choose **Bold** for the **Font Style** and **12** for the **Points,** and then click **OK.**

3 Click the **Right** box, type **YOUR NAME** in all caps, and click **OK.**

Step 11. Save your model with the file name **Project 1-1(your initials)** in at least two places.

Step 12. Print the drawing and turn it in.

Figure 1-14
Make Your Custom Header

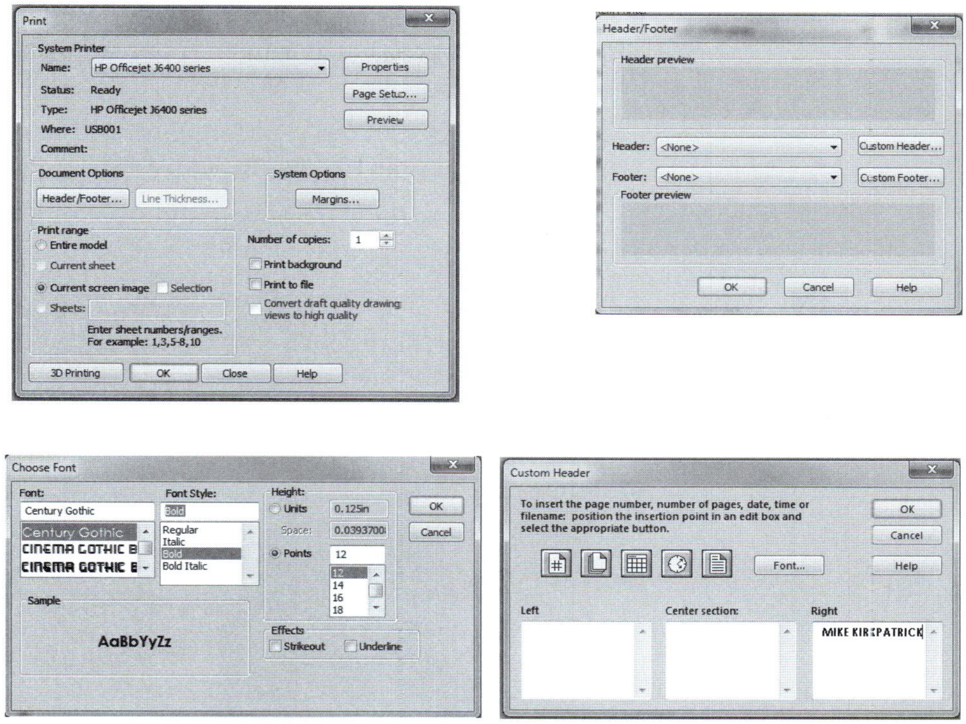

Project 1-2: **Open Project 1-1 and Modify It**

If you have exited SolidWorks, start it again and open *Project 1-1(your initials)* as shown in Figure 1-15.

Step 1. Click **Options** on the menu bar, select the **Document Properties** tab, and click **Units** in the list to the left.

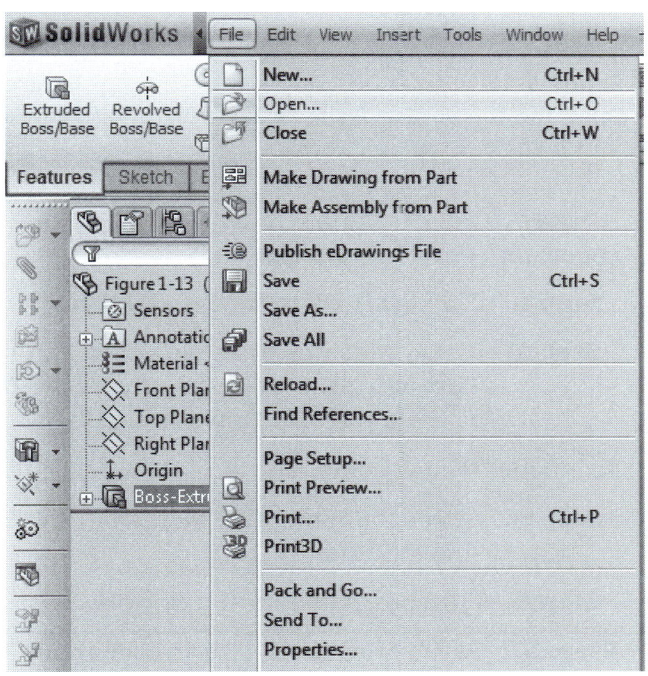

Figure 1-15
Open an Existing Drawing

Step 2. Make sure the **Units** are correctly set as shown in Figure 1-10 [notice that **IPS** (inch, pound, second) is checked], then click **OK.**

Step 3. Click the top vertical plane and select **Sketch** from the popup menu as shown in Figure 1-16.

Step 4. Click the **Circle** icon and use the **Circle** command to draw a circle on your sketch plane as shown in Figure 1-17.

Step 5. Use the **Smart Dimension** command to dimension the circle as shown in Figure 1-17.

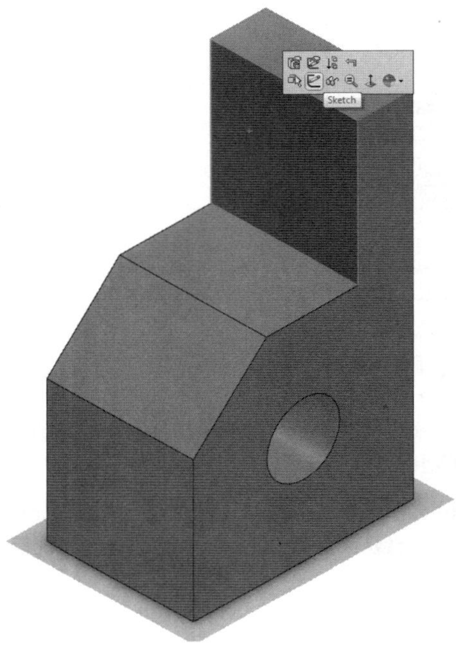

Figure 1-16
Select the Top Vertical Plane as the Sketch Plane

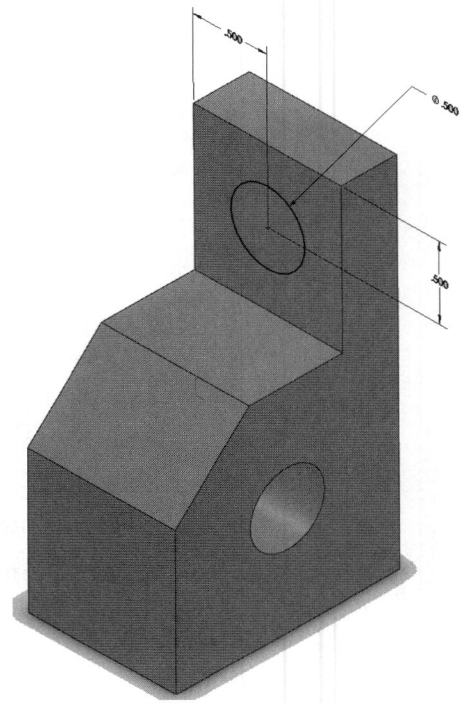

Figure 1-17
Draw a Circle and Dimension It

The sketch is now complete.

Step 6. Click **Exit Sketch,** then **Features,** and then **Extrude Cut.**

Step 7. Make the settings as shown in Figure 1-18, making sure that **Through All** is selected, then click **OK** (the green check mark at the top of the **Extrude Cut** box).

The model is now complete.

Step 8. Select the **Isometric** view from the **Heads-Up View** toolbar as shown in Figure 1-19.

Step 9. Put your name on the drawing.

Step 10. Save your model with the file name **Project 1-2(your initials)** in at least two places.

Step 11. Print the drawing and turn it in.

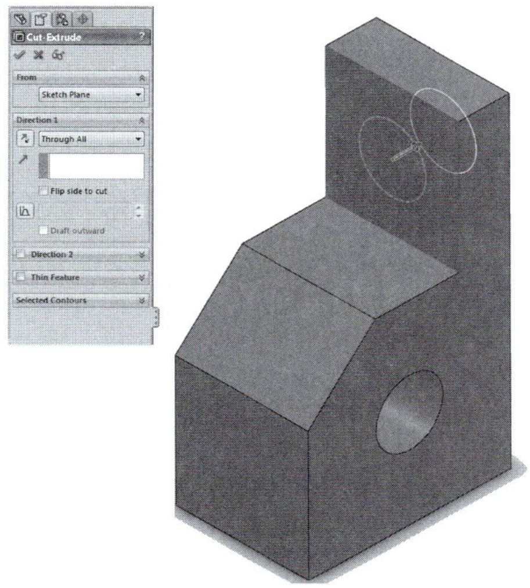

Figure 1-18
Make the Settings for the Extrude Cut

Figure 1-19
Project 1-2 Complete

Exercises

EXERCISE 1-1 Make a model from the sketch shown in Figure 1-20.

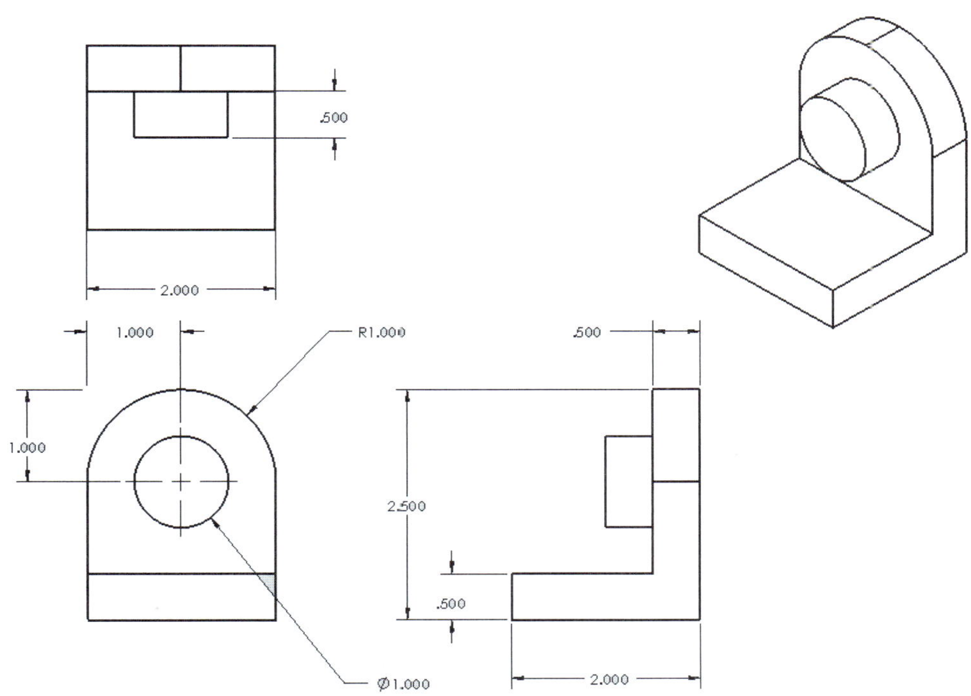

Figure 1-20
Exercise 1-1

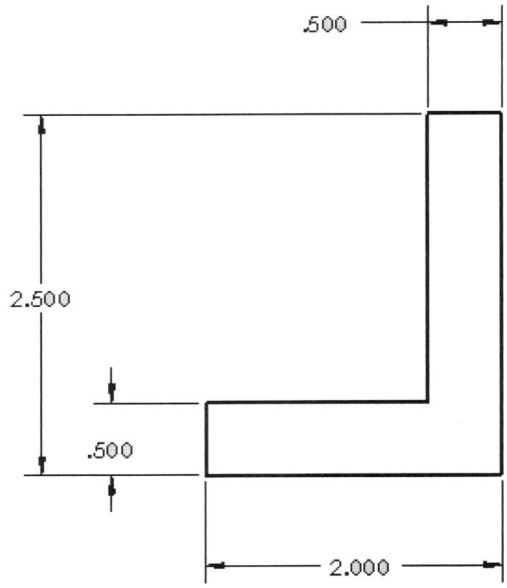

Figure 1-21
First Sketch for Exercise 1-1

Step 1. Draw and dimension the sketch on the front plane shown in Figure 1-21 and extrude it 2.000 inches.

Step 2. Click the front vertical 1.500 inches face and click **Sketch.**

Step 3. Draw a 1 inch diameter circle on that face, dimension it, and extrude it .500 inch.

Step 4. Click the **Fillet** command in the **Features** mode, set the value to **1 inch,** and click the two uppermost .500 inch horizontal lines to form the fillets on each side.

Step 5. Select the **Isometric** view from the **Heads-Up View** toolbar.

Step 6. Put your name on the drawing.

Step 7. Save your model with the file name **Exercise 1-1(your initials)** in at least two places.

Step 8. Print the drawing on an A-Size sheet and turn it in.

EXERCISE 1-2 **Make a model from the sketch shown in Figure 1-22.**

Step 1. Draw and dimension on the front plane the sketch shown in Figure 1-22 and extrude it .500 inch.

Step 2. Select the **Isometric** view from the **Heads-Up View** toolbar.

Step 3. Put your name on the drawing.

Step 4. Save your model with the file name **Exercise 1-2(your initials)** in at least two places.

Step 5. Print the drawing on an A-Size sheet and turn it in.

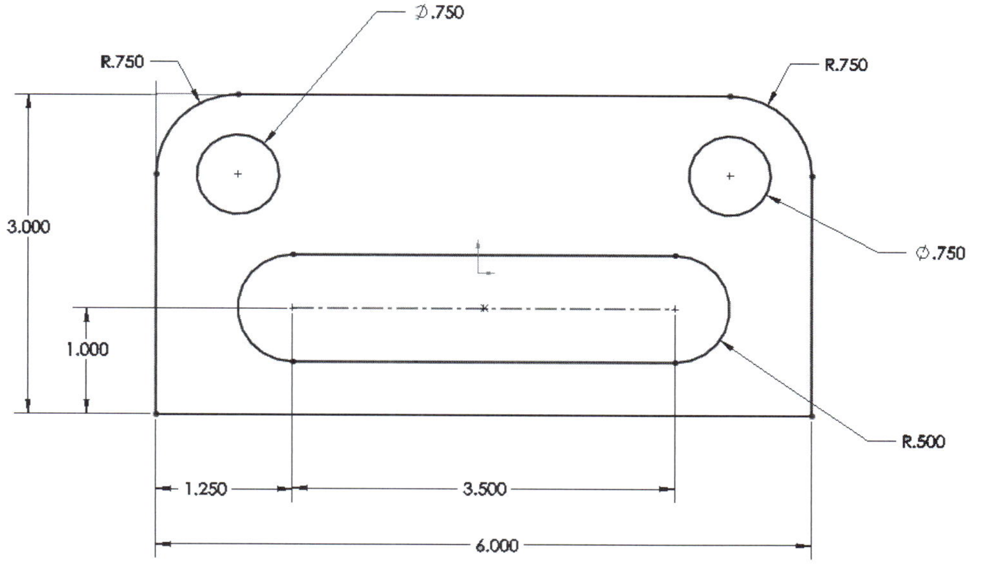

Figure 1-22
Exercise 1-2

EXERCISE 1-3 **Make a model from the sketch shown in Figure 1-23.**

Step 1. Draw and dimension the sketch on the top plane shown in Figure 1-24 and extrude it .375 inch.

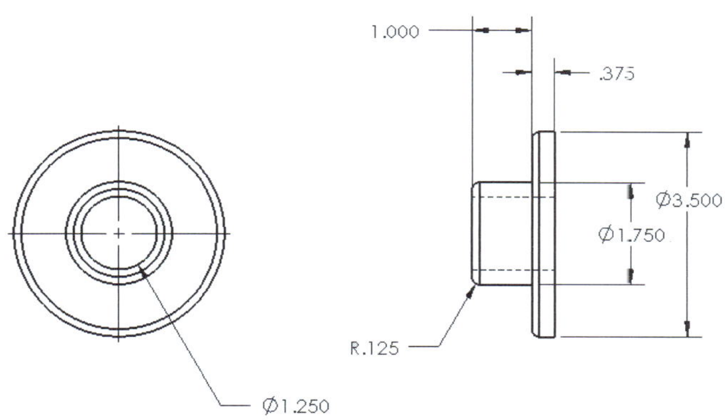

Figure 1-23
Exercise 1-3

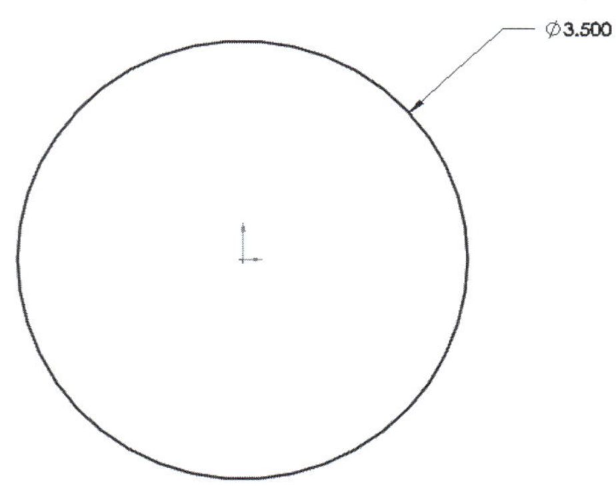

Figure 1-24
First Sketch for Exercise 1-3

Step 2. Click the top face and click **Sketch.**

Step 3. Draw a 1.750 inches diameter circle on that face, dimension it, and extrude it 1.000 inch.

Step 4. Click the top face of the 1.750 extruded circle, click **Sketch,** and draw a 1.250 diameter circle and dimension it.

Step 5. Click **Exit Sketch,** then **Features,** then **Extrude Cut.**

Step 6. Be sure **Through All** is selected, then click **OK** (the green check mark at the top of the **Extrude Cut** box).

Step 7. Click the **Fillet** command in the **Features** mode, set the value to **.125 inch,** and click the top edges of two extruded circles to make the fillets.

Step 8. Select the **Isometric** view from the **Heads-Up View** toolbar.

Step 9. Put your name on the drawing.

Step 10. Save your model with the file name **Exercise 1-3(your initials)** in at least two places.

Step 11. Print the drawing on an A-Size sheet and turn it in.

EXERCISE 1-4 Make a model from the sketch shown in Figure 1-25.

Step 1. Draw and dimension the sketch (of the front view) on the front plane shown in Figure 1-25 and extrude it 2.000 inches.

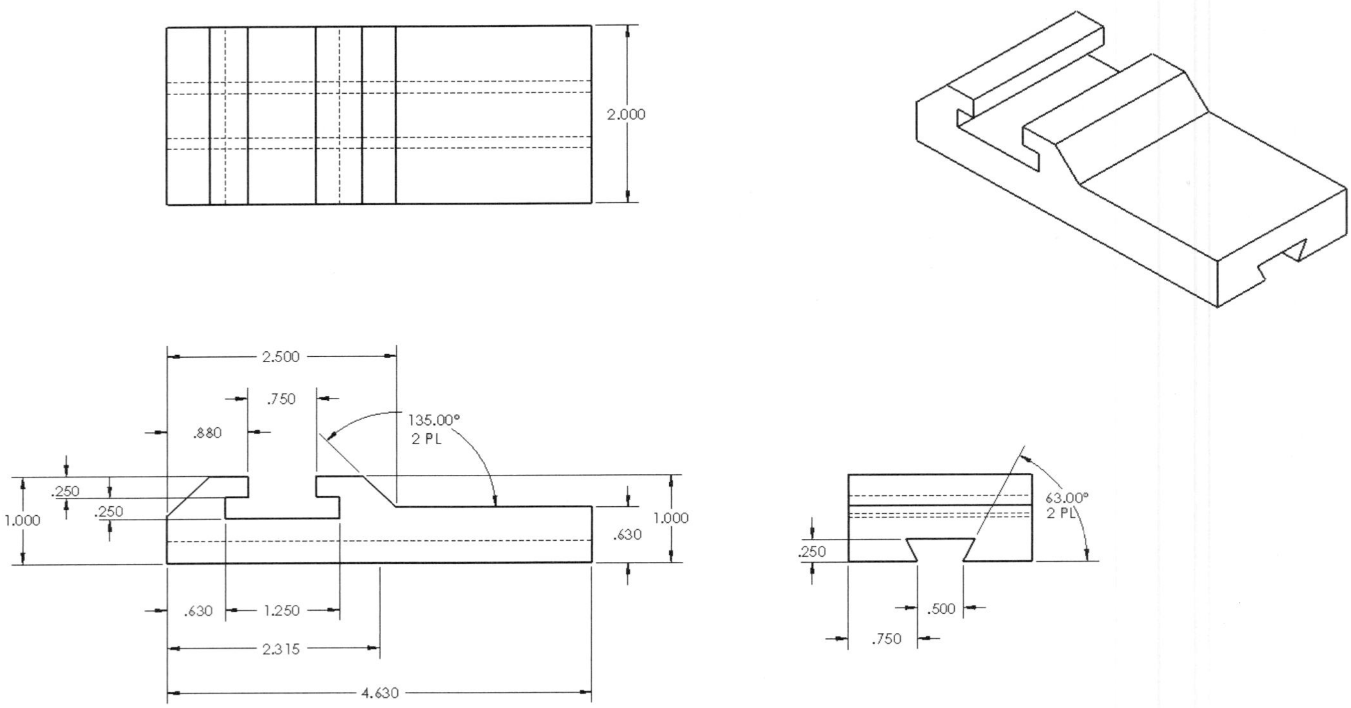

Figure 1-25
Exercise 1-4

Step 2. Click the right vertical .630 inch face and click **Sketch.**

Step 3. Draw and dimension the sketch of the cutout in the right side view. Be sure to add a .500 line at the bottom of the cutout so you can do an extruded cut.

Step 4. Click **Exit Sketch,** then **Features,** then **Extrude Cut.**

Step 5. Be sure **Through All** is selected, then click **OK** (the green check mark at the top of the **Extrude Cut** box).

Step 6. Select the **Isometric** view from the **Heads-Up View** toolbar.

Step 7. Put your name on the drawing.

Step 8. Save your model with the file name **Exercise 1-4(your initials)** in at least two places.

Step 9. Print the drawing on an A-Size sheet and turn it in.

1 one

Chapter Summary

This chapter contains the information necessary to build simple solid models. In addition, you have learned to put your name in the header and print a drawing. Now you have the skills and information necessary to produce simple solid models of sheet metal parts and print them.

Review Questions

Circle the most correct answer.

1. On which menu is the **Base Flange** command located?

   a. **Edit** c. **Insert**
   b. **View** d. **Tools**

2. The **Hole Wizard** allows you to make holes of which type from the **Type** tab?

   a. Through c. Countersink
   b. Counterbore d. All of these

3. Which of the following allows you to make a hole without using the **Hole Wizard?**

   a. **Extrude Boss/Base** c. **Extrude Cut**
   b. **Revolve Boss/Base** d. **Chamfer**

4. Which of these commands allows you to make a rectangular hole with a full radius on both ends?

   a. **Straight Slot** c. **Hole Wizard**
   b. **Simple Hole** d. **Extrude Cut**

5. What setting is used to make dimensions appear as inches?

   a. **Dimensions** c. **Units**
   b. **Annotations** d. **Detailing**

6. The **Edit Sheet** command is used to edit the sheet format.

   a. True b. False

7. The **Base Flange** command is used to make the overall shape of a sheet metal part.

   a. True b. False

8. The **Chamfer** command cannot be used in the features mode.

   a. True b. False

9. All counterbored holes must be done with the **Hole Wizard.**

   a. True b. False

10. All countersunk holes have the same angle for the countersink.

    a. True b. False

chapter one

Matching

Place the correct letter on the line to the right of the numbered item.

1. A command that gives you a feature that can be used to mirror holes _____
2. A command that allows you to edit a sheet format _____
3. A command that makes a counterbored hole _____
4. A command that can be used to duplicate a series of holes _____
5. A command that makes a rounded surface _____

a. **Hole Wizard**

b. **Fillet**

c. **Edit Sheet Format**

d. **Plane**

e. **Linear Pattern**

General Questions

Answer the following questions.

1. In what situations would you use the **Hole Wizard?**
2. When would you want to use the **Extrude Cut** command?
3. What commands are used to import dimensions into a drawing from a model?
4. How can the **Mirror** command be used to make a pattern of many holes?
5. When can **Smart Dimension** be used in making a drawing from a part?

2 chapter two

Making Drawings with Dimensions from the First Models

CHAPTER OBJECTIVES

After completing this chapter, you will be able to:

- Open an existing drawing and make a drawing from a part.
- Select a drawing format for the drawing.
- Drag drawing views from the **Task** pane onto the drawing format.
- Insert **Model Items** (dimensions) from the part to the drawing views.
- Arrange views and space dimensions correctly.
- Add notes to the drawing using **Edit Sheet** and to the drawing format using **Edit Sheet Format.**
- Correctly answer questions regarding SolidWorks commands used in this chapter.

Project 2-1: Opening an Existing Part and Making a Drawing from the Part

Step 1. Start SolidWorks.

Step 2. Click **Open....**

The **Open SolidWorks Document** dialog box appears.

Step 3. Click **Project 1-1(your initials)** and click **Open.**

The SolidWorks screen appears.

Step 4. Click **Make Drawing from Part** (Figure 2-1).

The **Sheet Format/Size** dialog box appears.

Step 5. Click **B (ANSI) Landscape** (Figure 2-2) and click **OK.**

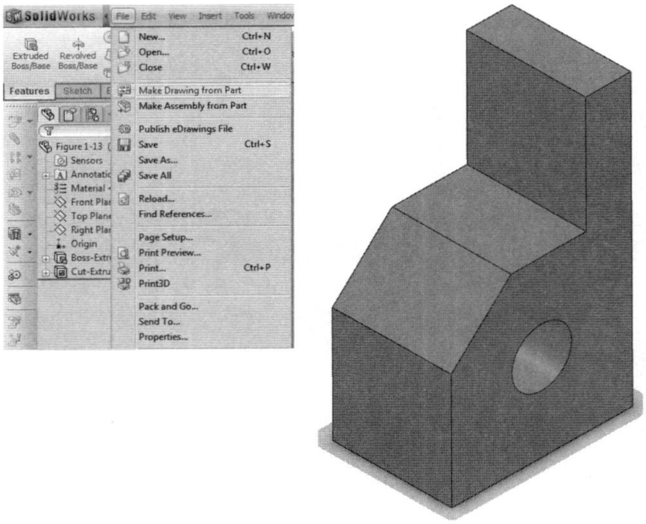

Figure 2-1
Click: Make Drawing from Part with Project 1-1 Open

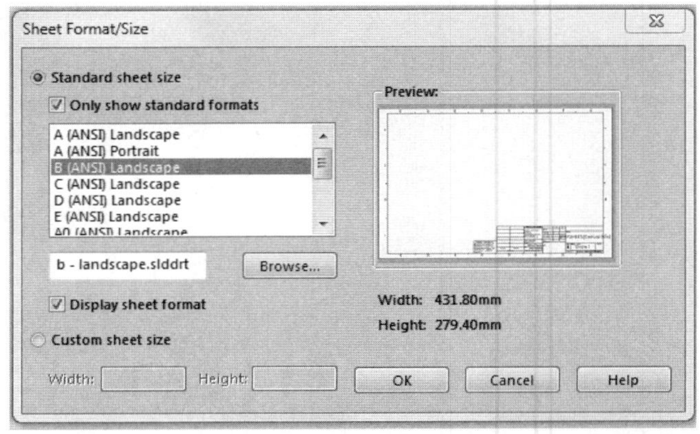

Figure 2-2
Click: B (ANSI) Landscape Sheet Format

Step 6. Right-click a point in the graphics area and click **Properties.** Then click **Third Angle,** making sure **Scale:** is **1:1** as shown in Figure 2-3, and click **OK.**

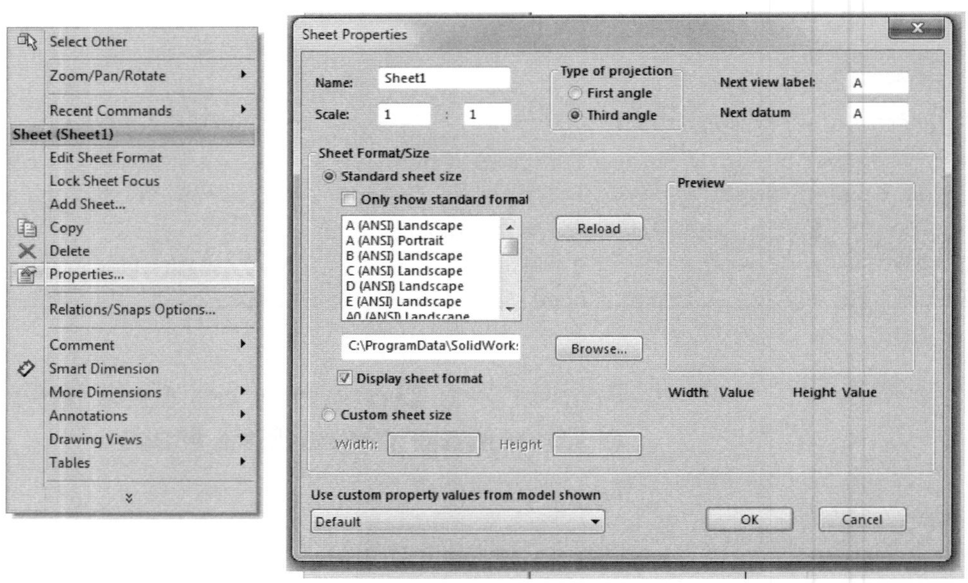

Figure 2-3
Make Sure You Are Using Third Angle Orthographic Projection and the Scale: Is 1:1

Step 7. Drag the **Front, Top, Right,** and **Isometric** drawing views from the **Task** pane as shown in Figure 2-4 and click the green check mark (better known as **OK**) in the **Drawing View Feature Manager.**

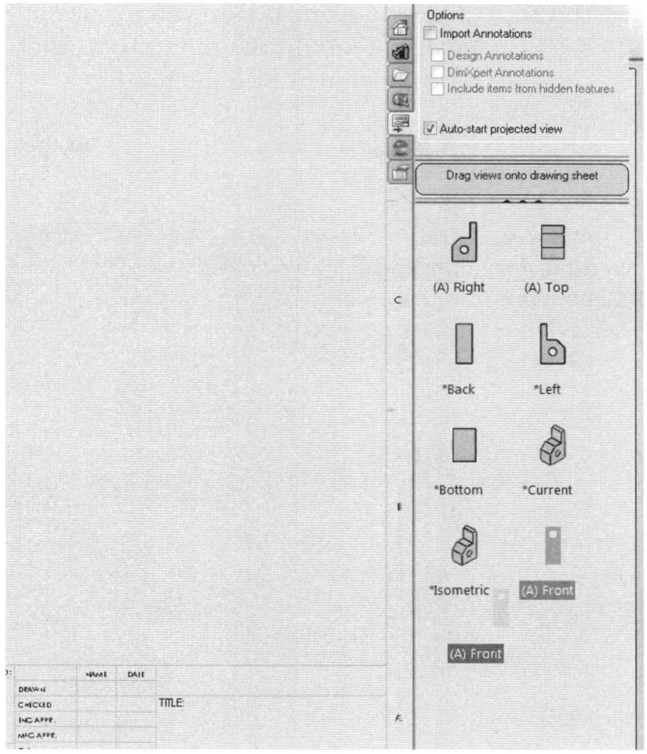

Figure 2-4
Drag Front, Top, Right and Isometric Views onto the Drawing Sheet

Step 8. Click each view and move it. You will have to click the dotted edge of each view to get the **Move** symbol, a plus with four arrows, as shown in Figure 2-5.

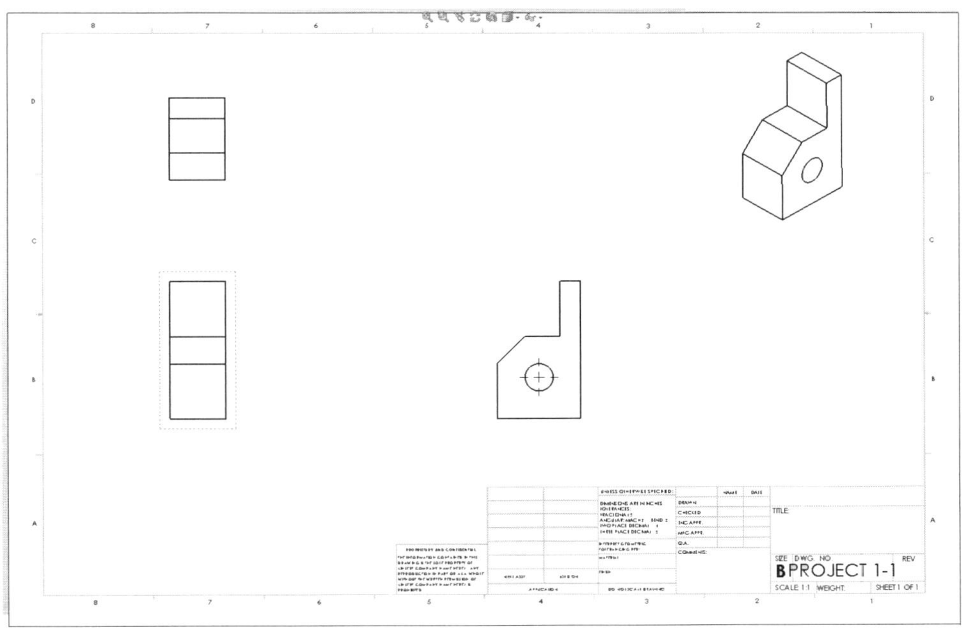

Figure 2-5
Arrange Views on the Sheet Format

Step 9. Click the **Annotation** tab (or select **Annotations** from the **Insert** menu on the menu bar), click **Model Items,** and make the settings shown in Figure 2-6.

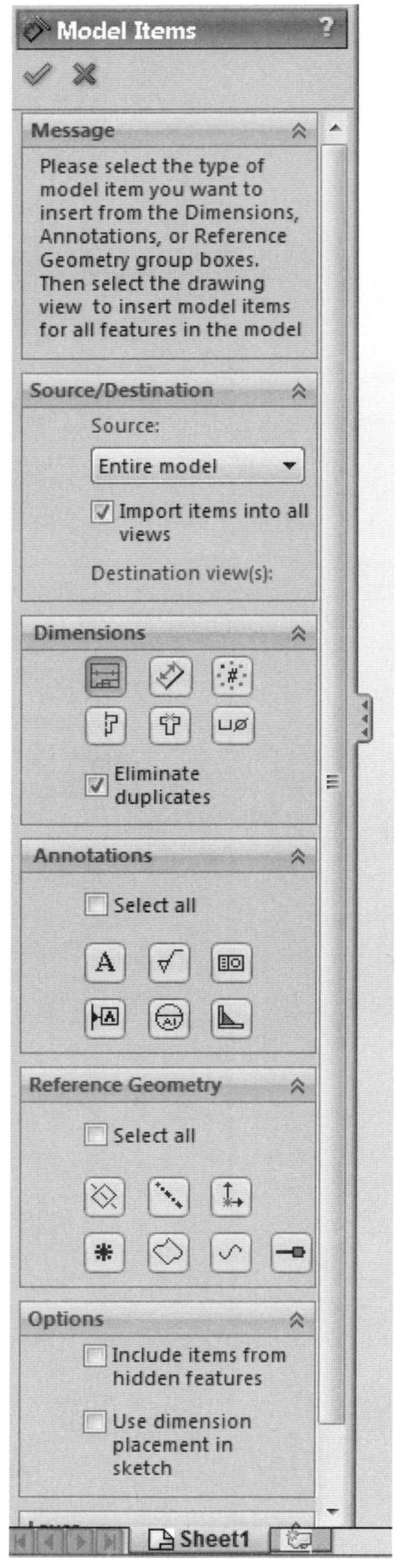

Figure 2-6
Settings for Importing Dimensions into Drawing Views

Be sure that the entire model is selected as the **Source** and the first icon under **Dimensions** is selected.

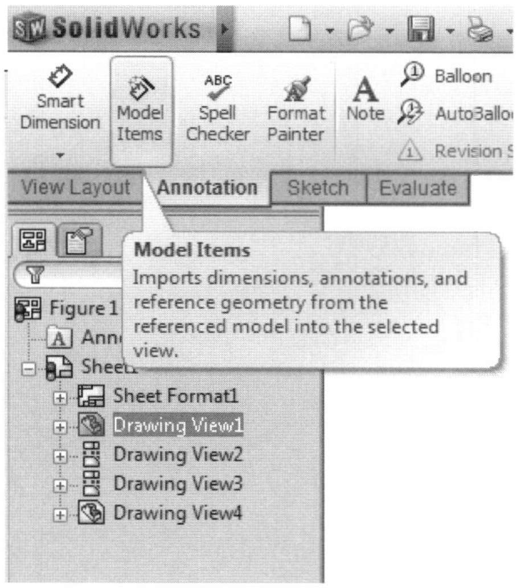

Step 10. Click a view and select **Hidden Lines Visible** as the **Display State** so that hidden lines appear in the **Front, Top,** and **Right** views.

Step 11. Arrange the views as shown in Figure 2-7. Once again, click each view and move it. (You will have to click the dotted edge of each view to get the **Move** symbol.)

Step 12. Arrange the dimensions as shown in Figure 2-8. **NOTICE THE SPACING OF ROWS OF DIMENSIONS AND THEIR PLACEMENT.** You may have to delete some dimensions that are not on the correct view and use **Smart Dimension** to put them on the correct view. You will find **Smart Dimension** on the **Tools** menu on the menu bar.

Step 13. Right-click any point on the drawing sheet. The popup menu appears. Click **Edit Sheet Format** as shown in Figure 2-9.

Step 14. Use the **Note** command to type your initials, today's date, your **School Name,** the **Project Number,** and the **Drawing Number** as shown in Figure 2-10. You will have to change the font size to **6** points in the **NAME** and **DATE** boxes, **12** points in the **Drawing Number** box, and **16** points in the **School Name** box and **Project Number** box. Change the font to **Arial.** You will also have to click on some of the items and move them into the correct space.

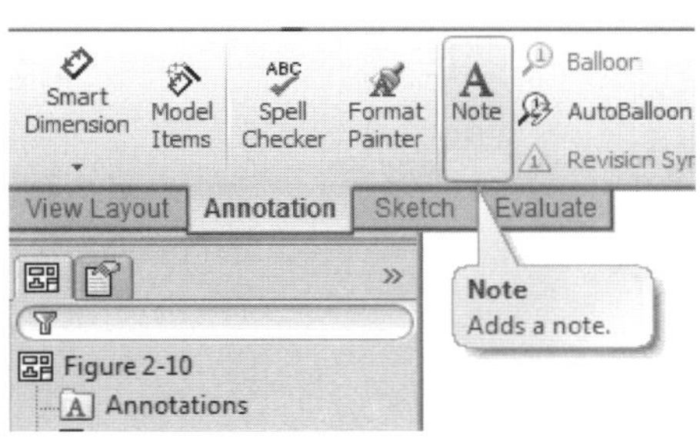

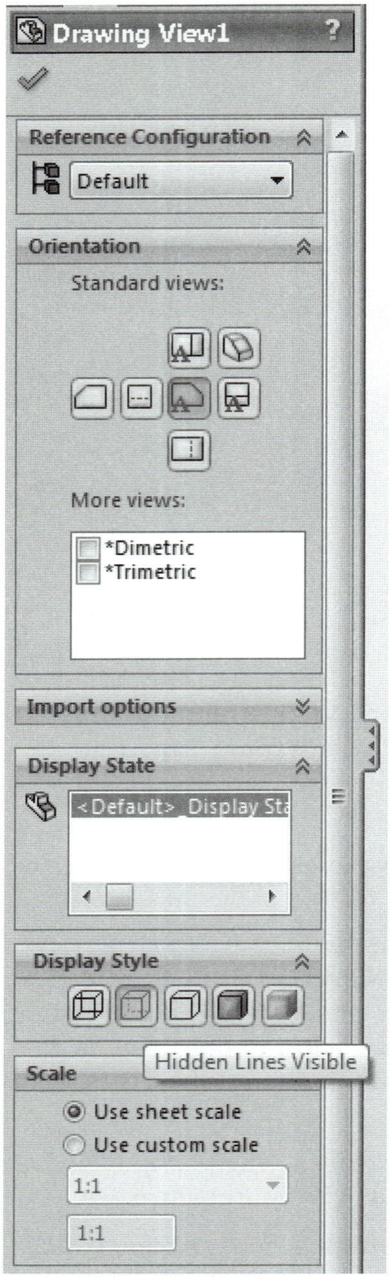

Figure 2-7
Select Hidden Lines Visible for Front, Top and Right Views

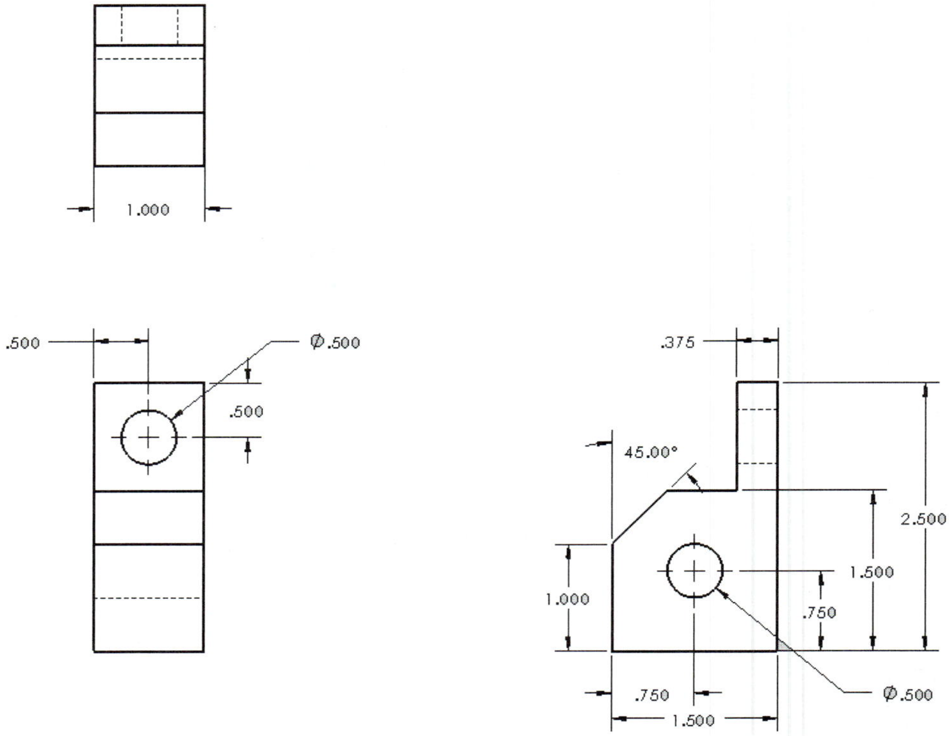

Figure 2-8
Views and Dimensions Correctly Arranged

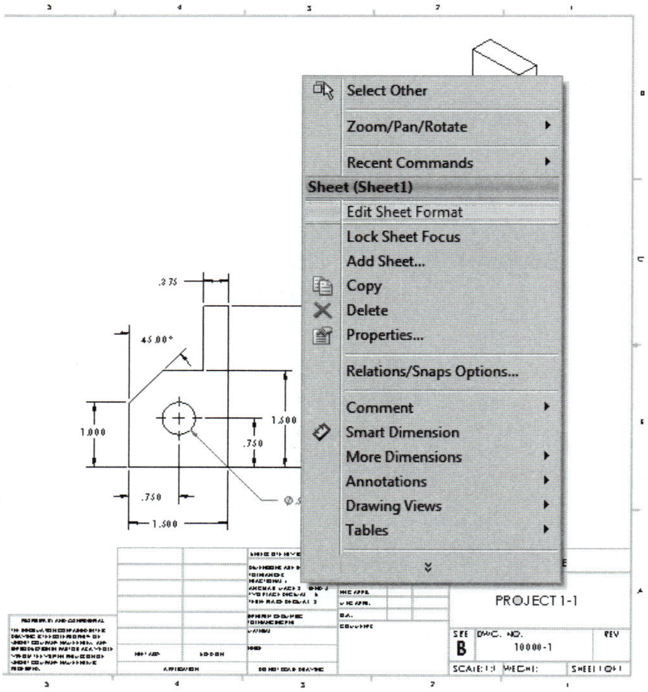

Figure 2-9
Click: Edit Sheet Format

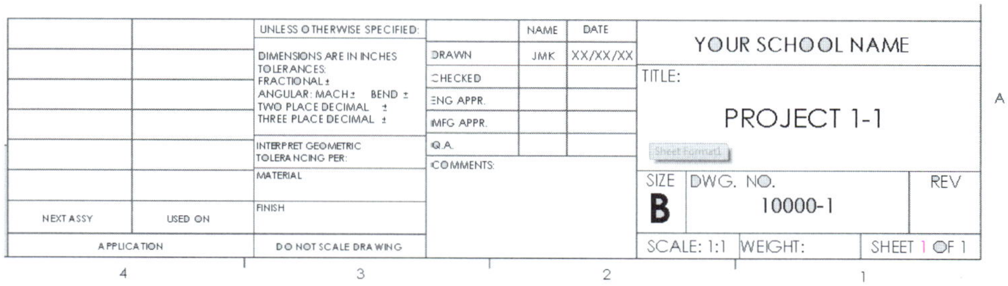

Figure 2-10
Edit Sheet Format

Step 15. Right-click any point on the drawing sheet. The popup menu appears. Click **Edit Sheet** as shown in Figure 2-11.

Step 16. Save your drawing with the file name **Project 2-1(your initials)** in at least two places.

Step 17. Print the drawing and turn it in.

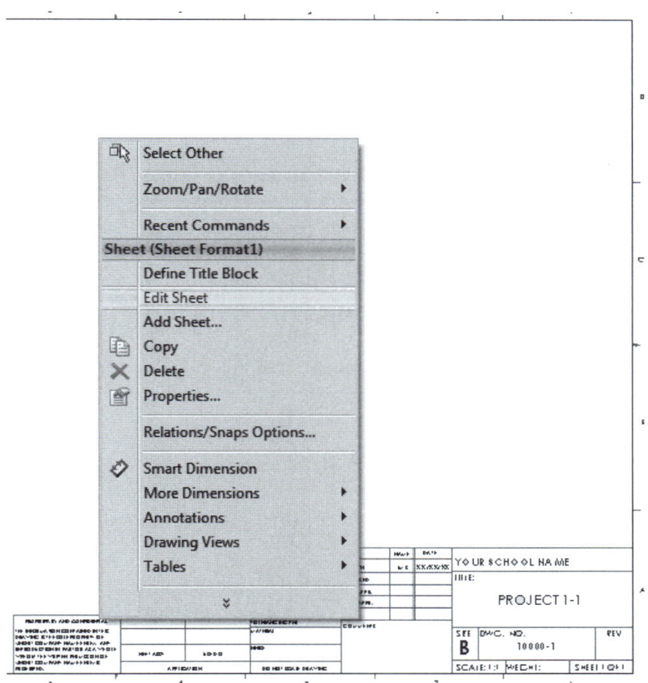

Figure 2-11
Click: Edit Sheet

Project 2-2: **Open Project 1-2 and Make a Drawing from the Part**

Use the steps described in Project 2-1 to make the drawing shown in Figure 2-12.

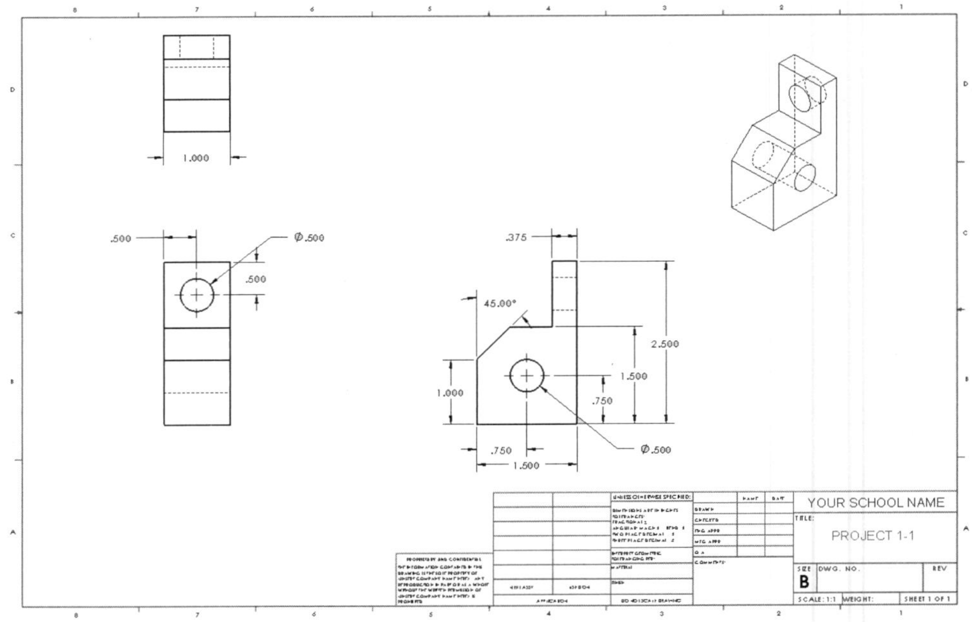

Figure 2-12
Project 2-2 Complete

Exercises

EXERCISE 2-1 **Make a drawing from the model shown in Figure 2-13.**

Step 1. Use the steps described in Project 2-1 to make the drawing shown in Figure 2-13.

Be sure to use the Drawing Number *10000-3* for this drawing.

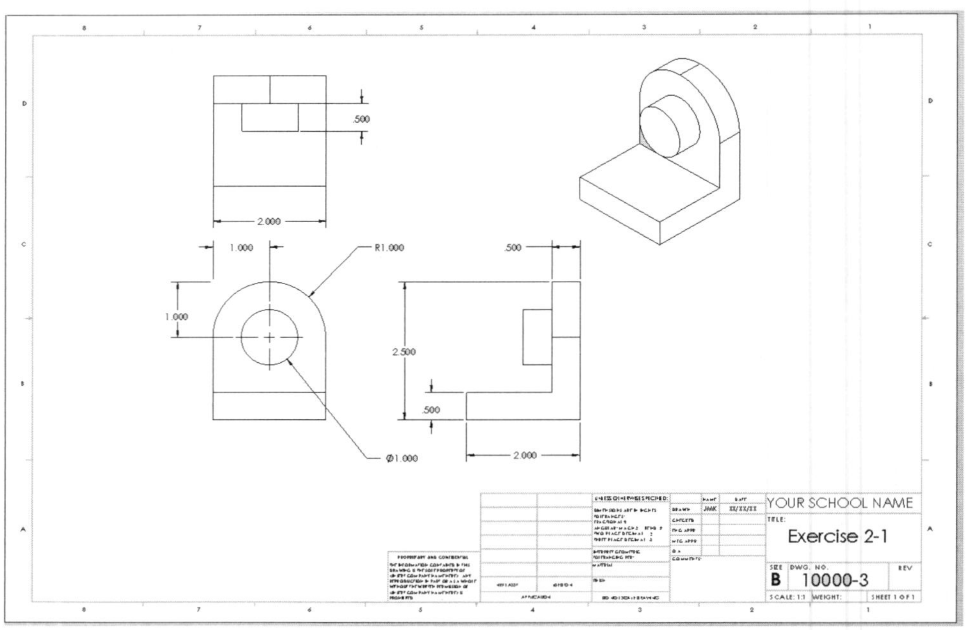

Figure 2-13
Exercise 2-1

Step 2. Save your drawing with the file name **Exercise 2-1(your initials)** in at least two places.

Step 3. Print the drawing and turn it in.

EXERCISE 2-2 Make a drawing from the model shown in Figure 2-14.

Step 1. Use the steps described in Project 2-1 to make the drawing shown in Figure 2-14.

Be sure to use the Drawing Number *10000-4* for this drawing.

Step 2. Save your drawing with the file name **Exercise 2-2(your initials)** in at least two places.

Step 3. Print the drawing and turn it in.

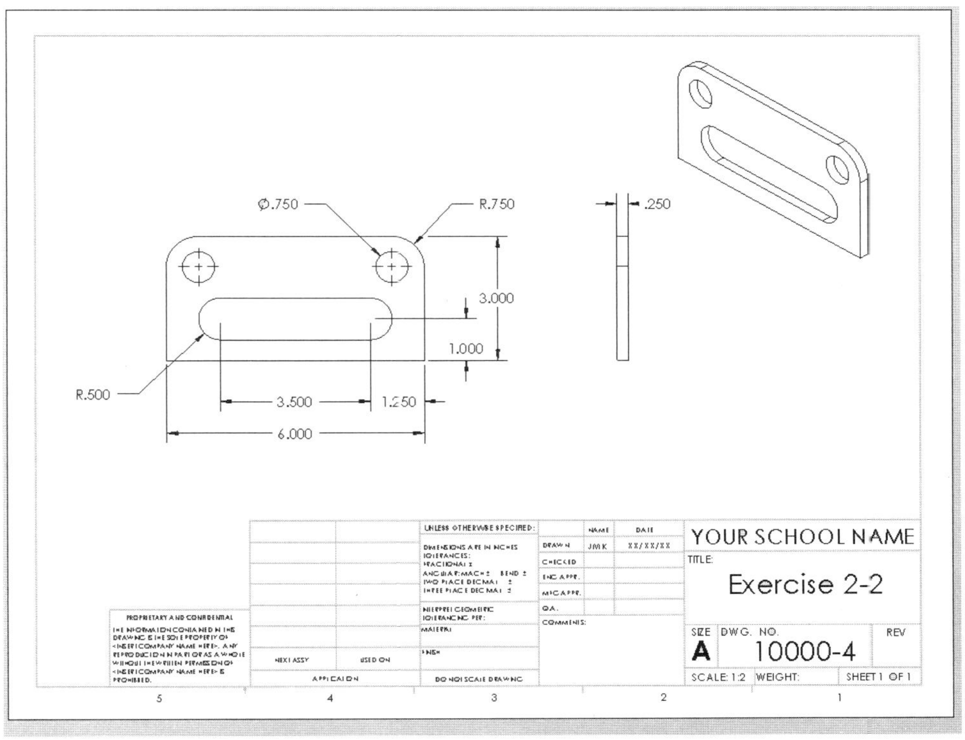

Figure 2-14
Exercise 2-2

EXERCISE 2-3 Make a drawing from the model shown in Figure 2-15.

Step 1. Use the steps described in Project 2-1 to make the drawing shown in Figure 2-15.

Be sure to use the Drawing Number *10000-5* for this drawing.

Step 2. Save your drawing with the file name **Exercise 2-3(your initials)** in at least two places.

Step 3. Print the drawing and turn it in.

Figure 2-15
Exercise 2-3

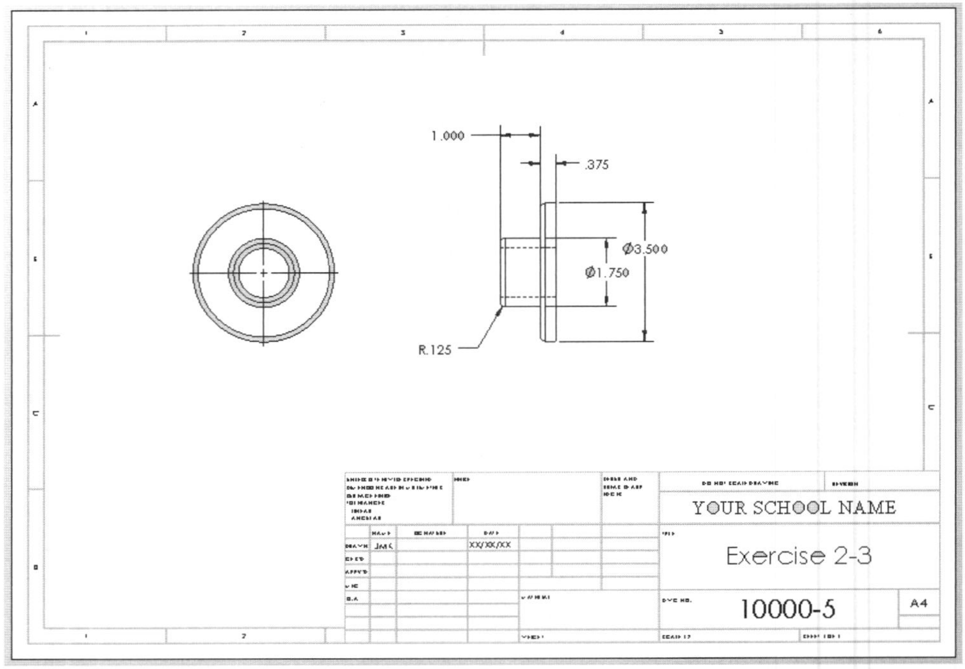

EXERCISE 2-4 **Make a drawing from the model shown in Figure 2-16.**

Step 1. Use the steps described in Project 2-1 to make the drawing as shown in Figure 2-16.

Be sure to use the Drawing Number *10000-6* for this drawing.

Step 2. Save your drawing with the name **Exercise 2-4(your initials)** in at least two places.

Step 3. Print the drawing and turn it in.

Figure 2-16
Exercise 2-4

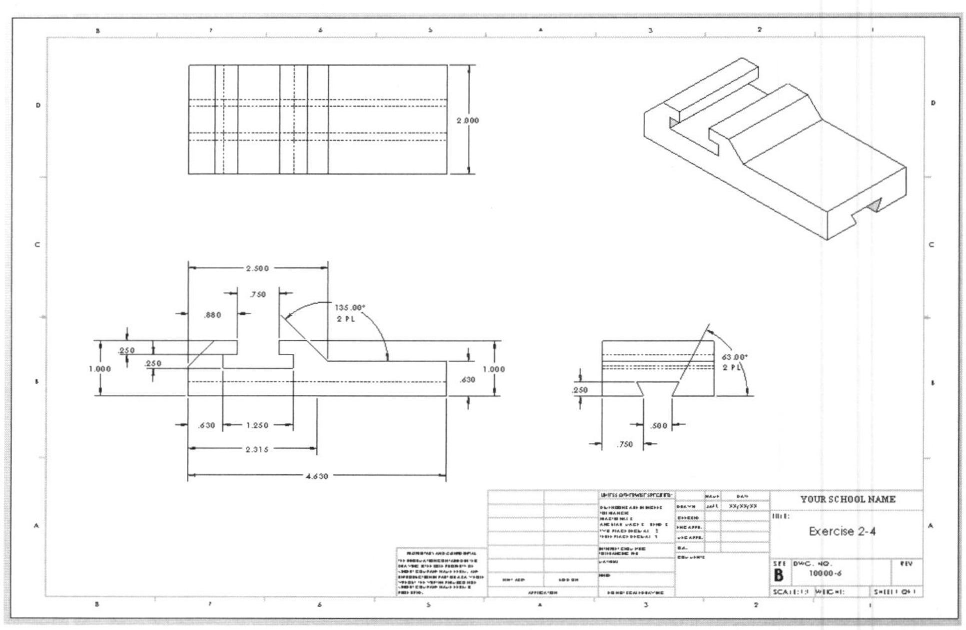

chapter two

Chapter Summary

This chapter contains the information needed to make drawings from simple solid models and sketches. Now you have the skills necessary to make drawings from simple solid models and sketches.

Review Questions

Circle the most correct answer.

1. Where is the **Make Drawing from Part** command located?
   a. **Edit** menu
   c. **Insert** menu
   b. **View** menu
   d. **File** menu
2. Which format was used in Project 2-1?
   a. A (ANSI) Portrait
   b. B (ANSI) Landscape
   c. A4 (ANSI) Landscape
   d. A4 (ANSI) Portrait
3. In which dialog box is the third angle projection method of drawing located?
   a. **Properties...**
   c. **Picture...**
   b. **Model Items...**
   d. **Object...**
4. From which icon on the **Task** pane are views of the part dragged and dropped?
   a. **Design Library**
   c. **SolidWorks Resources**
   b. **File Explorer**
   d. **View Palette**
5. What symbol must appear before views can be relocated?
   a. Edit
   c. Locate
   b. Move
   d. Origin
6. Model items are located on the **Annotation** tab.
   a. True
   b. False
7. An ANSI B Landscape format measures 8-1/2 inches × 11 inches.
   a. True
   b. False
8. Hidden lines cannot be shown in drawing views.
   a. True
   b. False
9. Dimensions cannot be moved after they have been imported into a drawing view.
   a. True
   b. False
10. The sheet format can be edited.
    a. True
    b. False

chapter two

Matching

Place the correct letter on the line to the right of the numbered item.

1. A command that is used to place the date in the **DATE** box _____
2. A setting in the **Properties** dialog box that allows you to change how views are placed _____
3. A command that allows you to view a title block _____
4. The area where the **View Palette** is located _____
5. A command that returns you from the **Sheet Format** to the drawing _____

a. **Task** pane
b. **Edit Sheet Format**
c. **Note** command
d. **Edit Sheet**
e. Third Angle Orthographic Projection

General Questions

Answer the following questions.

1. Why are drawing formats used?
2. What views would you use the most for the type of drawing work you choose?
3. What is the difference between first and third angle projection?
4. How can you relocate and add dimensions on a drawing?
5. Why are hidden views necessary on orthographic views?

3 chapter three

Making More Complex Models and Drawings with Dimensions from Those Models

CHAPTER OBJECTIVES

After completing this chapter, you will be able to:

- Make 2D sketches using drawing constructions and dimension them.
- Make complex extruded models.
- Make a sheet metal four-sided box.
- Open an existing drawing and make a drawing from a part.
- Select a drawing format for the drawing.
- Drag drawing views from the **Task** pane onto the drawing format.
- Insert **Model Items** (dimensions) from the part to the drawing views.
- Arrange views and space dimensions correctly.
- Add notes to the drawing using **Edit Sheet** and to the drawing format using **Edit Sheet Format.**
- Correctly answer questions regarding SolidWorks commands used in this chapter.

Project 3-1: Make a New Part (Figure 3-1)

Step 1. Start SolidWorks.

Step 2. Click **New....**

The **New SolidWorks Document** dialog box appears.

Step 3. Click **Part** and click **OK.**

The SolidWorks screen appears. Click **Options** on the menu bar, select the **Document Properties** tab, and click **Units** in the list to the left. Make sure the **Units** are correctly set [notice that **IPS** (inch, pound, second) is checked], then click **OK.**

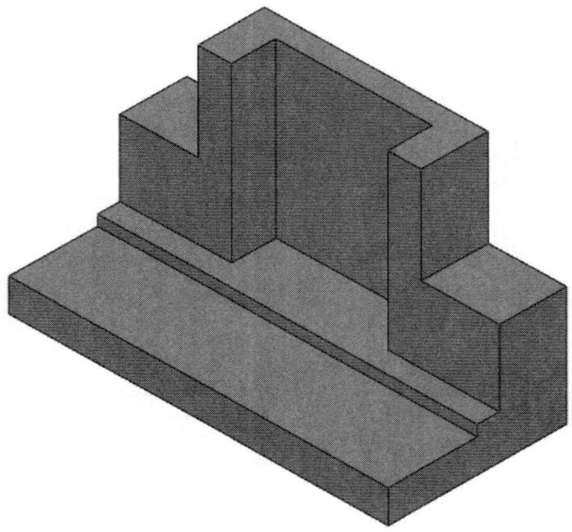

Figure 3-1
Project 3-1 Complete

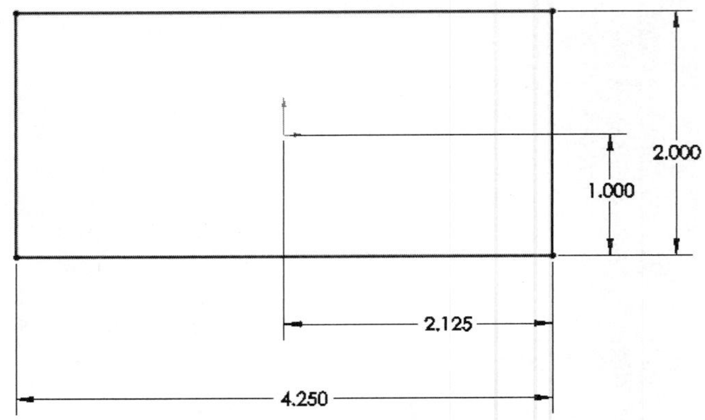

Figure 3-2
Base Sketch Complete

Step 4. Click the **Top Plane** and click **Sketch** so the Sketch mode is active.

Step 5. Draw a rectangle and dimension it as shown in Figure 3-2.

Step 6. Click **Exit Sketch.**

Step 7. Click **Extrude Boss/Base** and extrude the rectangle **.375** inch as shown in Figure 3-3.

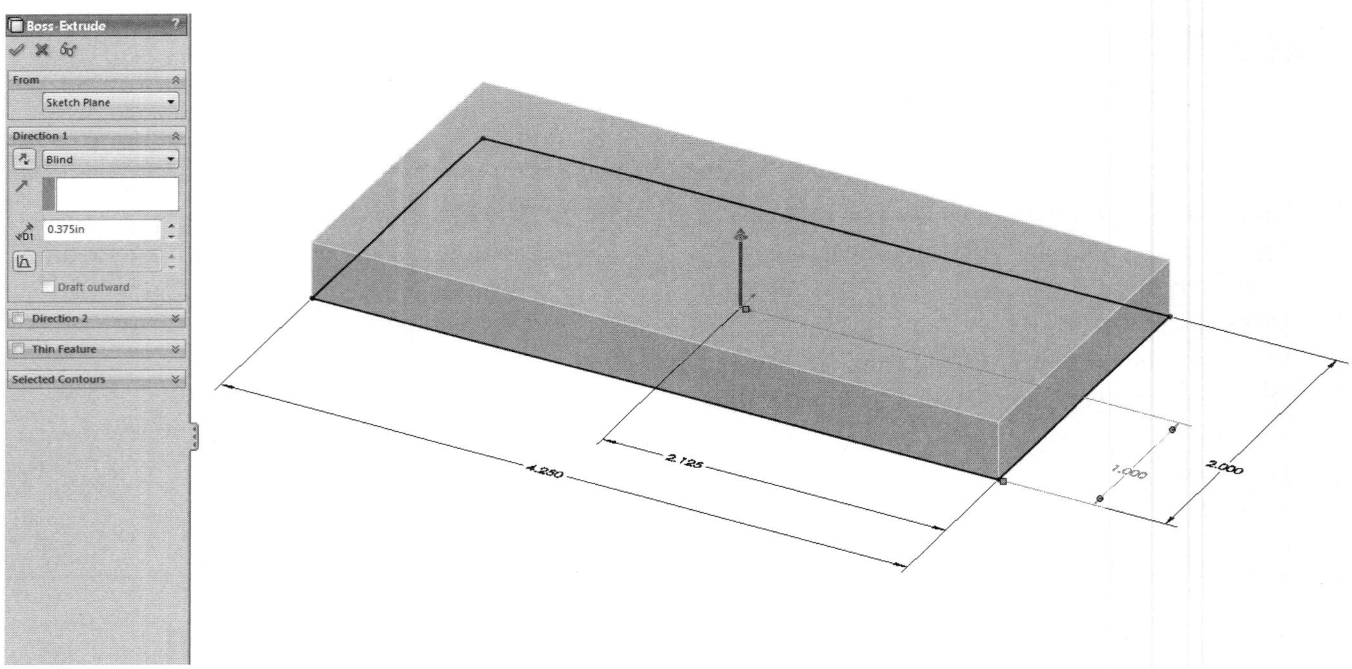

Figure 3-3
Base Extruded .375 Inch

Step 8. Click the top of the base and click **Sketch** as shown in Figure 3-4.

Step 9. Draw a rectangle and dimension it as shown in Figure 3-5.

Step 10. Click **Exit Sketch.**

Step 11. Click **Extrude Boss/Base** and extrude the rectangle **.125** inch as shown in Figure 3-6.

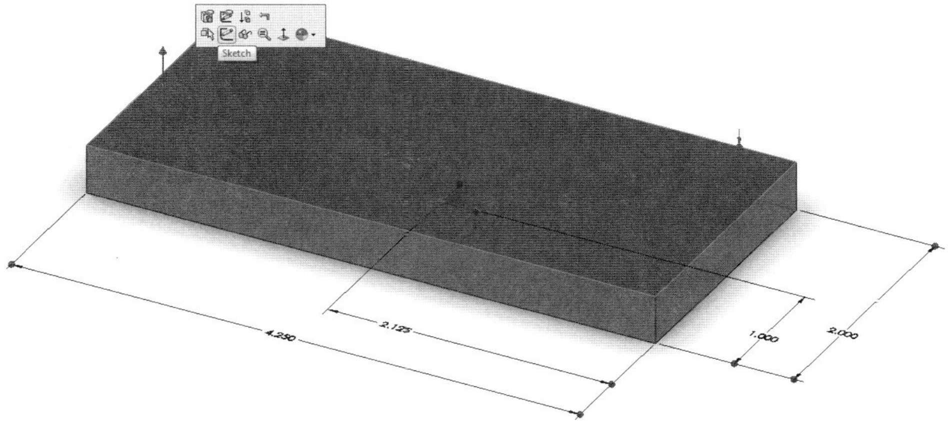

Figure 3-4
Select the Top of the Base as the Sketch Plane

Figure 3-5
Draw a Rectangle on the Sketch Plane and Dimension it

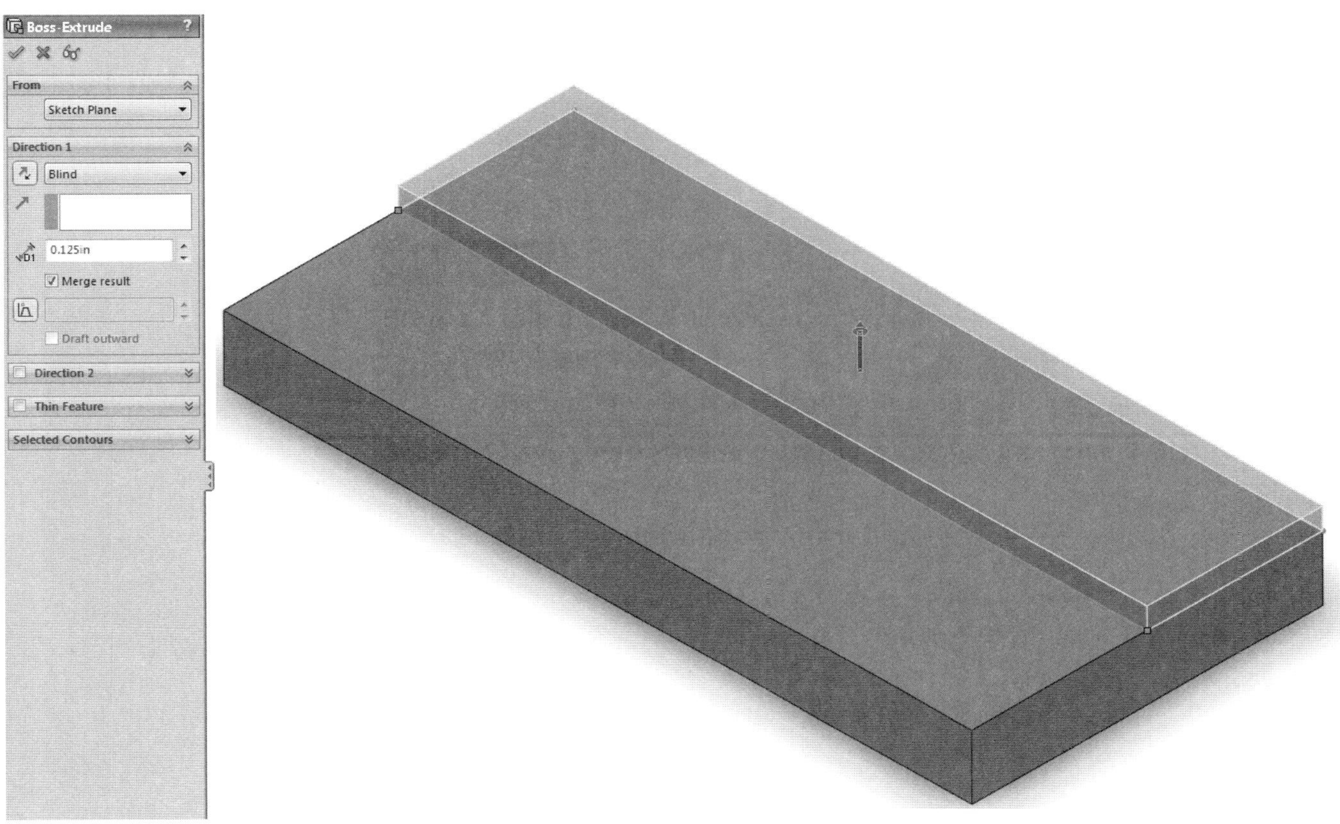

Figure 3-6
Extrude the Rectangle .125 Inch

Step 12. Draw two rectangles on the top of the second rectangle and dimension them as shown in Figure 3-7.

Figure 3-7
Draw 2 Rectangles on the Top of the Second Rectangle and Dimension Them

Step 13. Click **Exit Sketch.**

Step 14. Click **Extrude Boss/Base** and extrude the rectangles **.875** inch as shown in Figure 3-8.

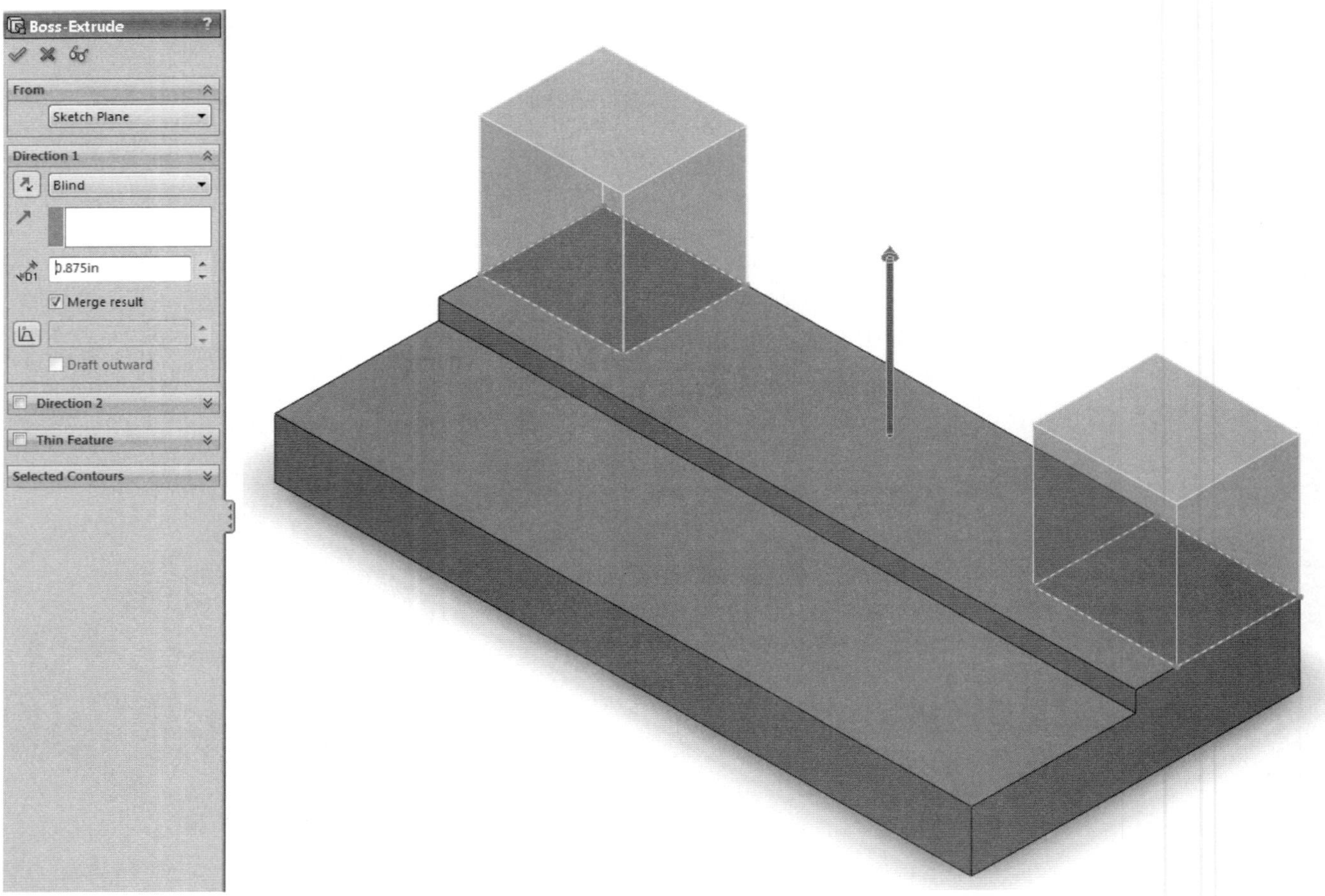

Figure 3-8
Two Rectangles Extruded .875 Inch

Step 15. Click the top of the second rectangle and click **Sketch** as shown in Figure 3-9.

Step 16. Click **Exit Sketch.**

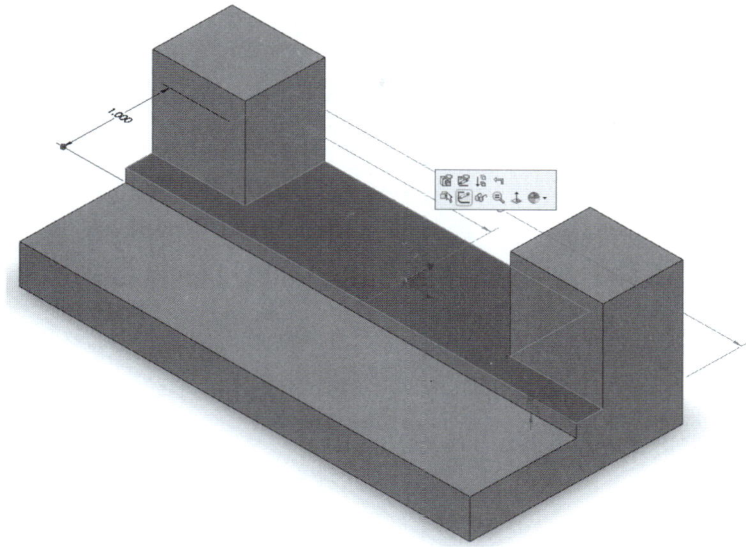

Figure 3-9
Select the Top Plane of the Second Rectangle as the Sketch Plane

Step 17. Draw a shape and dimension it, as shown in Figure 3-10.

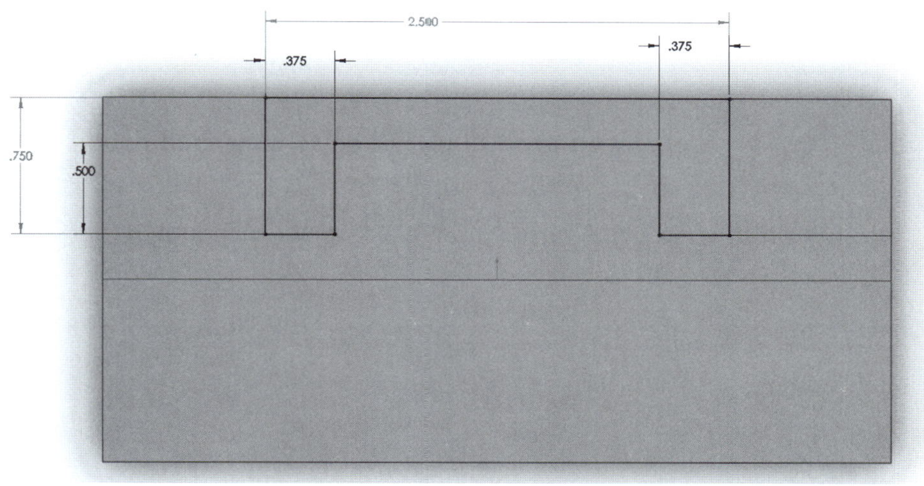

Figure 3-10
Draw a Shape and Dimension it

Step 18. Click **Extrude Boss/Base** and extrude the rectangles **2.00** inches as shown in Figure 3-11.

Step 19. Save your drawing (Figure 3-12) with the file name **Project 3-1(your initials)** in at least two places.

Step 20. Print the drawing and turn it in.

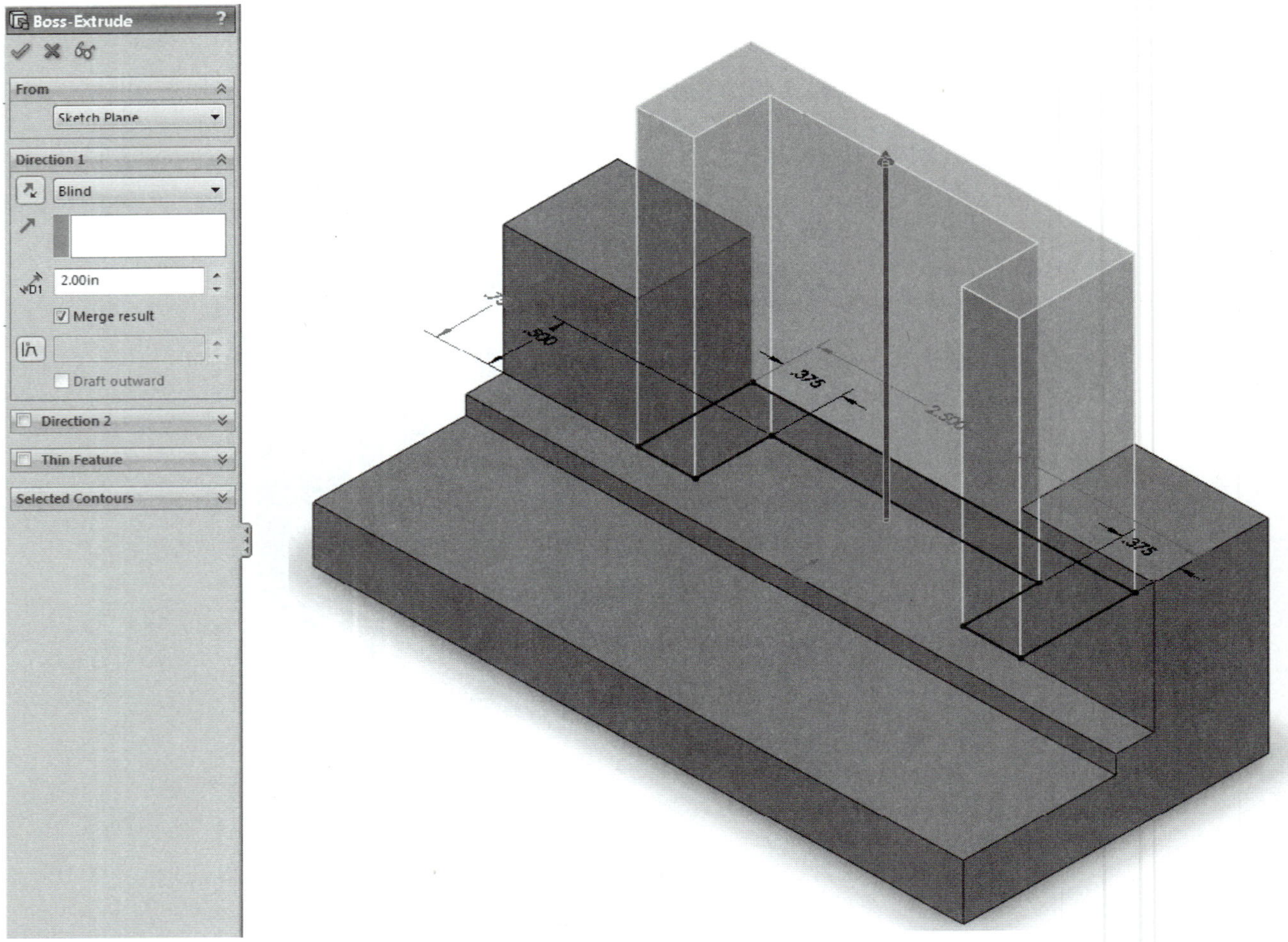

Figure 3-11
Extrude the Shape 2.00 Inches

Figure 3-12
Project 3-1 Complete

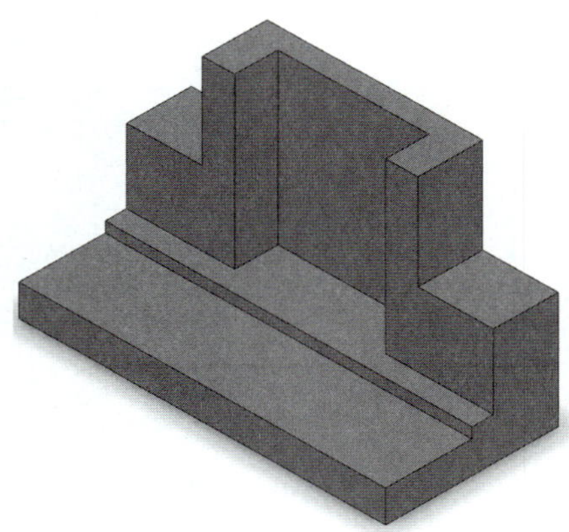

Project 3-2: **Open Project 3-1 and Make a Drawing from the Part**

Use the steps described in Project 3-1 to make the drawing as shown in Figure 3-13.

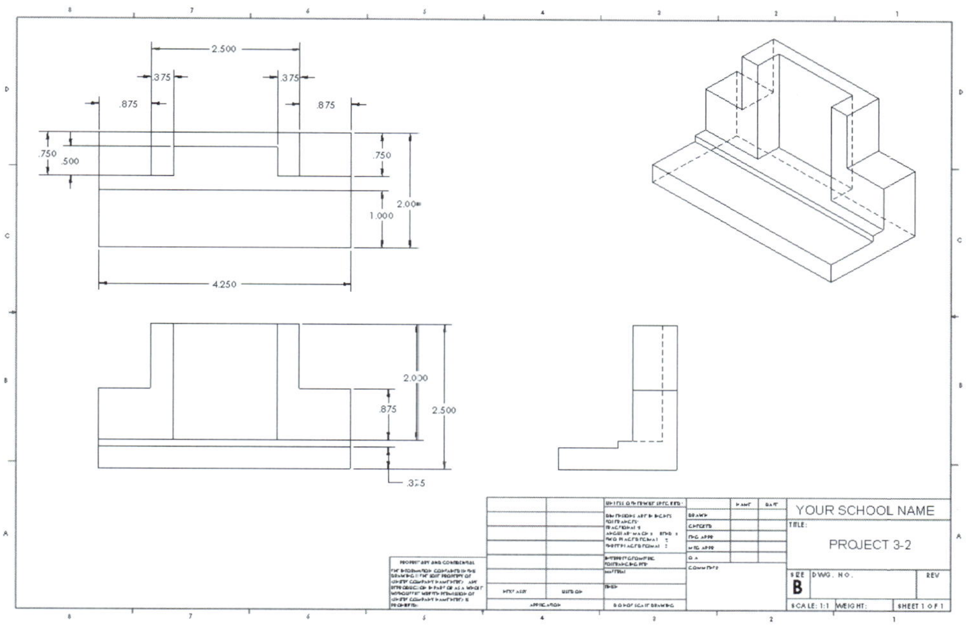

Figure 3-13
Project 3-2 Complete

Project 3-3: **Make a Four-Sided Sheet Metal Box (Figure 3-14)**

Step 1. Click **New....**

The **New SolidWorks Document** dialog box appears.

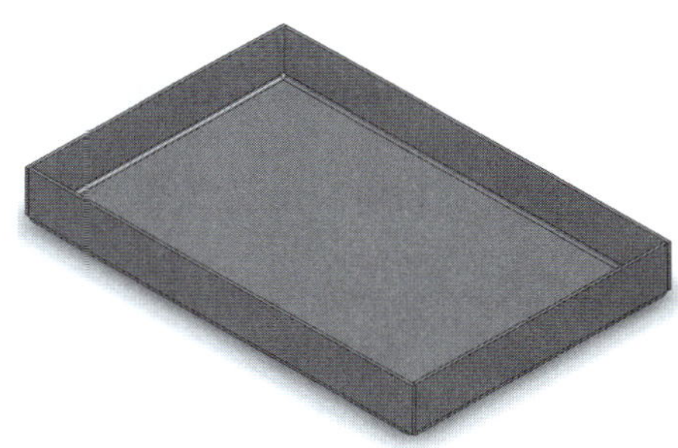

Figure 3-14
Project 3-3 Complete

Step 2. Click **Part** and click **OK.**

The SolidWorks screen appears.

Click **Options** on the menu bar, select the **Document Properties** tab, and click **Units** in the list to the left. Make sure the **Units** are correctly set [notice that **IPS** (inch, pound, second) is checked], then click **OK.**

Step 3. Click the top plane and click **Sketch** so that the Sketch mode is active.

Step 4. If the **Sheet Metal** toolbar is not visible, activate it from the **View** menu on the menu bar as shown in Figure 3-15.

Step 5. Draw a rectangle and dimension it as shown in Figure 3-16.

Step 6. Click **Exit Sketch.**

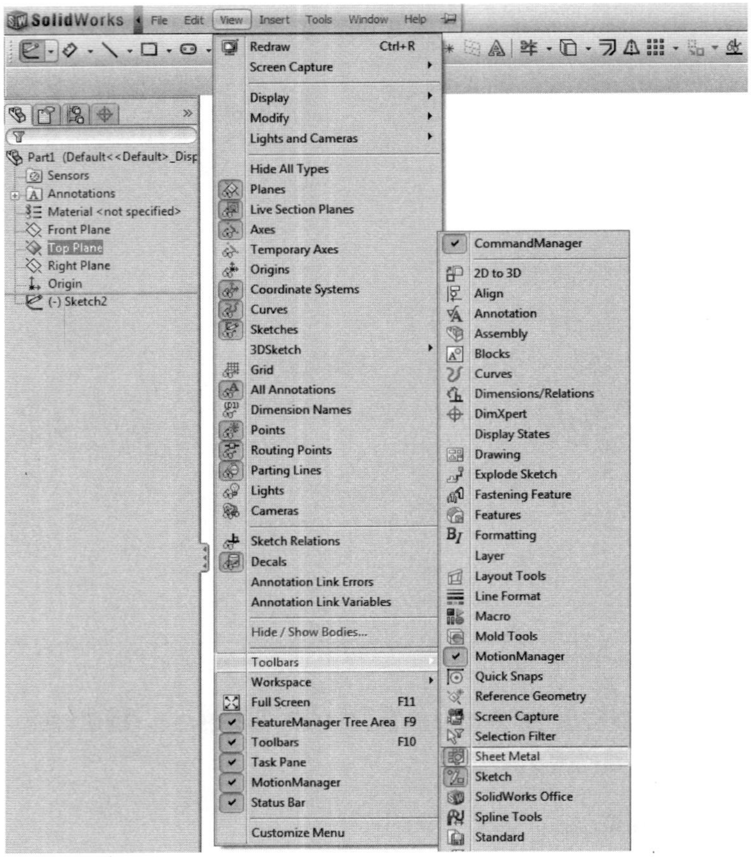

Figure 3-15
Activate the Sheet Metal Toolbar

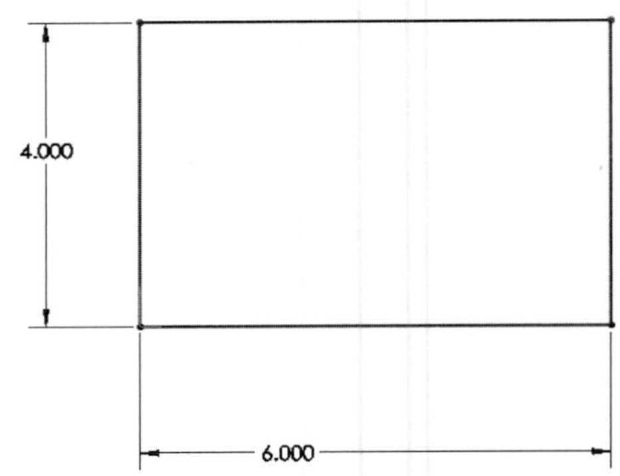

Figure 3-16
Draw and Dimension a Rectangle on the Top Plane

Step 7. Click **Base-Flange** from the **Sheet Metal** toolbar as shown in Figure 3-17.

Make sure the thickness box under **Sheet Metal Parameters** shows **0.048** and the **Reverse** direction is checked.

Figure 3-17
Make the Base Flange of the Four-Sided Box

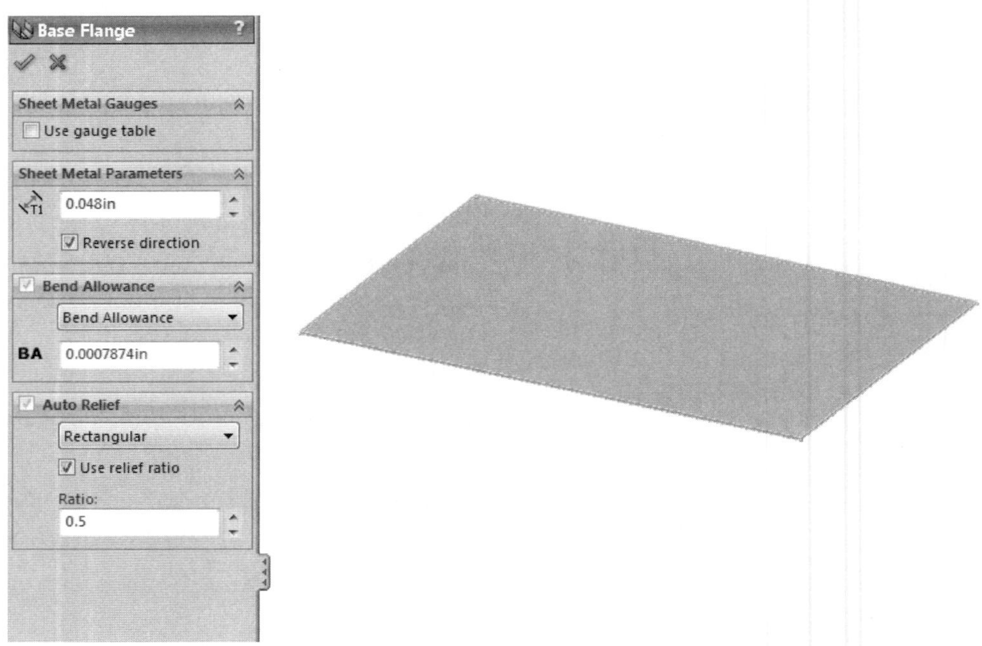

Step 8. Click **OK** (the green check mark at the top of the **Base Flange Manager**).

Step 9. In the **Feature Manager** right-click on the word **Sheet-Metal1** and click **Edit Feature** as shown in Figure 3-18.

Step 10. In the **Sheet-Metal1 Feature Manager** make sure the **Bend Radius** reads **0.05 in** as shown in Figure 3-19.

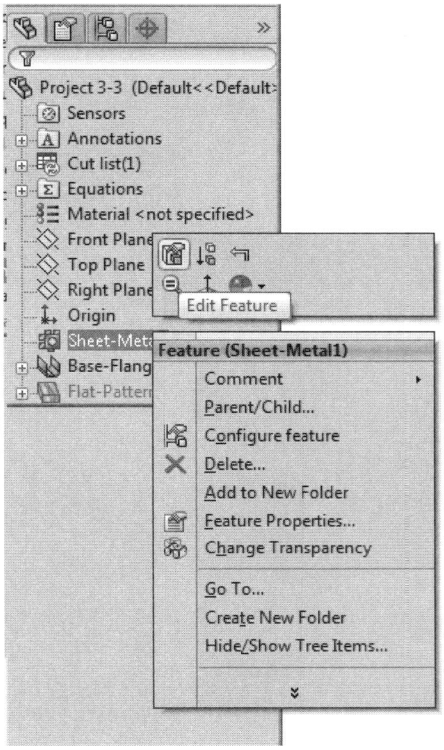

Figure 3-18
Right-Click: Sheet-Metal1 and Click: Edit Feature

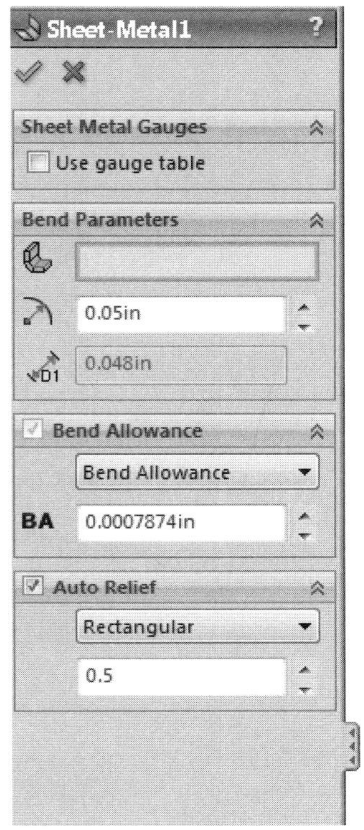

Figure 3-19
Set Bend Radius to 0.05 Inch

Step 11. Select **Edge Flange** on the **Sheet Metal** toolbar as shown in Figure 3-20.

Figure 3-20
Click: Edge Flange

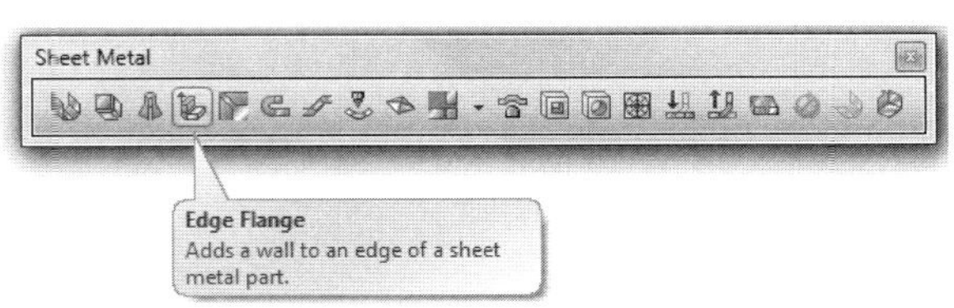

Step 12. Select the front, top edge of the base flange and move your mouse up, as shown in Figure 3-21.

With the direction set you are now able to click the top edge of the other three sides as shown in Figure 3-22.

Step 13. Set **Flange Length** to **.75** inch in the **Edge Flange Manager,** as shown in Figure 3-22.

Figure 3-21
Click the Front, Top Edge of the Base Flange

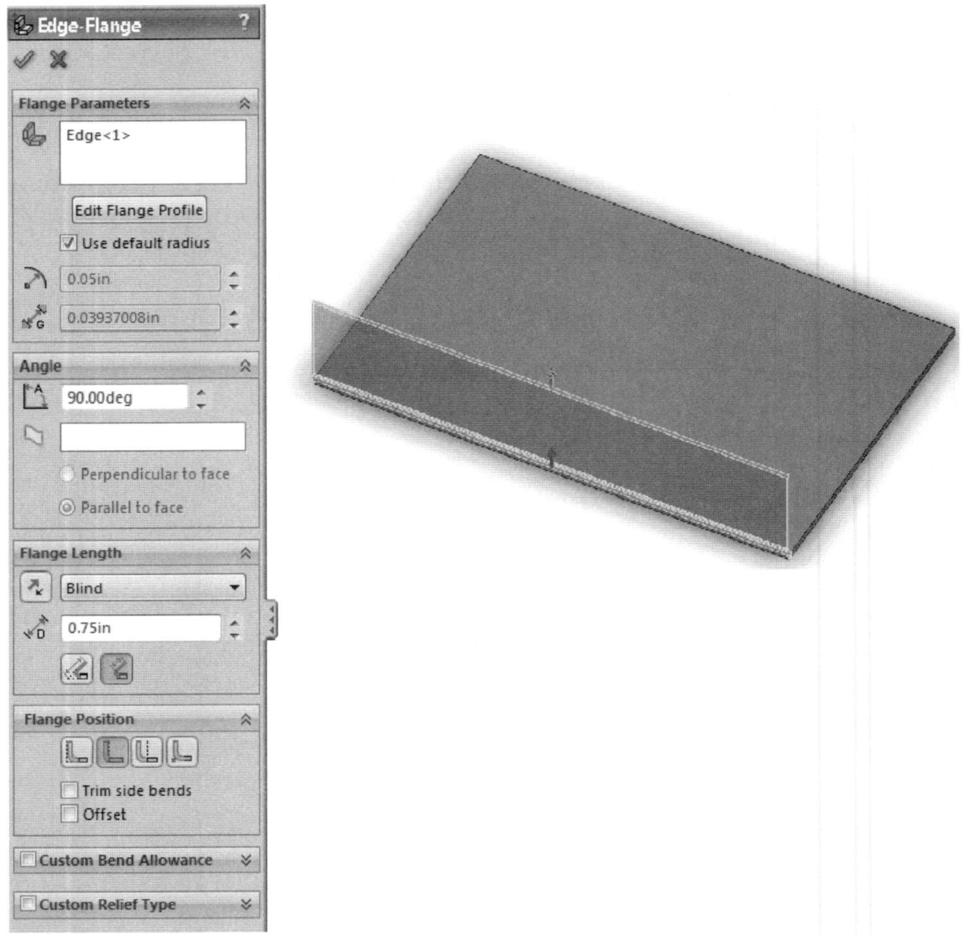

Figure 3-22
Make All Four Edge Flanges

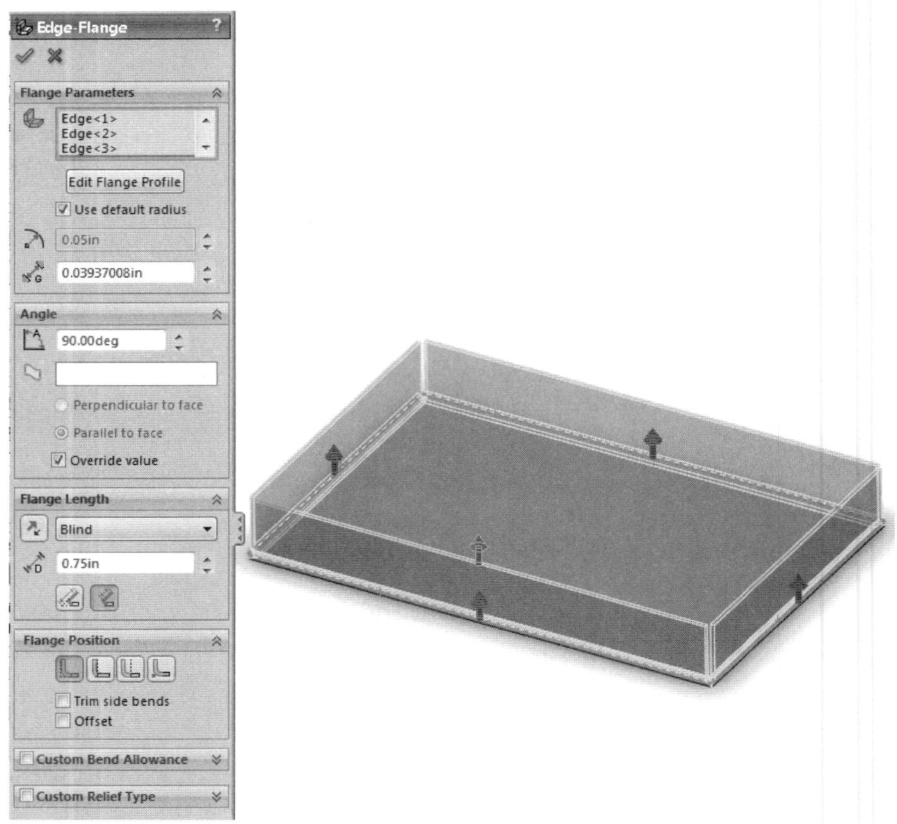

Step 14. Make sure the **Flange Position** is set to **Material Inside** and the **Outer Virtual Sharp** button is selected, as shown in Figure 3-23.

The term "Virtual Sharp" specifies the origin for the measurement of the length of the flange. It makes a sketch point at the exact intersection point of two sketch objects.

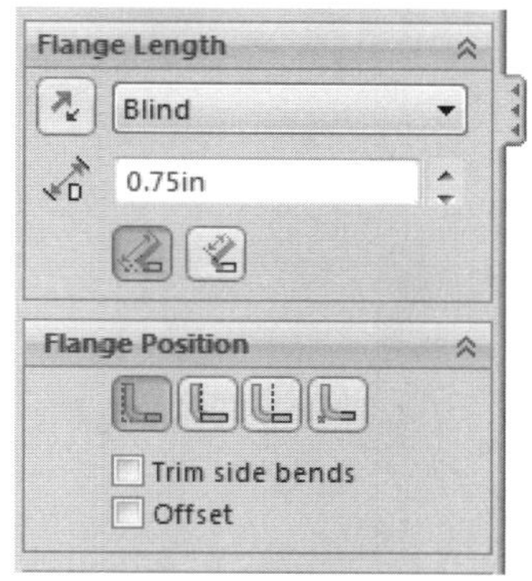

Figure 3-23
Flange Position Set to Inside and Outer Virtual Sharp is Depressed

Step 15. Click **OK** (the green check mark at the top of the **Edge-Flange Manager**).

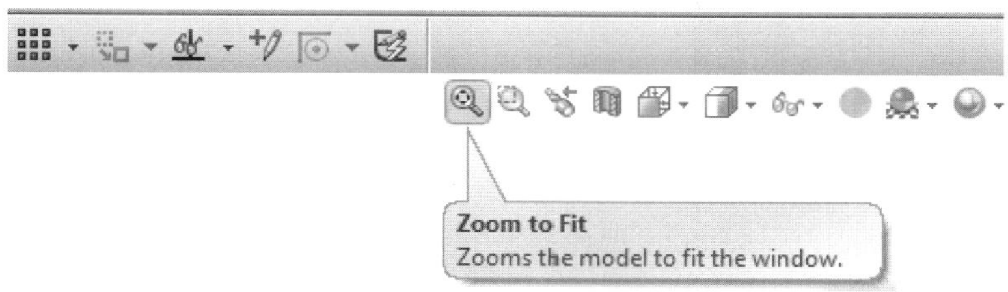

If your part is not where you want it, use the middle wheel on your mouse or the **Zoom to Fit** or **Zoom to Area** icons at the top of the SolidWorks screen.

Step 16. Close the corners of the box. To do this, click **Closed Corner** from **Corners** on the **Sheet Metal** toolbar and click the right edge of **Flange 1,** as shown in Figure 3-24.

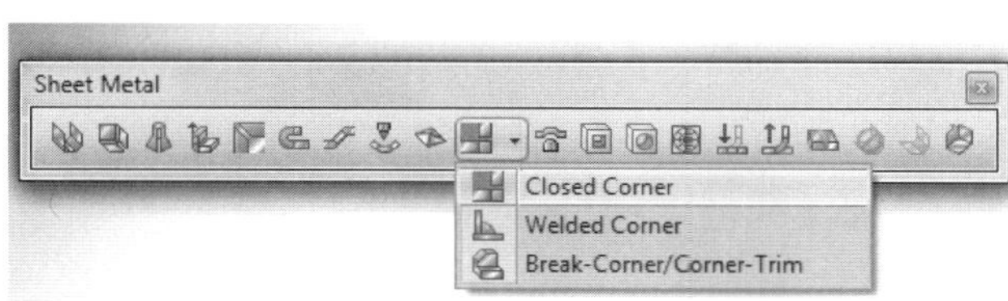

Step 17. Make the following settings: **Corner type - Overlap, 0.01** inch for the **Gap distance,** and make sure the **Open bend region** box is checked.

Step 18. Repeat Steps 16 and 17 for the remaining corners, then click **OK** to accept the settings and close the four corners. If you have trouble with the corners in selecting the correct face, right-click and click **Select Other,** or just try again.

Step 19. Put your name on the drawing.

Step 20. Save your sheet metal box with the file name **Project 3-3(your initials)** in at least two places. The final box is shown in Figure 3-25.

Step 21. Print the drawing on an A-Size sheet and turn it in.

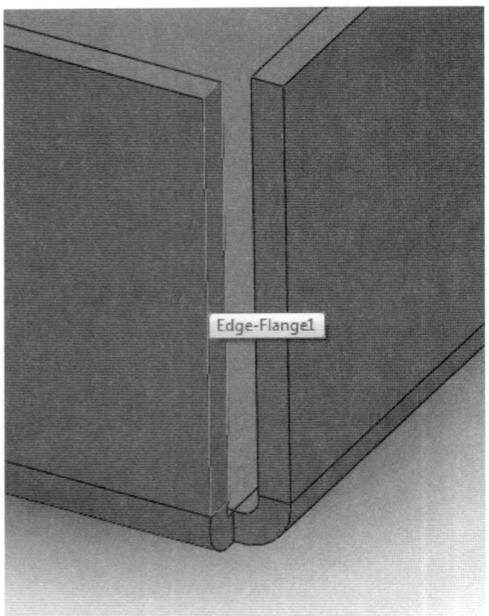

CLICK: THE EDGE AS SHOWN HERE.

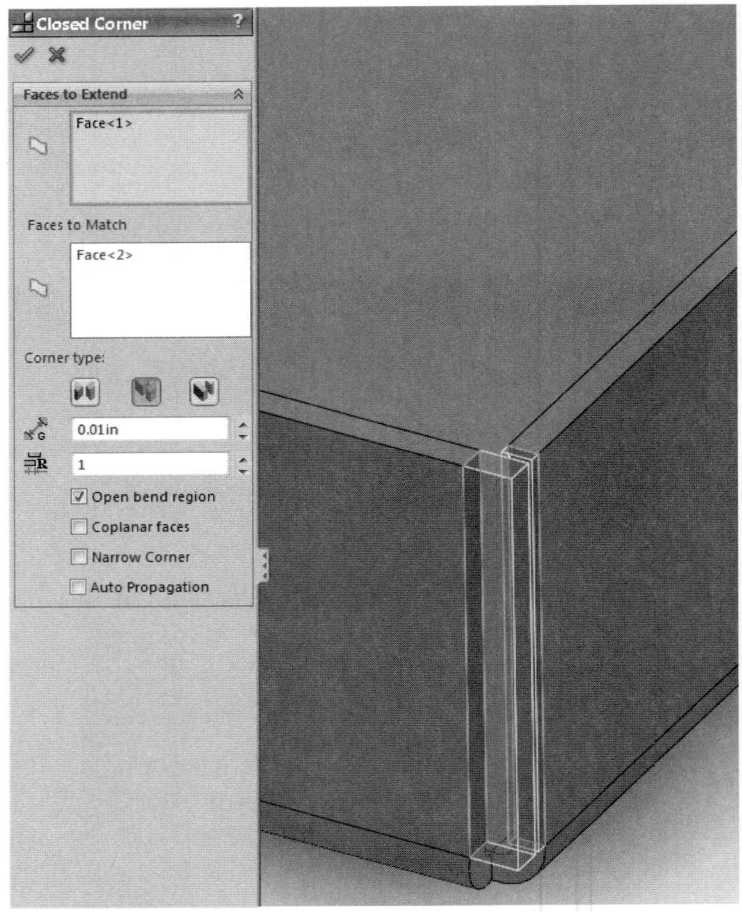

MAKE THE SETTINGS AS SHOWN HERE

Figure 3-24
Close the First Corner

Figure 3-25
Project 3-3 Complete

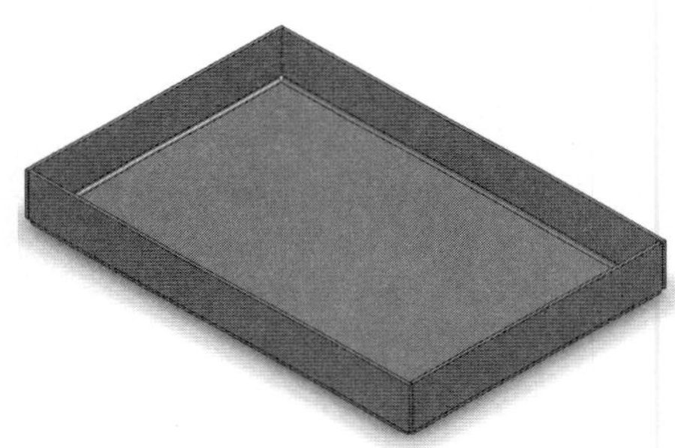

Exercises

EXERCISE 3-1 **Make a model from the sketch shown in Figure 3-26.**

Step 1. Draw and dimension the sketch on the top plane shown in Figure 3-27.

Step 2. Use the **Circular Sketch Pattern** tool to make the pattern shown in Figure 3-28. Be sure you click the .375 diameter circle so that **Arc2** shows in the **Entities to Pattern** box, **6** shows in the number box (the one to the right of the circular pattern), and

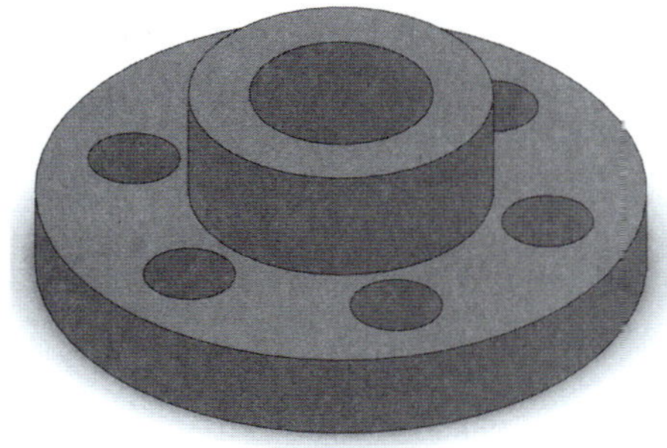

Figure 3-26
Exercise 3-1 Complete

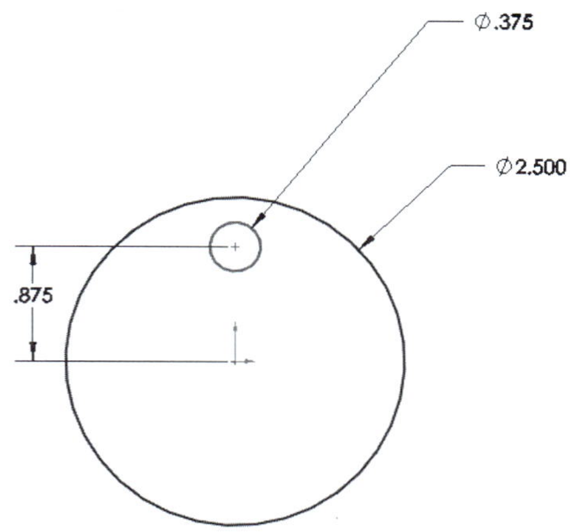

Figure 3-27
First Two Circles Dimensioned on the Top Plane

360 deg is shown in the angle box. When you get the pattern shown in Figure 3-16, click **OK.**

Figure 3-28
Make a Circular Pattern of 6 Holes

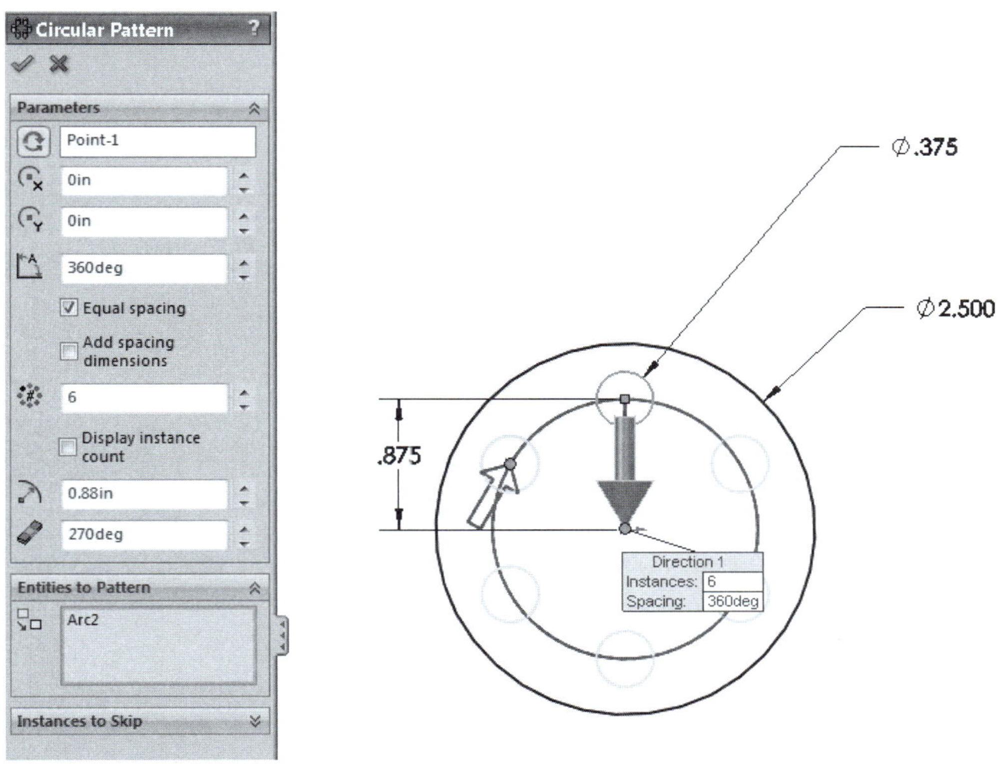

Step 3. Click **Exit Sketch.**

Step 4. Click **Extrude Boss/Base** and extrude the circle with the 6 holes .375 inch as shown in Figure 3-29.

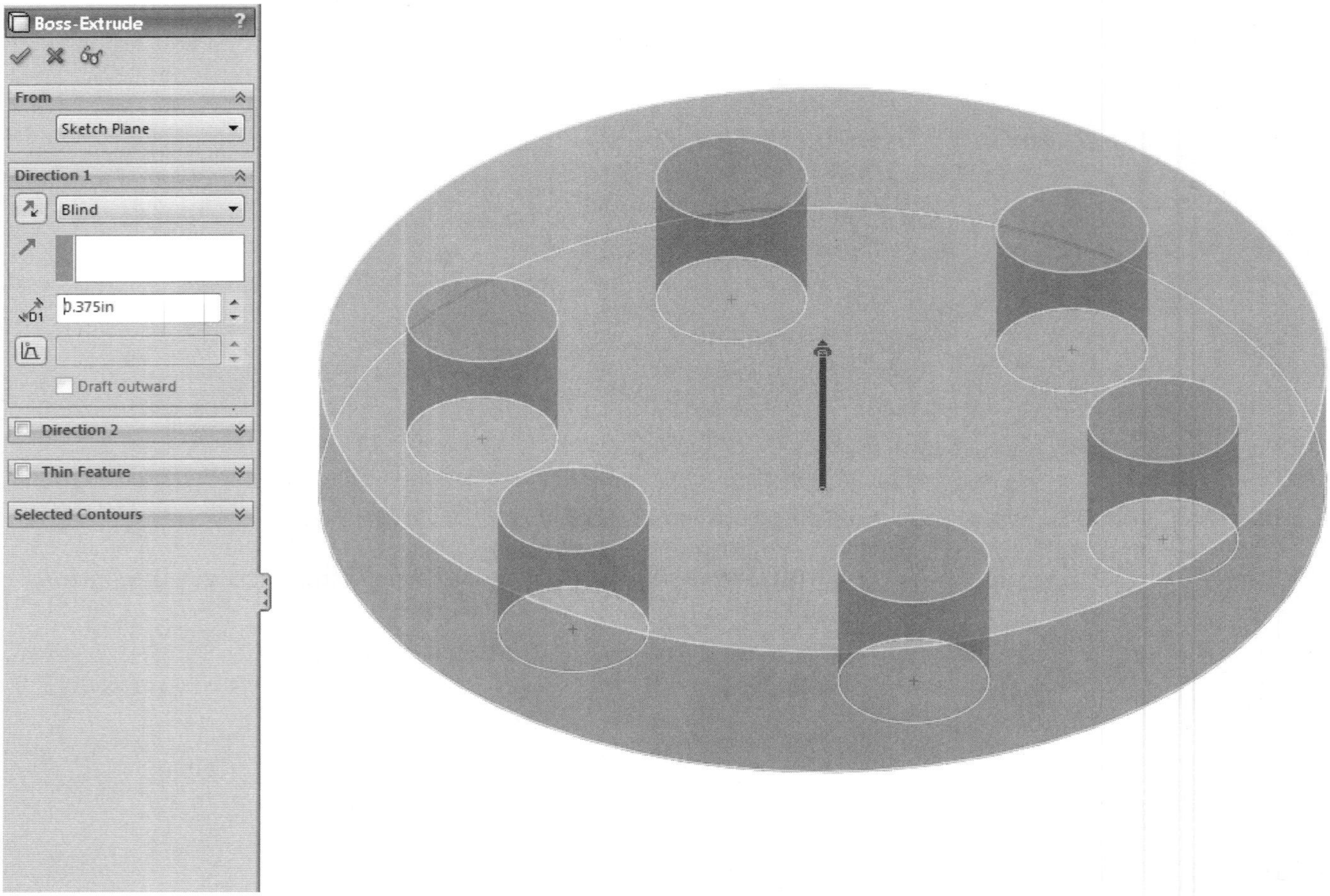

Figure 3-29
Extrude the Circle with 6 Holes .375 Inch

Step 5. Draw a 1.250 diameter circle with the center at the origin and dimension it as shown in Figure 3-30.

Step 6. Select the **Isometric** view from the **Heads-Up View** toolbar.

Step 7. Click **Exit Sketch.**

Figure 3-30
Draw a 1.250 Diameter Circle with the Center at the Origin and Dimension It

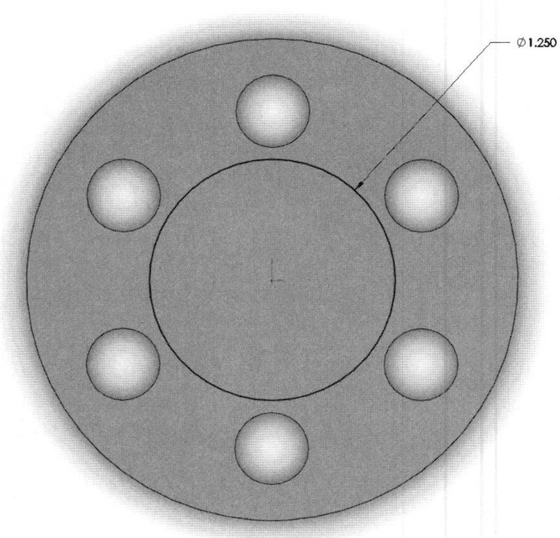

Step 8. Click **Extrude Boss/Base** and extrude the 1.250 diameter circle .500 inch as shown in Figure 3-31.

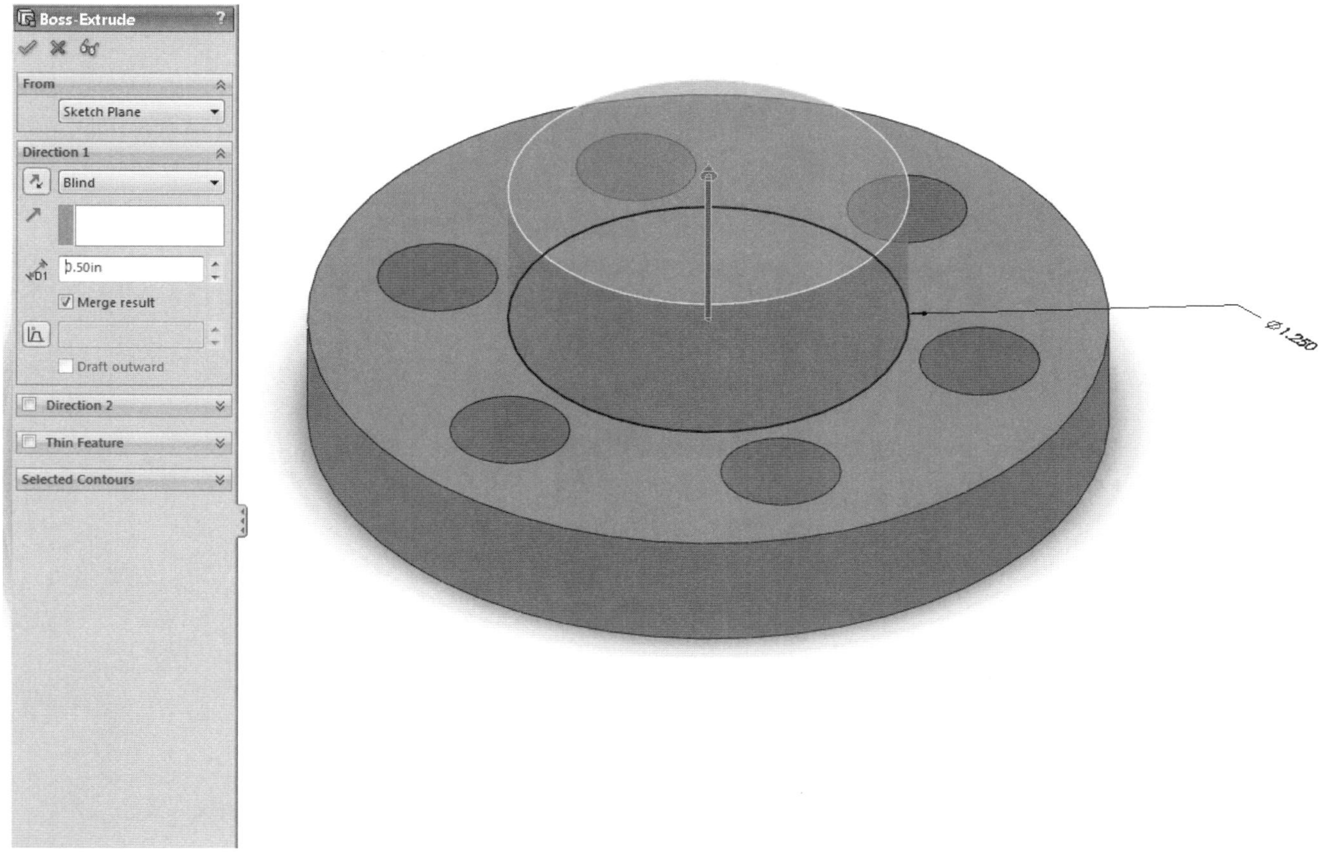

Figure 3-31
Extrude the 1.250 Diameter Circle 0.500 Inch

Step 9. Draw a .750 diameter circle on top of the 1.250 diameter extrusion and dimension it as shown in Figure 3-32.

Step 10. Click **Exit Sketch.**

Figure 3-32
Draw a .750 Diameter Circle on Top of the 1.250 Diameter Extrusion and Dimension It

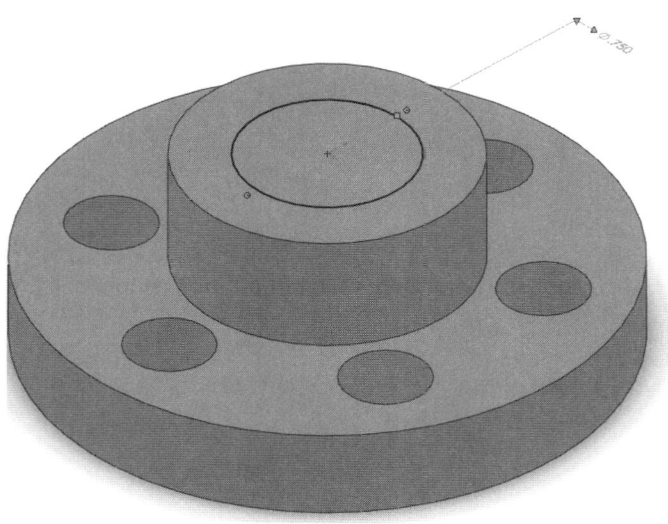

Step 11. Click **Extrude Cut** and cut the .750 diameter circle **Through All** as shown in Figure 3-33. (Be sure the box under **Direction 1** shows **Through All.** If the preview shows the wrong direction, click the icon with the two arrows to flip the direction 180 degrees.) The final model is shown in Figure 3-34.

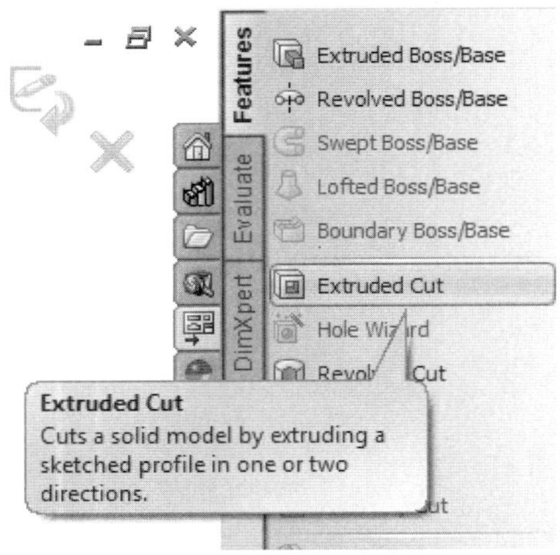

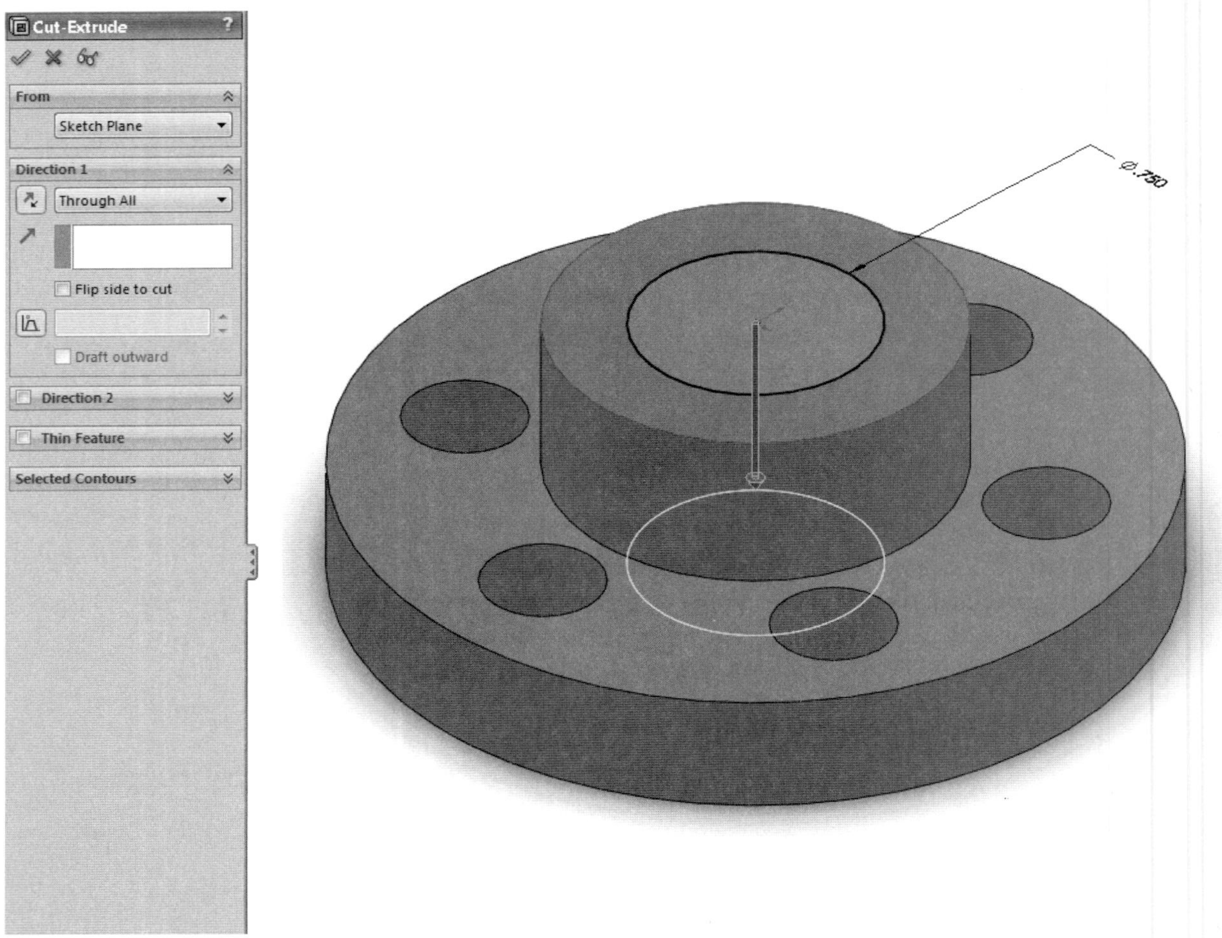

Figure 3-33
Do an Extrude Cut Through All

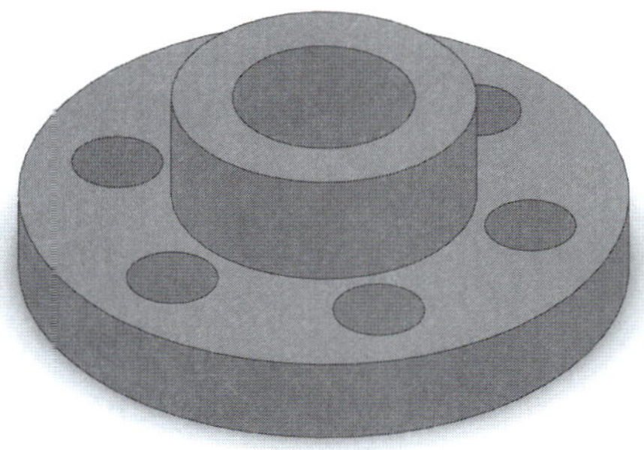

Figure 3-34
Exercise 3-1 Complete

Step 12. Put your name on the drawing.

Step 13. Save your model with the file name **Exercise 3-1(your initials)** in at least two places.

Step 14. Print the drawing on an A-Size or B-Size sheet and turn it in.

EXERCISE 3-2 Make the model shown in Figure 3-35.

Step 1. Click **New....**

The **New SolidWorks Document** dialog box appears.

Step 2. Click **Part** and click **OK.** The SolidWorks screen appears.

Click **Options** on the menu bar, select the **Document Properties** tab, and click **Units** in the list to the left. Make sure the **Units** are correctly set [notice that **IPS** (inch, pound, second) is checked], then click **OK.**

Step 3. Click the **Right Plane** and click **Sketch** so that the Sketch mode is active.

Step 4. If the **Sheet Metal** toolbar is not visible, activate it from the **View** menu on the menu bar.

Step 5. Draw a rectangle and dimension it as shown in Figure 3-36.

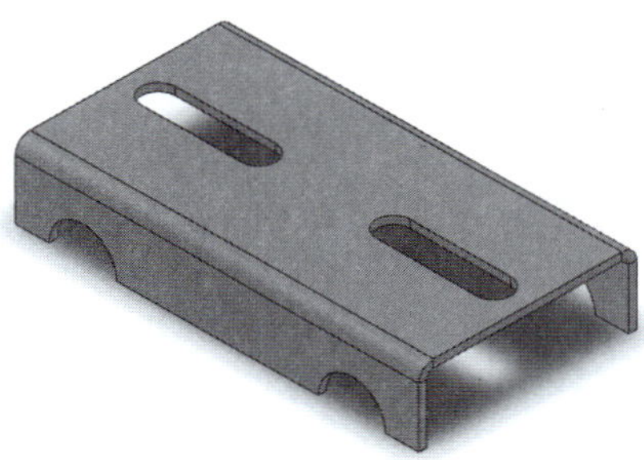

Figure 3-35
Exercise 3-2 Complete

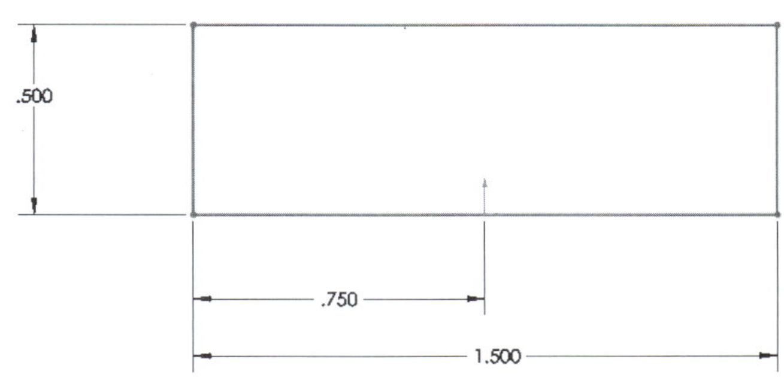

Figure 3-36
Draw and Dimension the Rectangle

Step 6. Trim the bottom of the rectangle. You may be able to trim it by holding down the left mouse button and dragging your cursor across the bottom line of the rectangle or you may have to click directly on the line.

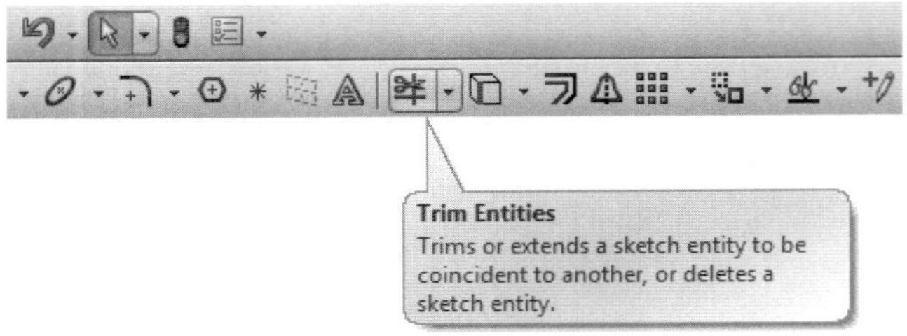

Step 7. Click **Exit Sketch.**

Step 8. Click **Base-Flange** from the **Sheet Metal** toolbar as shown in Figure 3-37.

Make sure that the thickness box under **Sheet Metal Parameters** shows **0.06 in** and **D1** near the top of the **Base Flange Manager** shows **3.00 in,** as shown in Figure 3-37.

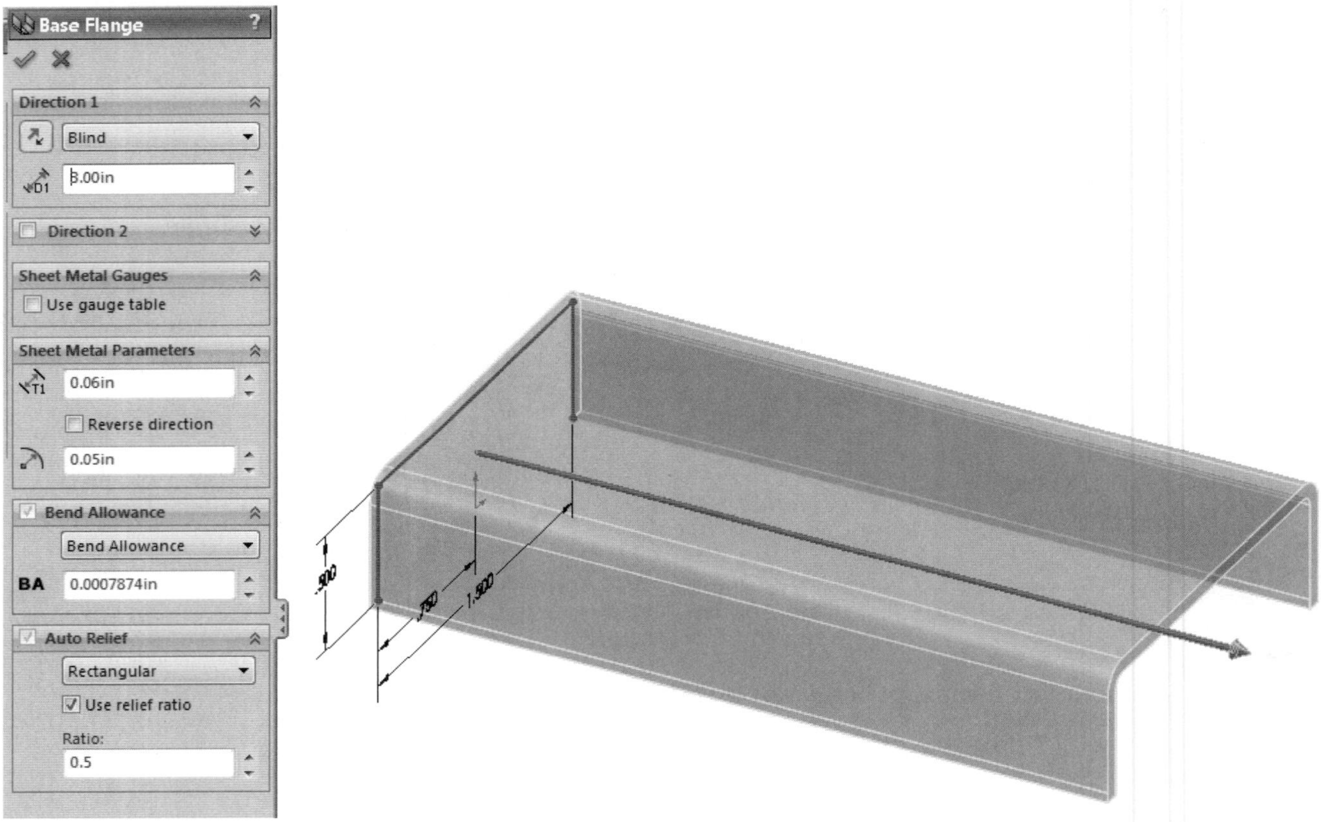

Figure 3-37
Make the Base Flange

Step 9. Click **OK** (the green check mark at the top of the **Base Flange Manager**).

Step 10. Click **Straight Slot** and draw and dimension the slot as shown in Figure 3-38. When you start the slot, move your mouse from the first point clicked (the .250 dimension) and move it directly to the right. Then, move your mouse up or down to make the slot.

Figure 3-38
Draw and Dimension the Straight Slot on the Top of the Base Flange

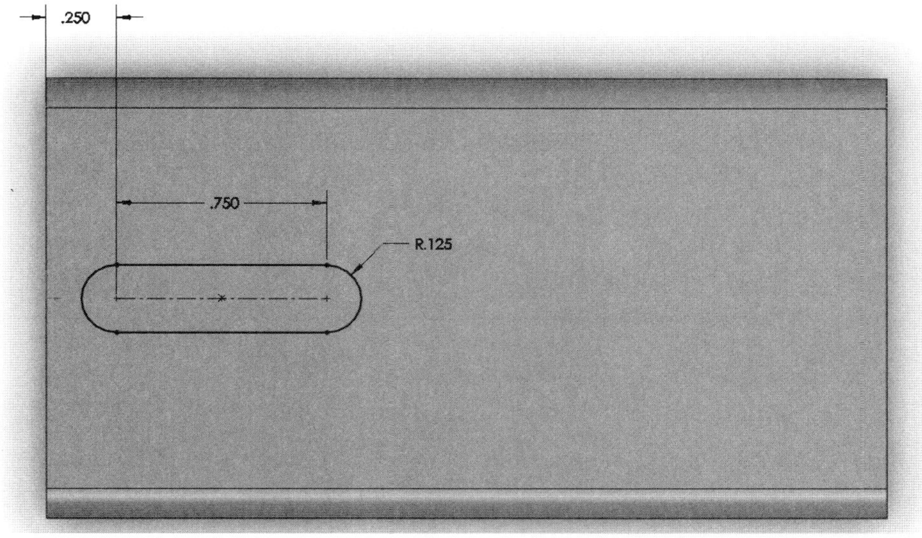

Step 11. Draw a line for the **MIRROR** command, as shown in Figure 3-39. The midpoint of the upper edge of the base flange shown in the figure must be the point where the line starts in the X-direction. The line can be any length. The **MIRROR** command in SolidWorks requires a line or other object for the item to mirror about.

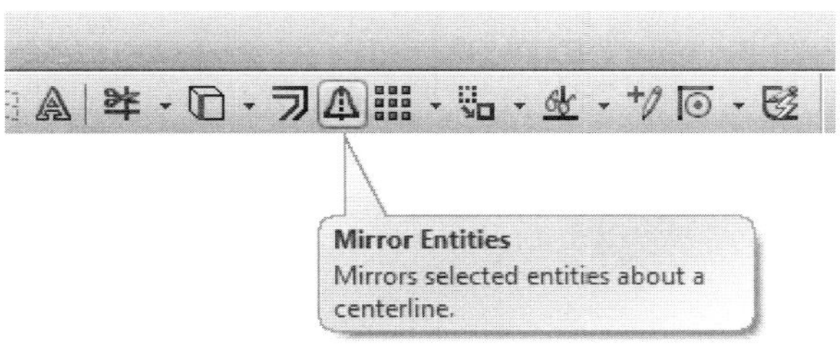

Figure 3-39
Draw a Line for the Mirror Command

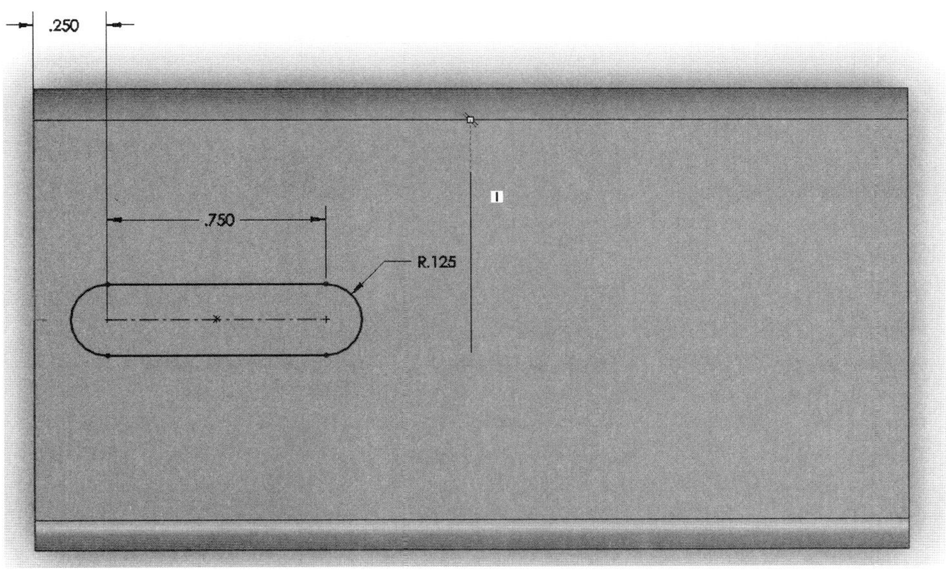

Step 12. Select **Mirror** and use a window (click left and drag right) to select the slot on the left side so that **Slot1** shows in the **Entities to mirror:** box, as shown in Figure 3-40. Then, click the box under **Mirror about**: and click the line you just drew. Then, click **OK.**

Figure 3-40
Mirror the Slot

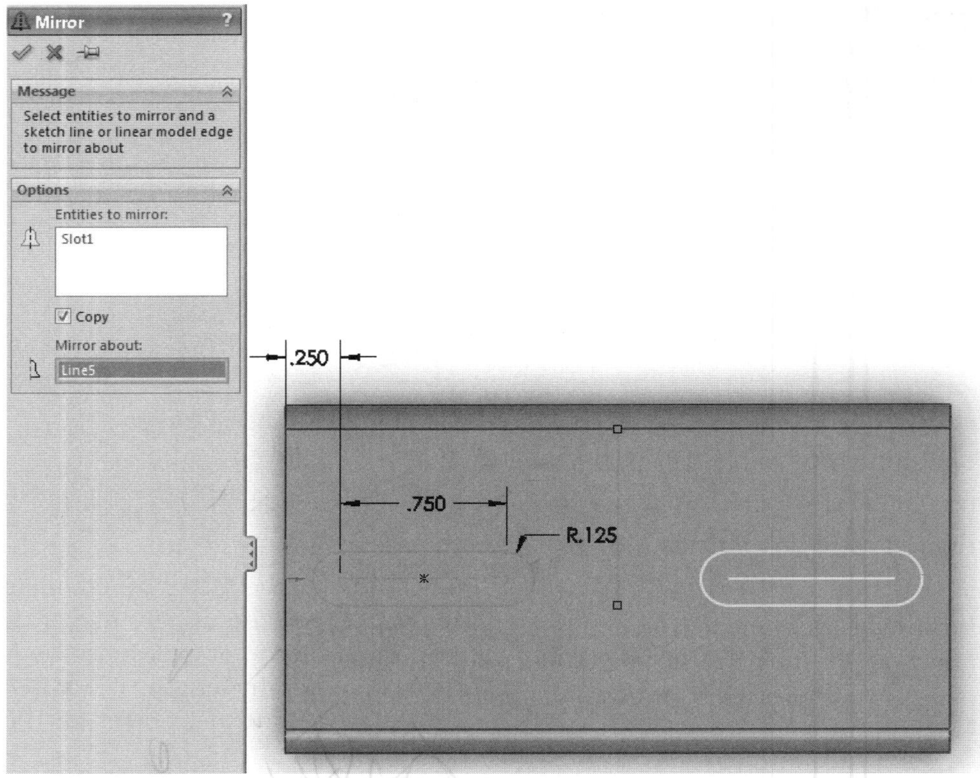

Step 13. Click **Exit Sketch,** select **Isometric view,** and click **Extrude Cut,** as shown in Figure 3-41 (make sure **Through All** shows in the box under **Direction 1**). Then, click **OK.**

Figure 3-41
Make the Extruded Cut

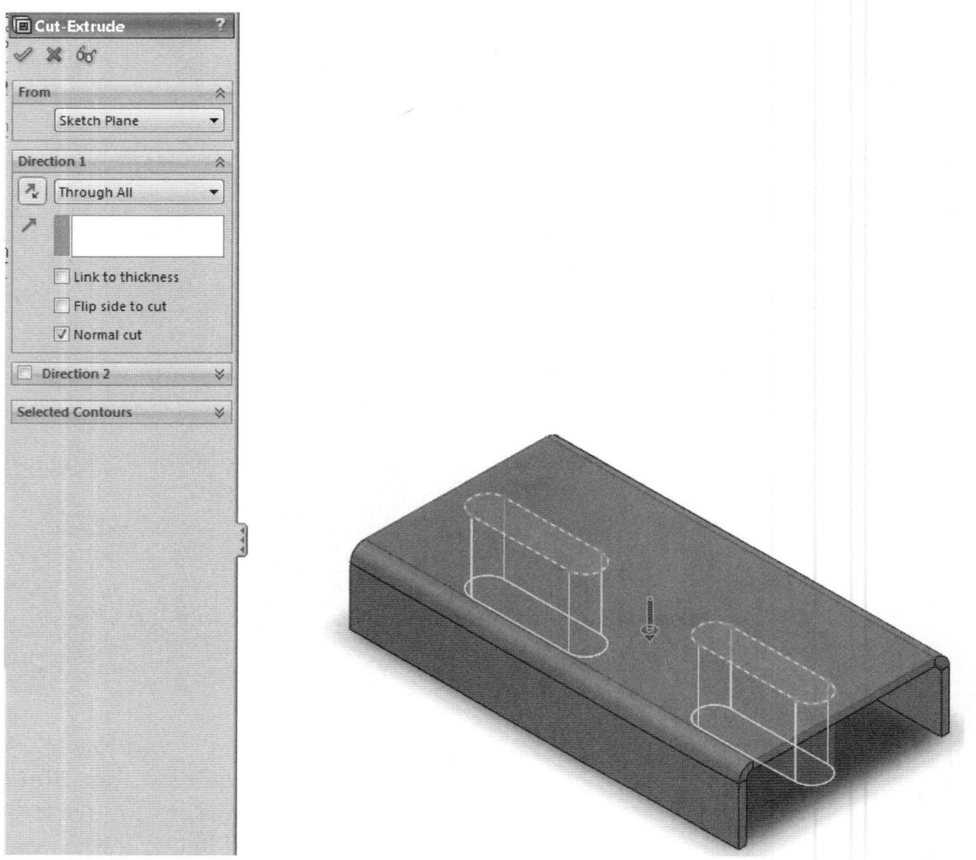

Step 14. Select **3 Point Arc** and draw the arc shown in Figure 3-42 as described next:

1. Click **P1** and **P2.**
2. Drag the arc and click **P3.**
3. Draw a line from **P1** to **P2.**

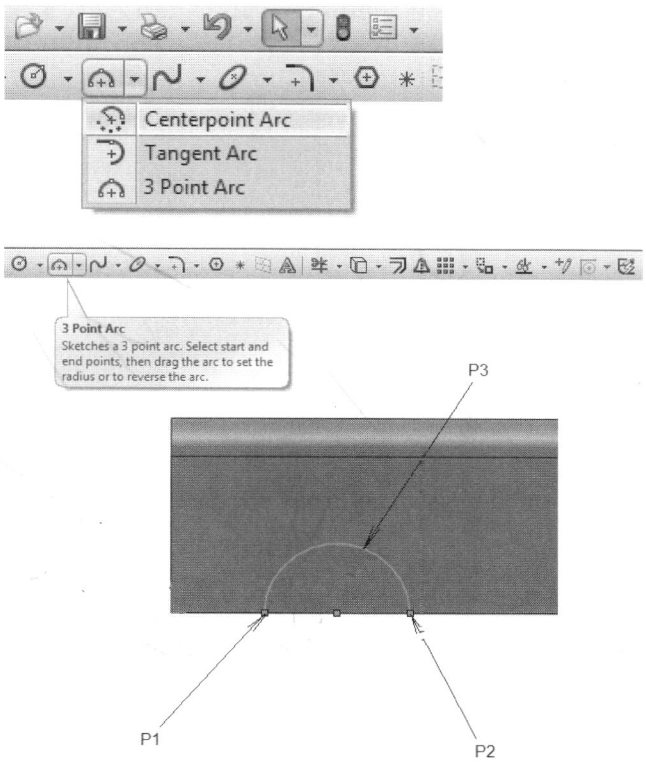

Figure 3-42
Draw a 3 Point Arc on the Side of the Base Flange

Step 15. Dimension the arc as shown in Figure 3-43.

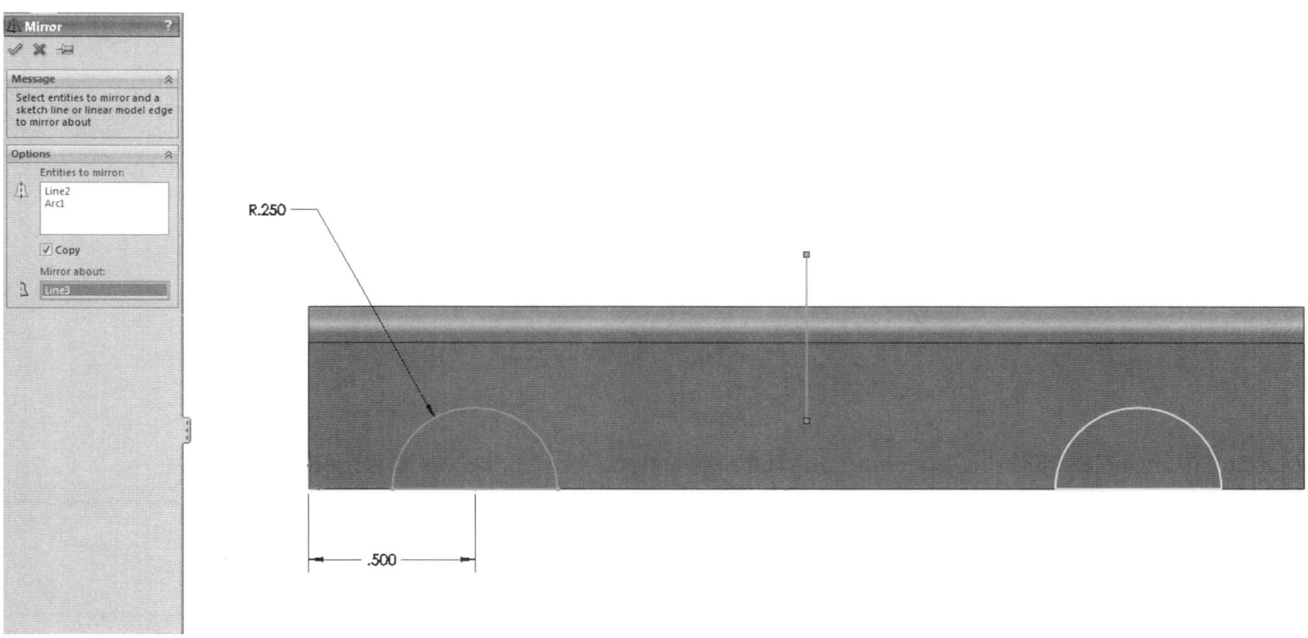

Figure 3-43
Dimension and Mirror the Arc

Step 16. Draw a line in the midpoint of the model as shown in Figure 3-43 (the line can be any length).

Step 17. Select **Mirror** and use a window to select the arc and the line connecting the ends of the arc so **Line2** and **Arc1** show in the **Entities to mirror:** box. Then, click the box under **Mirror about:** and click the line you just drew. Then, click **OK.**

Step 18. Select **Isometric view** and click **Extrude Cut** as shown in Figure 3-44 (make sure **Through All** shows in the box under **Direction 1**). Then, click **OK.**

The final model is shown in Figure 3-45.

Figure 3-44
Make the Extruded Cut

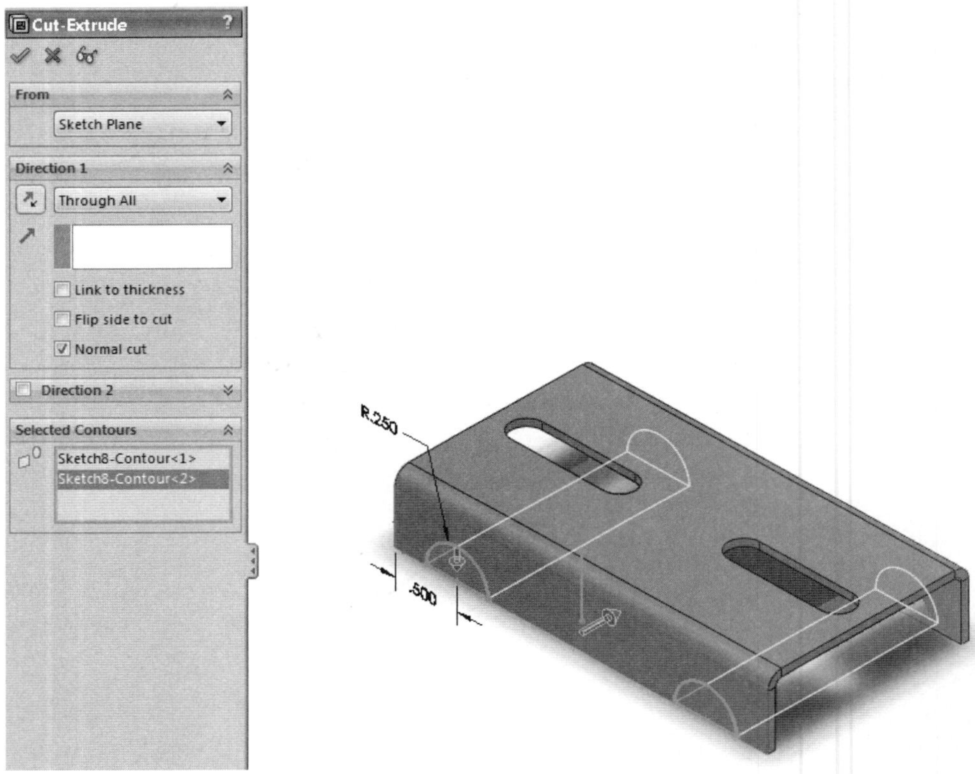

Figure 3-45
Exercise 3-2 Complete

EXERCISE 3-3 **Make the model shown in Figure 3-46.**

Step 1. Click **New....**

The **New SolidWorks Document** dialog box appears.

Figure 3-46
Exercise 3-3 Complete

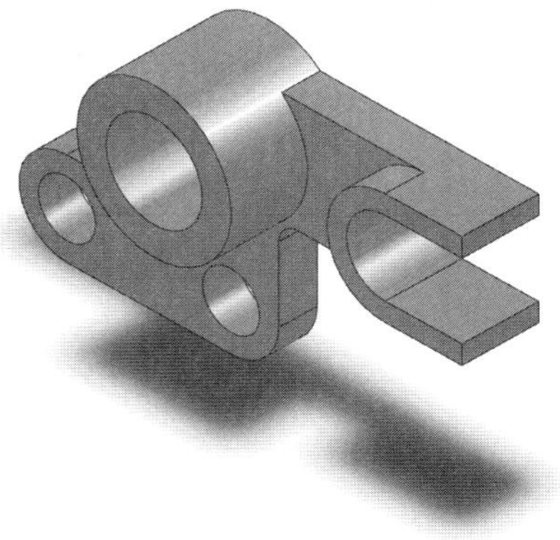

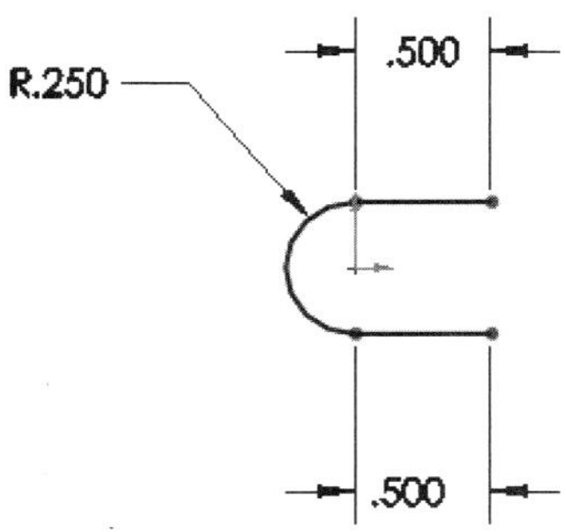

Figure 3-47
Use Centerpoint Arc and Line to Draw the Part on the Front Plane and Dimension It

Step 2. Click **Part** and click **OK.**

The SolidWorks screen appears.

Click **Options** on the menu bar, select the **Document Properties** tab, and click **Units** in the list to the left. Make sure the **Units** are correctly set [notice that **IPS** (inch, pound, second) is checked], then click **OK.**

Step 3. Click the **Front Plane** and click **Sketch** so that the Sketch mode is active.

Step 4. Click **Centerpoint Arc** and draw the arc, as shown in Figure 3-47. To draw the arc, click the center, then drag the arc around to get a semicircle as shown.

Step 5. Draw two lines from the endpoints of the two arcs and dimension the arc and the two lines as shown.

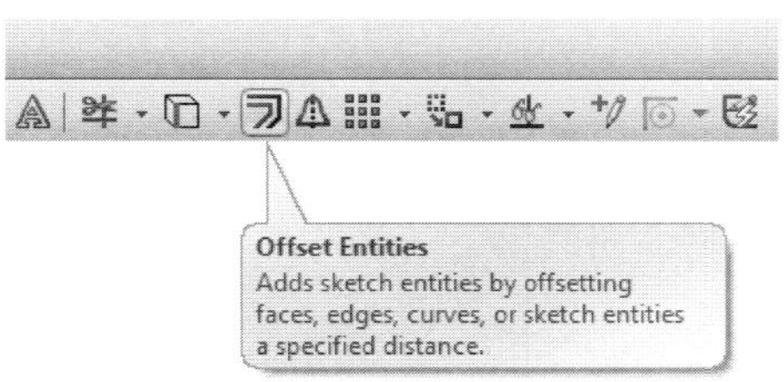

Step 6. Click **Offset** and click the arc and the two lines. If the offset is to the inside, click **Reverse** as shown in Figure 3-48. Make sure **.125** (the size of the offset) shows in the box beneath the word **Parameters,** as shown in Figure 3-48.

Figure 3-48
Use Offset to Offset the Part .125 Inch

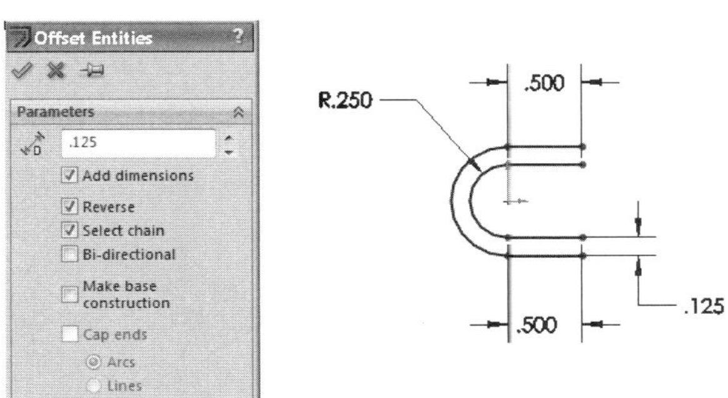

Step 7. Draw lines across the ends of the sketch.

Step 8. Click **Exit Sketch,** select **Isometric view,** and click **Extrude** as shown in Figure 3-49 (make sure **0.05 in** shows in the box next to **D1**). Then, click **OK.**

Step 9. Click the **Front Plane** and click **Sketch** so that the Sketch mode is active as shown in Figure 3-50.

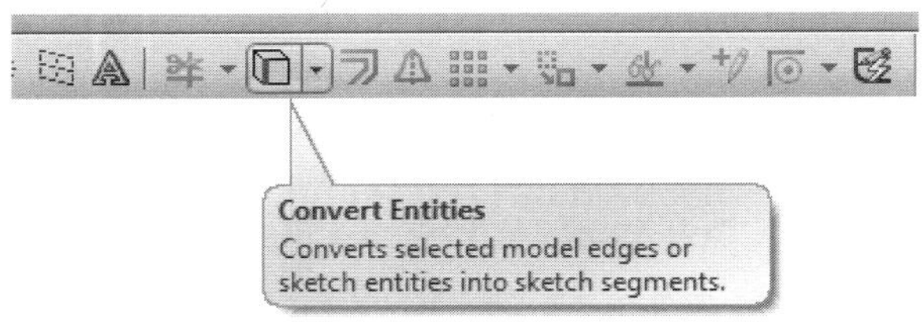

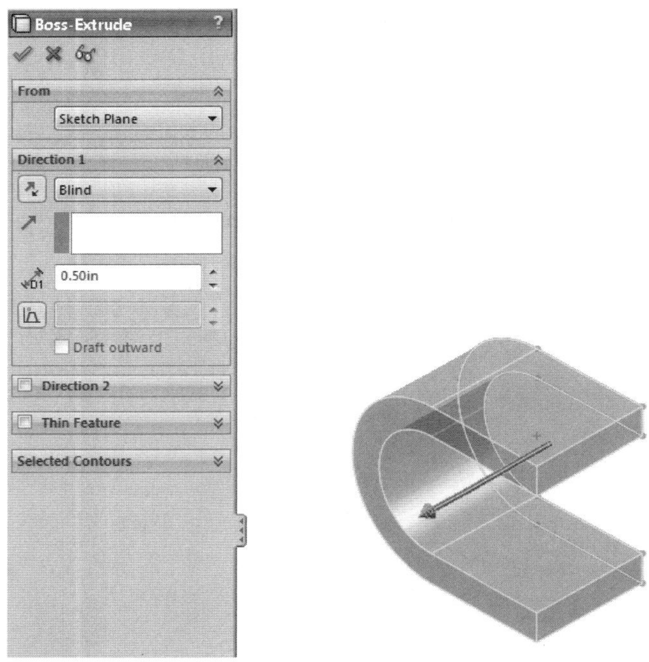

Figure 3-49
Extrude the Sketch 0.50 Inch

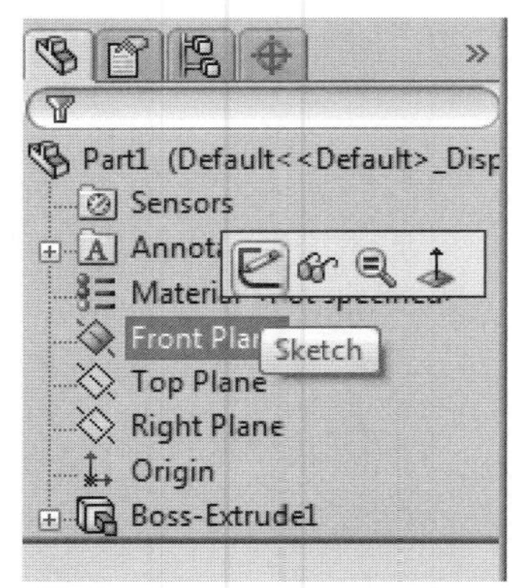

Figure 3-50
Click: the Front Plane and Click: Sketch

Step 10. Click **Convert Entities** and click **P1, P2,** and **P3,** as shown in Figure 3-51.

Step 11. Draw and dimension the circle as shown in Figure 3-51.

Step 12. Extend the two converted lines by clicking **P1** and **P2.**

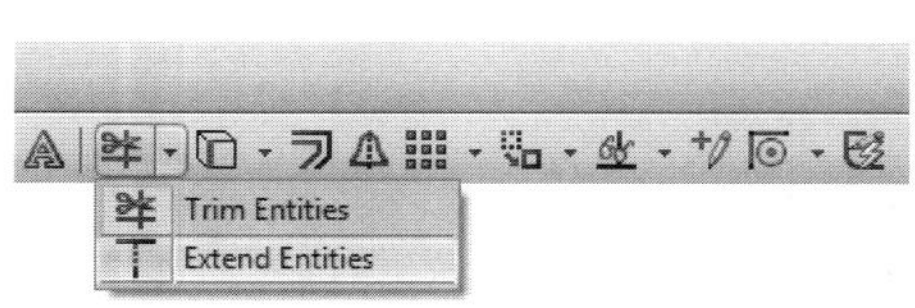

Figure 3-51
Convert and Extend Entities

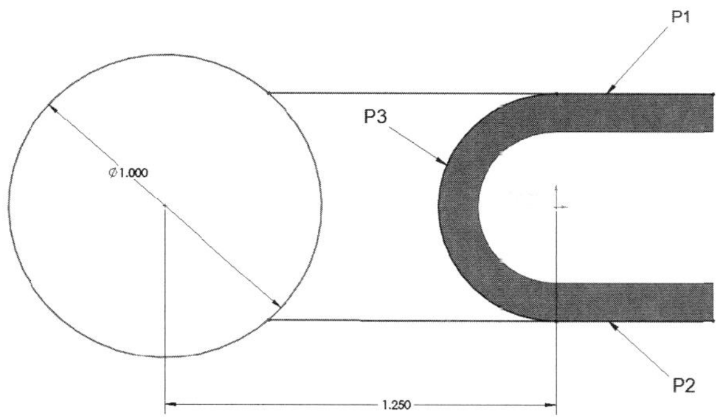

Step 13. Draw and dimension the two smaller circles as shown in Figure 3-52.

Figure 3-52
Copy the Two Circles

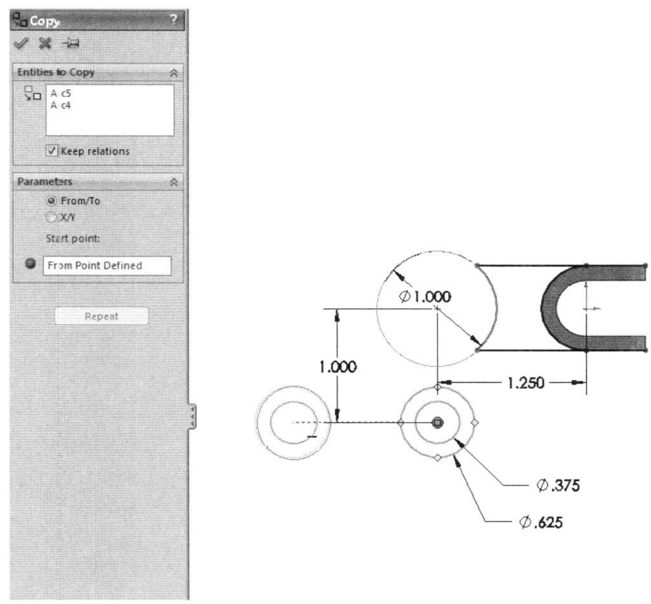

Step 14. Click **Copy Entities** and click the two smaller circles so that **Arc5** and **Arc4** appear in the **Entities to Copy** box, then click the **From/To** button. Then, click the center of the two smaller circles and copy them to the left, as shown in Figure 3-52.

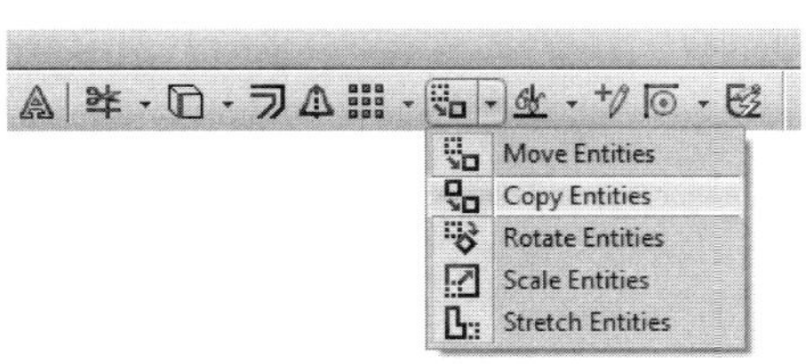

Step 15. Complete the sketch as shown in Figure 3-53. The only command you have not used so far is the sketch fillet. To do that, click **Sketch Fillet** and set the **Fillet Radius** to **0.125 inch** and click the two lines for that fillet. You will do this twice for this sketch.

Step 16. Trim the sketch as shown in Figure 3-54.

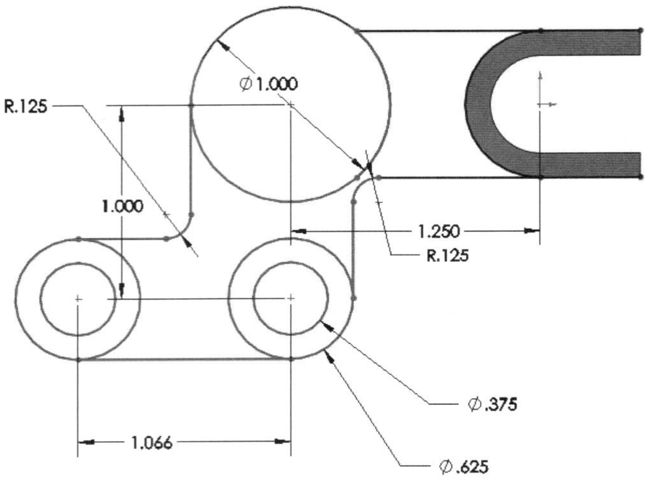

Figure 3-53
Complete the Sketch Before Trimming

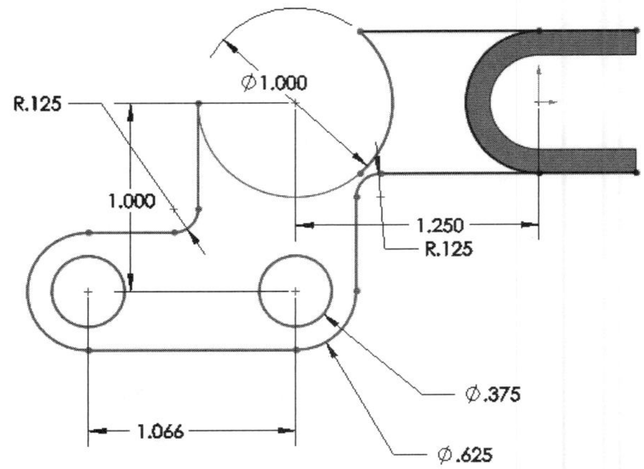

Figure 3-54
Trim the Sketch

Step 17. Click **Exit Sketch,** select **Isometric view,** and click **Extrude** as shown in Figure 3-55 (make sure **0.25 in** shows in the box next to **D1**). Then, click **OK.**

Step 18. Click the front plane and click **Sketch** so that the Sketch mode is active.

Step 19. Draw a 1.000 diameter circle and a .626 diameter circle with their centers at the center of the 1.000 diameter arc from the previous sketch and dimension them as shown in Figure 3-56.

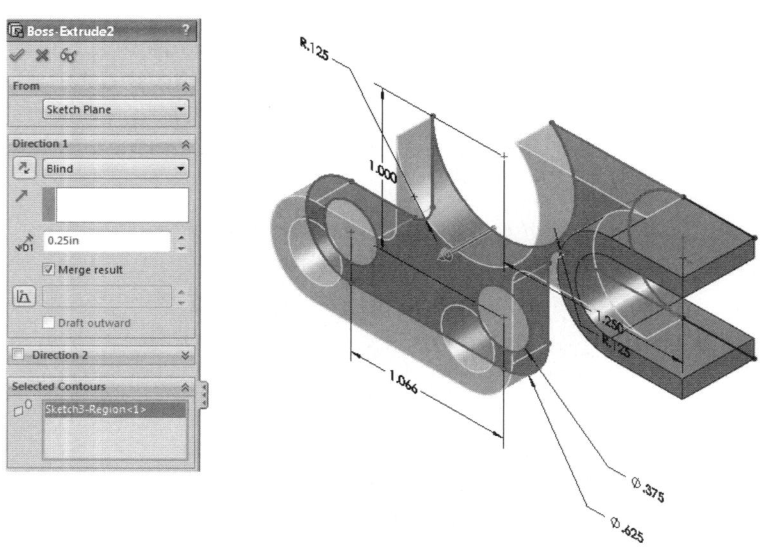

Figure 3-55
Extrude the Sketch 0.25 Inch

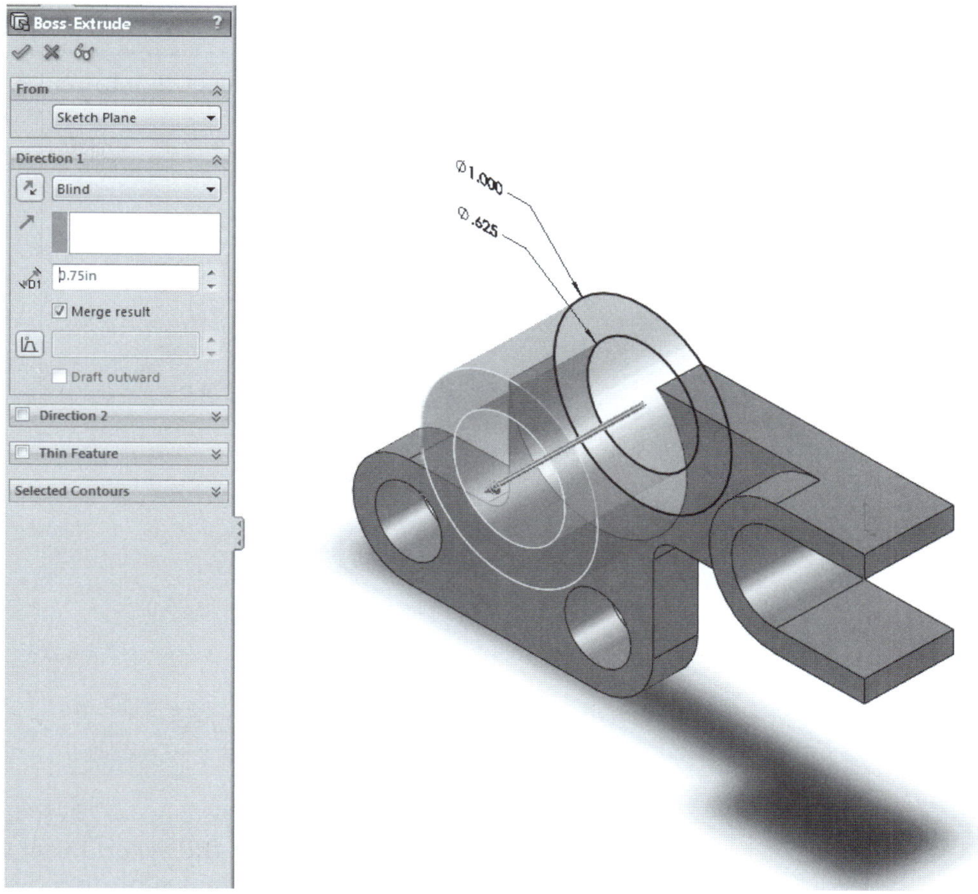

Figure 3-56
Extrude the Circles 0.75 Inch

Step 20. Select the **Isometric** view from the **Heads-Up View** toolbar.

Step 21. Click **Exit Sketch.**

Step 22. Click **Extrude Boss/Base** and extrude the two circles **0.75 inch** as shown in Figure 3-56.

The final model is shown in Figure 3-57.

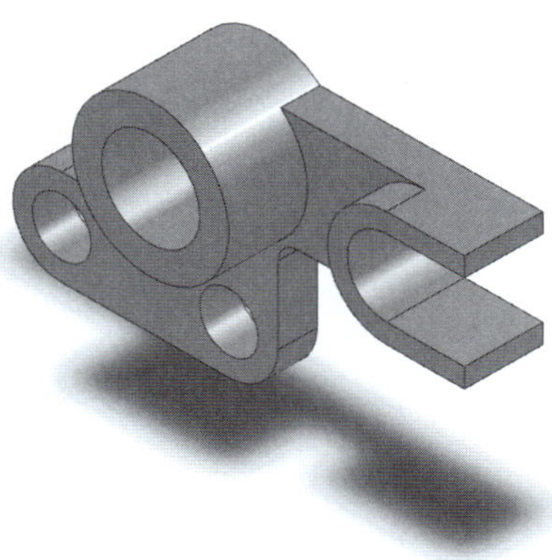

Figure 3-57
Exercise 3-3 Complete

EXERCISE 3-4 Open Exercise 3-3 and make a drawing from the part.

Use the steps described in Project 3-1 to make the drawing as shown in Figure 3-58.

Figure 3-58
Exercise 3-4 Complete

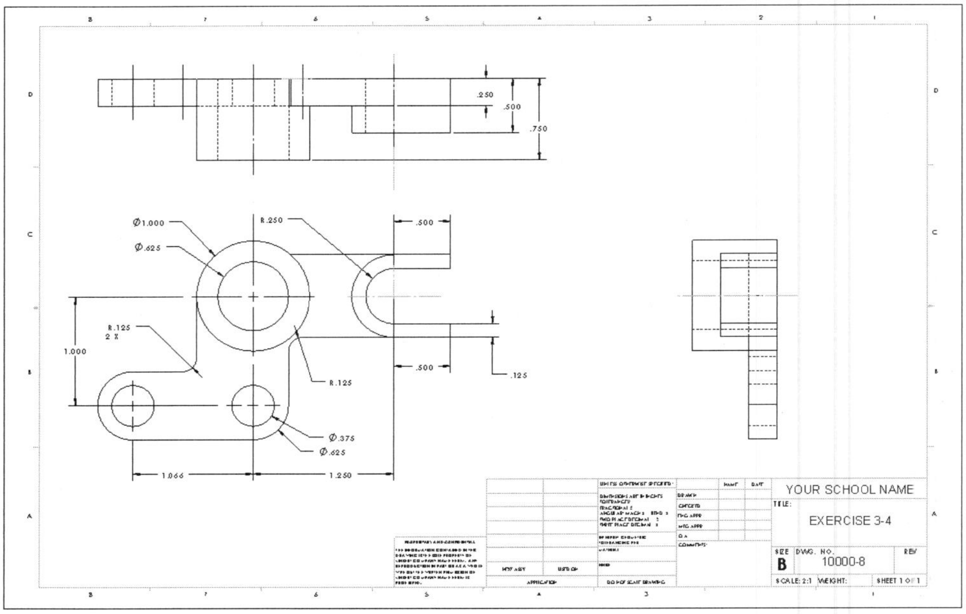

Make sure to arrange the dimensions so that the first row of dimensions is farther from the drawing than the second row is from the first row, as shown. Be sure also that no dimension touches an object line, the drawing format, or an extension line or dimension line.

3 chapter three

Chapter Summary

This chapter contains the information needed to make more complex models and drawings from those models. Now you have the skills and information necessary to make more complex models and drawings of complex models.

Review Questions

Circle the most correct answer.

1. Which of the following commands makes a rectangle most quickly?
   a. **Rectangle** c. **Line**
   b. **Square** d. **Offset**
2. Which of the following settings is needed to make sure inches is used instead of millimeters?
   a. **Notes** c. **Units**
   b. **Annotations** d. **Dimensions**
3. In which dialog box is the third angle projection method of drawing located?
   a. **Properties...** c. **Picture...**
   b. **Model Items...** d. **Object...**
4. Which is the first object drawn in a sheet metal part?
   a. Front-Flange c. Side-Flange
   b. Edge-Flange d. Base-Flange
5. In what menu on the menu bar is the **Sheet Metal** toolbar located?
   a. **File** c. **Tools**
   b. **Insert** d. **View**
6. Third angle projection is used on drawings in the United States.
   a. True b. False
7. The **Sheet Metal** toolbar does not have a command for closing the corners of flanges.
   a. True b. False
8. The **Circular Sketch Pattern** tool makes rows and columns of holes.
   a. True b. False
9. The **Straight Slot** command is a faster method of making a slot than using two circles and two lines.
   a. True b. False
10. The **Mirror** command requires a line or a plane to use as a **Mirror about:** an object.
    a. True b. False

Matching

Place the correct letter on the line to the right of the numbered item.

1. A command that is used to make a copy of lines and arcs at a larger size _____
2. A setting in the **Properties** dialog box that allows you to change how views are shown _____
3. Where the **Edge-Flange** command is located _____
4. A command that requires a line or a plane to work _____
5. A command that makes lines from the edge of a model _____

a. **Sheet Metal** toolbar
b. **Mirror** command
c. **Convert Entities** command
d. **Offset** command
e. Third angle orthographic projection

General Questions

Answer the following questions.

1. How can you use the **Sheet Metal** toolbar?
2. When would you use the inch system of units instead of metric units?
3. Why is third angle projection used here instead of first angle projection?
4. Why is it important to be consistent with how you arrange rows of dimensions?
5. Why are drawing formats necessary and what information do they have that a model drawing does not?

chapter four

Making Complex Extruded and Revolved Features, Assigning Materials to the Models, and Making Drawings with Dimensions from Those Models

CHAPTER OBJECTIVES

After completing this chapter, you will be able to:

- Make complex extruded models.
- Make complex revolved models.
- Assign materials to models.
- Correctly answer questions regarding SolidWorks commands used in this chapter.

Project 4-1: **Make a Table (Figure 4-1)**

Step 1. Start SolidWorks.

Step 2. Click **New....**

The **New SolidWorks Document** dialog box appears.

Step 3. Click **Part** and click **OK.**

Figure 4-1
Project 4-1 Complete

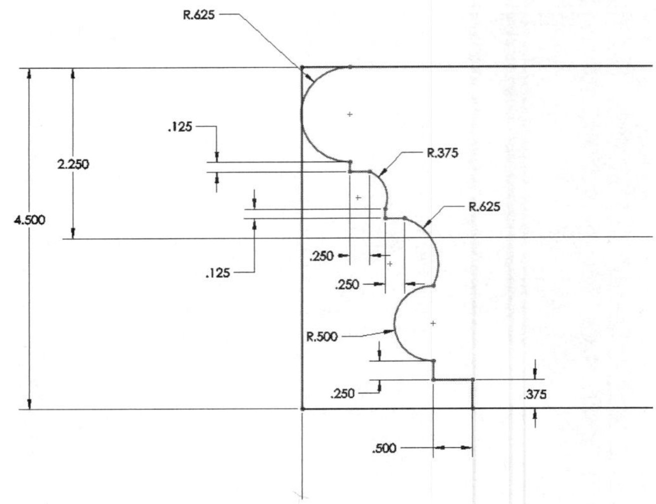

Figure 4-2
Draw and Dimension the Sketch on the Front Plane and Trim Unnecessary Lines

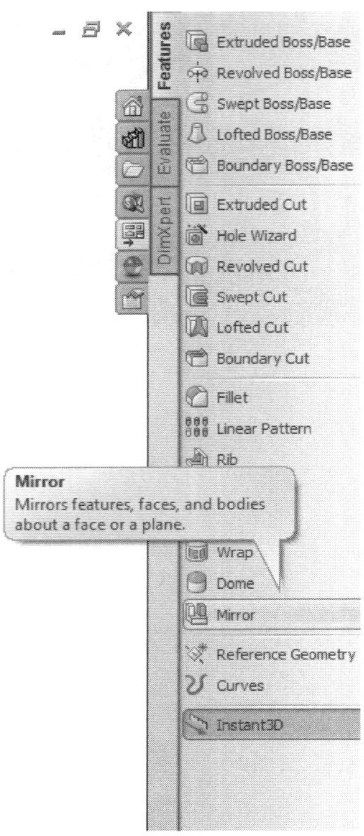

The SolidWorks screen appears. Click **Options** on the menu bar, select the **Document Properties** tab, and click **Units** in the list to the left.

Make sure the **Units** are correctly set [notice that **IPS** (inch, pound, second) is checked], then click **OK.**

Step 4. Click the **Front Plane** and click **Sketch** so that the Sketch mode is active.

Step 5. Draw a rectangle measuring 56 inches long × 4.5 inches high and draw the sketch on the left end and dimension it as shown in Figure 4-2. Trim any unnecessary lines.

Step 6. Use a window to select the sketch you just drew and trimmed and use **Copy** (press **<Ctrl>** and the **<C>** key or select **Copy** from the **Edit** menu) to copy the sketch for the **Extrude Cut** in Step 13.

Step 7. Draw a line from the origin as shown on the right end of Figure 4-3 (the line can be any length).

Step 8. Select **Mirror** and use a window to select the arcs and the lines of the sketch of Figure 4-2. Then, click the **Box** under **Mirror about:** and click the line you just drew. Then, click **OK.**

Step 9. Click **Exit Sketch.**

Step 10. Click **Extrude Boss/Base** and extrude the sketch **17 inches** as shown in Figure 4-4.

Step 11. Click the **Right** plane, click **Sketch,** press **<Ctrl>** and the **<V>** key to copy the sketch, and draw additional lines as shown in Figure 4-5. (You may have to move the sketch so it appears as shown in Figure 4-5.)

Step 12. Click **Exit Sketch.**

Step 13. Click **Extrude Cut** and cut the model with the sketch **60 inches** from the mid plane as shown in Figure 4-6.

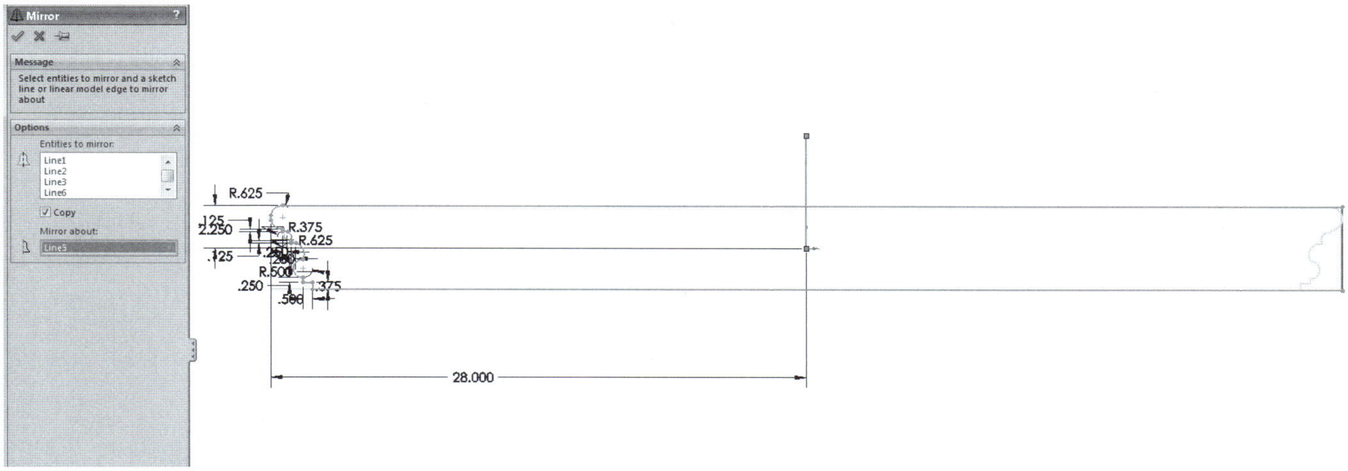

Figure 4-3
Mirror the Sketch to the Right End of the Rectangle

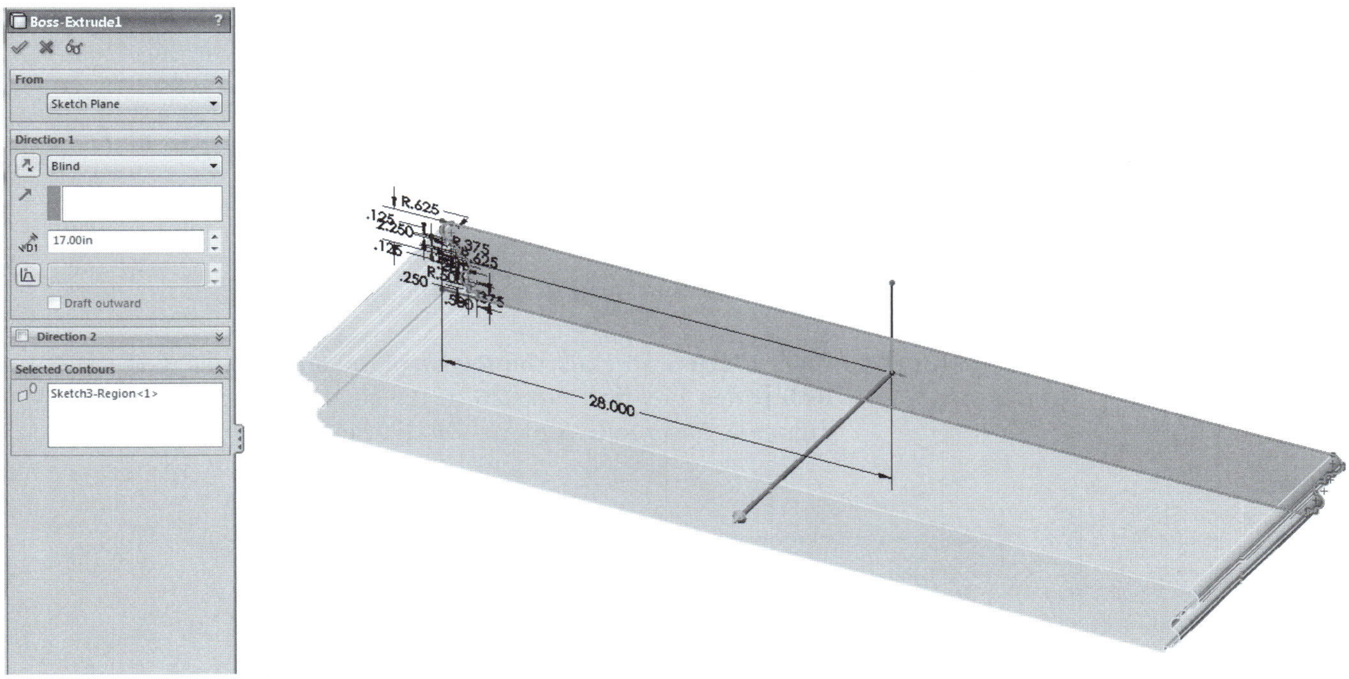

Figure 4-4
Extrude the Sketch 17 Inches

Figure 4-5
Paste the Sketch from Figure 4-2 on the Right Plane and Draw Additional Lines

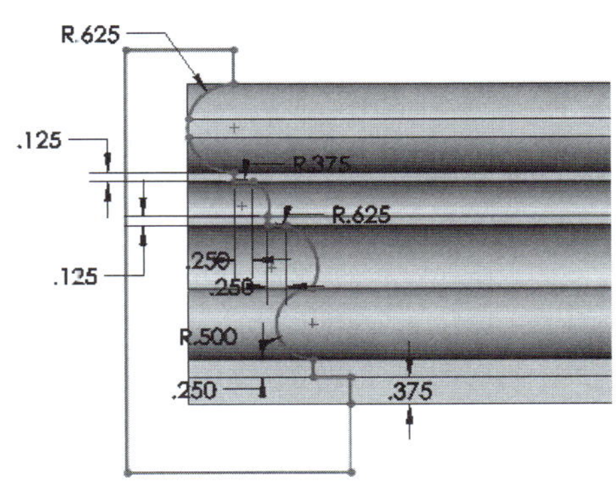

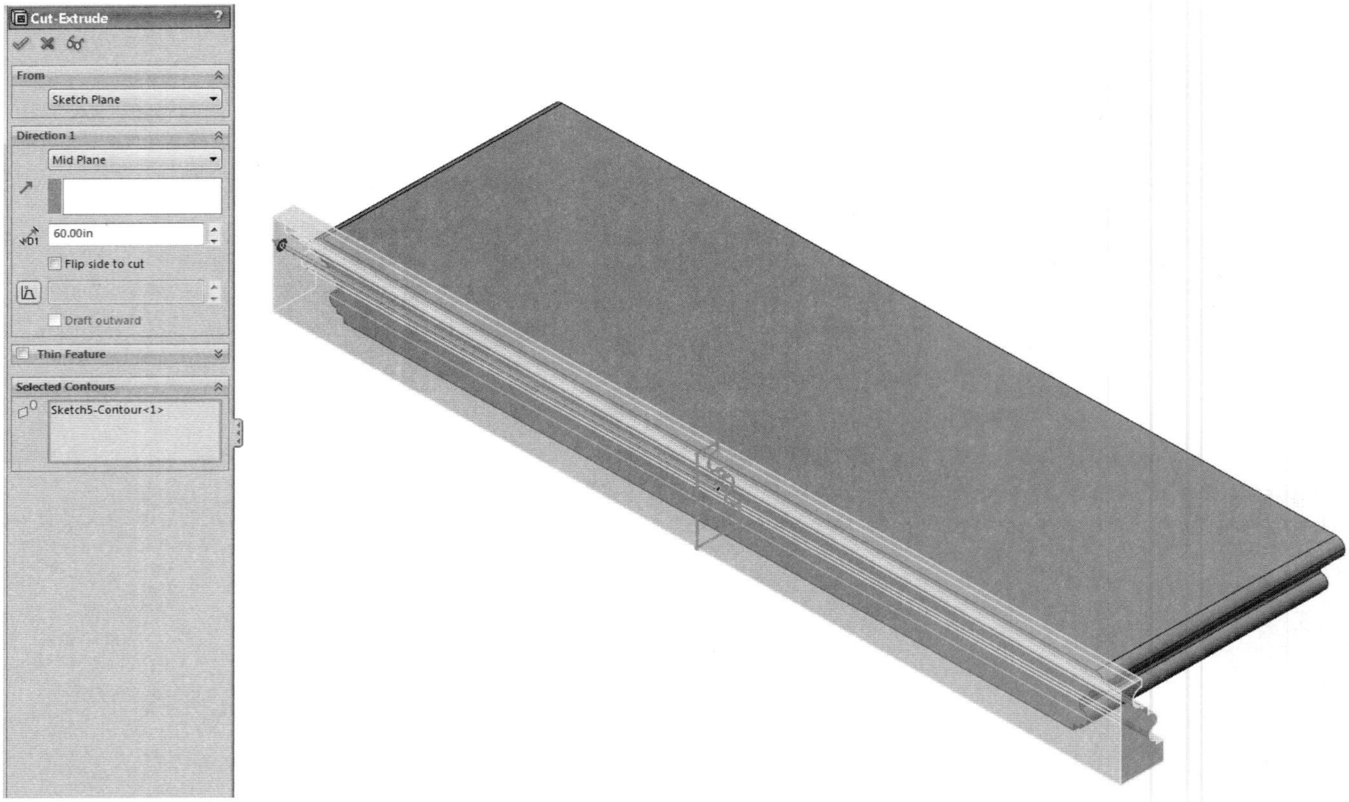

Figure 4-6
Make the Extrude Cut from the Mid Plane 60.00 Inches Long

Step 14. Click **Plane** and click the **Right** plane so it appears in the **First Reference** box and type **26** in the distance box as shown in Figure 4-7 to make a plane for the next sketch.

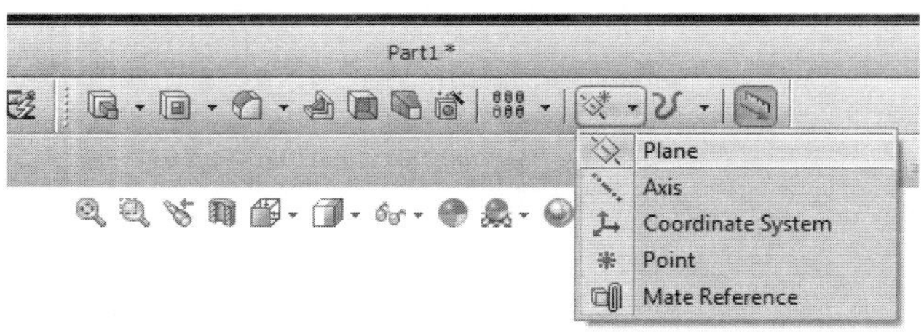

Step 15. Click **Plane1** and click **Sketch** so that the Sketch mode is active.

Step 16. Draw and dimension the sketch as shown in Figure 4-8.

Step 17. Click **Exit Sketch.**

Step 18. Click **Extrude Boss/Base** and extrude the sketch **1.375 inches** as shown in Figure 4-9.

Step 19. Select **Mirror** and select **Boss-Extrude 2** (the feature you just extruded). Then, click **Box** under **Mirror Mirror Face/Plane** and click **Right** plane as shown in Figure 4-10. Then, click **OK.**

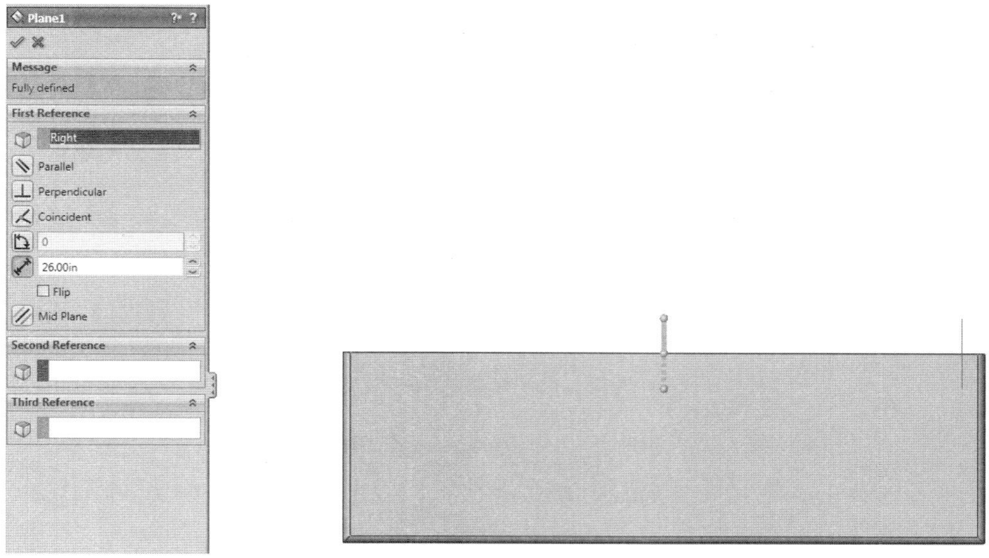

Figure 4-7
Make a New Plane 26 Inches from the Right Plane

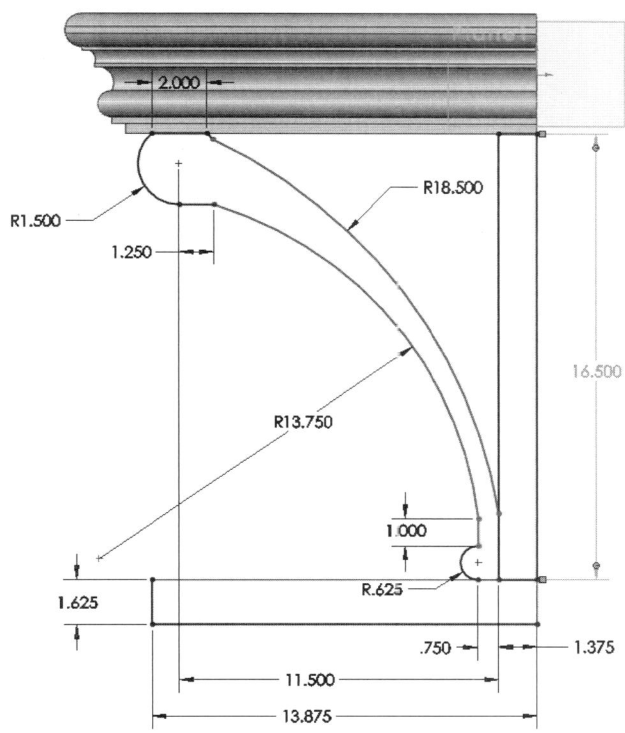

Figure 4-8
Draw and Dimension the Sketch on Plane1

Step 20. Select **Fillet,** set the fillet radius to **.50 inch** as shown in Figure 4-11, and click the front vertical edges of the feature you extruded in Figure 4-9 and mirrored in Figure 4-10.

Step 21. Click the bottom of the feature you just extruded so that the Sketch mode is active and draw and dimension the sketch as shown in Figure 4-12.

Figure 4-9
Extrude the Sketch 1.375 Inches

Figure 4-10
Mirror the Second Feature

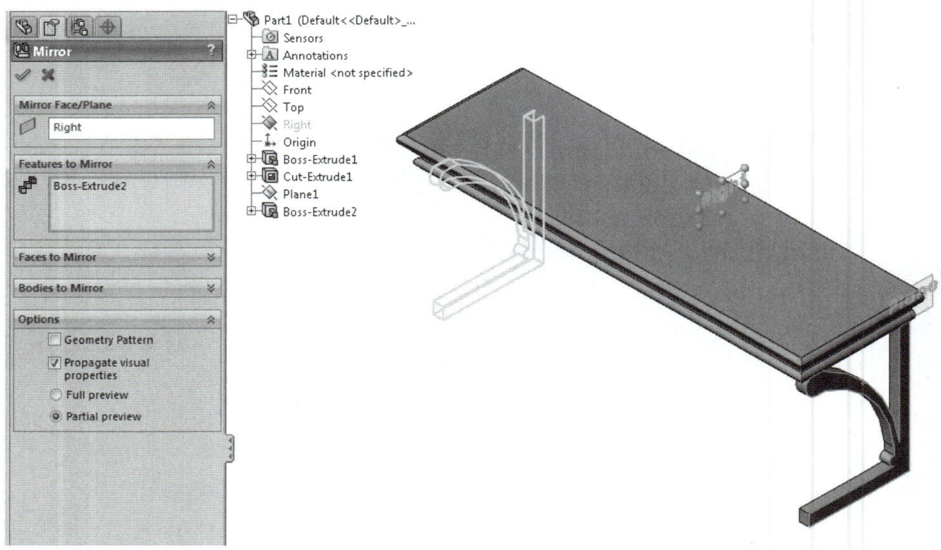

Step 22. Click **Extrude Boss/Base** and extrude the sketch **1.75 inches** as shown in Figure 4-13.

Step 23. Click **Plane,** click the **Front** plane so that it appears in the **Second Reference** box, and type **1.50** in the distance box as shown in Figure 4-14 to make a plane for the next sketch.

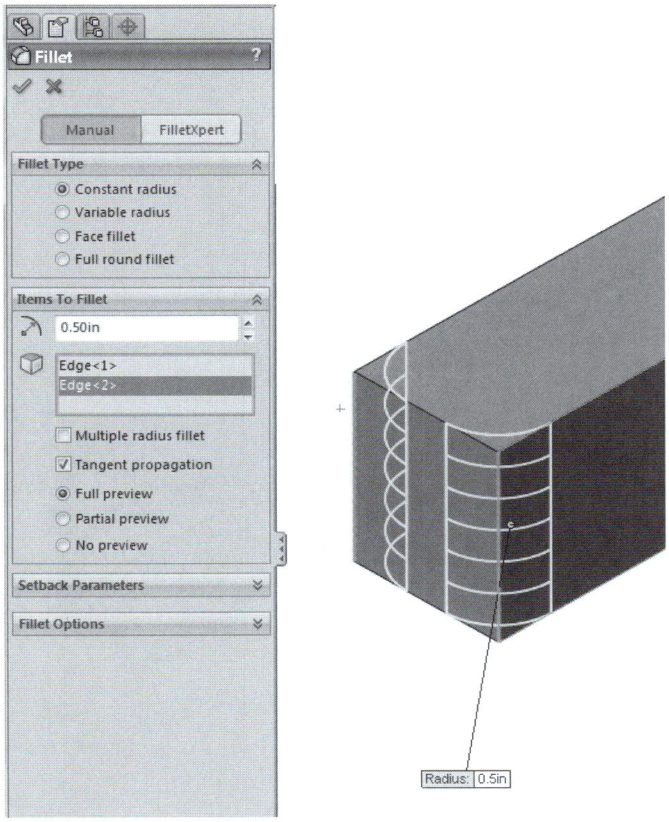

Figure 4-11
Fillet Four Vertical Edges .50 Inch

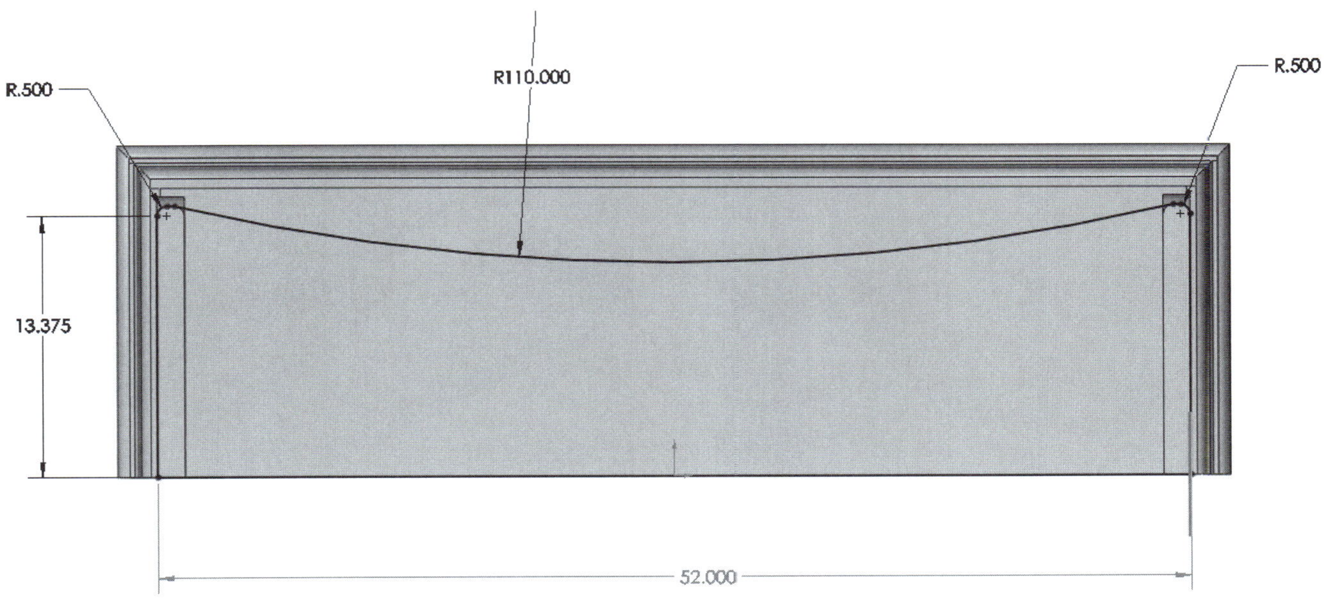

Figure 4-12
Draw and Dimension the Sketch on the Bottom of the Feature Just Extruded

Step 24. Draw and dimension the sketch as shown in Figure 4-15.

Step 25. Click **Revolved Base/Base** and click the **3.500 inches** line of the sketch to revolve the sketch as shown in Figure 4-15.

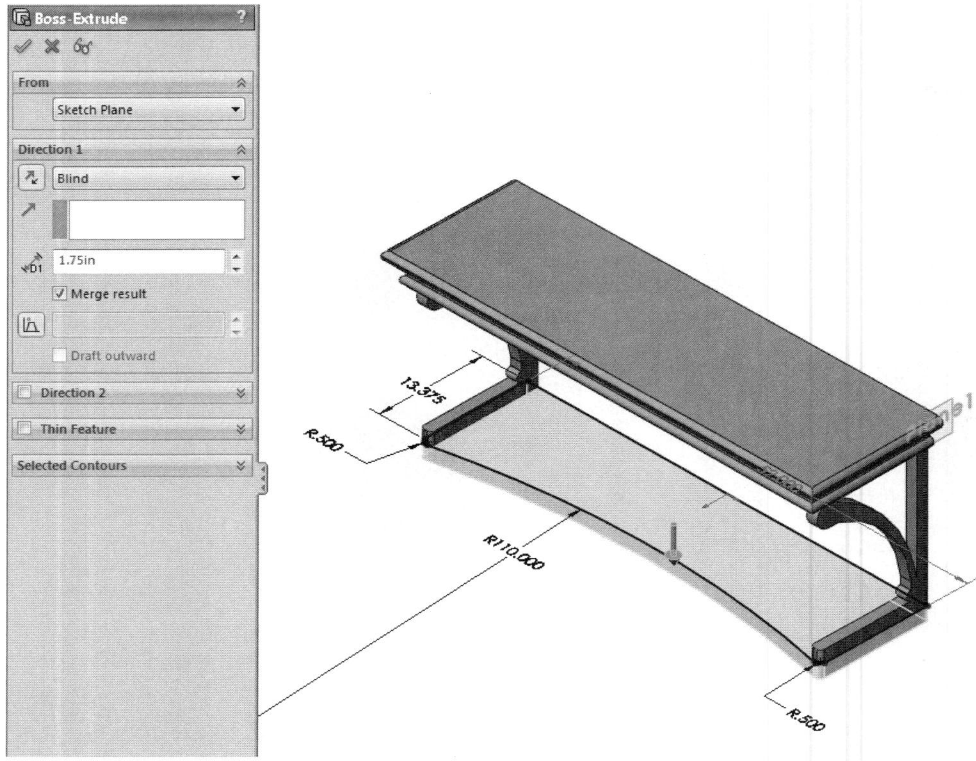

Figure 4-13
Extrude the Sketch 1.75 Inches

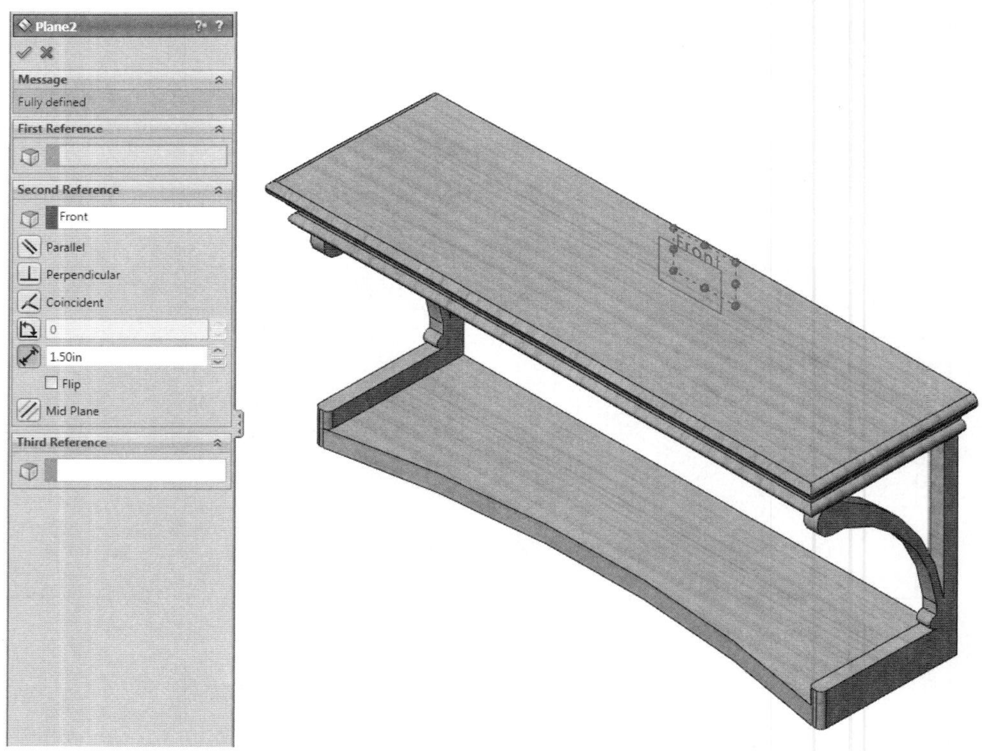

Figure 4-14
Make a Plane 1.50 Inches from the Front Plane

Step 26. Click **Bottom view** and click **Linear Pattern** as shown in Figure 4-16.

Click the revolved sketch you just made so it appears in the **Features to Pattern** box.

Figure 4-15
Make the Sketch for the Revolve Boss/Base on Plane 2 and Revolve It

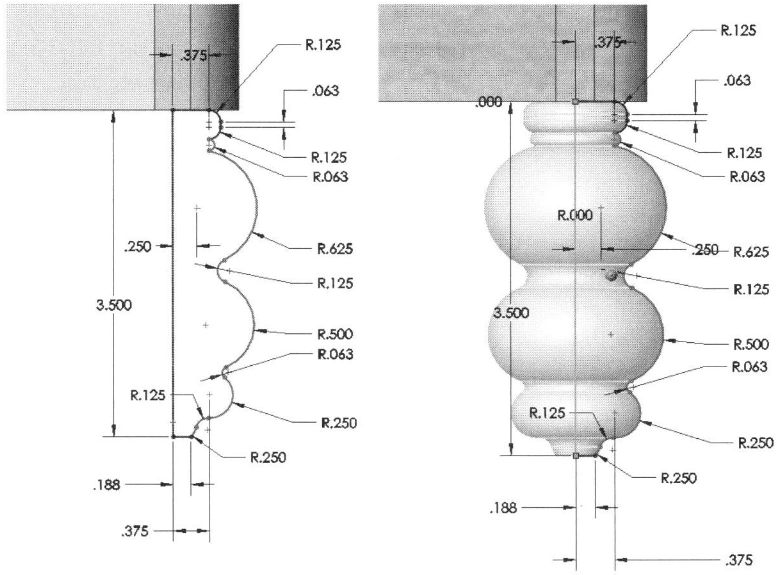

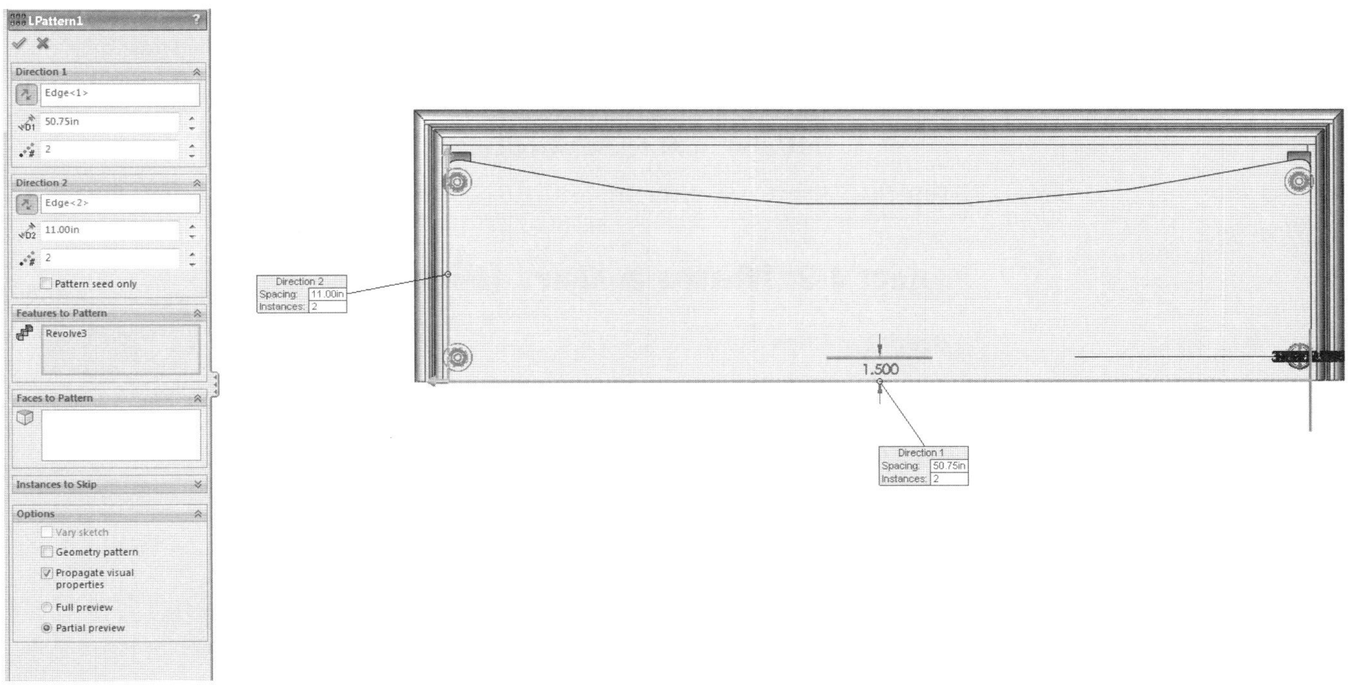

Figure 4-16
Make Settings for the Linear Pattern and Make the Pattern

Click the long edge of the model so it appears in the box under **Direction 1** and type **50.75 inches** in the **D1** box and **2** in the **Number of Instances** box.

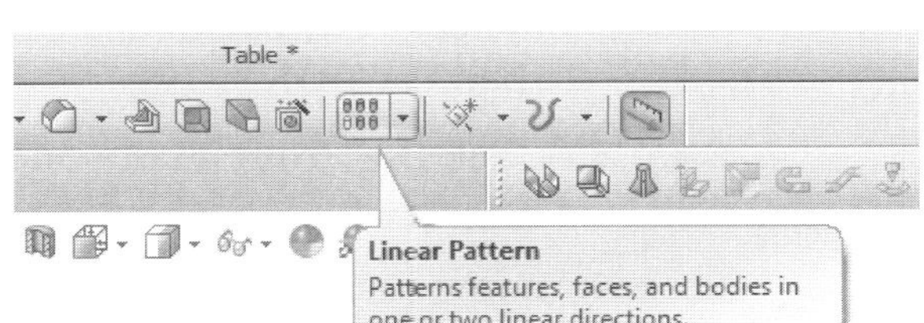

Click a vertical edge of the model so that it appears in the box under **Direction 2** and type **11.00 inches** in the **D2** box and **2** in the **Number of Instances** box.

Step 27. Click **OK.**

Step 28. Click **Isometric view** so that the model appears as shown in Figure 4-17.

Figure 4-17
Project 4-1 Complete

Step 29. Save your drawing (Figure 4-17) with the file name **Project 4-1(your initials)** in at least two places.

Step 30. Print the drawing and turn it in.

Project 4-2: **Make a Dart (Figure 4-18)**

Step 1. Click **New....**

The **New SolidWorks Document** dialog box appears.

Figure 4-18
Project 4-2 Complete

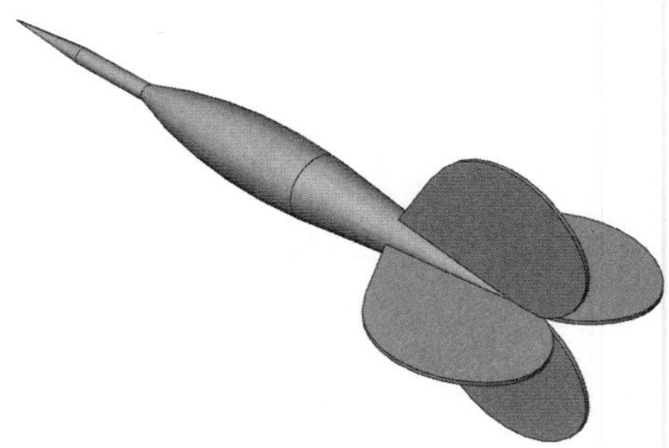

Step 2. Click **Part** and click **OK.**

The SolidWorks screen appears.

Click **Options** on the menu bar, select the **Document Properties** tab, and click **Units** in the list to the left. Make sure the **Units** are correctly set [notice that **IPS** (inch, pound, second) is checked], then click **OK.**

Step 3. Click the **Front Plane** and click **Sketch** so that the Sketch mode is active.

Step 4. Draw the sketch for a **Revolve Boss/Base** operation and dimension it as shown in Figure 4-19.

Figure 4-19
Make a Sketch for a Revolve Boss/Base on the Front Plane and Revolve it

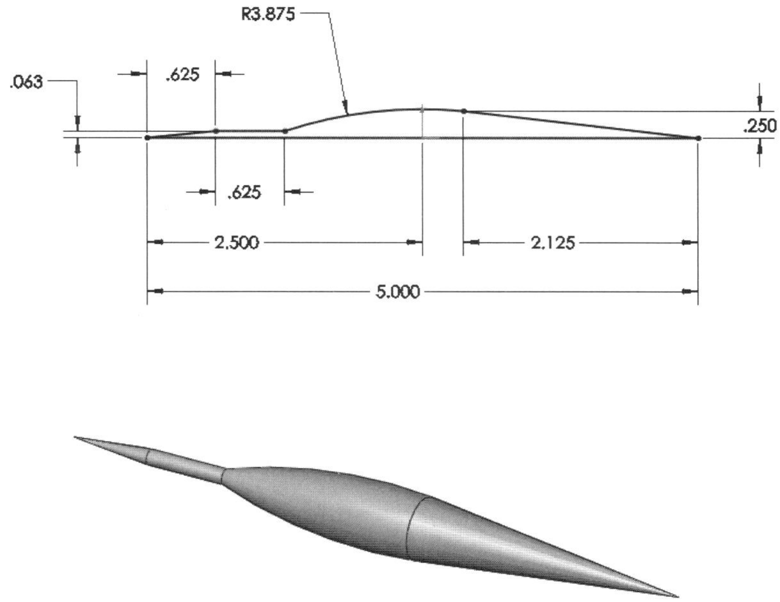

Step 5. Click **Exit Sketch.**

Step 6. Click **Revolve Boss/Base** and click the **5.000** line as shown in Figure 4-19.

Step 7. Click **OK.**

Step 8. Draw a sketch using **Line, 3 Point Arc,** and **Spline** with the dimensions shown in Figure 4-20. Draw the **3 Point Arc** at the lower right corner of the dart, then use **Spline** to complete the curved part of the sketch. Be sure you have a line connecting the end of the 3 point arc and the left end of the spline. You will have to approximate some measurements because the spline cannot be dimensioned.

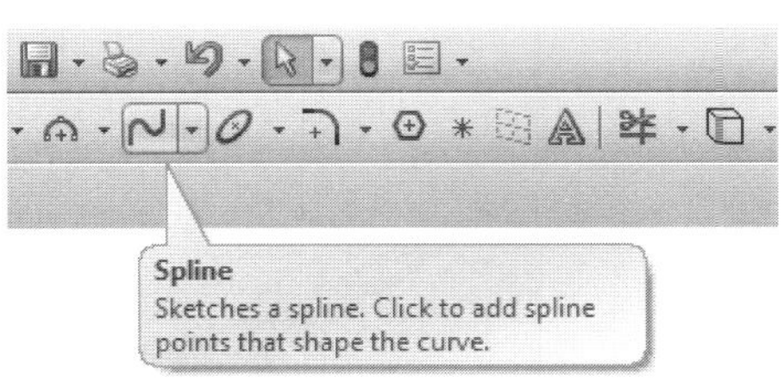

Figure 4-20
Draw a Sketch Using Line, 3 Point Arc, and Spline Using the Dimensions Shown. (Approximate Any Other Measurements)

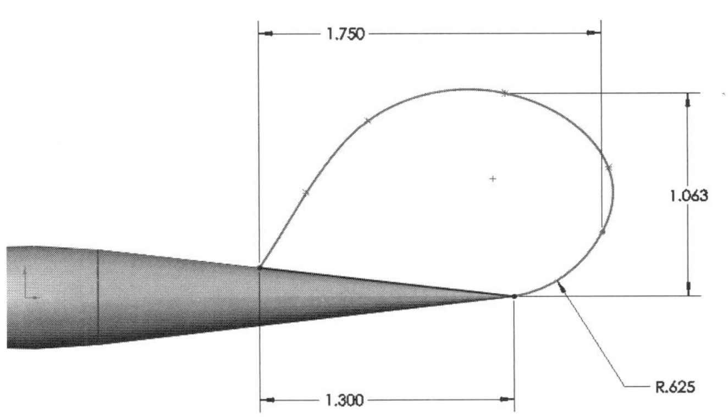

Step 9. Draw a horizontal line of any length from the origin for the **Mirror** command. Click the line of the sketch, the 3 point arc, and the spline so that they appear in the **Entities to mirror:** box and click the horizontal line from the origin so that it appears in the **Mirror about:** box as shown in Figure 4-21. Then, click **OK.**

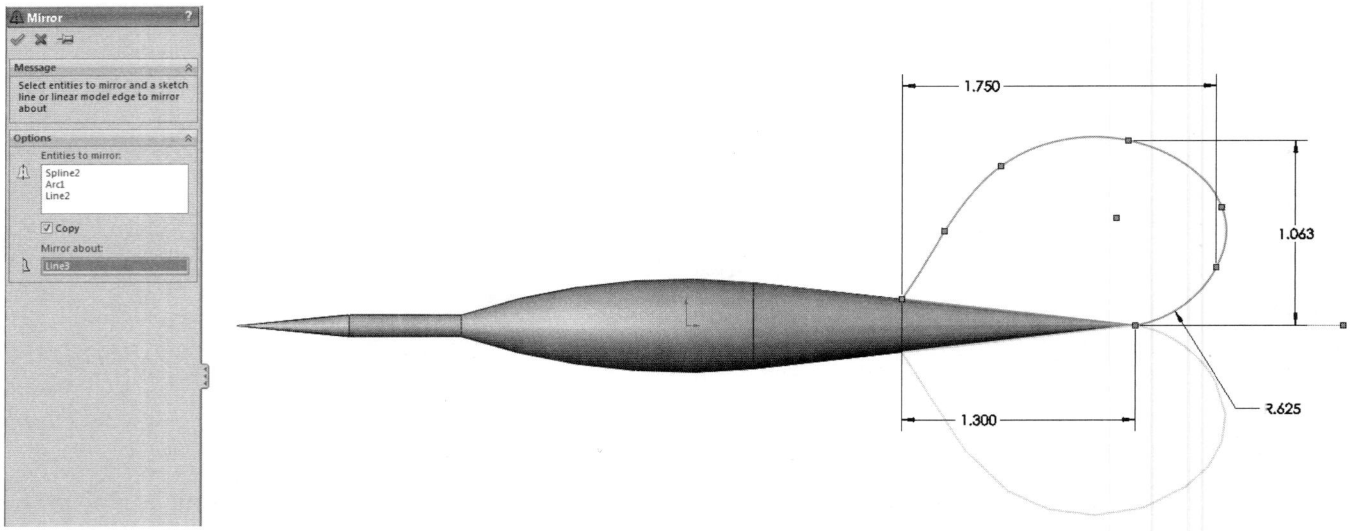

Figure 4-21
Mirror the Sketch

Step 10. Use a window to select the sketch you just drew and mirrored and use **Copy** (press **<Ctrl>** and the **<C>** key or select **Copy** from the **Edit** menu) to copy the sketch for the **Extrude** in Step 11.

Step 11. **Extrude** the sketch **.03 inch** from the **Mid Plane** as shown in Figure 4-22.

Figure 4-22
Extrude the Sketch .03 Inch from the Mid Plane

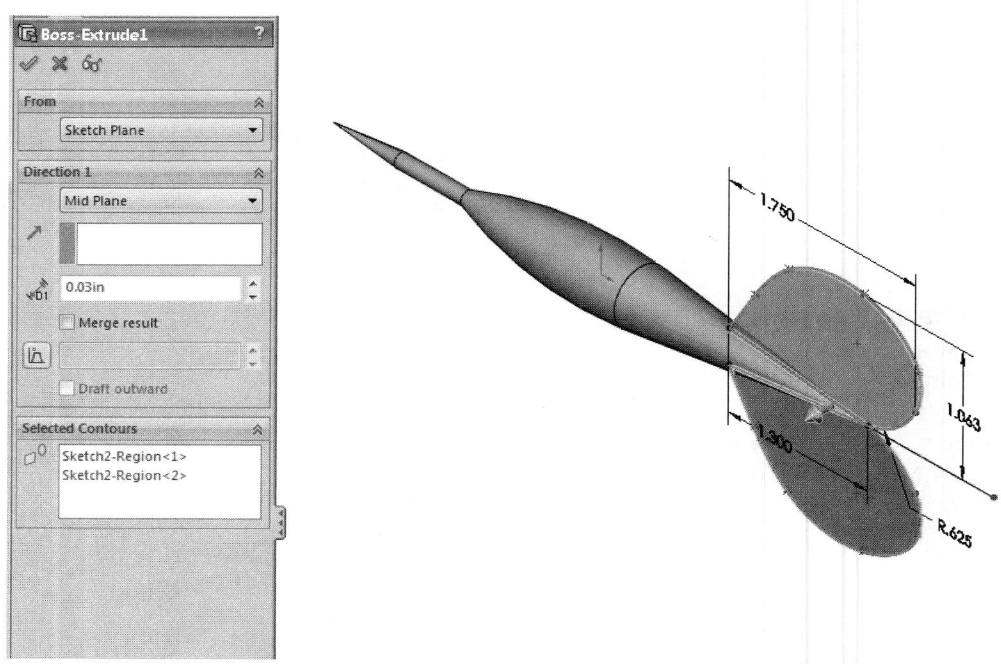

Step 12. Click the top plane, click **Sketch,** and press **<Ctrl>** and the **<V>** key to copy the sketch as shown in Figure 4-23. (You may have to move the sketch so that it appears as shown in Figure 4-23.)

Step 13. Right-click **Material <not specified>** in the **Project Manager** and click **Edit Material** as shown in Figure 4-24.

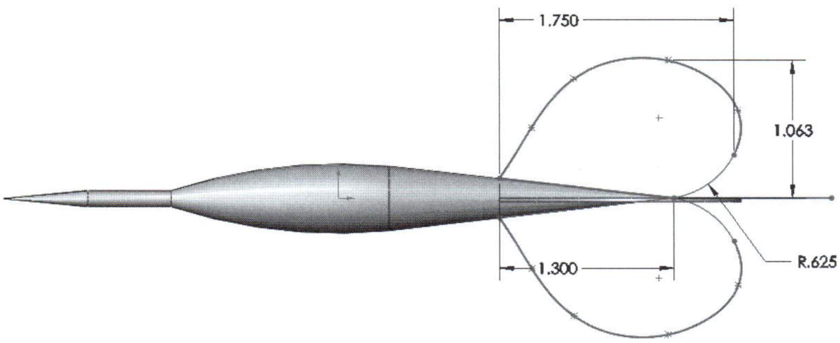

Figure 4-23
Click: the Top Plane, Click: Sketch, Press: <Ctrl> and the <V> Key to Copy the Sketch from Step 10. (You May Have to Move the Sketch so It Appears as Shown)

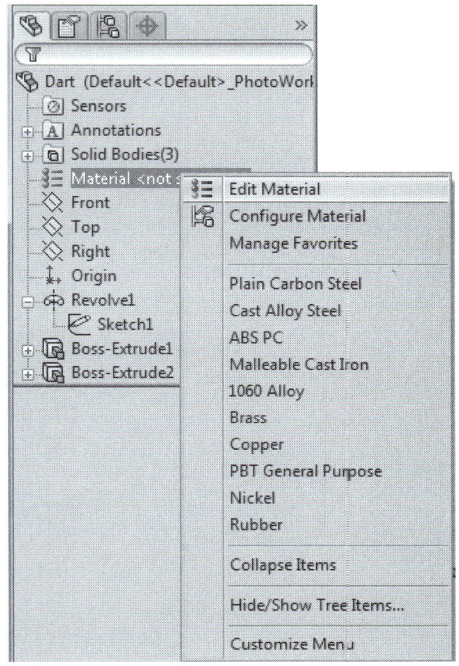

Figure 4-24
Right Click: Material <not specified> in the Project Manager and Click: Edit Material

Step 14. Use **Edit Material** to assign material to the model as shown in Figure 4-25.

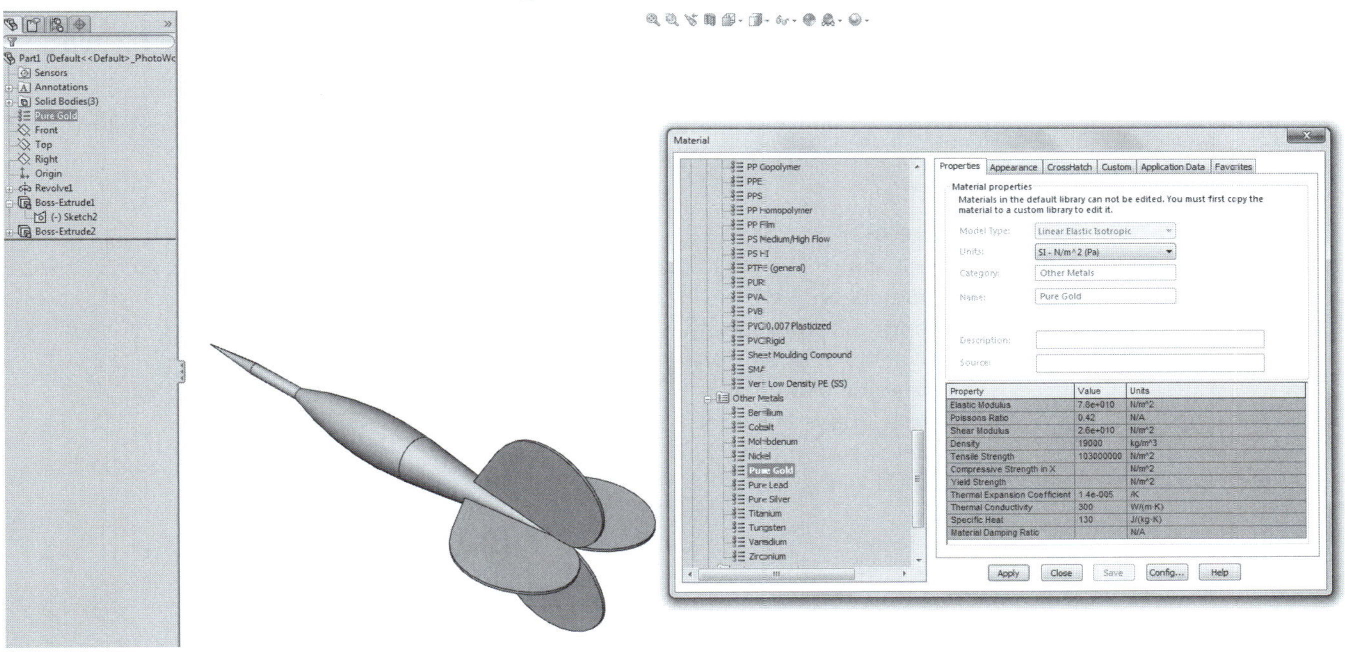

Figure 4-25
Assign Material to the Model Using Edit Material

Step 15. Click **Apply** and then click **Close.**

Step 16. Click **Isometric view** so that the model appears as shown in Figure 4-26.

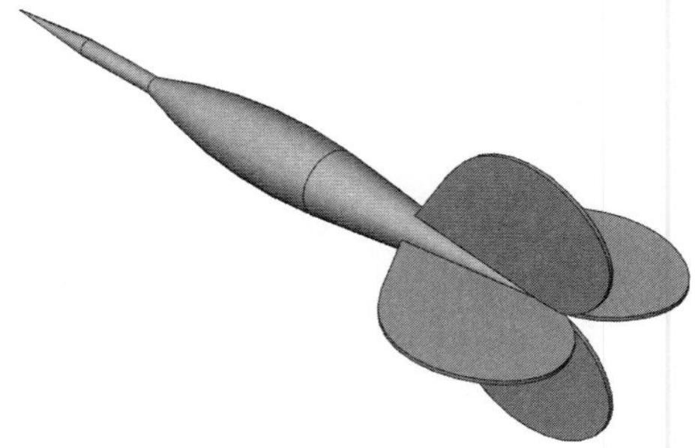

Figure 4-26
Project 4-2 Complete

Step 17. Save your drawing (Figure 4-26) with the file name **Project 4-2(your initials)** in at least two places.

Step 18. Print the drawing and turn it in.

Project 4-3: **Make a New Part in Millimeters**

Step 1. Click **New....**

The **New SolidWorks Document** dialog box appears.

Step 2. Click **Part** and click **OK.**

The SolidWorks screen appears.

Click **Options** on the menu bar, select the **Document Properties** tab, and click **Units** in the list to the left. Make sure the **Units** are correctly set [notice that **MMGS** (millimeter, gram, second) is checked], then click **OK.**

Step 3. Click the **Front Plane** and click **Sketch** so that the Sketch mode is active.

Step 4. Draw the sketch for a **Revolve Boss/Base** and dimension it as shown in Figure 4-27.

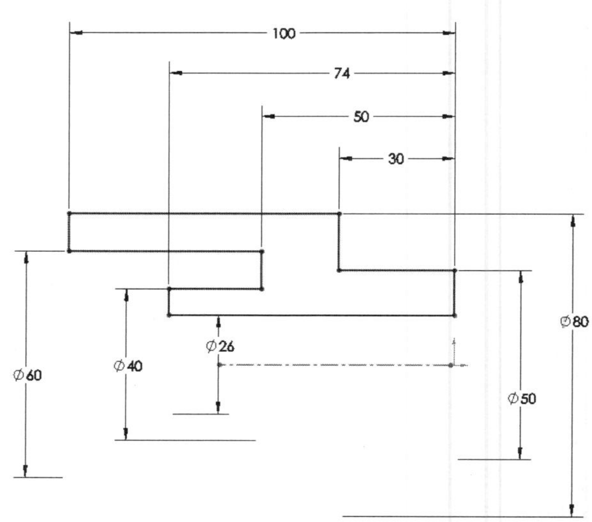

Figure 4-27
Draw a Sketch in Millimeters on the Front Plane for a Revolve and Revolve It. Be Sure to Draw a Centerline and Click Below It so the Diameter Symbol Shows in the Diameter Dimension

Make sure to use the **Centerline** command when you make the sketch, that the line for the revolve is made with the **Centerline** command, and that when you make the diameter dimensions you click below the centerline so that the dimension shows a diameter symbol as shown in Figure 4-27.

Step 5. Click **Exit Sketch.**

Step 6. Click **Revolve Boss/Base** and click the centerline as shown in Figure 4-27.

Step 7. Click **OK.**

Step 8. Click **Fillet** on the **Features** tab, set the radius to **5.00** and click the three edges shown in Figure 4-28.

Figure 4-28
Make 5.00 Millimeter Fillets on Three Edges. Use Edit Material to Assign a Material.
Project 4-3 Complete

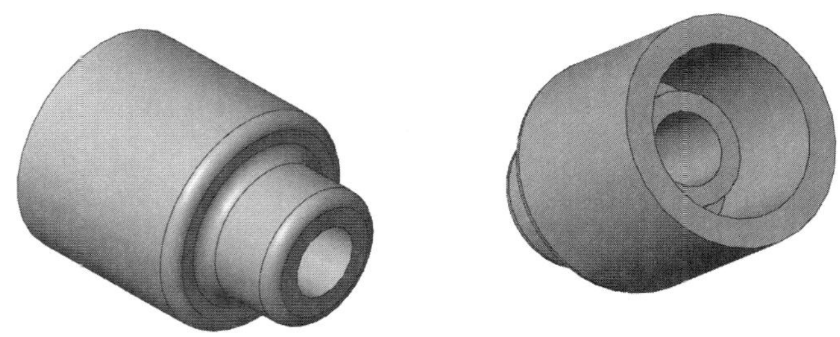

Step 9. Right-click **Material <not specified>** in the **Project Manager** and click **Edit Material.**

Step 10. Use **Edit Material** to assign material to the model as shown in Figure 4-28.

Step 11. Click **Apply** and then click **Close.**

Step 12. Click **Isometric view** so that the model appears as shown in the left side of Figure 4-28.

Step 13. Save your drawing (Figure 4-28) with the file name **Project 4-3(your initials)** in at least two places.

Step 14. Print the drawing and turn it in.

Project 4-4: Make New Part in Inches

Step 1. Click **New....**

The **New SolidWorks Document** dialog box appears.

Step 2. Click **Part** and click **OK.**

The SolidWorks screen appears.

Click **Options** on the menu bar, select the **Document Properties** tab, and click **Units** in the list to the left. Make sure the **Units** are correctly set [notice that **IPS** (inch, pound, second) is checked], then click **OK.**

Step 3. Click the **Front Plane** and click **Sketch** so that the Sketch mode is active.

Step 4. Draw the sketch for a **Revolve Boss/Base** and dimension it as shown in Figure 4-29.

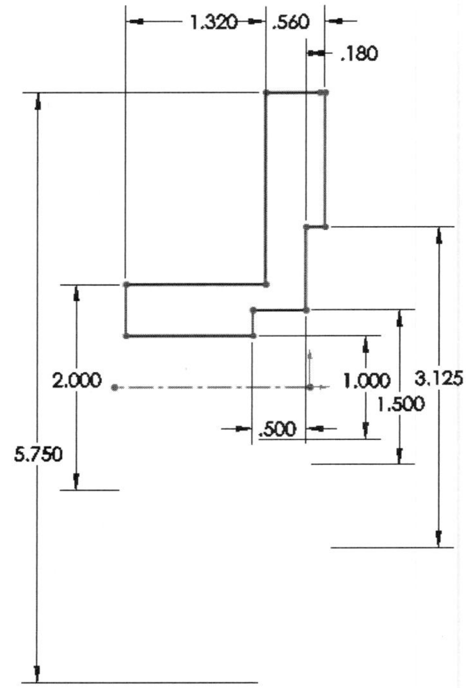

Figure 4-29
Make a Sketch for a Revolved Boss/Base on the Front Plane and Revolve It

Make sure that when you make the sketch, the line for the revolve is made with the **Centerline** command and that when you make the diameter dimensions you click below the centerline so that the diameter dimensions are as shown in Figure 4-29.

Step 5. Click **Exit Sketch.**

Step 6. Click **Revolve Boss/Base** and click the centerline as shown in Figure 4-29.

Step 7. Click **OK.**

Step 8. Draw a circle for an **Extrude Cut** on the 5.750 diameter surface and dimension it as shown in Figure 4-30.

Step 9. Make a circular pattern of the circle using 8 instances, **Arc1** (the .500 diameter circle) as the **Entities to Pattern,** and **Point-1** or a circular edge of the model as the center (the first text box under **Parameters** in the **Circular Pattern Manager**) as shown in Figure 4-30.

Make the **Extrude Cut Through All.**

Step 10. Click **Fillet** on the **Features** tab, set the radius to **.125,** and click the three edges shown in Figure 4-31.

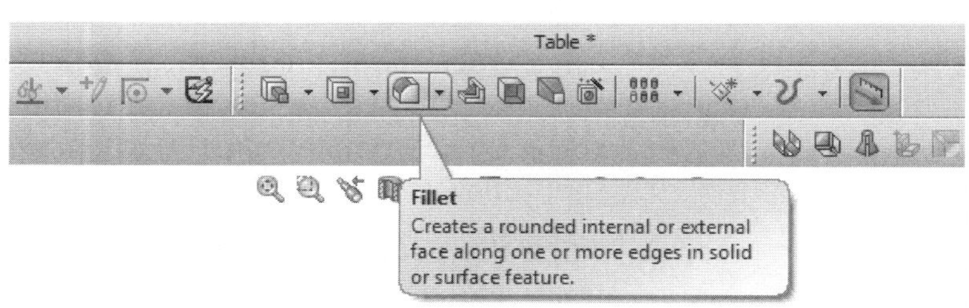

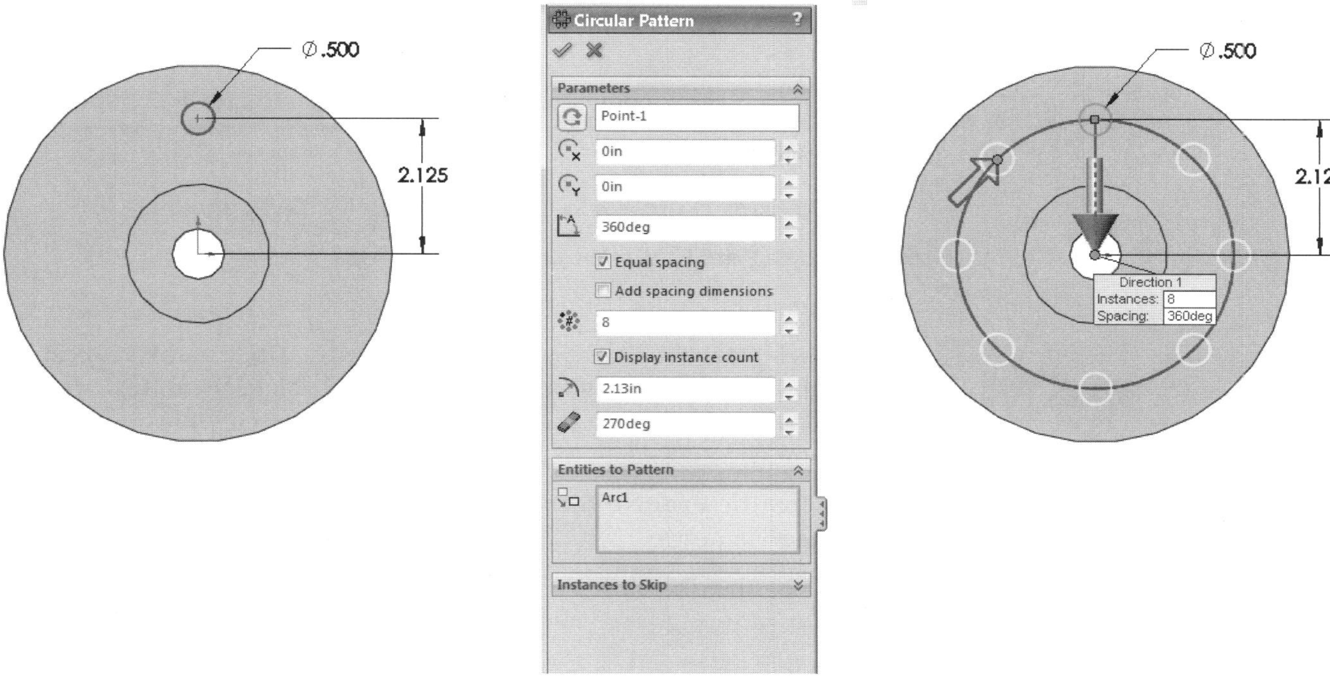

Figure 4-30
Draw a Circle for an Extrude Cut on the 5.750 Diameter Surface and Dimension It. Make a Circular Pattern of the Circle as Shown in the Circular Pattern Manager. Make the Extrude Cut Through All

Figure 4-31
Make .125 Radius Fillets on Three Edges and Use Edit Material to Assign a Material.
Project 4-4 Complete

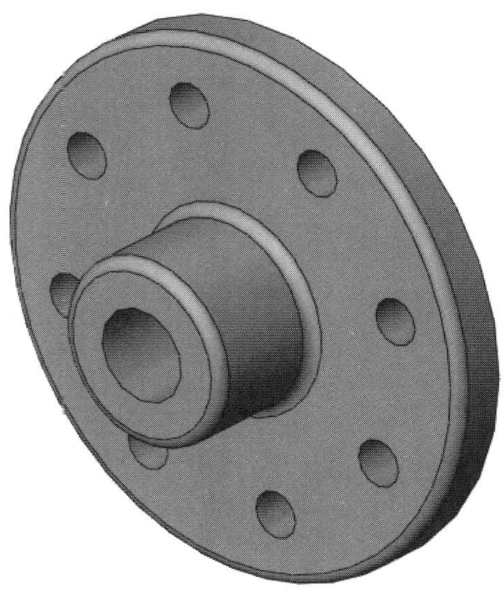

Step 11. Right-click **Material <not specified>** in the **Project Manager** and click **Edit Material.**

Step 12. Use **Edit Material** to assign material to the model as shown in Figure 4-31.

Step 13. Click **Apply** and then click **Close.**

Step 14. Click **Isometric view** so that the model appears as shown in Figure 4-31.

Step 15. Save your drawing (Figure 4-31) with the file name **Project 4-4(your initials)** in at least two places.

Exercises

EXERCISE 4-1 **Make a model from the sketch shown in Figures 4-32 through 4-36.**

Step 1. Click **New....**

The **New SolidWorks Document** dialog box appears.

Step 2. Click **Part** and click **OK.**

The SolidWorks screen appears.

Click **Options** on the menu bar, select the **Document Properties** tab, and click **Units** in the list to the left. Make sure the **Units** are correctly set [notice that **IPS** (inch, pound, second) is checked], then click **OK.**

Step 3. Click the **Front Plane** and click **Sketch** so that the Sketch mode is active.

Step 4. Draw the sketch for a **Revolve Boss/Base** and dimension it as shown in Figure 4-32.

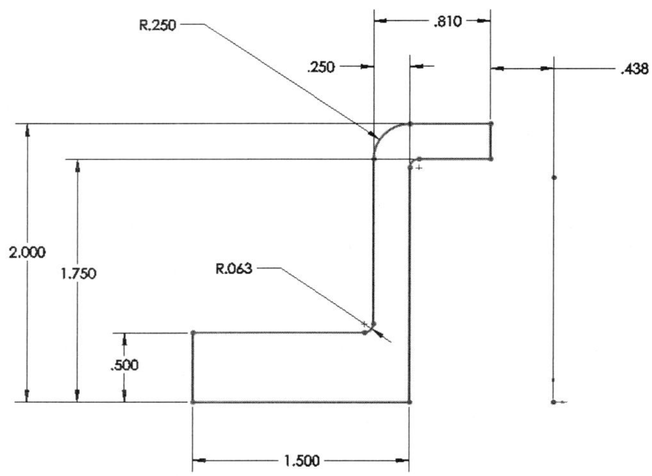

Figure 4-32
Draw a Sketch for a Revolved Boss/Base on the Front Plane and Revolve It

Make sure that when you make the sketch, the line for the revolve is made with the **Centerline** command and that when you make the diameter dimension you click to the right of the centerline so that the diameter dimension is as shown in Figure 4-32.

Perform the **Revolve Boss/Base.**

Step 5. Make a circular pattern of the circle using 4 instances, as shown in Figure 4-33.

Step 6. Draw four circles at the centers of the circles, as shown in Figure 4-34, and dimension one of them.

Step 7. Click **Add Relation** and click the dimensioned circle. Then, click another circle and click **Equal** in the **Add Relations Manager.** (You may need to do this three times.)

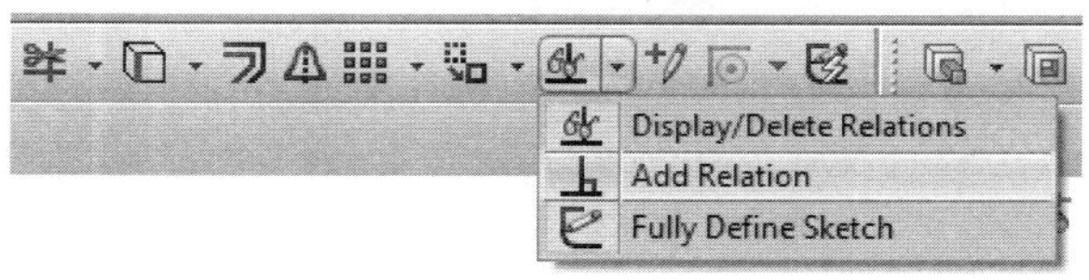

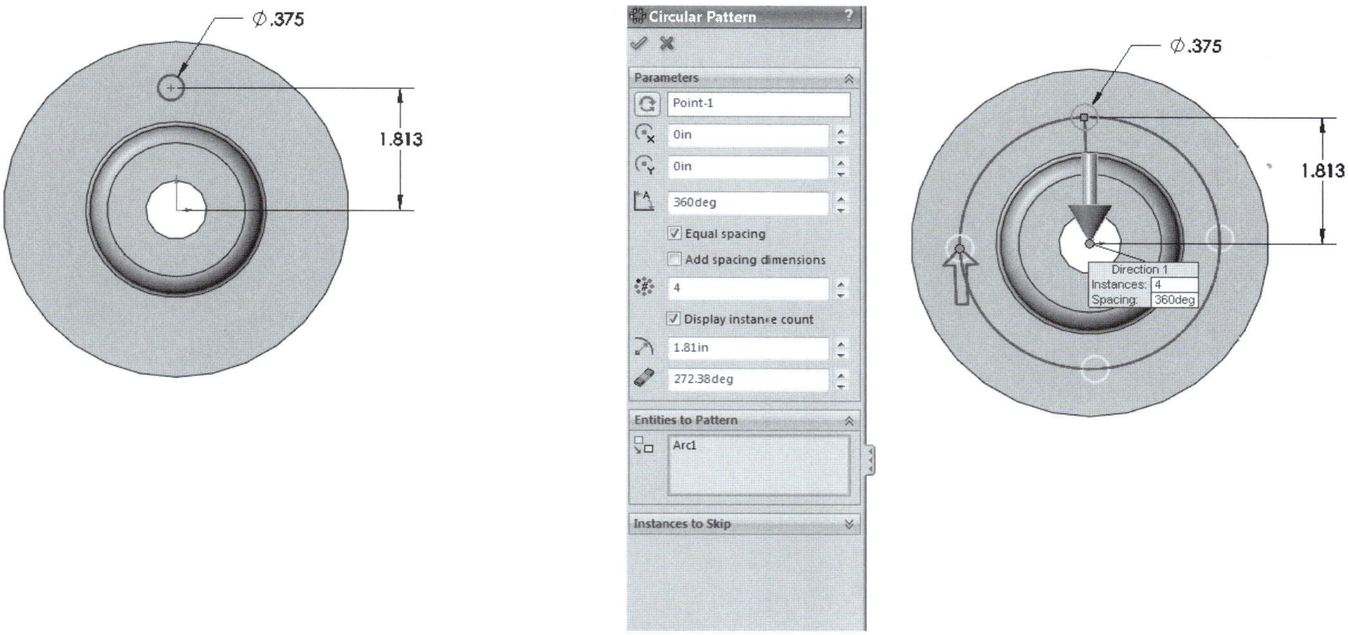

Figure 4-33
Draw a Circle for an Extrude Cut and Dimension It. Make a Circular Pattern of the Circle as Shown in the Circular Pattern Manager

Figure 4-34
Draw Four Circles at the Centers of the Circles as Shown. Dimension One of Them and Add the Equal Relation to the Other Three

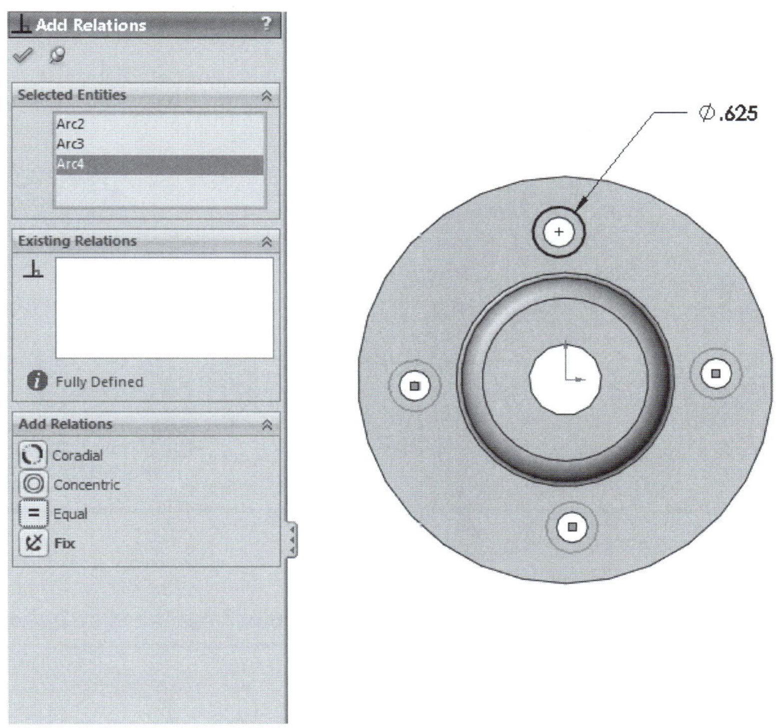

Step 8. Make an **Extrude Cut .125 inch** deep as shown in Figure 4–35.

Step 9. Right-click **Material <not specified>** in the **Project Manager** and click **Edit Material.**

Step 10. Use **Edit Material** to assign material to the model as shown in Figure 4-35.

Figure 4-35
Make an Extrude Cut
.125 Inch Deep

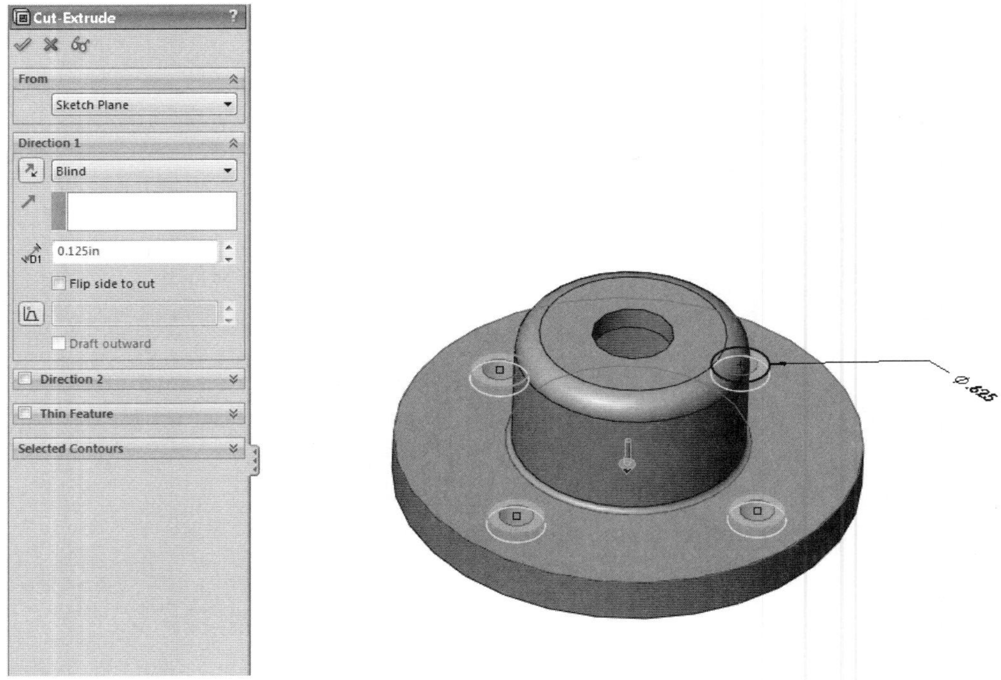

Step 11. Click **Apply** and then click **Close.**

Step 12. Click **Isometric view** so that the model appears as shown in Figure 4-35.

Step 13. Save your drawing (Figure 4-36) with the file name **Exercise 4-1(your initials)** in at least two places.

Figure 4-36
Use Edit Material to Assign a Material.
Exercise 4-1 Complete

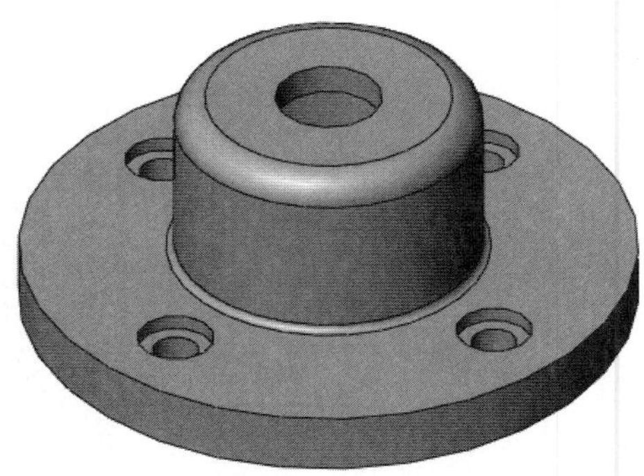

EXERCISE 4-2 Make a model from the sketch shown in Figures 4-37 through 4-42.

Step 1. Click **New....**

The **New SolidWorks Document** dialog box appears.

Step 2. Click **Part** and click **OK.**

The SolidWorks screen appears.

Click **Options** on the menu bar, select the **Document Properties** tab, and click **Units** in the list to the left. Make sure the **Units** are correctly set [notice that **IPS** (inch, pound, second) is checked], then click **OK.**

Step 3. Click the **Front Plane** and click **Sketch** so that the Sketch mode is active.

Step 4. Draw the sketch for a **Revolve Boss/Base** and dimension it as shown in Figure 4-37.

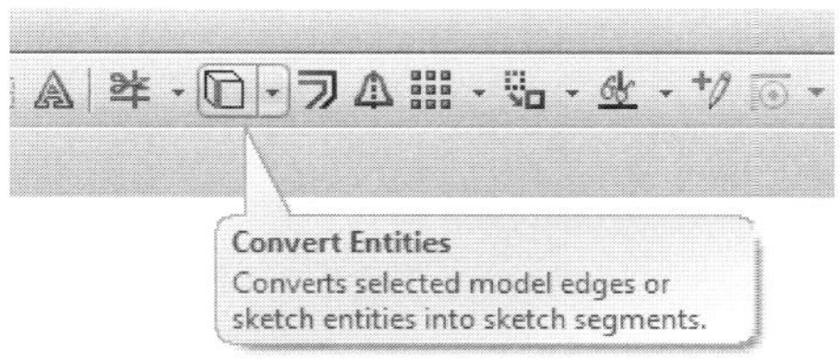

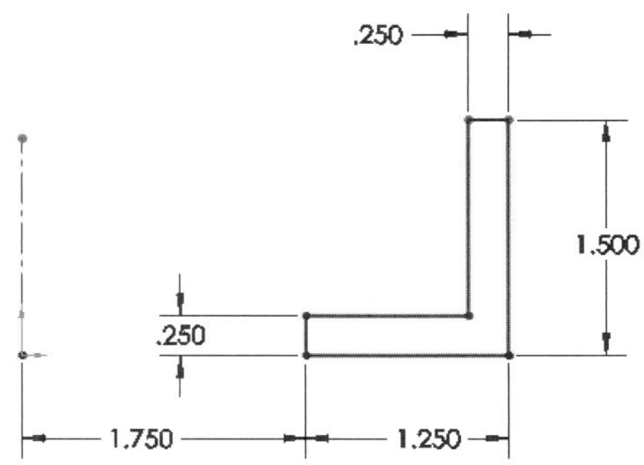

Figure 4-37
Draw a Sketch for a Revolved Boss/Base on the Front Plane and Revolve It

Make sure that when you make the sketch, the line for the revolve is made with the **Centerline** command and that when you make the diameter dimension you click to the left of the centerline so that the diameter dimension is as shown in Figure 4-37.

Perform the **Revolve Boss/Base**.

Step 5. Click **Convert Entities** and click the inner outside edge of the revolution to the sketch as shown in Figure 4-38. This changes the edge to a circle that can be used in the next sketch.

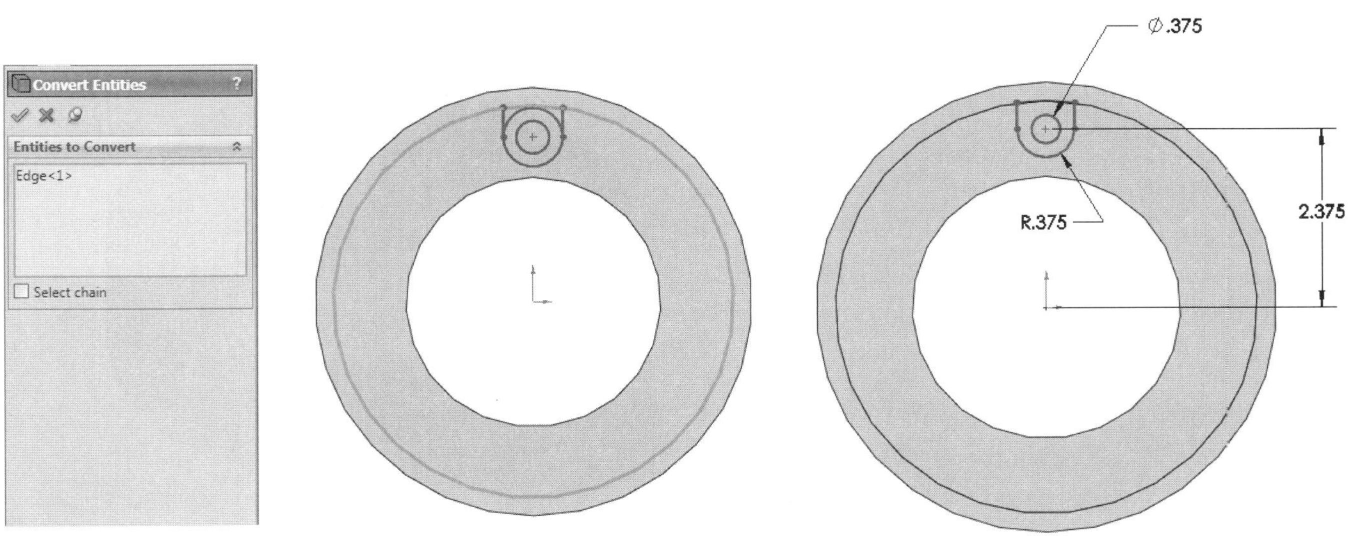

Figure 4-38
Use Convert Entities to Convert the Inner Outside Edge of the Revolution to a Circle. Trim and Dimension the Sketch as Shown

Step 6. Trim and dimension the sketch as shown in Figure 4-38.

Step 7. Make a circular pattern of the sketch using 3 instances, the entire sketch as the **Entities to Pattern,** and **Point-1** or a circular edge of the model as the center (the first text box under **Parameters** in the **Circular Pattern Manager**) as shown in Figure 4-39.

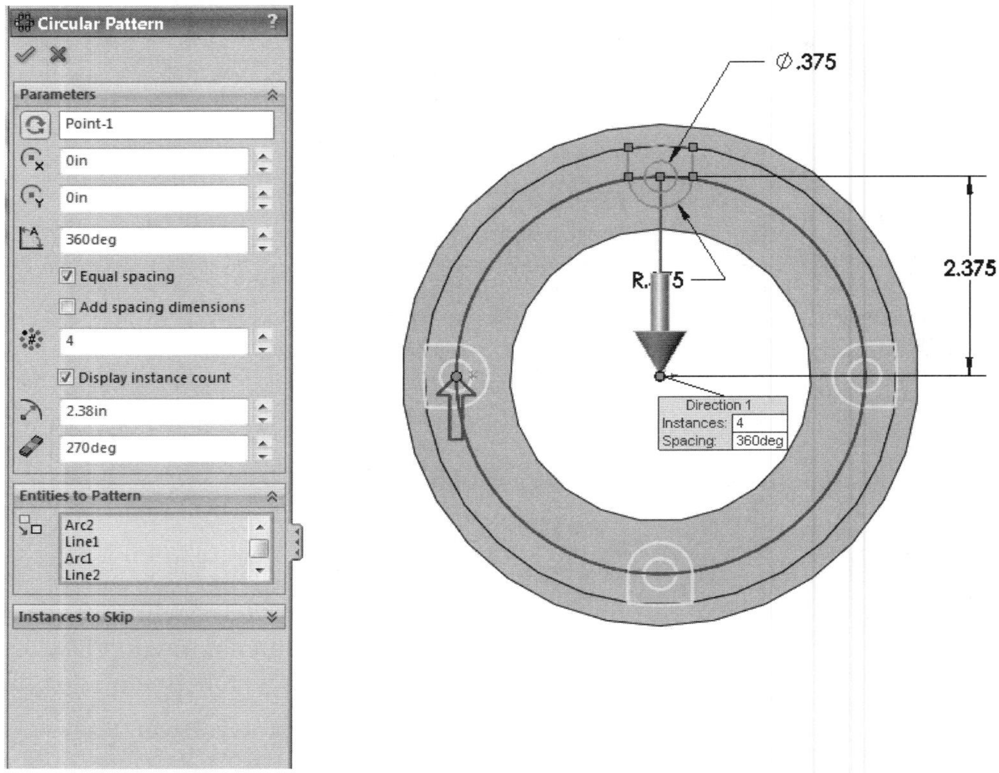

Figure 4-39
Make a Circular Pattern of the Sketch

Step 8. Make an **Extrude Boss/Base** of the patterned sketch .25 inch deep as shown in Figure 4-40.

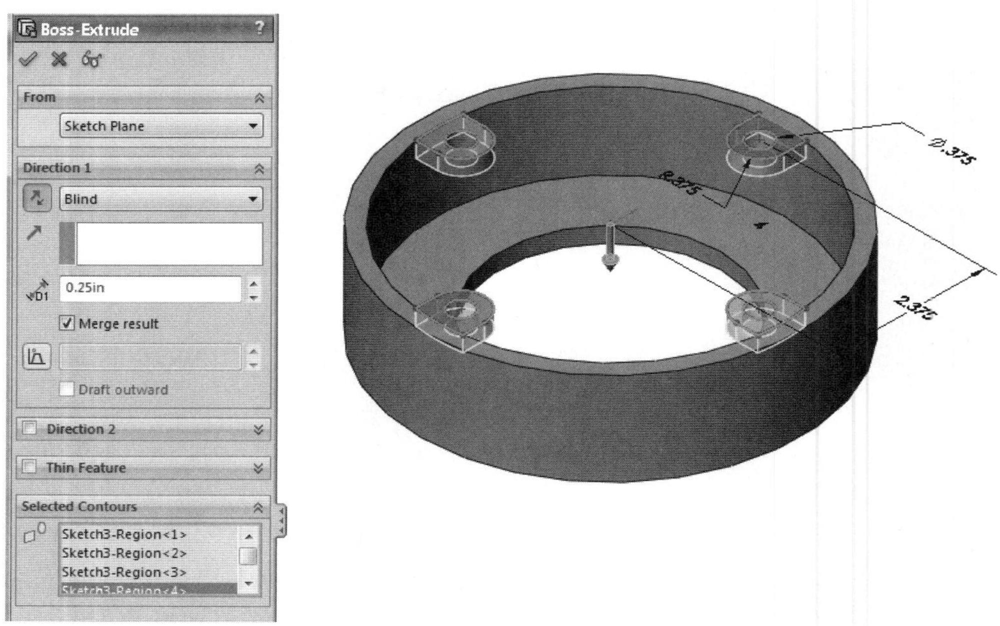

Figure 4-40
Make an Extrude Boss/Base of the Patterned Sketch .25 Inch Deep

Step 9. Click **OK** (the green check mark at the top of the **Boss-Extrude Manager**).

Step 10. Draw a circle for an **Extrude Cut** and dimension it as shown in Figure 4-41.

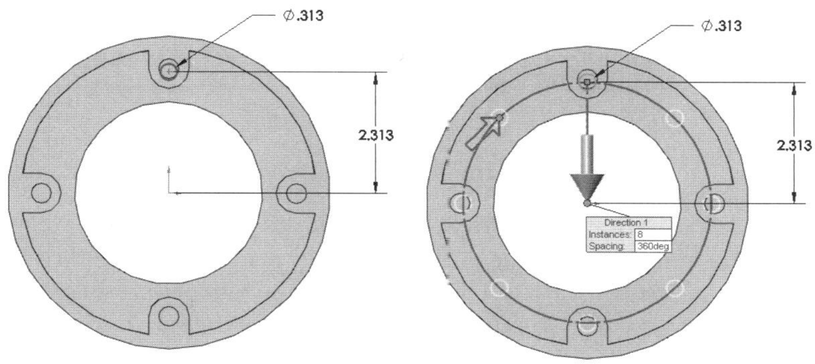

Figure 4-41
Draw a Circle for an Extrude Cut and Dimension It. Make a Circular Pattern of the Circle Using 8 Instances

Step 11. Make a circular pattern of the circle using 8 instances as shown in Figure 4-41.

Make the **Extrude Cut Through All** as shown in Figure 4-42.

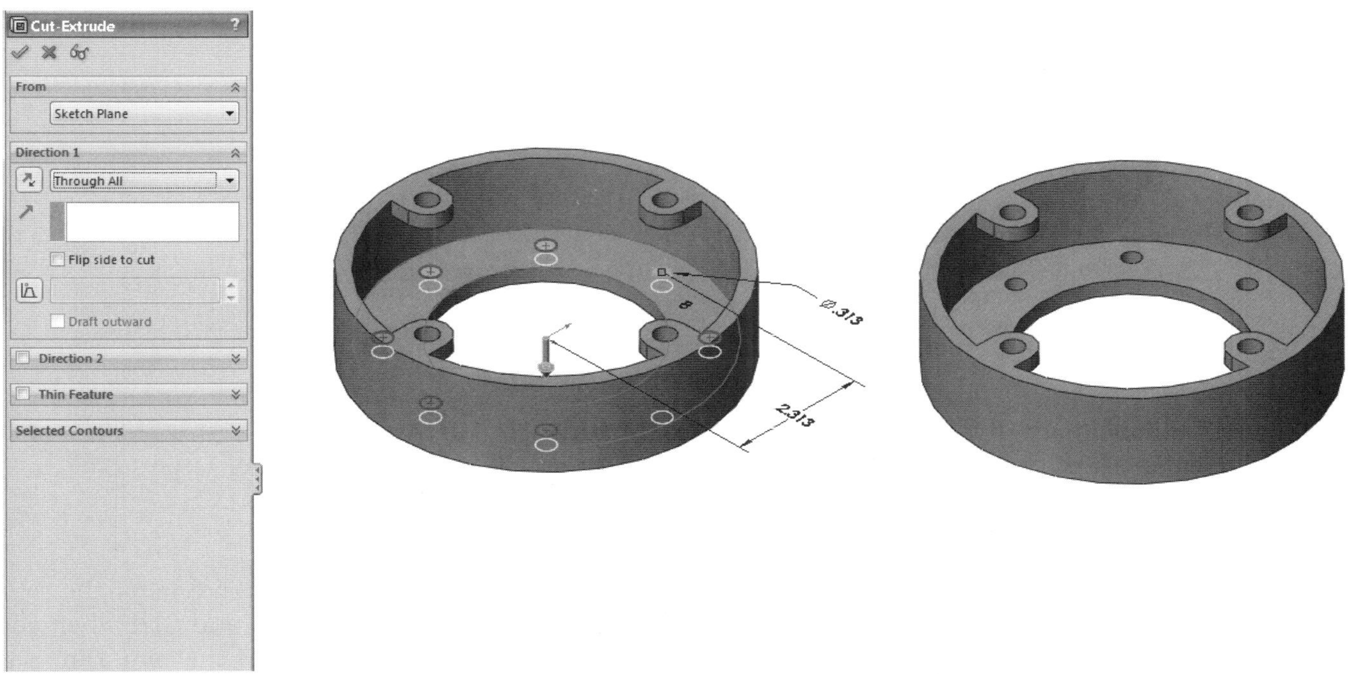

Figure 4-42
Make an Extrude Cut Through All. Use Edit Material to Assign a Material. Exercise 4-2 Complete

Step 12. Right-click **Material <not specified>** in the **Project Manager** and click **Edit Material.**

Step 13. Use **Edit Material** to assign material to the model as shown in Figure 4-42.

Step 14. Click **Apply** and then click **Close.**

Step 15. Click **Isometric view** so that the model appears as shown in the right side of Figure 4-42.

Step 16. Save your drawing (Figure 4-42) with the file name **Project 4-2(your initials)** in at least two places.

EXERCISE 4-3 **Make a model from the sketch shown in Figures 4-43 through 4-46.**

Step 1. Click **New....**

The **New SolidWorks Document** dialog box appears.

Step 2. Click **Part** and click **OK.**

The SolidWorks screen appears.

Click **Options** on the menu bar, select the **Document Properties** tab, and click **Units** in the list to the left. Make sure the **Units** are correctly set [notice that **IPS** (inch, pound, second) is checked], then click **OK.**

Step 3. Click the **Right Plane** and click **Sketch** so that the Sketch mode is active.

Step 4. Draw the sketch for a **Revolve Boss/Base** and dimension it as shown in Figure 4-43.

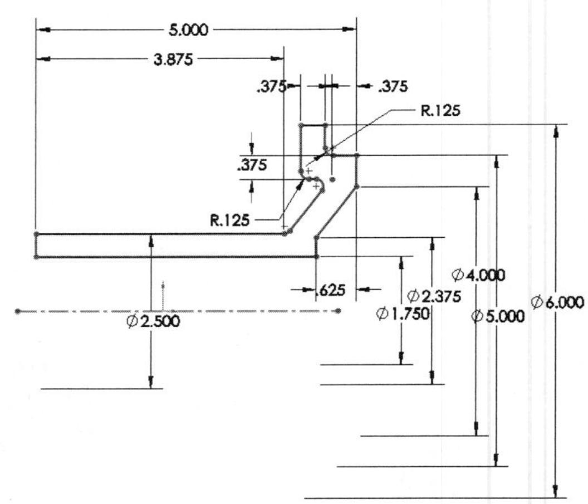

Figure 4-43
Draw a Sketch for a Revolve Boss/Base on the Right Plane and Revolve It

Make sure that when you make the sketch, the line for the revolve is made with the **Centerline** command and that when you make the diameter dimension you click below the centerline so that the diameter dimensions are as shown in Figure 4-43.

Perform the **Revolve Boss/Base.**

Step 5. Draw a circle for an **Extrude Cut** on the front plane of the model and make the **Extrude Cut** 1.00 inch deep with a 45 degree draft angle as shown in Figure 4-44.

Step 6. Click **Rectangle,** draw the rectangle, and dimension it as shown in Figure 4-45.

Step 7. Make a **Circular Pattern** of 4 instances as shown in Figure 4-45.

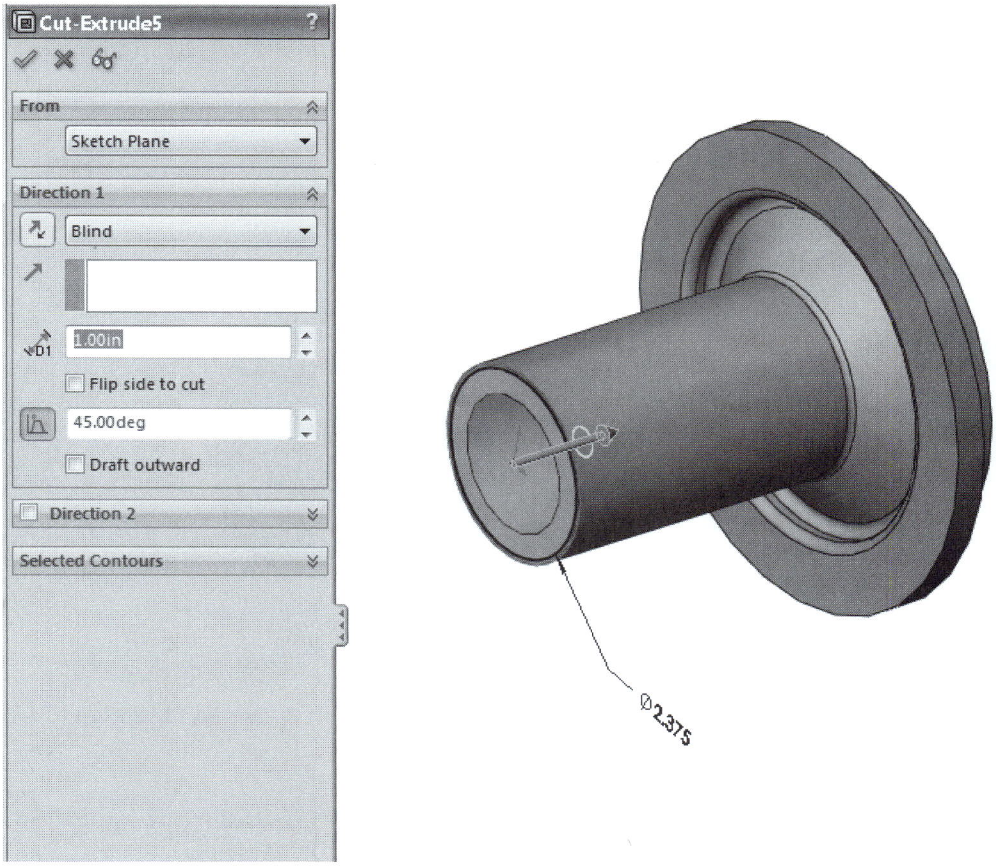

Figure 4-44
Draw a Circle for an Extrude Cut on the Front Plane of the Model and Make the Cut 1.00 Inch Deep with a 45 Degree Draft Angle

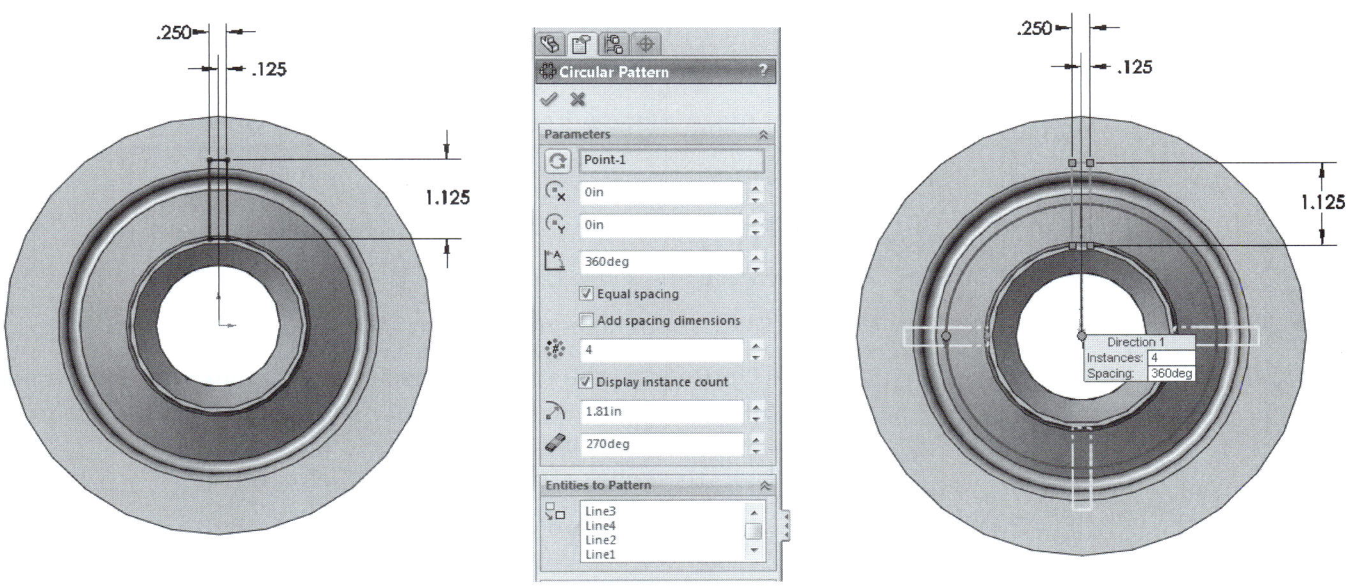

Figure 4-45
Draw a Rectangle for an Extrude Boss/Base, Dimension It, and Make a Circular Pattern of 4 Instances

Step 8. Make an **Extrude Boss/Base** of the circular pattern of rectangles **Up to next** as shown in Figure 4-46.

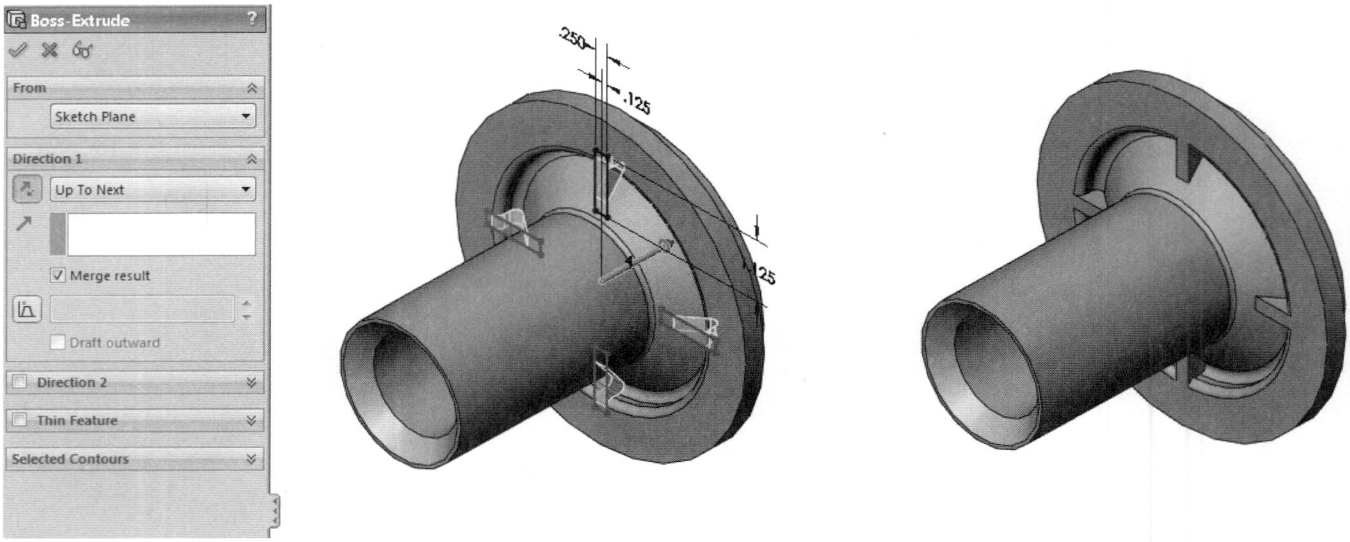

Figure 4-46
Make an Extrude Boss/Base of the Circular Pattern Up to Next. Use Edit Material to Assign a Material. Exercise 4-3 Complete

Step 9. Right-click **Material <not specified>** in the **Project Manager** and click **Edit Material.**

Step 10. Use **Edit Material** to assign material to the model as shown in Figure 4-46.

Step 11. Click **Apply** and then click **Close.**

Step 12. Click **Isometric view** so that the model appears as shown in the right side of Figure 4-46.

Step 13. Save your drawing (Figure 4-46) with the file name **Project 4-3(your initials)** in at least two places.

EXERCISE 4-4 Make a model from the sketch shown in Figures 4-47 and 4-48.

Step 1. Click **New....**

The **New SolidWorks Document** dialog box appears.

Step 2. Click **Part** and click **OK.**

The SolidWorks screen appears.

Click **Options** on the menu bar, select the **Document Properties** tab, and click **Units** in the list to the left. Make sure the **Units** are correctly set [notice that **IPS** (inch, pound, second) is checked], then click **OK.**

Step 3. Click the **Front Plane** and click **Sketch** so that the Sketch mode is active.

Step 4. Draw the sketch for a **Revolve Boss/Base** and dimension it as shown in Figure 4-47.

Figure 4-47
Make a Sketch for a Revolve Boss/Base on the Front Plane and Revolve It

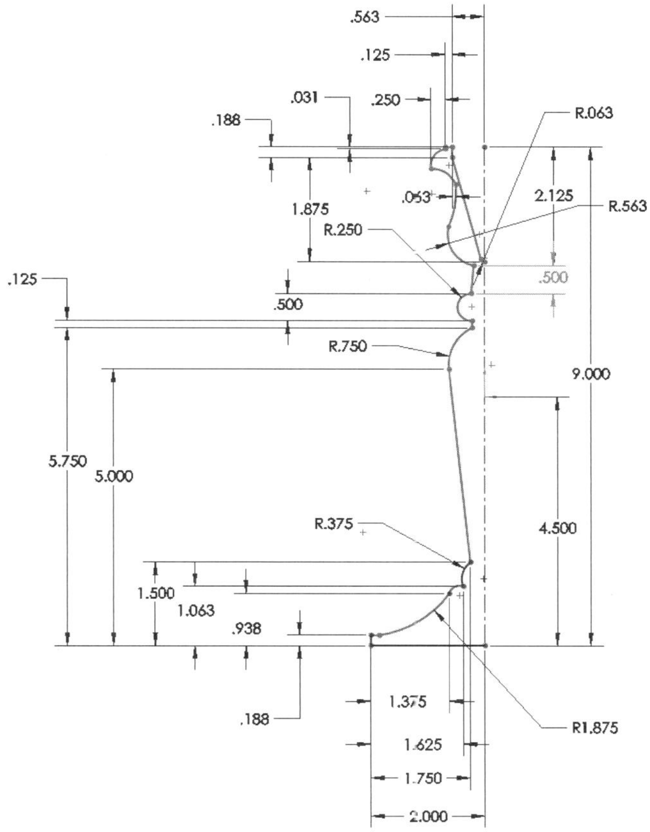

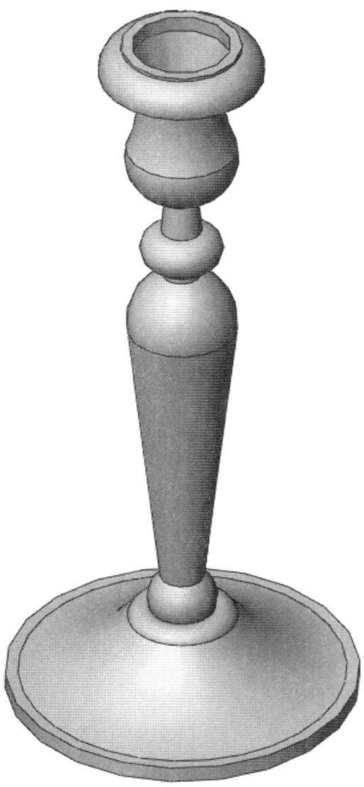

Figure 4-48
Use Edit Material to Assign a Material.
Exercise 4-4 Complete

Make sure that when you make the sketch, the line for the revolve is made with the **Centerline** command as shown in Figure 4-47.

Perform the **Revolve Boss/Base**

Step 5. Right-click **Material** in the **Project Manager** and click **Edit Material.**

Step 6. Use **Edit Material** to assign material to the model as shown in Figure 4-48.

Step 7. Click **Apply** and then click **Close.**

Step 8. Click **Isometric view** so that the model appears as shown in Figure 4-48.

Step 9. Save your drawing (Figure 4-48) with the file name **Project 4-4(your initials)** in at least two places.

chapter four

Chapter Summary

This chapter contains the information needed to make more complex models with revolved features and extrude cuts. You also have learned how to assign materials to these models. Now you have the skills and information necessary to make more complex models with revolved features and extrude cuts and apply materials to those models.

Review Questions

Circle the most correct answer.

1. Which of the following commands makes a rectangle most quickly?
   a. **Rectangle** c. **Line**
   b. **Square** d. **Offset**
2. Which of the following settings is needed to make sure inches is used instead of millimeters?
   a. **Notes** c. **Units**
   b. **Annotations** d. **Dimensions**
3. Which of the following is needed as the axis of revolution to make sure you get a **Revolved Boss/Base**?
   a. **Line** c. **Centerline**
   b. **3 Point Arc** d. **Front Plane**
4. On which item do you click to get the **Edit Material** command?
   a. **Material <not specified>** c. **Edit Material**
   b. **Material** d. **Annotations**
5. Which command is used to convert solid edges into circles?
   a. **Fully Define Sketch** c. **Add Relation**
   b. **Intersection Curve** d. **Convert Entities**
6. An **Extrude Cut** can only be made at 90 degree angles.
   a. True b. False
7. The **Convert Entities** command can change a solid edge into a line.
   a. True b. False
8. There is only one way to make a circle equal to another circle.
   a. True b. False
9. A plane can be created at a specific distance from the **Front Plane.**
   a. True b. False
10. A fillet cannot be made after the sketch has been turned into a solid.
    a. True b. False

chapter four

Matching

Place the correct letter on the line to the right of the numbered item.

1. A category in the **Material** dialog box _____	a. **Spline**
2. A command that makes a surface on which a sketch can be made _____	b. **Add Relation**
3. A command that can be used to make a complex curve _____	c. **Plane**
4. A command that can make a circle the same size as another circle _____	d. **Centerline**
5. A command that is often used as the center of revolution _____	e. **ANSI 1020**

General Questions

Answer the following questions.

1. How can you use the **Revolve Boss/Base** command?
2. When would you use the draft angle on an **Extruded Cut**?
3. Why is it necessary to assign materials to solid models?
4. What is the best command to make the center of revolution for a **Revolve Boss/Base**?
5. Where would you use the **Circular Pattern** command?

5 chapterfive

Using Extrude Cut, Hole Simple, Hole Wizard, and Rib to Make Complex Models and Drawings with Dimensions Including Tolerances

CHAPTER OBJECTIVES

After completing this chapter, you will be able to:

- Make complex extruded models.
- Make models with counterbored, countersunk, and threaded holes.
- Make models with ribs.
- Make drawings from complex models with dimensions and tolerances.
- Assign materials to models
- Correctly answer questions regarding SolidWorks commands used in this chapter.

In this chapter you will use many of the same commands you have used in previous chapters. The one area that may require some explanation is tolerances.

Uses of Tolerances

Because machines do not build perfect parts, drawings must have tolerances on all dimensions. A tolerance is the amount the part can vary from the perfect dimensions shown on the drawing.

Usually the format block of a drawing has a list of the standard tolerances that are used on all dimensions. For any dimension with a tolerance different from the standard one, the tolerance is shown with the dimension. A tolerance of +.005 and −.005 on a dimension of .500 means that the part when inspected can fall between .505 and .495 and pass as a good part. Any other dimension, and the part must be corrected or scrapped.

Tolerances beyond three decimal places are often shown by placing the upper and lower limits of the dimension directly on the drawing. In the case of the .500 dimension with a tolerance of + and − .005 those limits are shown as .505 and .495.

Project 5-3 is an assembly of two cylinders and a part with two holes. These parts demonstrate the use of clearance and interference fits.

Clearance Fits

Many tolerances are used to allow parts to fit together and function as they should no matter which pair of the thousands of parts are mated. For example, a cylinder with a circular dimension of 1.500 and a tolerance of +.004 and −.004 has an upper limit of 1.504 and a lower limit of 1.496. When it is inserted into a part with a hole that has a dimension of 1.510 and a tolerance of +.004 and −.004, the minimum clearance is .002. The hole has an upper limit of 1.514 and a lower limit of 1.506; therefore, no matter which cylinder is placed in which hole, there will always be at least .002 clearance between the cylinder and the hole.

Interference Fits

Some parts must be fitted together so that they will not come apart easily. This fit is made under heat or pressure and is called an interference fit. In such cases, the shaft is always bigger than the hole. For example, a cylinder with a circular dimension of 1.500 and a tolerance of +.004 and −.004 has an upper limit of 1.504 and a lower limit of 1.496. When it is inserted into a part with a hole that has a dimension of 1.490 and a tolerance of +.004 and −.004, the minimum interference is .002. The hole has an upper limit of 1.494 and a lower limit of 1.486; therefore, no matter which cylinder is placed in which hole, there will always be at least .002 interference between the cylinder and the hole.

Project 5-1: Make a Part with Simple and Complex Holes (Figure 5-1)

Step 1. Start SolidWorks.

Step 2. Click **New....**

The **New SolidWorks Document** dialog box appears.

Step 3. Click **Part** and click **OK.**

The SolidWorks screen appears. Click **Options** on the menu bar, select the **Document Properties** tab, and click **Units** in the list to the left. Make sure the **Units** are correctly set [notice that **IPS** (inch, pound, second) is checked], then click **OK.**

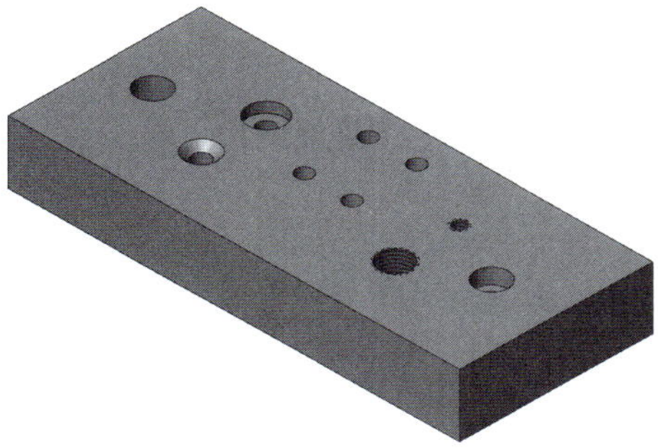

Figure 5-1
Project 5-1 Complete

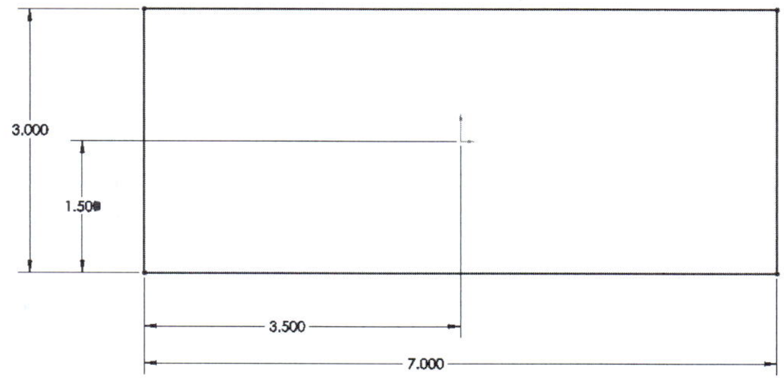

Figure 5-2
Draw a Sketch for an Extrude Boss/Base on the Top Plane and Extrude It 1.00 Inch

Step 4. Click the **Top Plane** and click **Sketch** so the Sketch mode is active.

Step 5. Draw a rectangle measuring 7.000 inches long × 3.000 inches high and extrude it 1.00 inch (Figure 5-2).

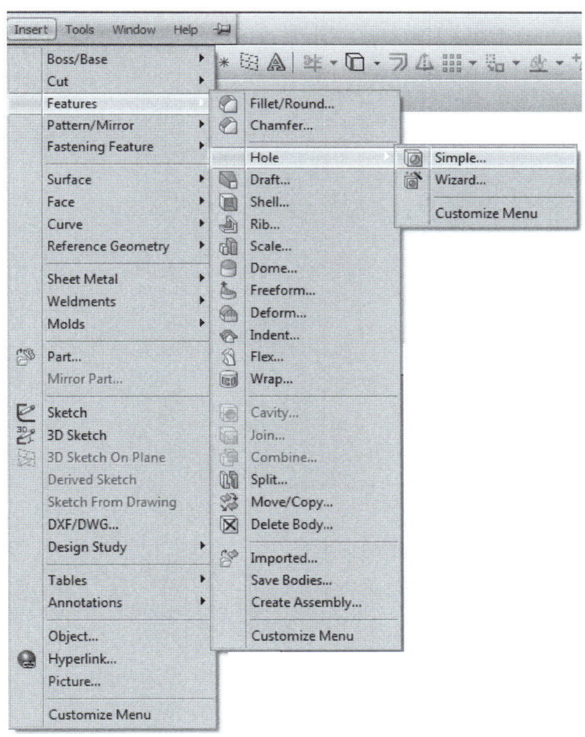

Step 6. Use **Hole Simple** to make a .500 diameter hole **Through All** and use **Smart Dimension** to locate the hole and click **OK** as shown in Figure 5-3.

Step 7. Use **Hole Wizard** to make a 1/4 inch diameter counterbored hole. Choose the **Hole Type** specifications under the **Type** tab as shown in Figure 5-4. Then, click the **Positions** tab and click a point to locate the hole. Use **Smart Dimension** to locate the hole and click **OK.**

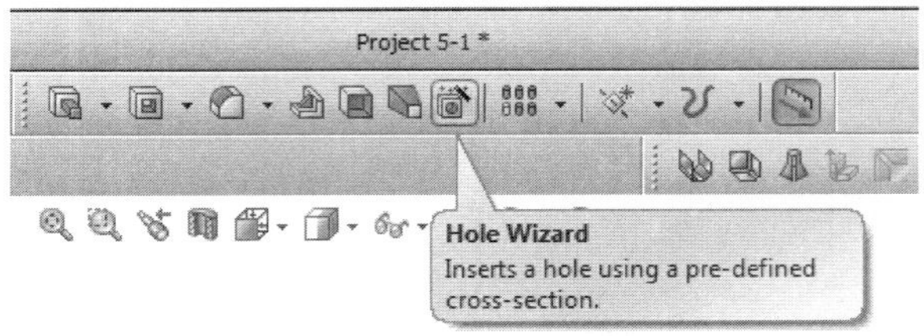

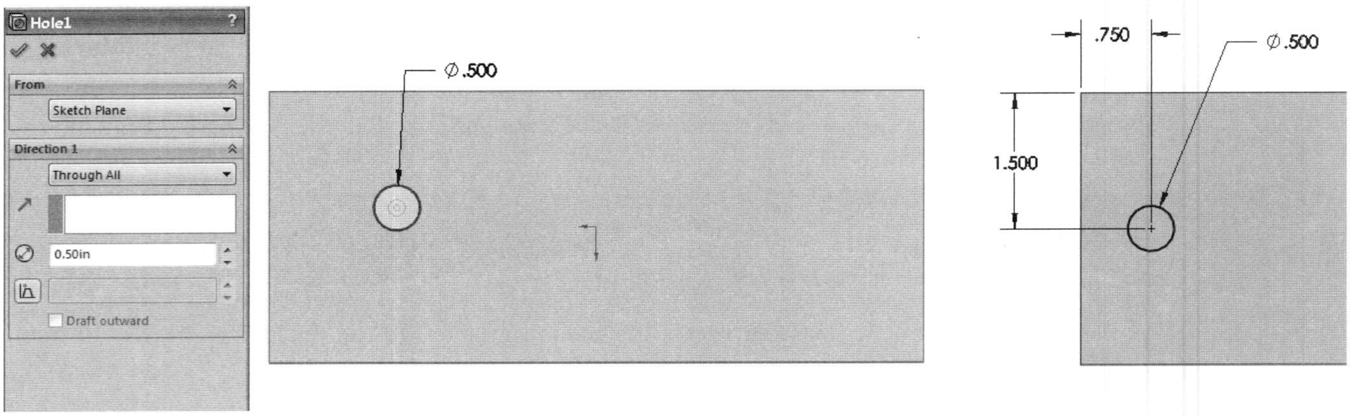

Figure 5-3
Use Hole Simple to Make a .500 Diameter Hole Through All. Use Smart Dimension to Locate the Hole

Figure 5-4
Use Hole Wizard to Make a 1/4 inch Diameter Counterbored Hole. Make Settings Per the Type Tab. Click: Positions Tab and Locate the Hole. Dimension the Hole.
Click: OK

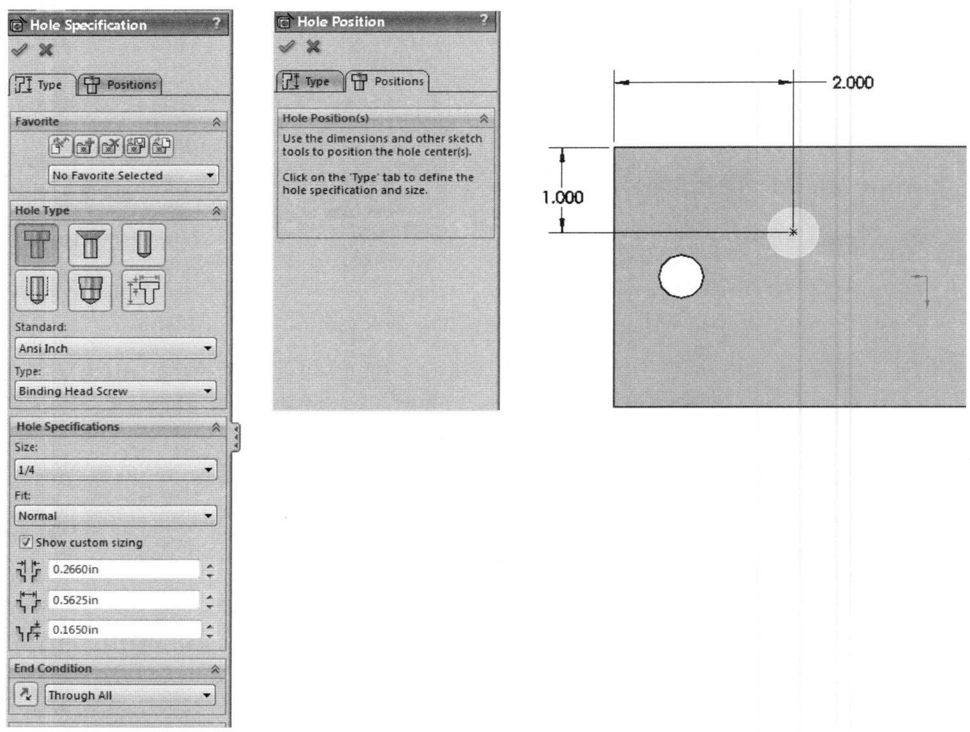

Step 8. Use **Hole Wizard** to make a 1/4 inch diameter countersunk hole. Choose the **Hole Type** specifications under the **Type** tab as shown in Figure 5-5. Then, click the **Positions** tab and click a point to locate the hole. Use **Smart Dimension** to locate the hole and click **OK.**

Step 9. Use **Hole Wizard to** make four 1/4 inch diameter holes without threads. Choose the **Hole Type** specifications under the **Type**

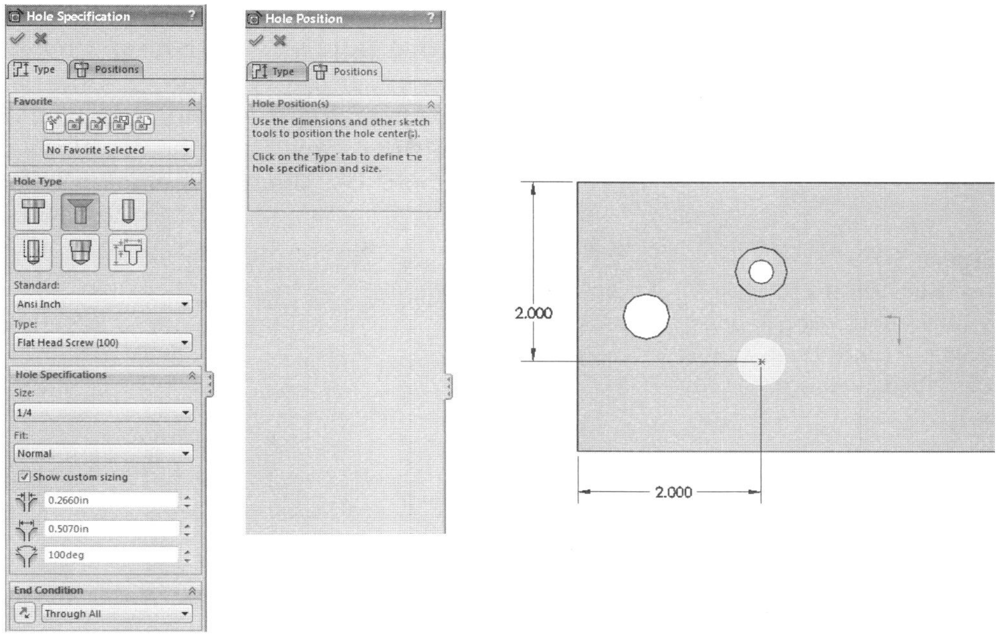

Figure 5-5
Use Hole Wizard to Make a 1/4 inch Diameter Countersunk Hole as Described in Figure 5-4

tab as shown in Figure 5-6. Then, click the **Positions** tab and click four points to locate the holes. Use **Smart Dimension** to locate the first three holes, then use **Add Relation - Horizontal** to locate the fourth hole and click **OK.**

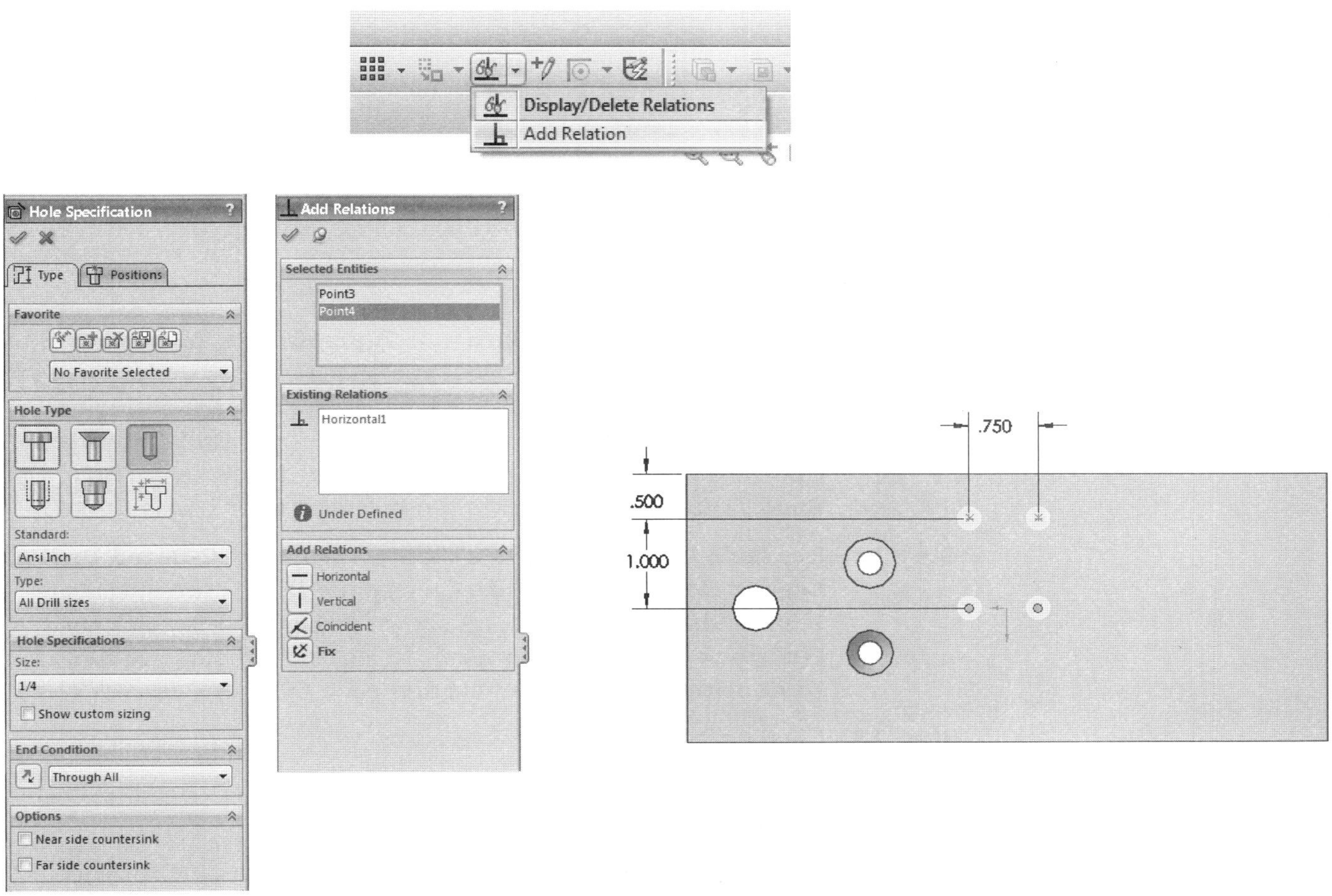

Figure 5-6
Use Hole Wizard to Make 4 1/4 inches Holes. Dimension the First 3 and Use Add Relation to Locate the Other Hole

Step 10. Use **Hole Wizard** to make a 1/4-20 inch diameter **Bottoming Tapped Hole** x .75 inch deep. Choose the **Hole Type** specifications under the **Type** tab as shown in Figure 5-7. Check **Show custom sizing** as shown, then type **.75** in the depth box under **End Condition.** Click the **Positions** tab and click a point to locate the hole. Use **Smart Dimension** to locate the hole and click **OK.**

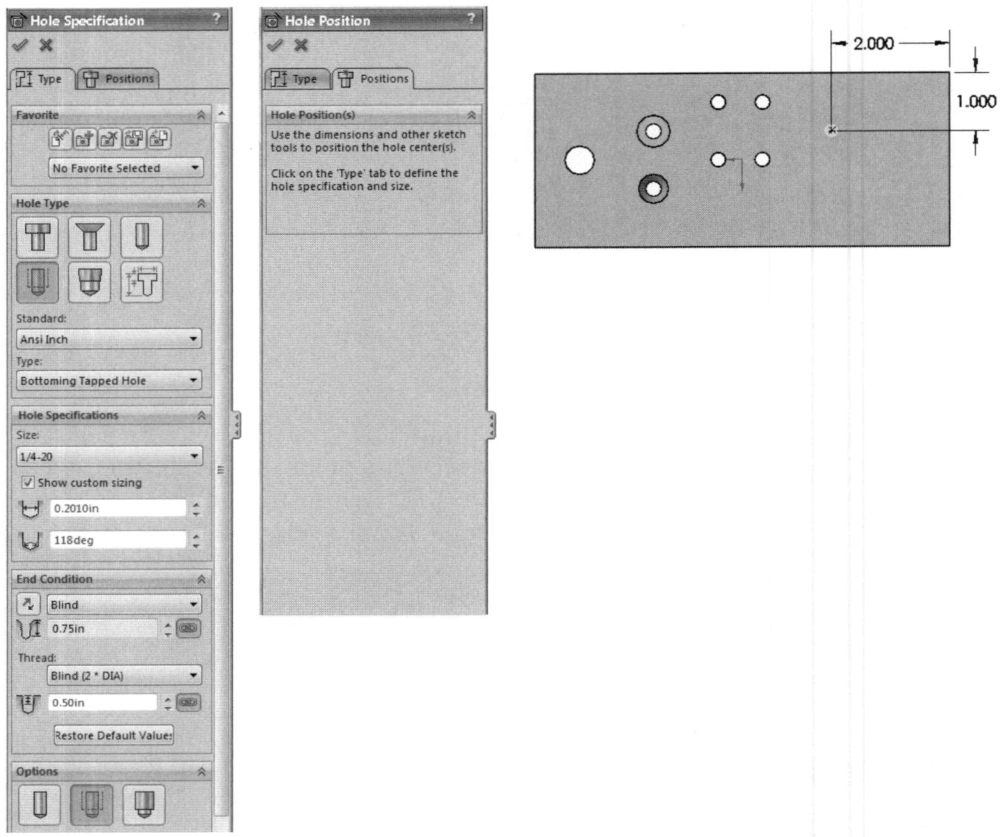

Figure 5-7
Use Hole Wizard to Make a 1/4-20 Bottoming Tapped Hole x .75 inch Deep as Described in Figure 5-4

Step 11. Use **Hole Wizard** to make a 1/4 **Tapered Pipe Tap** hole x .67 inch deep. Choose the **Hole Type** specifications under the **Type** tab as shown in Figure 5-8. Check **Show custom sizing** as shown, then type **.67** in the depth box under **End Condition.** Click the **Positions** tab and click a point to locate the hole. Use **Smart Dimension** to locate the hole and click **OK.**

Step 12. Use **Hole Wizard** to make a .25 inch diameter **C-Bored Drilled Legacy Hole Through All.** Choose the **Hole Type** specifications under the **Type** tab as shown in Figure 5-9. Click the **Positions** tab and click a point to locate the hole. Use **Smart Dimension** to locate the hole and click **OK.**

Step 13. Click **Exit Sketch.**

Step 14. Click **Isometric view** so that the model appears as shown in Figure 5-10.

Step 15. Save your drawing (Figure 5-10) with the file name **Project 5-1(your initials)** in at least two places.

Step 16. Print the drawing and turn it in.

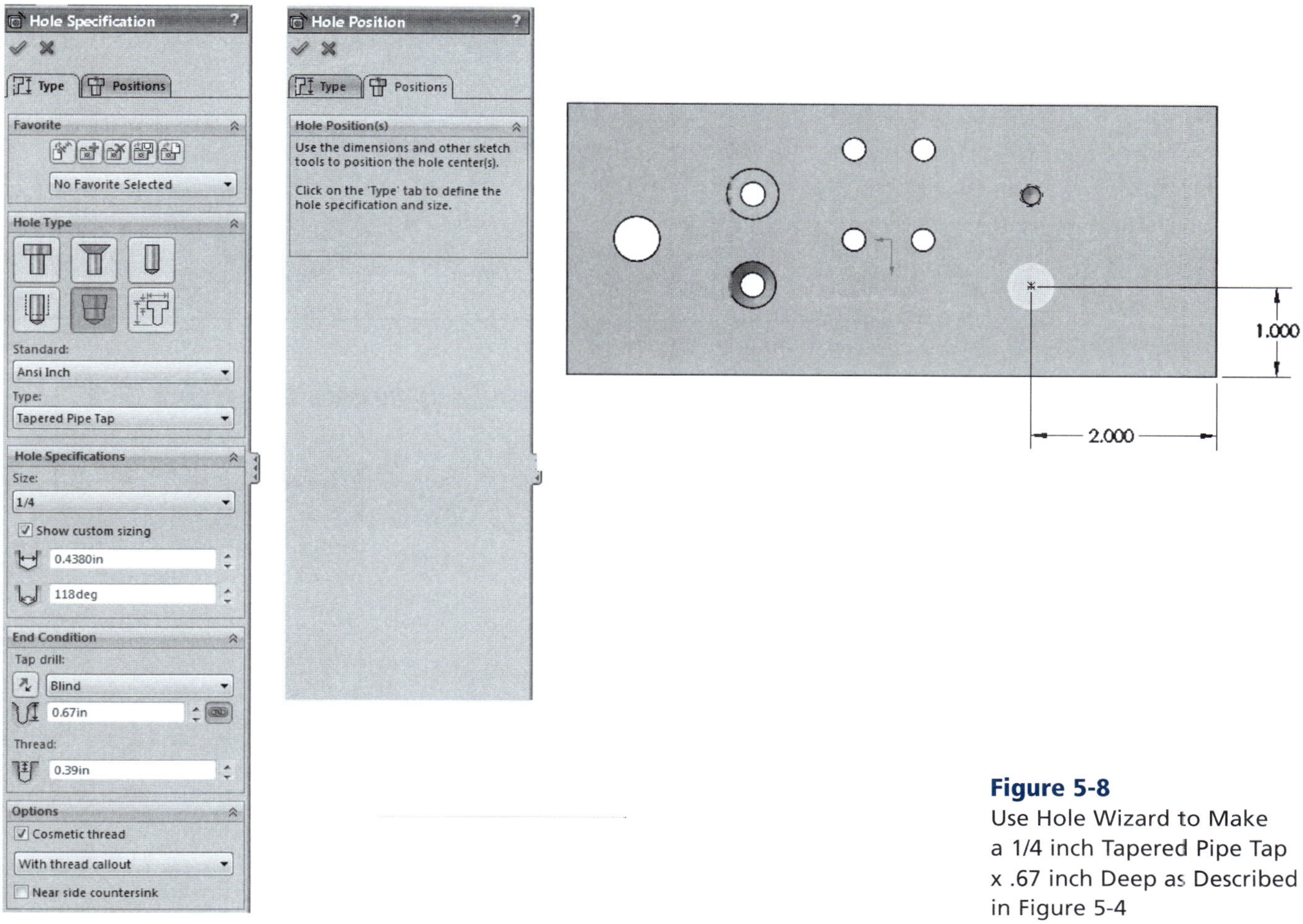

Figure 5-8
Use Hole Wizard to Make a 1/4 inch Tapered Pipe Tap x .67 inch Deep as Described in Figure 5-4

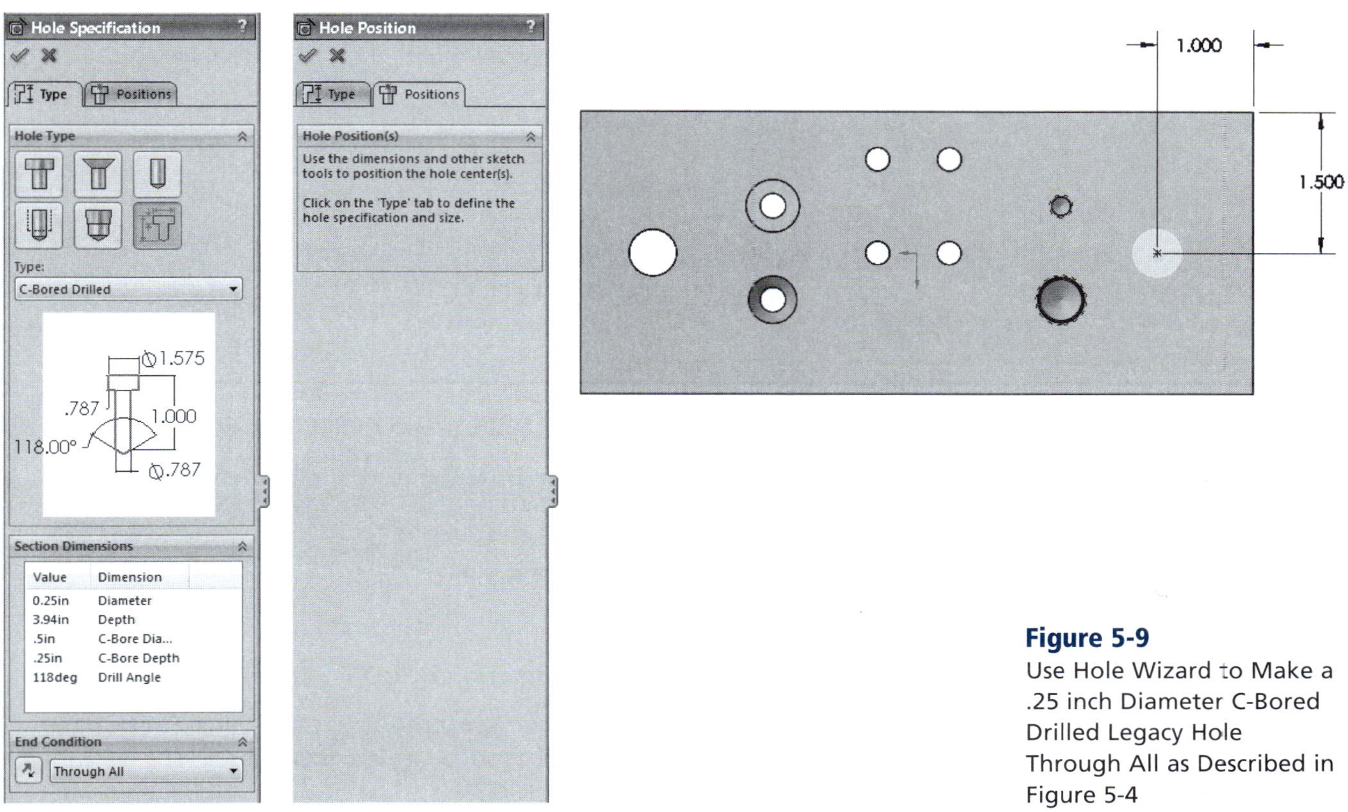

Figure 5-9
Use Hole Wizard to Make a .25 inch Diameter C-Bored Drilled Legacy Hole Through All as Described in Figure 5-4

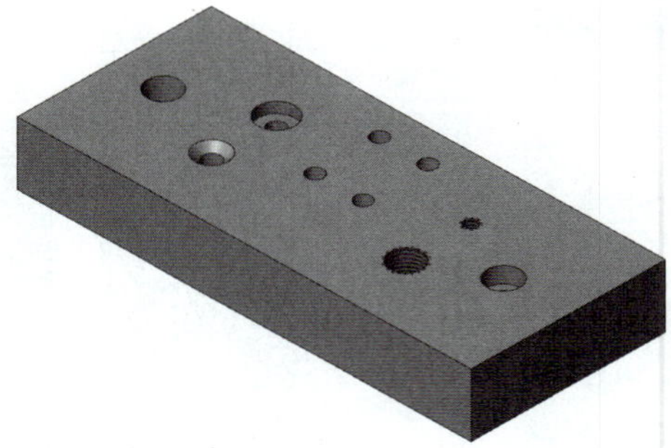

Figure 5-10
Project 5-1 Complete

Project 5-2: **Make a Drawing from Project 5-1 (Figure 5-11)**

Step 1. Start SolidWorks.

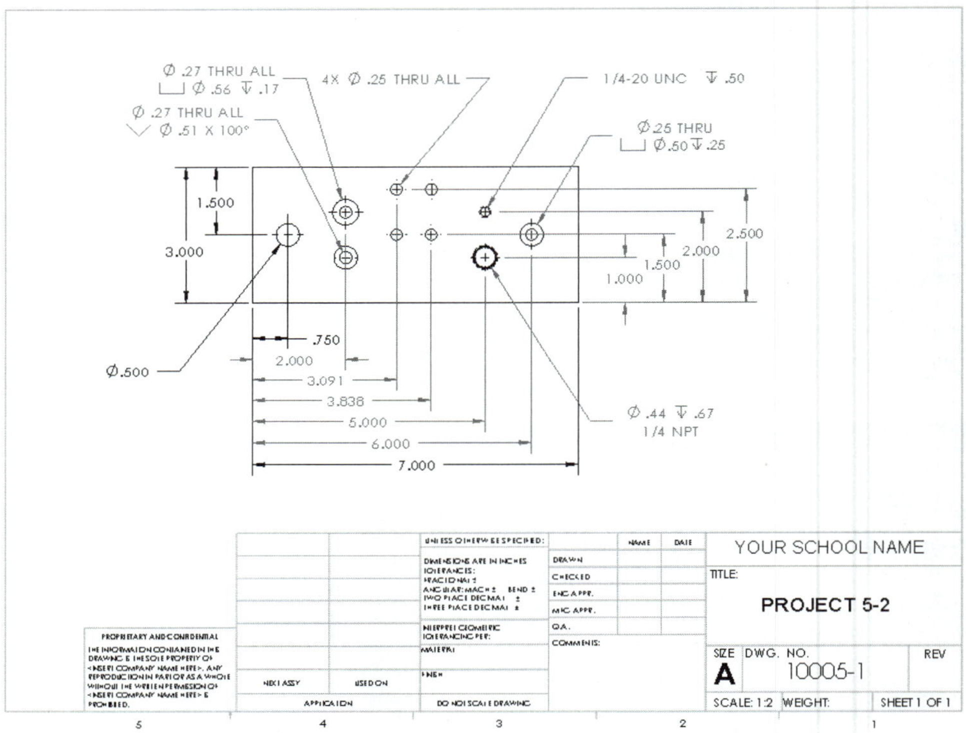

Figure 5-11
Project 5-2 Complete

Step 2. Click **Open....**

The **Open SolidWorks Document** dialog box appears.

Step 3. Click **Project 5-1(your initials)** and click **Open.**

The SolidWorks screen appears.

Step 4. Click **Make Drawing from Part.**

The **Sheet/Format** dialog box appears.

Step 5. Click **(ANSI) Landscape** and click **OK.**

Step 6. Right-click a point in the graphics area and click **Properties.** Then click **Third Angle** and make sure **Scale:** is **1:2** as shown in Figure 5-11 and click **OK.**

Step 7. Drag **Front view** from the **Task** pane, arrange it so that it is centered in the drawing area, and click the green check mark in the **Drawing View Feature Manager.**

Step 8. Click the **Annotation** tab (or select **Annotations** from the **Insert** menu on the menu bar), click **Model Items,** and make the settings as shown in Figure 2-6. (You will have to turn to Chapter 2 to view this figure.)

Be sure the **Entire** model is selected as the **Source:** and the first icon under **Dimensions** is selected.

Step 9. Arrange the dimensions as shown in Figure 5-11. **NOTICE THE SPACING OF ROWS OF DIMENSIONS AND THEIR PLACEMENT.** You may have to delete some dimensions that are not on the correct view and use **Smart Dimension** to put them on the correct view.

Step 10. Right-click any point on the drawing sheet.

The popup menu appears. Click **Edit Sheet Format.**

Step 11. Use the **Note Command** to type your **Initials,** today's date, your **School Name,** the **Project Number,** and the **Drawing Number** as shown in Figure 5-11. You will have to change the font size to **6** points in the **NAME** and **DATE** boxes, **12** points in the **Drawing Number** box, and **16** points in the **School Name** and **Project Number** boxes. Change the font to **Arial.** You will also have to click on some of the items and move them into the correct space.

Step 12. Right-click any point on the drawing sheet.

The popup menu appears. Click **Edit Sheet.**

Step 13. Save your drawing with the file name **Project 5-2(your initials)** in at least two places.

Step 14. Print the drawing and turn it in.

Project 5-3: Make a New Assembly of Three Parts with Bilateral Tolerances on All Three Parts (Figure 5-12)

Step 1. Start SolidWorks.

Step 2. Click **New....**

The **New SolidWorks Document** dialog box appears.

Step 3. Click **Part** and click **OK.**

The SolidWorks screen appears. Click **Options** on the menu bar, select the **Document Properties** tab, and click **Units** in the list to the left. Make sure the **Units** are correctly set [notice that **IPS** (inch, pound, second) is checked], then click **OK.**

Step 4. Click the **Top Plane** and click **Sketch** so that the Sketch mode is active.

Figure 5-12
Project 5-3 Complete

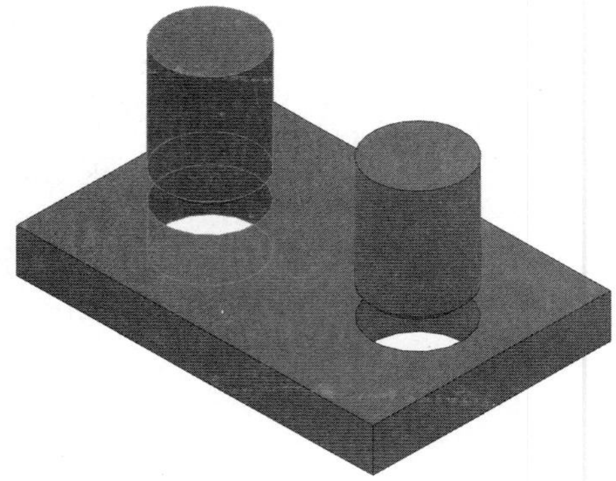

Step 5. Draw a circle and dimension it with a tolerance of +.005 and −.005 as shown in Figure 5-13. Then extrude it 2.00 inches.

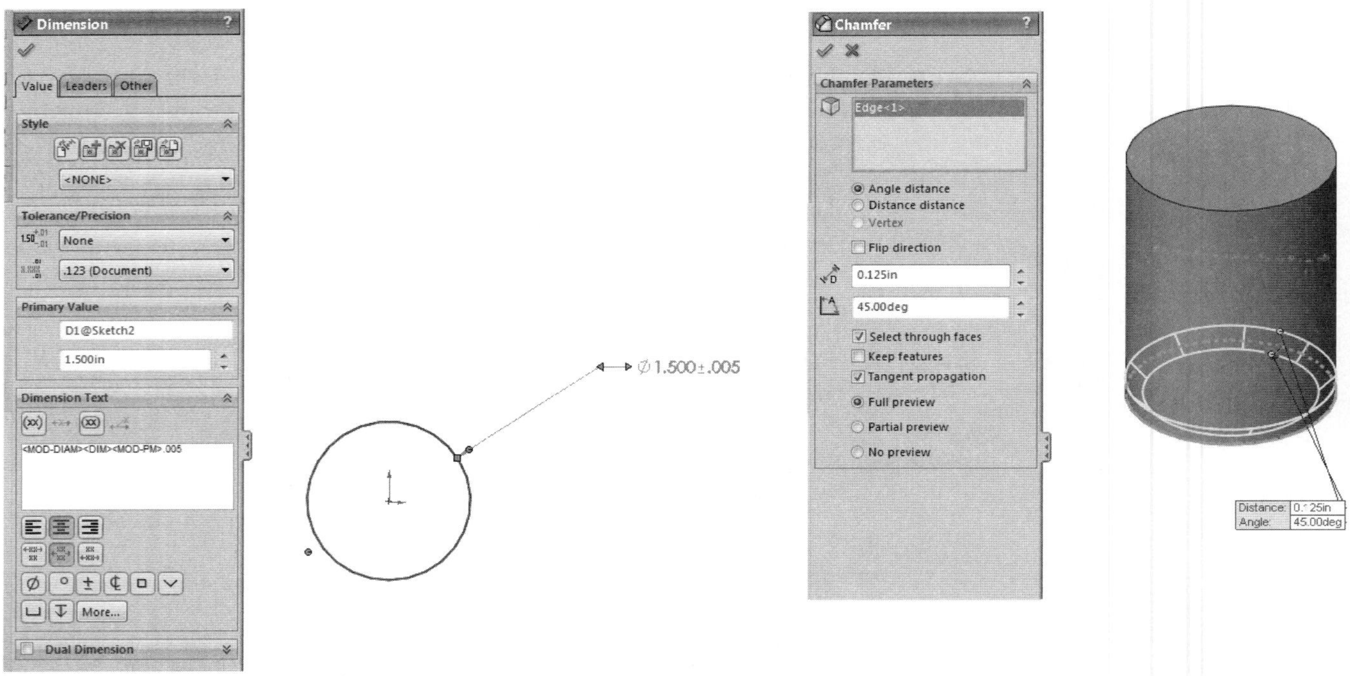

Figure 5-13
Draw and Dimension a Circle on the Top Plane with a Tolerance of ±.005 and Extrude It 2.000 Inches. Draw a .125 Chamfer on the Bottom of the Cylinder

Step 6. Use **Chamfer** from the **Features** toolbar to make a .125 chamfer on the bottom of the cylinder as shown in Figure 5-13.

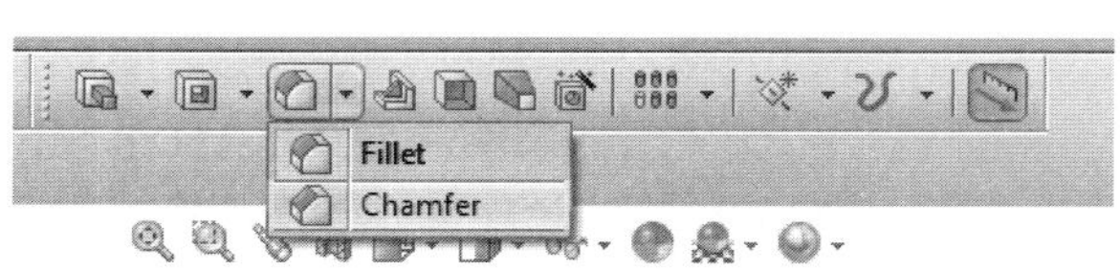

Step 7. Click **Exit Sketch.**

Step 8. Click **Isometric view** so that the model appears as shown in Figure 5-13.

Step 9. Save your drawing (Figure 5-13) with the file name **Cylinder1** in at least two places.

Step 10. Click **New....**

The **New SolidWorks Document** dialog box appears.

Step 11. Click **Part** and click **OK.**

The SolidWorks screen appears. Click **Options** on the menu bar, select the **Document Properties** tab, and click **Units** in the list to the left. Make sure the **Units** are correctly set [notice that **IPS** (inch, pound, second) is checked], then click **OK.**

Step 12. Click the **Top Plane** and click **Sketch** so that the Sketch mode is active.

Step 13. Click **Exit Sketch.**

Step 14. Click **Isometric view** so that the model appears as shown in Figure 5-13.

Step 15. Save your drawing (Figure 5-13) with the file name **Cylinder1** in at least two places.

Step 16. Click **New....**

The **New SolidWorks Document** dialog box appears.

Step 17. Click **Part** and click **OK.**

The SolidWorks screen appears. Click **Options** on the menu bar, select the **Document Properties** tab, and click **Units** in the list to the left. Make sure the **Units** are correctly set [notice that **IPS** (inch, pound, second) is checked], then click **OK.**

Step 18. Click the **Top Plane** and click **Sketch** so that the Sketch mode is active.

Step 19. Draw and dimension a rectangle and two circles with + and − tolerances on the circle diameters, as shown in Figure 5-14, and extrude the part .75 inch.

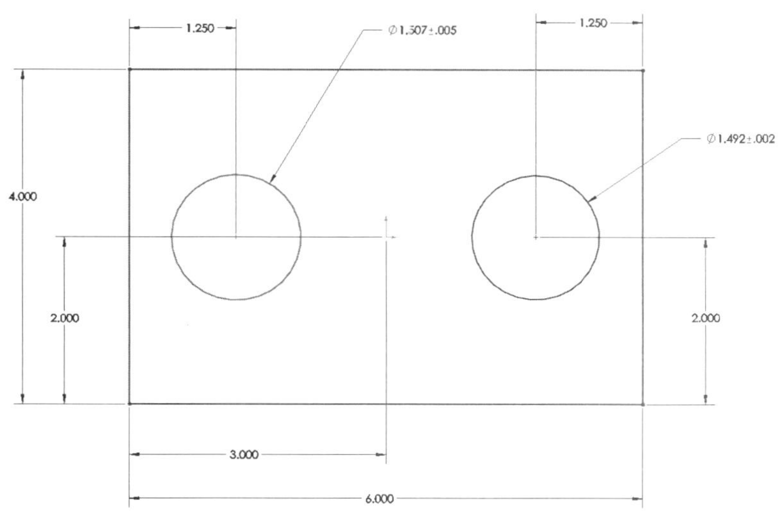

Figure 5-14
Draw and Dimension a Rectangle and Two Circles with ± Tolerances on the Top Plane and Extrude the Part .75 Inch

Step 20. Save your drawing with the file name **Project 5-3(your initials)** in at least two places.

Step 21. Click **New....**

The **New SolidWorks Document** dialog box appears.

Step 22. Click **Part** and click **OK.**

The SolidWorks screen appears. Click **Options** on the menu bar, select the **Document Properties** tab, and click **Units** in the list to the left. Make sure the **Units** are correctly set [notice that **IPS** (inch, pound, second) is checked], then click **OK.**

Step 23. Click the **Top Plane** and click **Sketch** so that the Sketch mode is active.

Step 24. Draw a circle and dimension it with a tolerance of + and − .004 as shown in Figure 5-15. Then extrude it 2.00 inches.

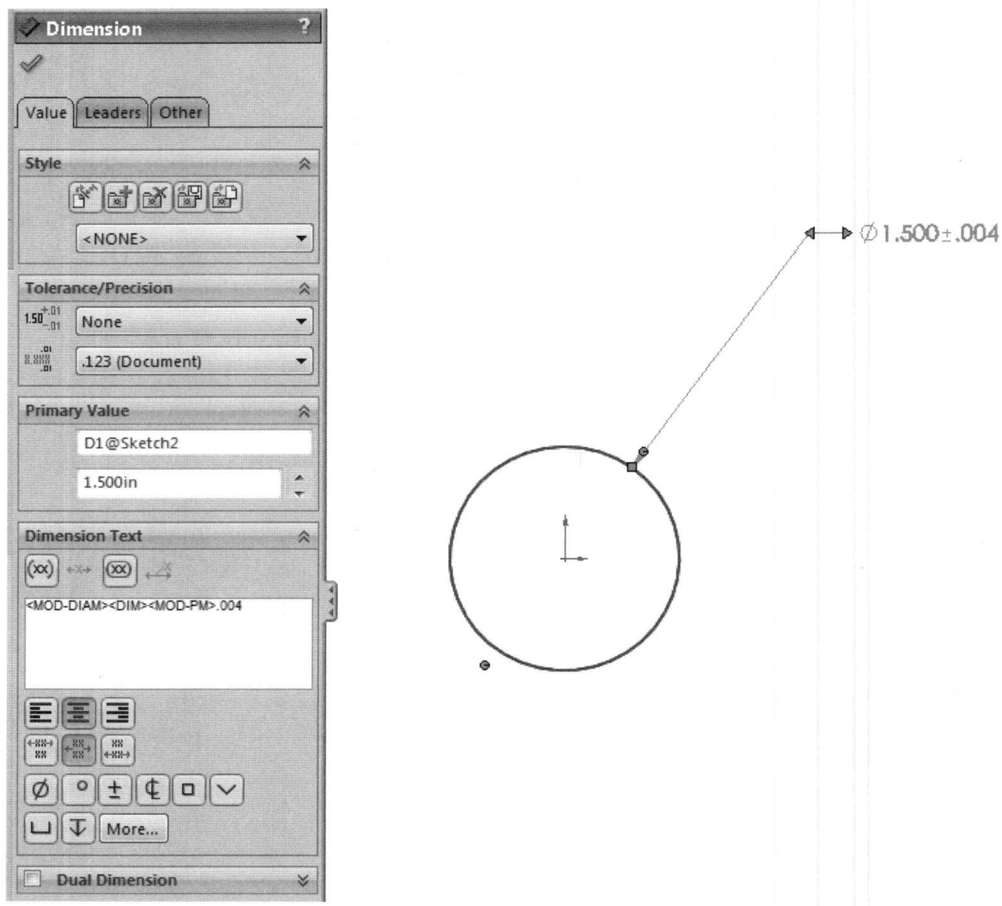

Figure 5-15
Draw and Dimension a Circle on the Top Plane with a Tolerance of ±.004 and Extrude It 2.00 Inches. Draw a .125 Chamfer on the Bottom of the Cylinder

Step 25. Use **Chamfer** from the **Features** toolbar to make a .125 chamfer on the bottom of the cylinder.

Step 26. Click **Exit Sketch.**

Step 27. Click **Isometric view** so that the model appears as shown in Figure 5-13.

Step 28. Save your drawing with the file name **Cylinder2** in at least two places.

Step 29. Open **Project 5-3(your initials)** and click **Isometric view.**

Step 30. Click **Make Assembly from Part** as shown in Figure 5-16.

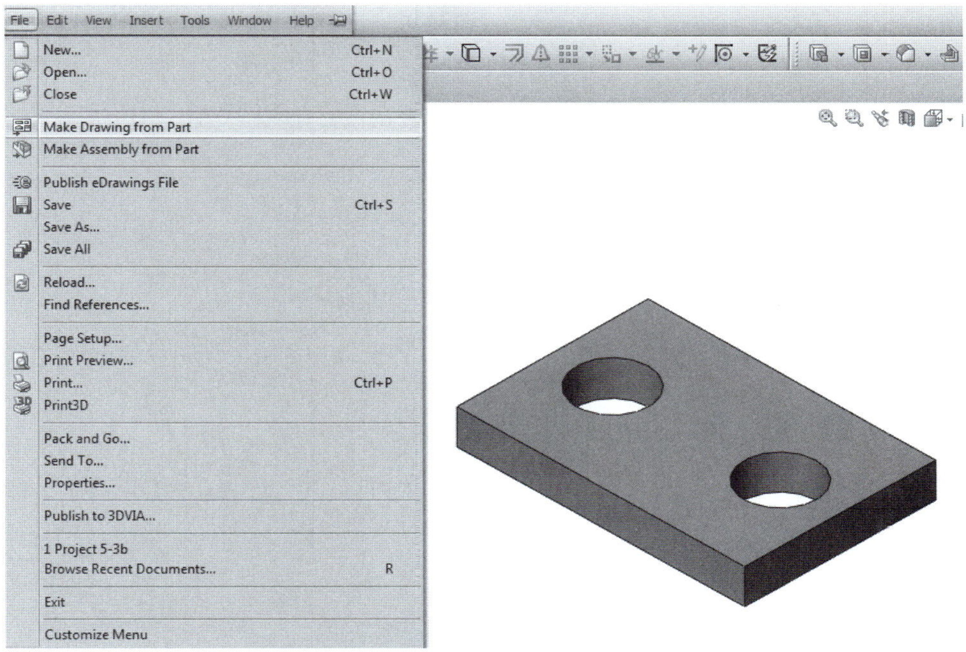

Figure 5-16
Click: Make Assembly from Part and Click: OK. Then Use Insert Components to Insert Both Cylinders

Step 31. Use **Insert Components** to insert **Cylinder1** and **Cylinder2.**

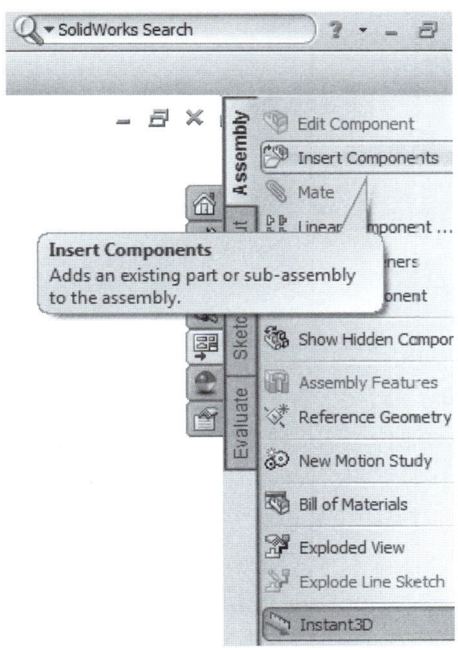

Step 32. Use **Mate - Concentric** (click the cylindrical edge of the cylinders, then click the cylindrical edge of the holes) to position the cylinders above the holes.

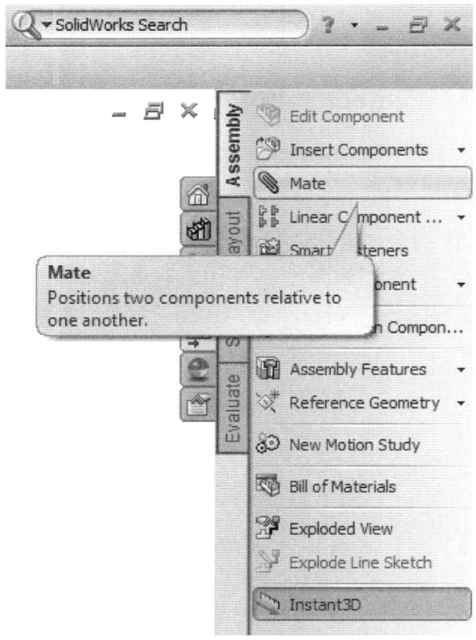

Step 33. Move the cylinders vertically (click on the cylinder and move it) so that they are in the positions shown in Figure 5-17.

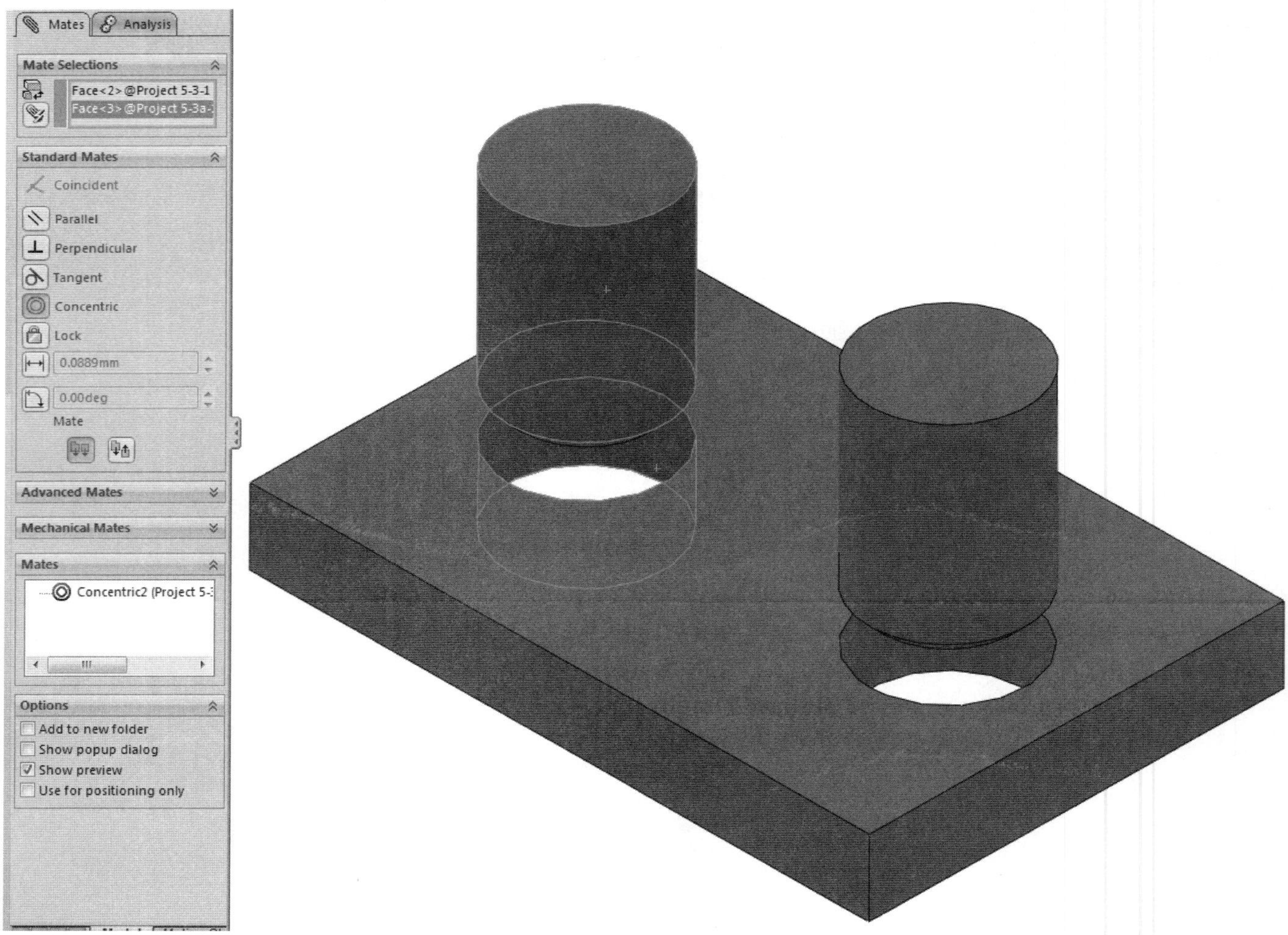

Figure 5-17
Apply Concentric Mates to Both Cylinders and Both Holes. Project 5-3 Complete

Step 34. Save your drawing (Figure 5-17) with the file name **Project 5-3(your initials)** in at least two places.

Step 35. Print the drawing and turn it in.

Project 5-4: **Make a Drawing from Project 5-3 (Figure 5-17)**

Step 1. Start SolidWorks.

Step 2. Click **Open....**

The **Open SolidWorks Document** dialog box appears.

Step 3. Click **Project 5-3(your initials)** and click **Open.**

The SolidWorks screen appears.

Step 4. Click **Make Drawing from Part.**

The **Sheet/Format** dialog box appears.

Step 5. Click **A (ANSI) Landscape** and click **OK.**

Step 6. Right-click a point in the graphics area and click **Properties.** Then click **Third Angle** and make sure **Scale:** is **1:2** as shown in Figure 5-18 and click **OK.**

Figure 5-18
Project 5-4 Complete

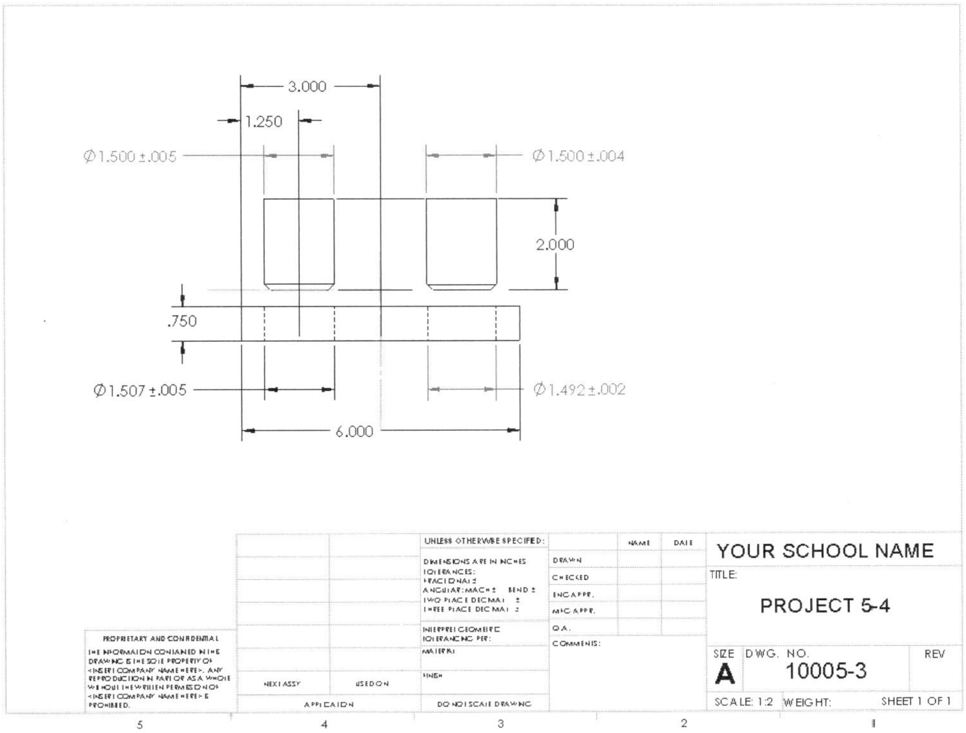

Step 7. Drag **Front view** from the **Task** pane, arrange it so that it is centered in the drawing area, and click the green check mark in the **Drawing View Feature Manager.**

Step 8. Click the **Annotation** tab (or select **Annotations** from the **Insert** menu on the menu bar), click **Model Items,** and make the settings as shown in Figure 2-6. (You will have to turn to Chapter 2 to view this figure.) Be sure the **Entire** model is selected as the **Source:** and the first icon under **Dimensions** is selected.

Step 9. Arrange the dimensions as shown in Figure 5-18. **NOTICE THE SPACING OF ROWS OF DIMENSIONS AND THEIR PLACEMENT.** You may have to delete some dimensions that are not on the correct view and use **Smart Dimension** to put them on the correct view.

Step 10. Right-click any point on the drawing sheet.

The popup menu appears. Click **Edit Sheet Format.**

Step 11. Use the **Note** command to type your **Initials,** today's date, your **School Name,** the **Project Number,** and the **Drawing Number** as shown in Figure 5-18. You will have to change the font size to **6** points in the **NAME** and **DATE** boxes, **12** points in the **Drawing Number** box, and **16** points in the **School Name** and **Project Number** boxes. Change the font to **Arial.** You will also have to click on some of the items and move them into the correct space.

Step 12. Right-click any point on the drawing sheet.

The popup menu appears. Click **Edit Sheet.**

Step 13. Save your drawing with the file name **Project 5-4(your initials)** in at least two places.

Step 14. Print the drawing and turn it in.

Project 5-5: **Make a New Part (Figure 5-19)**

Step 1. Start SolidWorks.

Step 2. Click **New....**

The **New SolidWorks Document** dialog box appears.

Step 3. Click **Part** and click **OK.**

The SolidWorks screen appears. Click **Options** on the menu bar, select the **Document Properties** tab, and click **Units** in the list to the left. Make sure the **Units** are correctly set [notice that **IPS** (inch, pound, second) is checked], then click **OK.**

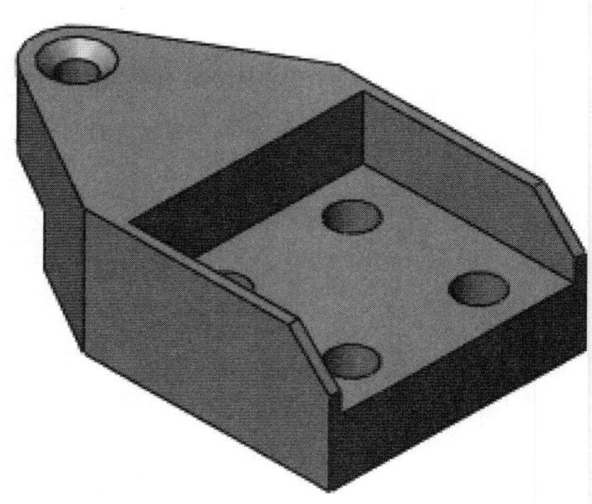

Figure 5-19
Project 5-5 Complete

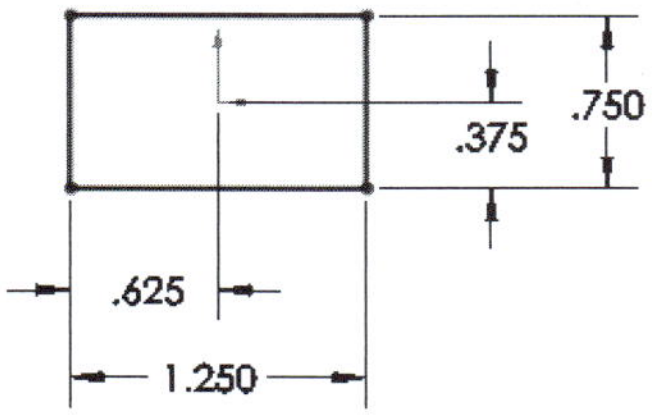

Figure 5-20
Draw a Sketch on the Top Plane Using Rectangle and Dimension It. Extrude the Sketch .375 Inch

Step 4. Click the **Top Plane** and click **Sketch** so that the Sketch mode is active.

Step 5. Draw a rectangle measuring 1.250 inches long × .750 inches high, dimension it, and extrude it .375 inch as shown in Figure 5-20.

Step 6. Click the top plane of the extruded rectangle, make a sketch for an **Extrude Cut,** and make the cut **Through All** as shown in Figure 5-21.

Step 7. Draw a rectangle for an **Extrude Cut** on the top of the model, dimension it, and make the cut .188 inch deep as shown in Figure 5-22.

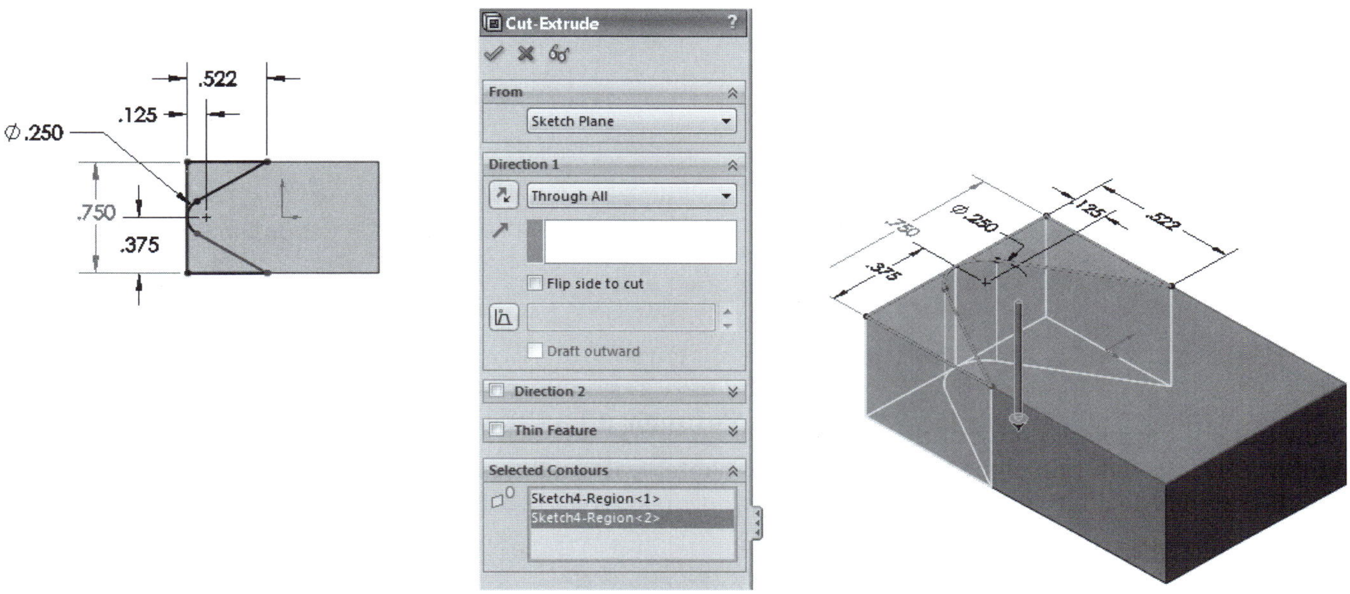

Figure 5-21
Make a Sketch for an Extrude Cut on the Top Plane of the Model, Dimension It and Make the Cut Through All

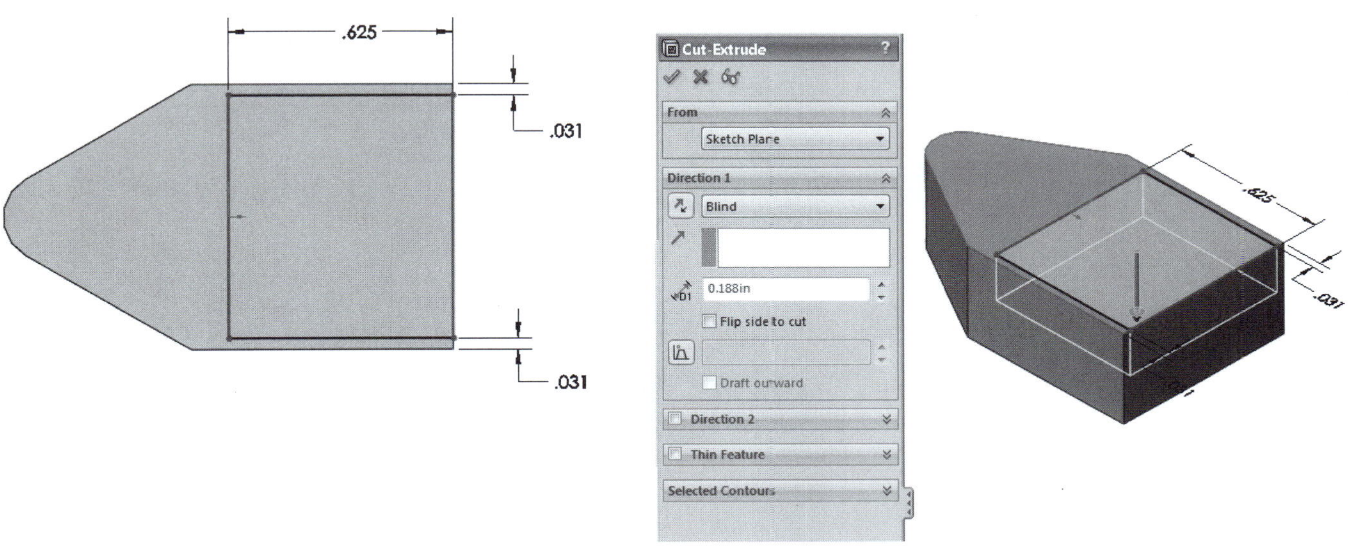

Figure 5-22
Draw a Rectangle for an Extrude Cut on the Top of the Model, Dimension It and Make the Cut .188 Inch Deep

Step 8. Use **Chamfer** (from the **Features** toolbar) to make a .125 inch chamfer on two edges as shown in Figure 5-23.

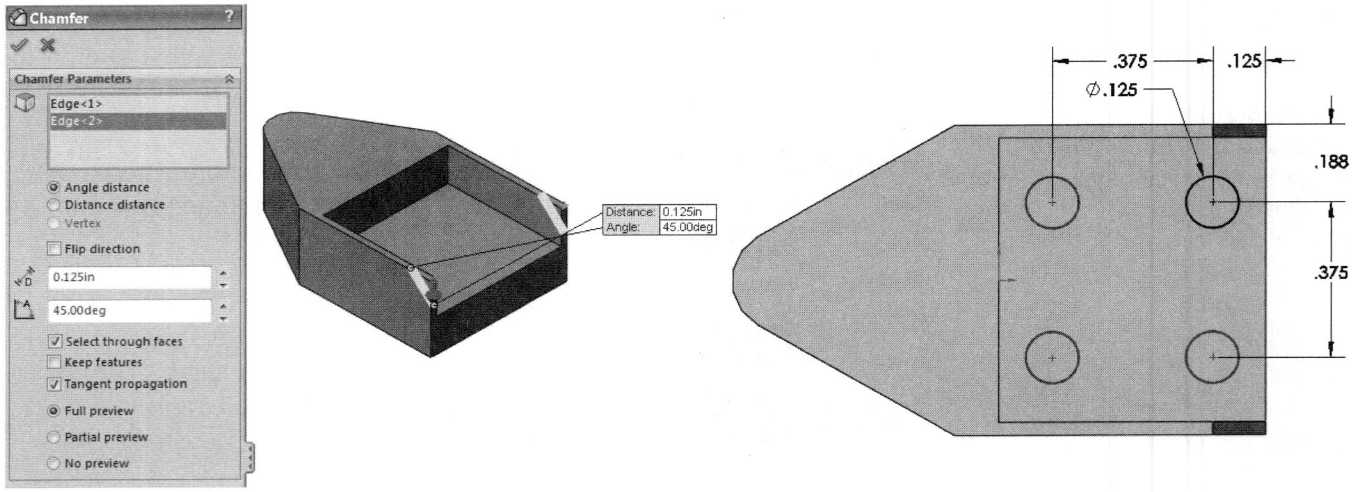

Figure 5-23
Use Chamfer to Make a .125 Inch Chamfer on Two Edges. Draw a Sketch for an Extrude Cut on the Surface Shown and Make the Cut Through All

Step 9. Draw a sketch of four circles for an **Extrude Cut** and dimension it on the base of the **Extrude Cut** from Step 7 and make the sketch **Through All** as shown in Figure 5-23.

Step 10. Make a sketch for an **Extrude Cut** on the front plane of the model and make the cut **Through All** as shown in Figure 5-24.

Figure 5-24
Make a Sketch for an Extrude Cut on the Front Plane of the Model and Make the Extrude Cut Through All

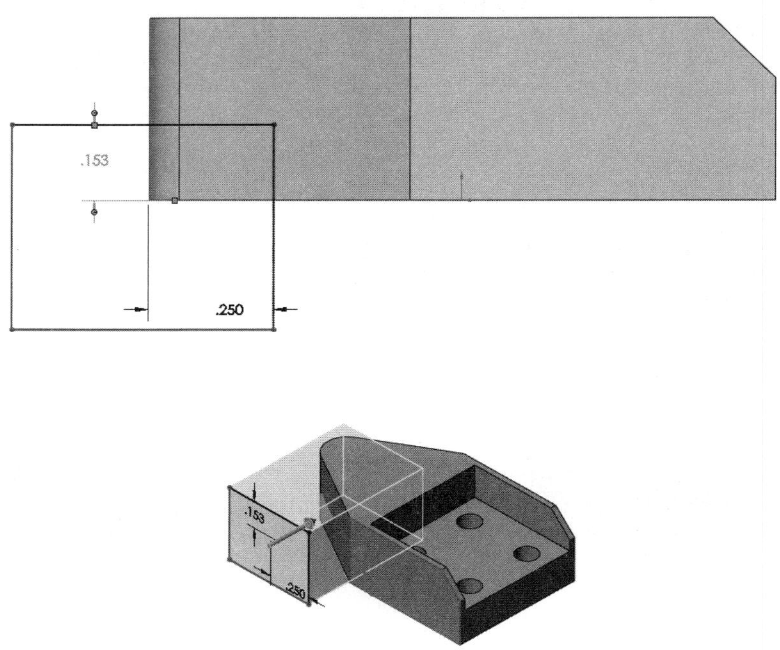

Step 11. Use **Hole Wizard** to make a countersunk hole for a #2 flat head machine screw as shown in Figure 5-25.

Figure 5-25
Use Hole Wizard to Make a Countersunk Hole for a #2 Flat Head Machine Screw.

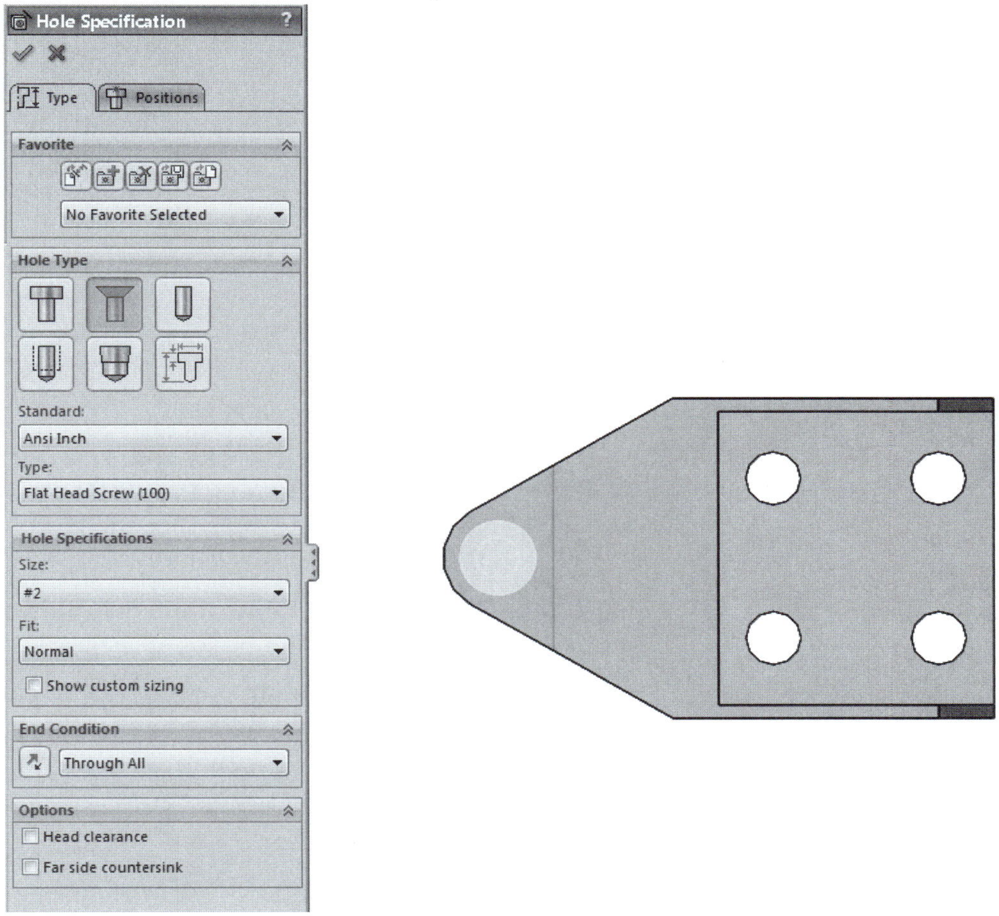

Step 12. Save your drawing (Figure 5-26) with the file name **Project 5-5(your initials)** in at least two places.

Step 13. Print the drawing and turn it in.

Figure 5-26
Project 5-5 Complete

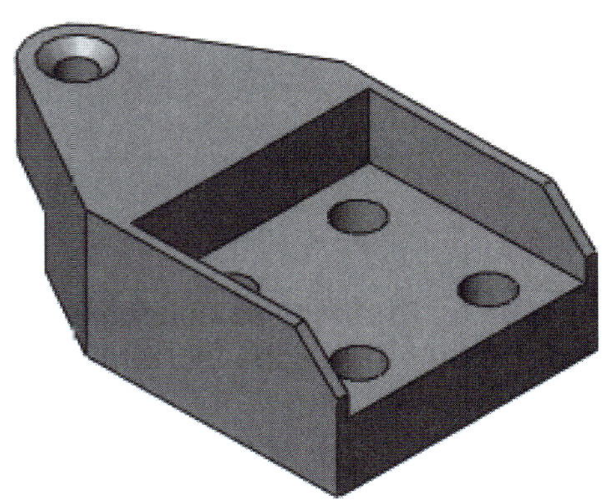

Project 5-6: Make New Parts (Figure 5-27)

Step 1. Start SolidWorks.

Step 2. Click **New....**

The **New SolidWorks Document** dialog box appears.

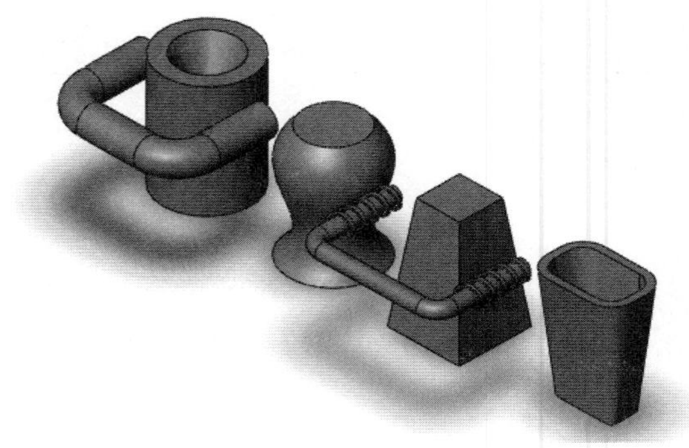

Figure 5-27
Project 5-6 Complete

Step 3. Click **Part** and click **OK.**

The SolidWorks screen appears. Click **Options** on the menu bar, select the **Document Properties** tab, and click **Units** in the list to the left. Make sure that the **Units** are correctly set [notice that **IPS** (inch, pound, second) is checked], then click **OK.**

Step 4. Click the **Top Plane** and click **Sketch** so that the Sketch mode is active.

Step 5. Draw a 1.500 diameter circle and extrude it 2.000 inches as shown in Figure 5-28.

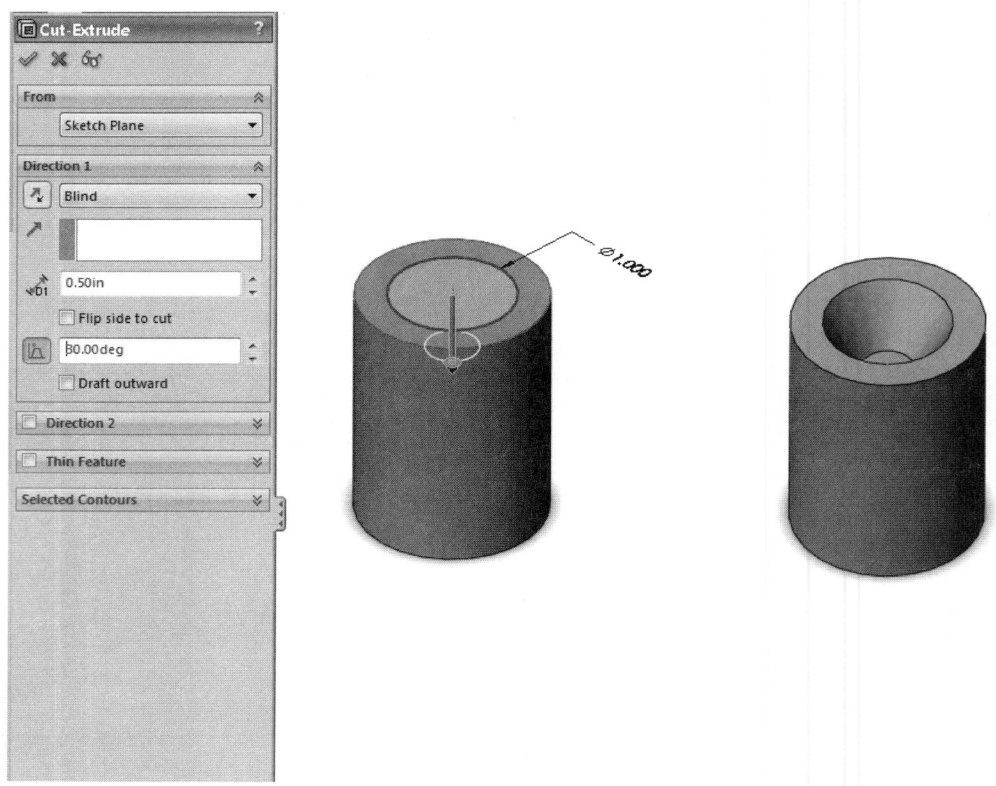

Figure 5-28
Draw a 1.500 Diameter Circle on the Top Plane and Extrude It 2.000 Inches. Draw a 1.00 Diameter Circle on Top of the Extruded Circle and Use Extrude Cut to Make a .50 Inch Deep Cut with a 90 Degree Draft Angle

Step 6. Click the top plane of the extruded circle and use **Extrude Cut** to make a .50 inch deep cut with a 90° **Draft angle** as shown in Figure 5-28.

Step 7. Draw a 1.500 diameter circle on the top plane.

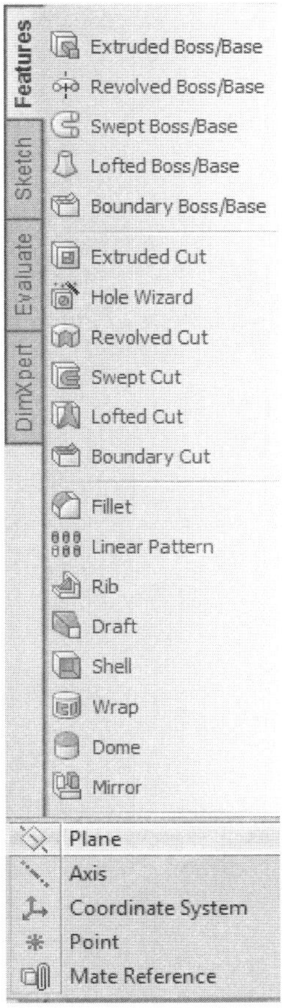

Step 8. Make a **Plane** .50 inch from the top plane.

Step 9. Draw a 1.000 diameter circle on **Plane1.**

Step 10. Make a **Plane** 1.00 inch from **Plane1.**

Step 11. Draw a 1.500 diameter circle on **Plane2.**

Step 12. Make a **Plane** .50 inch from **Plane2.**

Step 13. Draw a 1.000 diameter circle on **Plane3.**

Step 14. Use **Lofted Boss/Base** to make the **Lofted Solid** shown in Figure 5-29.

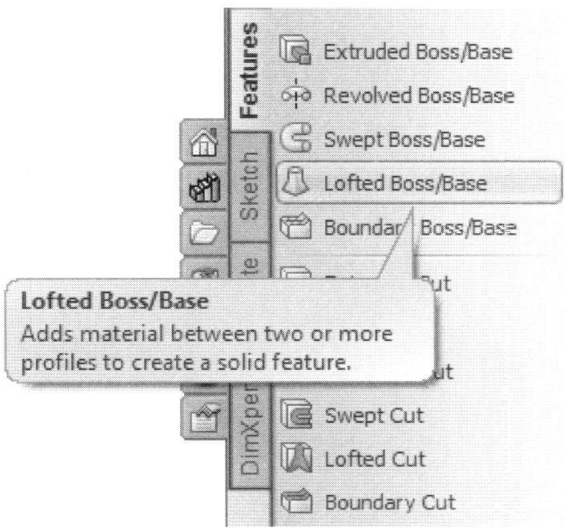

Step 15. Draw a rectangle on the top plane and dimension it as shown on the left side of Figure 5-30.

Step 16. Draw a rectangle on **Plane3** and dimension it as shown in the center of Figure 5-30.

Step 17. Use **Boundary Boss/Base** and click the two corners as shown on the right side of Figure 5-30 to make the solid **Boundary Boss/Base.**

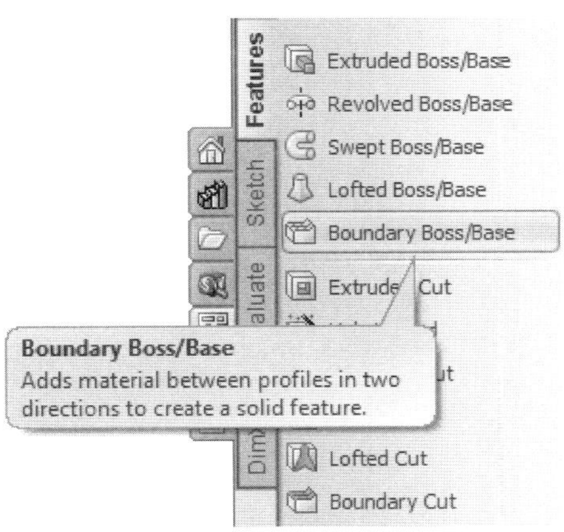

Step 18. Draw a rectangle on **Plane3** and dimension it as shown in Figure 5-31.

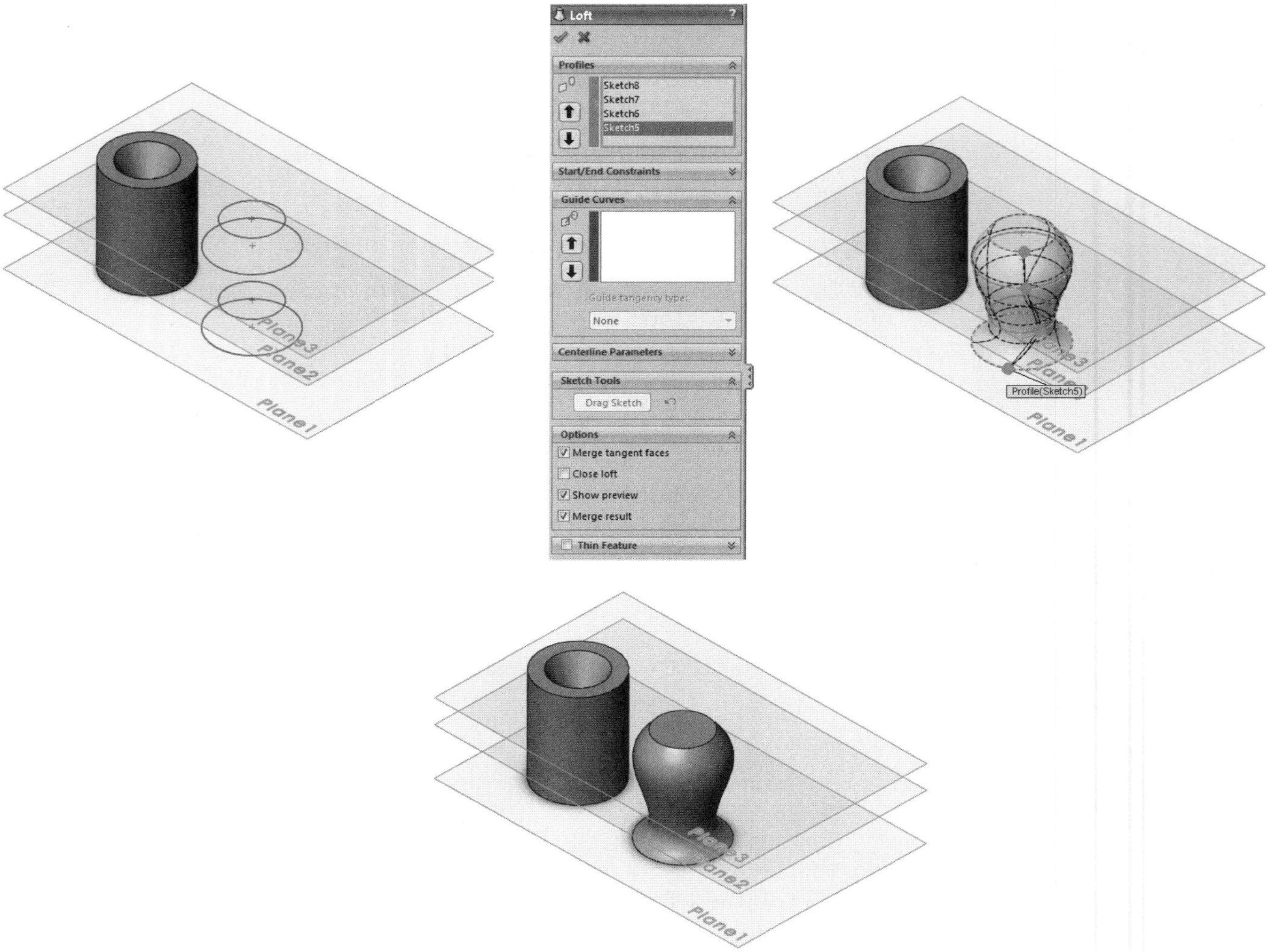

Figure 5-29
Draw a 1.500 Diameter Circle on the Top Plane. Make a Plane .50 Inch from the Top Plane and Draw a 1.000 Diameter Circle on Plane1. Make a Plane 1.00 Inch on Plane1 and Draw a 1.500 Diameter Circle on Plane2. Make a Plane .50 Inch from Plane2 and Draw a 1.000 Diameter Circle on Plane3. Use Lofted Boss/Base to Make a Lofted Solid.

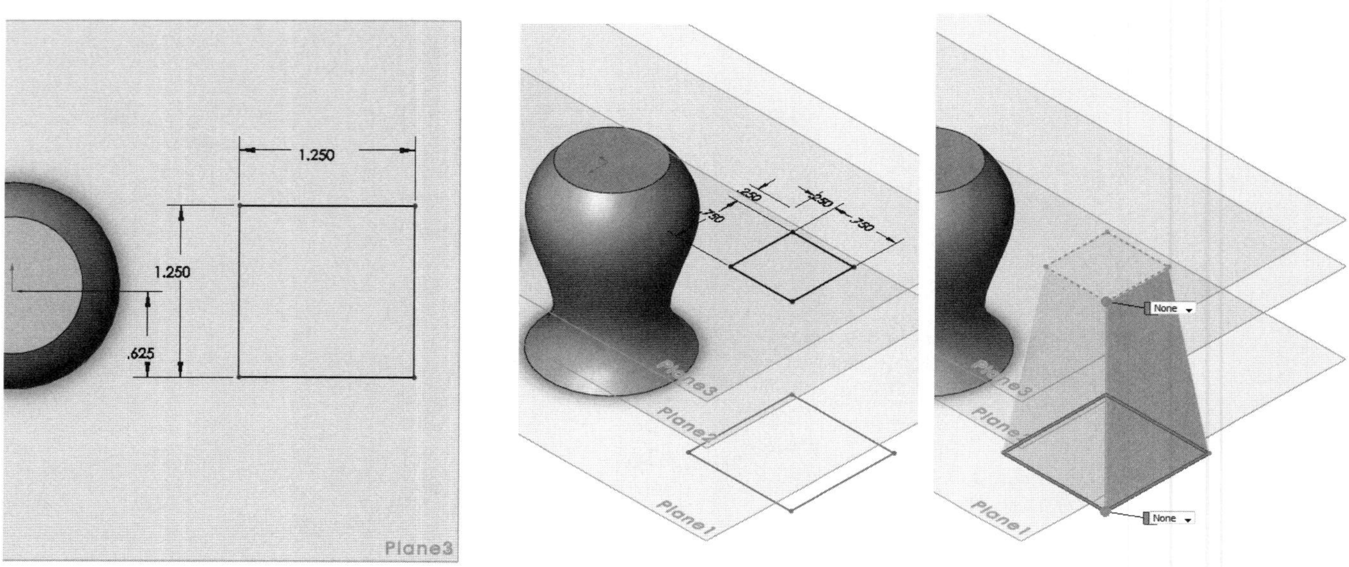

Figure 5-30
Draw a Rectangle on the Top Plane and Dimension It. Draw a Rectangle on Plane3 and Dimension It. Use Boundary Boss/Base and Click: the Two Corners to Make It

Figure 5-31
Draw a Rectangle on Plane3 and Dimension It.

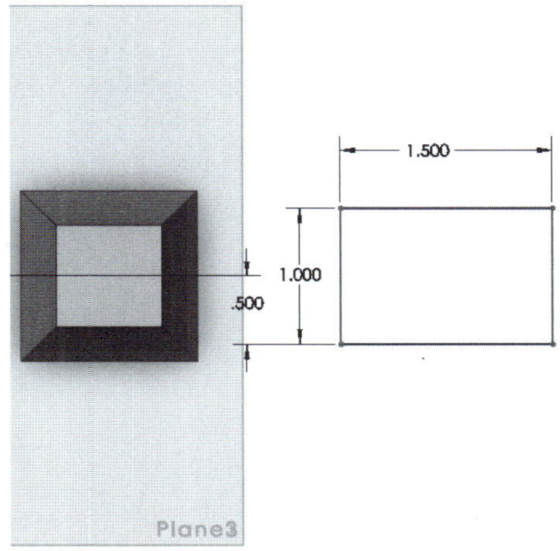

Step 19. Use **Sketch Fillet** on all corners of the rectangle to make a .375 radius on all corners as shown in the upper right of Figure 5-32.

Figure 5-32
Use Fillet to Make a .375 Radius on All Corners of the First Rectangle. Draw a .500 × 1.000 Rectangle on the Top Plane Centered in the First Rectangle and Make a .63 Radius on All Four Corners

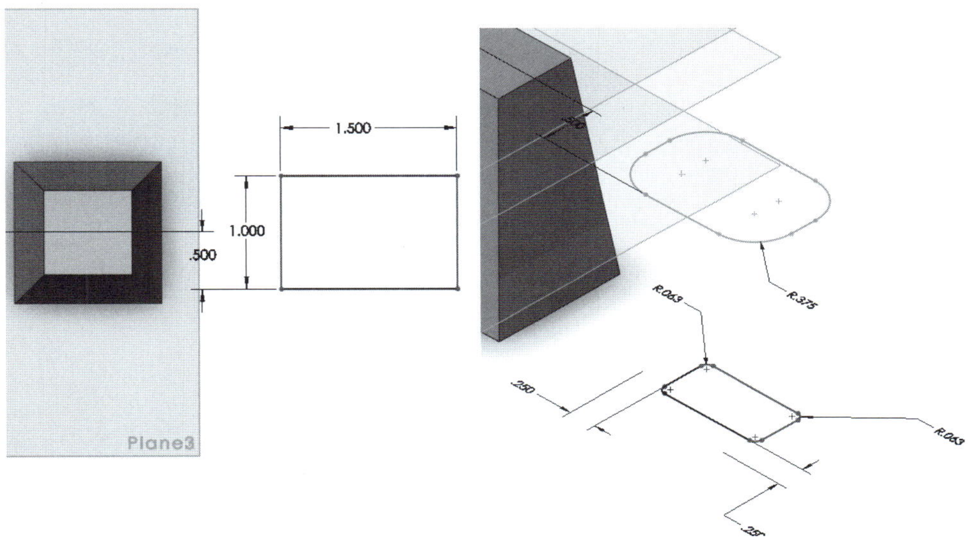

Step 20. Draw a .500 × 1.000 rectangle on the top plane from the top view so that it is centered in the first rectangle and make a .063 radius on all four corners as shown in the lower right of Figure 5-32.

Step 21. Use **Boundary Boss/Base** and click an edge on the same side of each rectangle to make the **Boundary Boss/Base** as shown in Figure 5-33.

Step 22. Use **Convert Entities** to change the edges of the top surface of the **Boundary Boss/Base** to lines and use **Offset** to make a .125 offset as shown in the top left side of Figure 5-34.

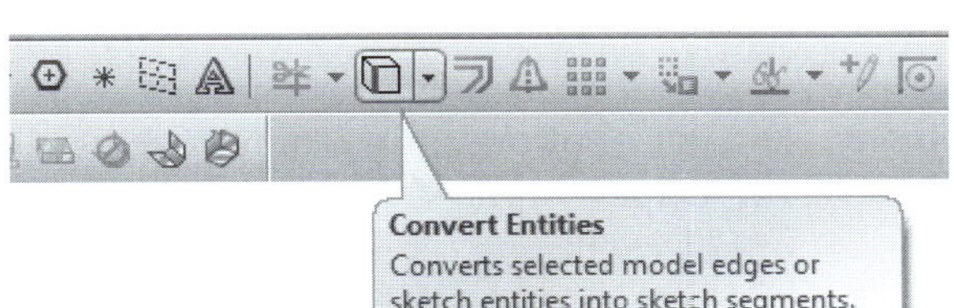

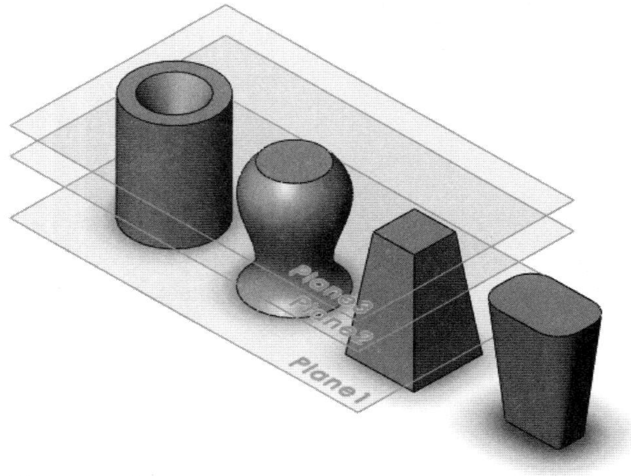

Figure 5-33
Use Boundary Boss/Base and Click: An Edge on the Same Side of Each Rectangle to Make the Solid

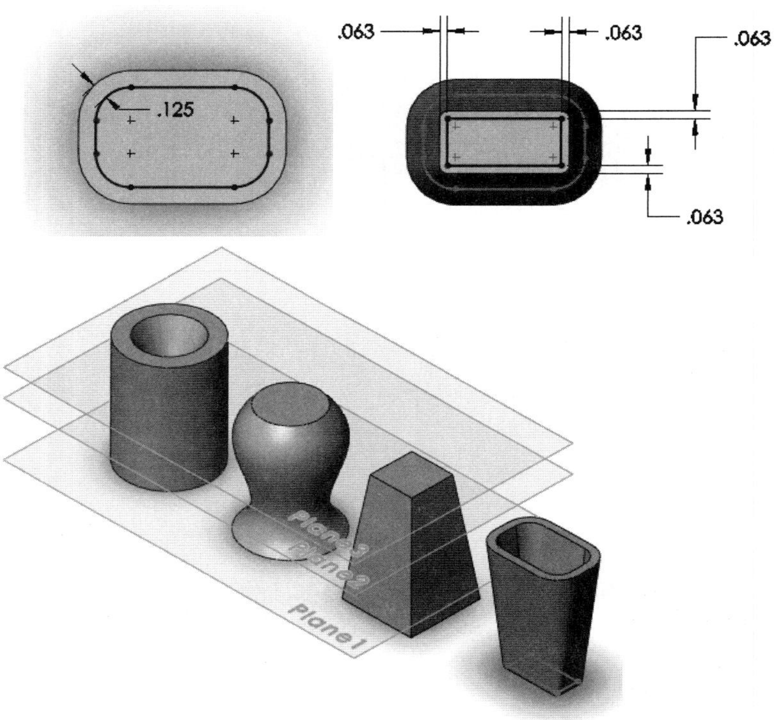

Figure 5-34
Use Convert Entities to Change the Edges of the Top Surface of the Boundary Boss/Base. Use Offset to Make a .125 Offset on Plane3 as Shown from the Top View. Draw a Rectangle on the Top Plane and Dimension It. (Make Sure It Is Centered in the Offset Rectangle.) Use Boundary Cut to Make the Cut

Step 23. Draw a rectangle on the top plane and dimension it as shown in the upper right of Figure 5-34. (Make sure it is centered in the offset rectangle as shown.)

Step 24. Use **Boundary Cut** to make the cut as shown in Figure 5-34.

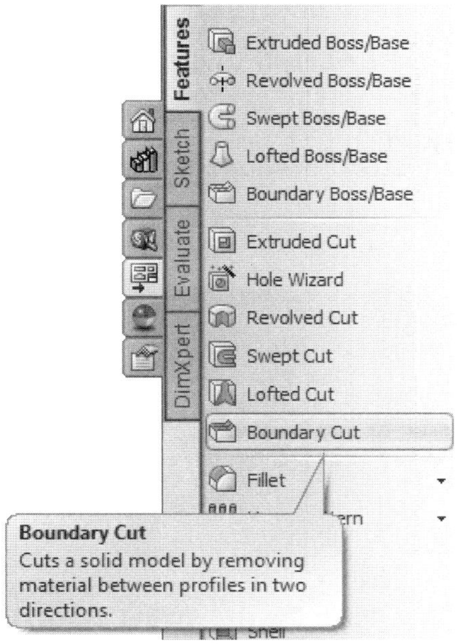

Step 25. Draw a **Path** for the **Swept Boss/Base** on **Plane2** so that one end of the path intersects with the front plane as shown on the left side of Figure 5-35.

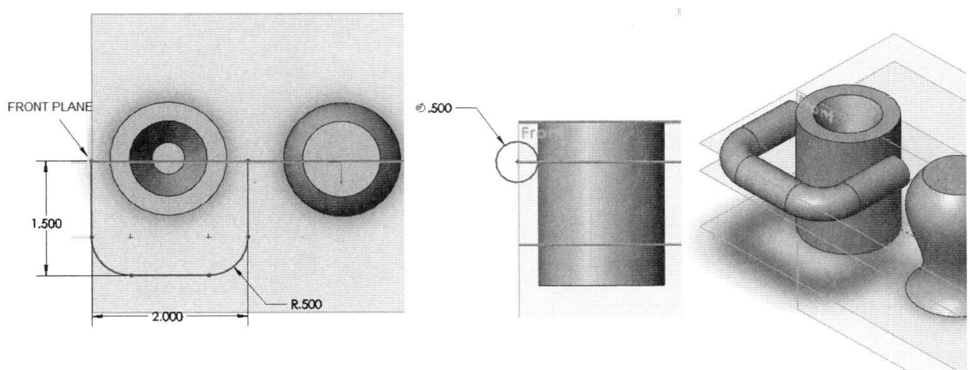

Figure 5-35
Draw a Path for the Swept Boss/Base on Plane2 So One End of the Path Intersects with the Front Plane as Shown. On the Front Plane Draw a .500 Diameter Circle for the Profile. Use Swept Boss/Base to Make the Sweep

Step 26. On the front plane draw a .500 diameter circle for the **Profile** as shown in the center of Figure 5-35.

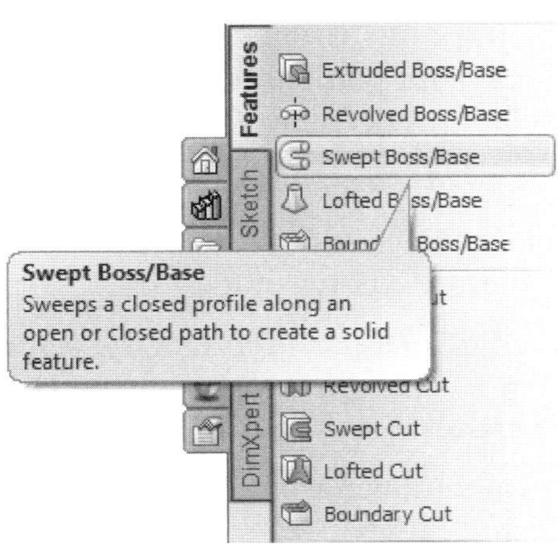

Step 27. Use **Swept Boss/Base** to make the sweep.

Step 28. Use **Edit Sketch** to draw a .375 diameter circle in the center of the .500 diameter circle of the **Profile** of the **Swept Boss/Base,** then click **Exit Sketch,** and the **Swept Cut** is miraculously completed, as shown in the right side of Figure 5-36.

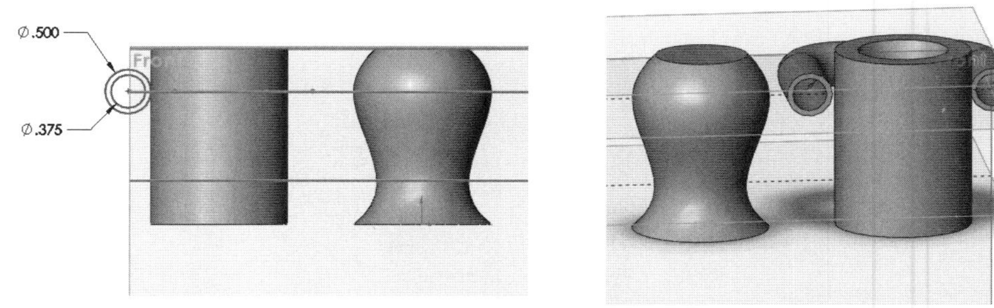

Figure 5-36
Use Edit Sketch to Draw a .375 Diameter Circle on the Swept Contour Sketch, Then Exit Sketch. Swept Cut Complete

Step 29. Draw a **Path** for the **Swept Boss/Base** on **Plane2** so that one end of the path intersects with the front plane as shown on the left side of Figure 5-37.

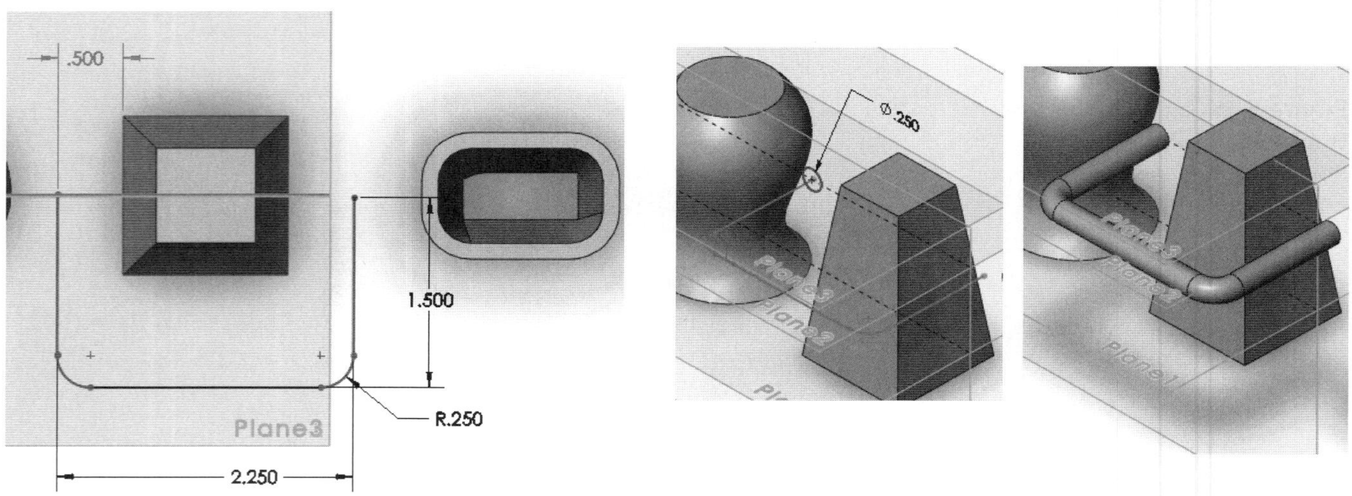

Figure 5-37
Draw a Path for a Swept Boss/Base on Plane2 and Dimension It So One End Intersects with the Front Plane. Draw a .250 Diameter Circle on the Front Plane at the End of the Path for the Contour. Use Swept Boss/Base

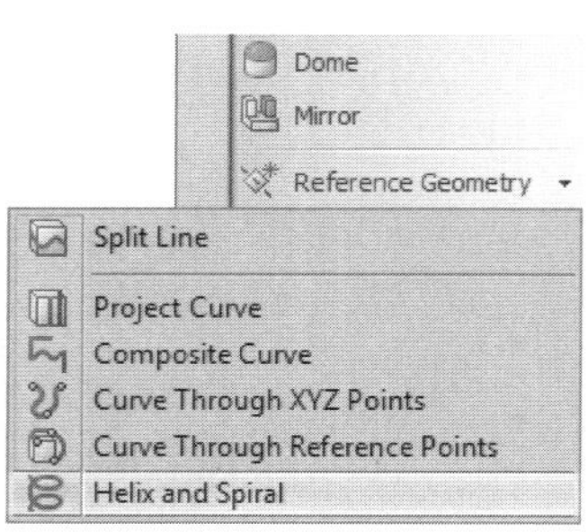

Step 30. On the front plane draw a .250 diameter circle for the **Profile** as shown in the center of Figure 5-37.

Step 31. Use **Swept Boss/Base** to make the sweep.

Step 32. Click the back face of the new sweep and draw a .250 diameter circle in the center of it.

Step 33. Click **Helix/Spiral** and use the settings in the **Helix/Spiral2 Feature Manager** (yours may be **Helix/Spiral1)** to make the **Helix/Spiral** as shown in the left side of Figure 5-38.

Step 34. Draw and dimension a rectangle on **Plane2** as shown in the center of Figure 5-38.

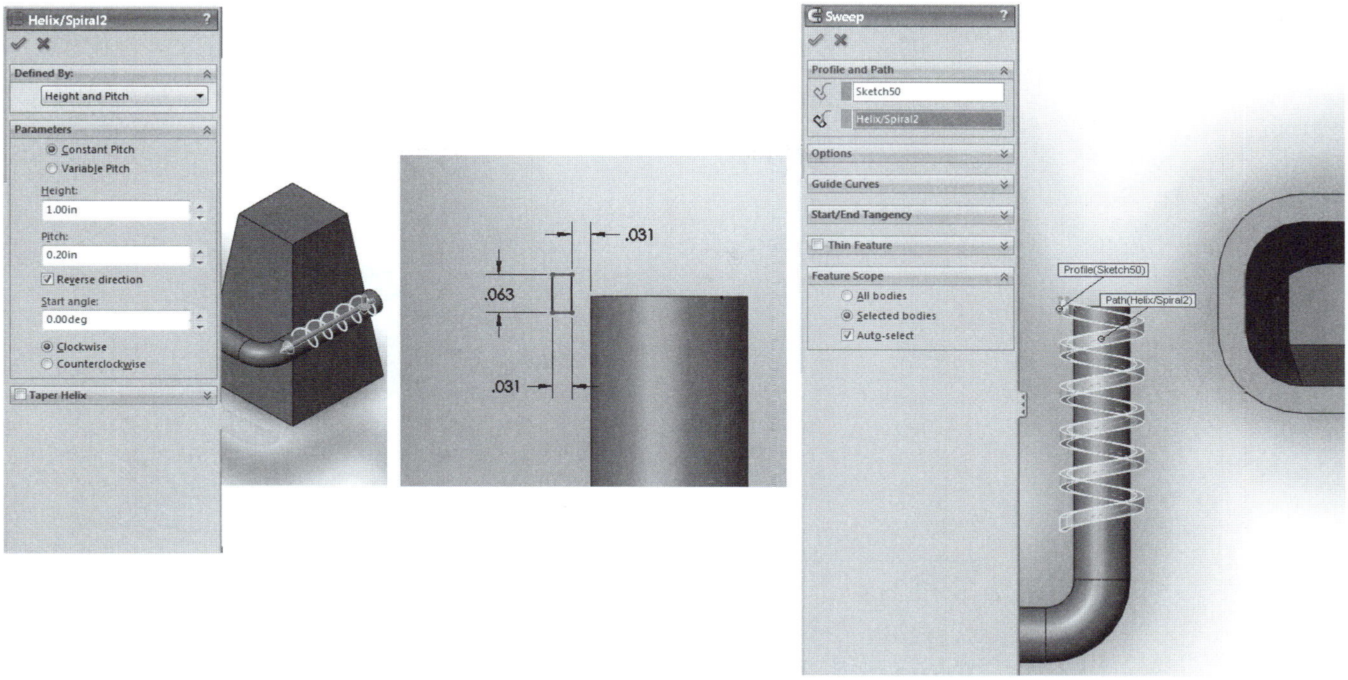

Figure 5-38
Click the Back Face of the Sweep and Draw a .250 Diameter Circle in the Center of It. Click: Helix/Spiral and Make the Settings in Helix/Spiral2 to Make It. Draw and Dimension a Rectangle on Plane2. Click: Swept Boss/Base. Make Sure the Sketch Is Shown in Profile and the Helix/Spiral Is in the Path Box

Step 35. Click **Exit Sketch.**

Step 36. Click **Swept Boss/Base.** Make sure that the rectangle sketch, **Sketch 50** (yours may be a different sketch number), is shown in the **Profile Box** and that the **Helix/Spiral** is shown in the **Path Box** as shown on the right side of Figure 5-38, then click **OK.**

Step 37. Right-click **Material <not specified>** and click **Edit Material** to assign a material as shown in Figure 5-39. Then click **Isometric view.**

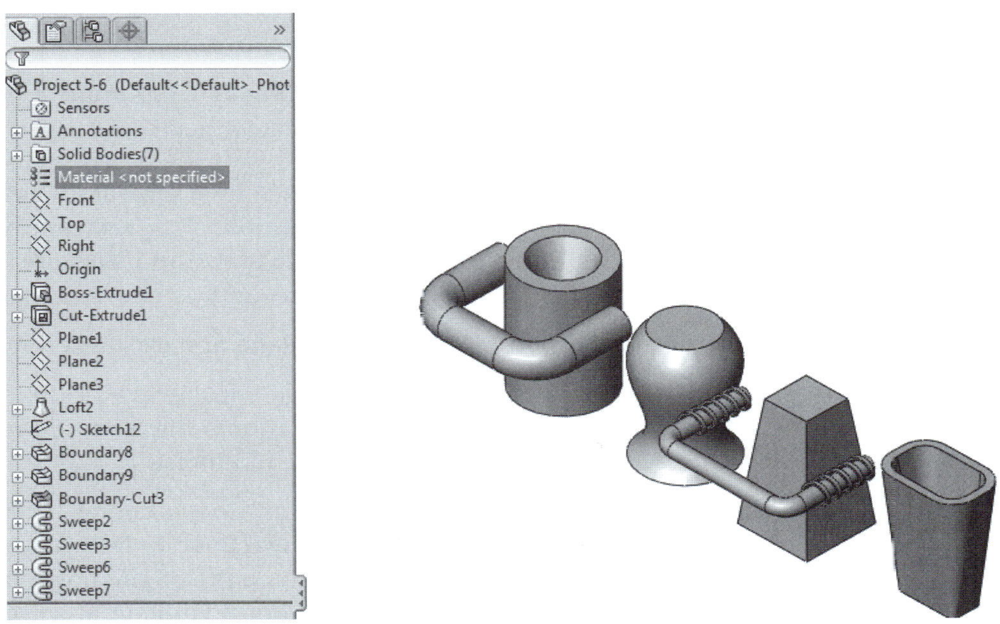

Figure 5-39
Right Click: Material <not specified> and Click: Edit Material to Assign a Material

Step 38. Save your drawing (Figure 5-40) with the file name **Project 5-6(your initials)** in at least two places.

Step 39. Print the drawing and turn it in.

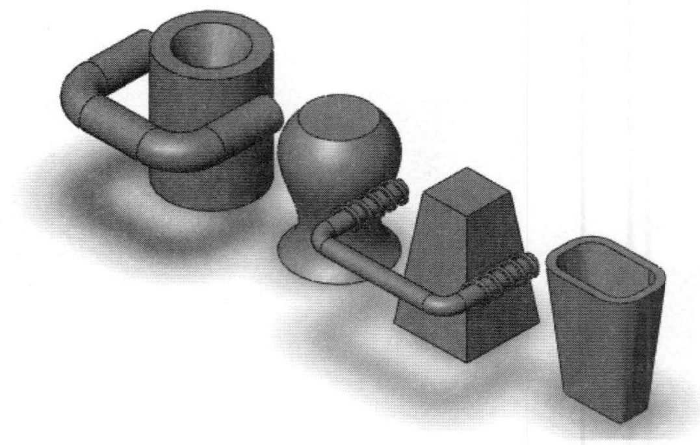

Figure 5-40
Project 5-6 Complete

Project 5-7: **Make a New Part with a Rib on Each End (Figure 5-41)**

Step 1. Start SolidWorks.

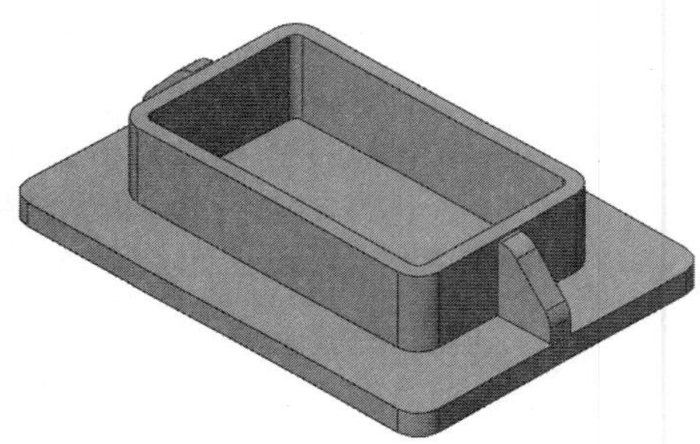

Figure 5-41
Project 5-7 Complete

Step 2. Click **New....**

The **New SolidWorks Document** dialog box appears.

Step 3. Click **Part** and click **OK.**

The SolidWorks screen appears. Click **Options** on the menu bar, select the **Document Properties** tab, and click **Units** in the list to the left. Make sure the **Units** are correctly set [notice that **IPS** (inch, pound, second) is checked], then click **OK.**

Step 4. Click the **Top Plane** and click **Sketch** so the Sketch mode is active.

Step 5. Draw a rectangle measuring 7.500 inches long × 5.250 inches high with .375 radius fillets on all four corners, dimension it, and extrude it .375 inch.

Step 6. Use **Convert Entities** to change the edges of the top surface of the **Extrude Boss/Base** to lines and arcs and use **Offset** to make a 1.000 inch offset as shown in the top left side of Figure 5-42. (You will not be able to offset the arcs.)

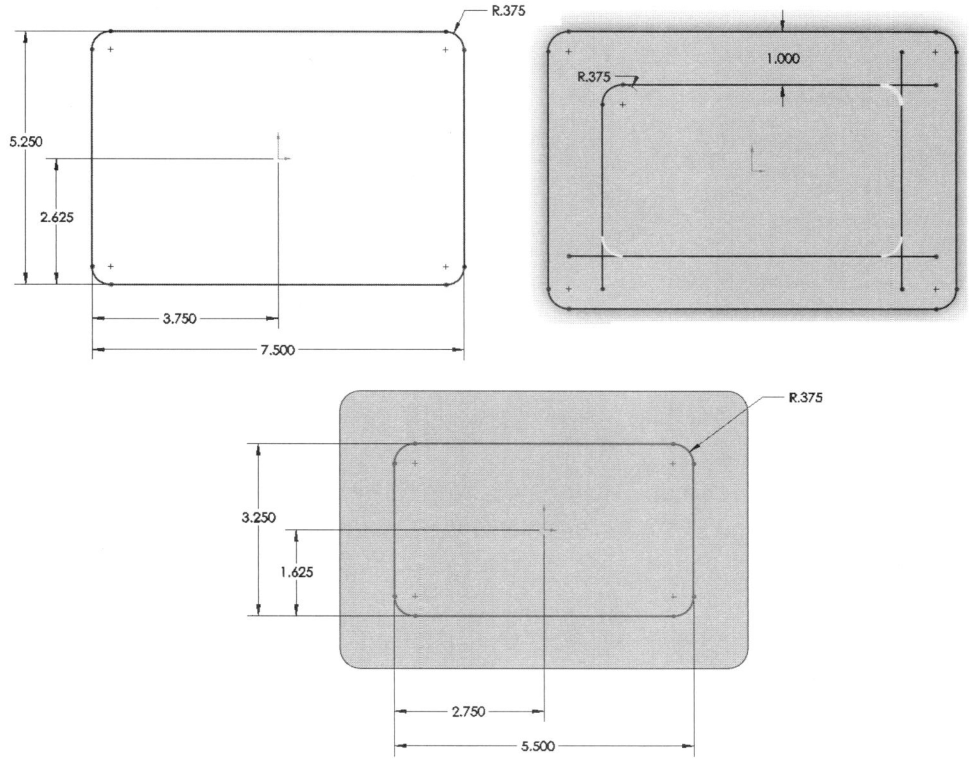

Figure 5-42
Make a Sketch for an Extrude Boss/Base on the Top Plane and Extrude It .375 Inch Use Convert Entities to Change Outside Edges to Lines and Arcs and Offset the Entities 1.000 Inch Inside. Make a .375 Inch Fillet at Each Corner and Dimension the Sketch as Shown, then Delete the Converted Entities, Extrude the Sketch 1.25 Inches

Step 7. Make a .375 inch fillet at each corner and dimension the sketch as shown in Figure 5-42.

Step 8. Delete the converted entities and extrude the sketch 1.25 inches.

Step 9. Use the **Shell** command to shell the top extrusion as shown in Figure 5-43. (Be sure to click the top of the extrusion so that **Face<1>** shows in the **Faces to Remove** box and **.250 inch** shows in the **Thickness** box (next to **D1**).)

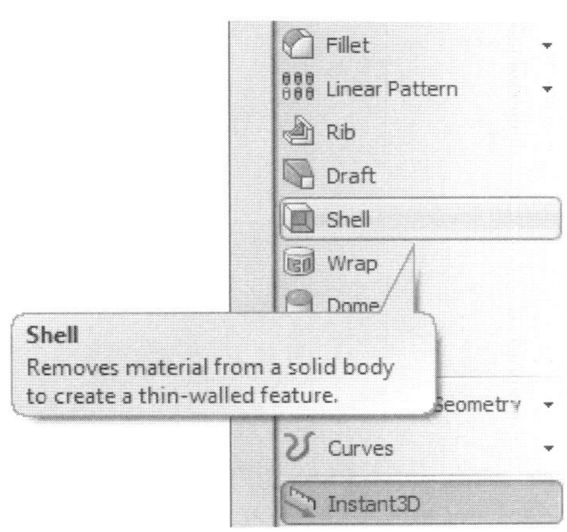

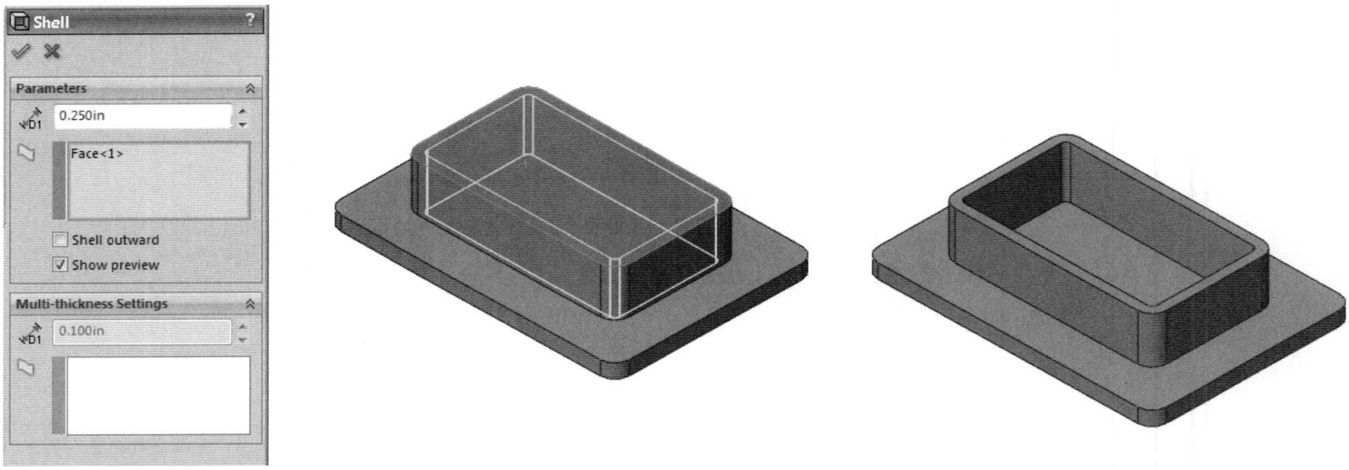

Figure 5-43
Use the Shell Command as Shown. Be Sure to Click: the Top of the Extrusion

Step 10. Make a sketch for a **Rib** on the front plane as shown in Figure 5-44 and make the **Rib** using the settings shown in the **Rib Manager.**

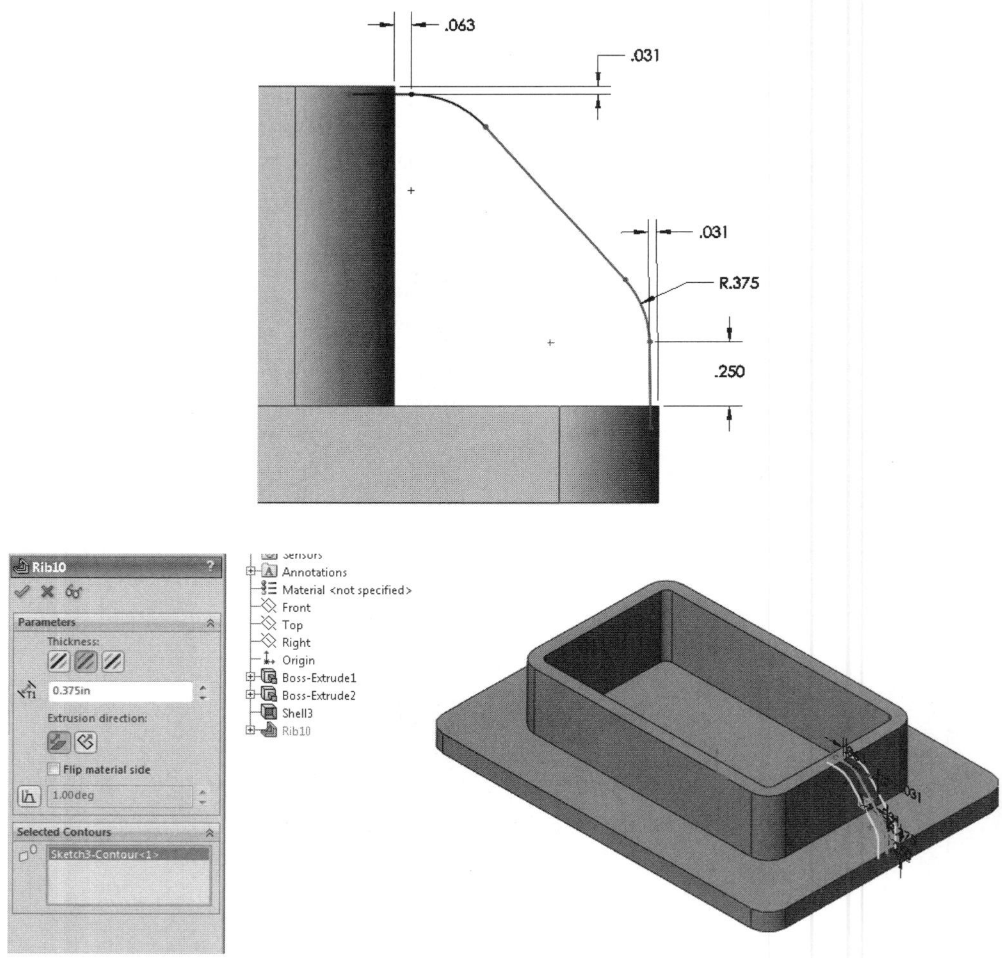

Figure 5-44
Make a Sketch for a Rib on the Front Plane and Make the Rib Using the Settings Shown in the Rib Manager

Step 11. Mirror the **Rib** using the right plane as the **Mirror Face/Plane** and **Rib10** as the **Features to mirror** as shown in Figure 5-45.

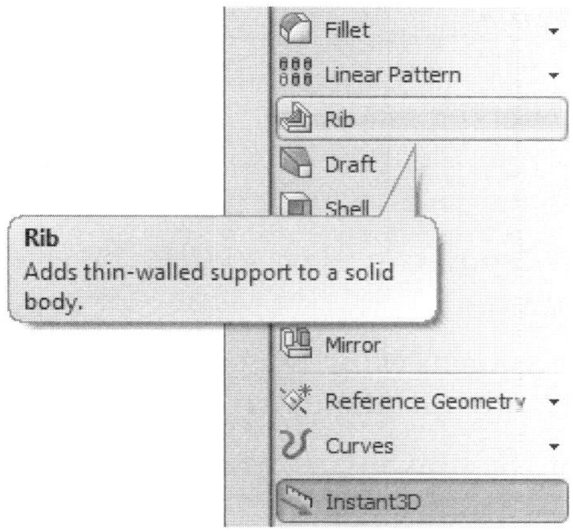

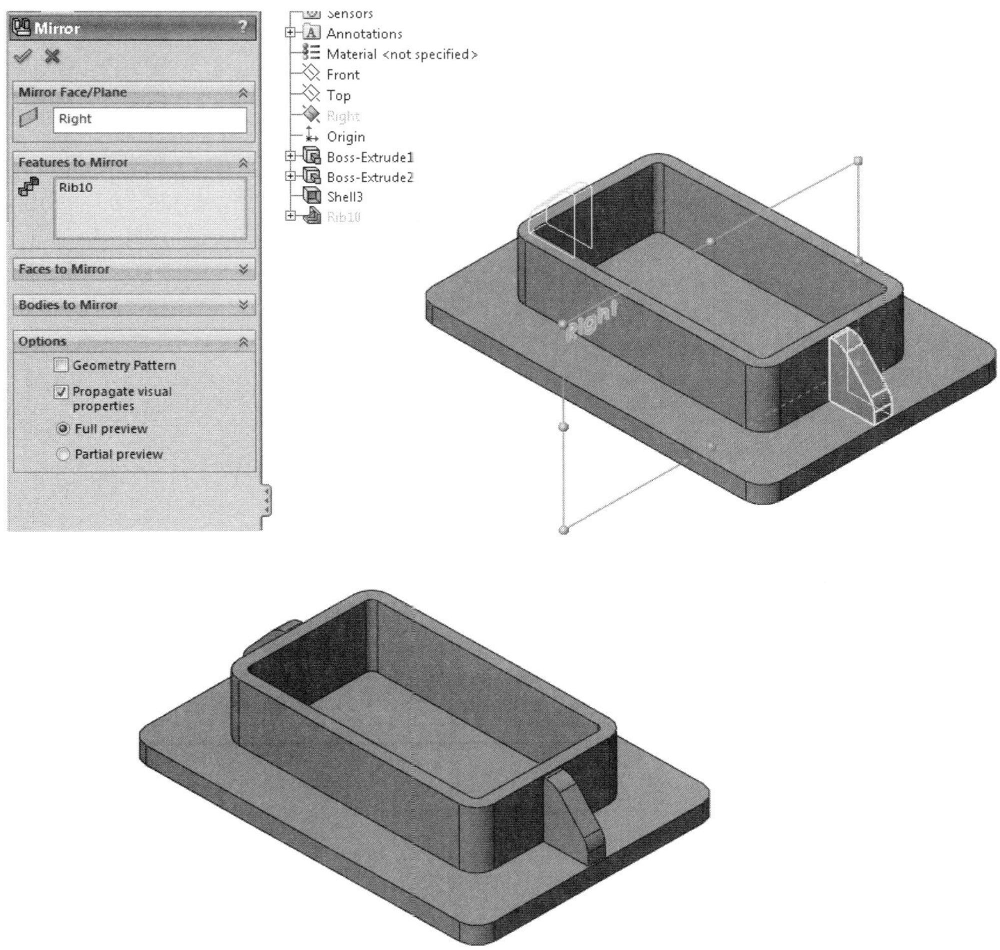

Figure 5-45
Mirror the Rib Using the Right Plane as the Mirror Face/Plane. Project 5-7 Complete

Step 12. Save your drawing (Figure 5-45) with the file name **Project 5-7(your initials)** in at least two places.

Step 13. Print the drawing and turn it in.

Exercises

EXERCISE 5-1 **Make a model (Figure 5-46) from the drawing shown in Figure 5-47.**

Step 1. Use the commands and settings from the Projects in this chapter to make the model shown in Figure 5-46. Use the dimensions shown in Figure 5-47 to make the model. Add 1/16 radius filet to the edges shown in Figure 5-46.

Step 2. Click **Isometric view** so that the model appears as shown in Figure 5-46.

Step 3. Save your drawing (Figure 5-46) with the file name **Exercise 5-1(your initials)** in at least two places.

Step 4. Print the drawing and turn it in.

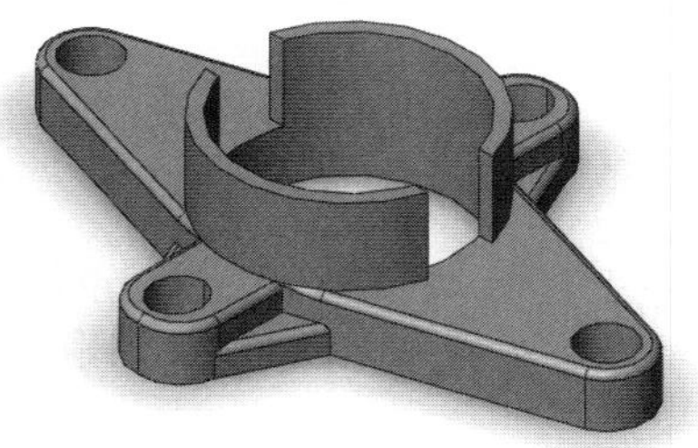

Figure 5-46
Exercise 5-1 Complete

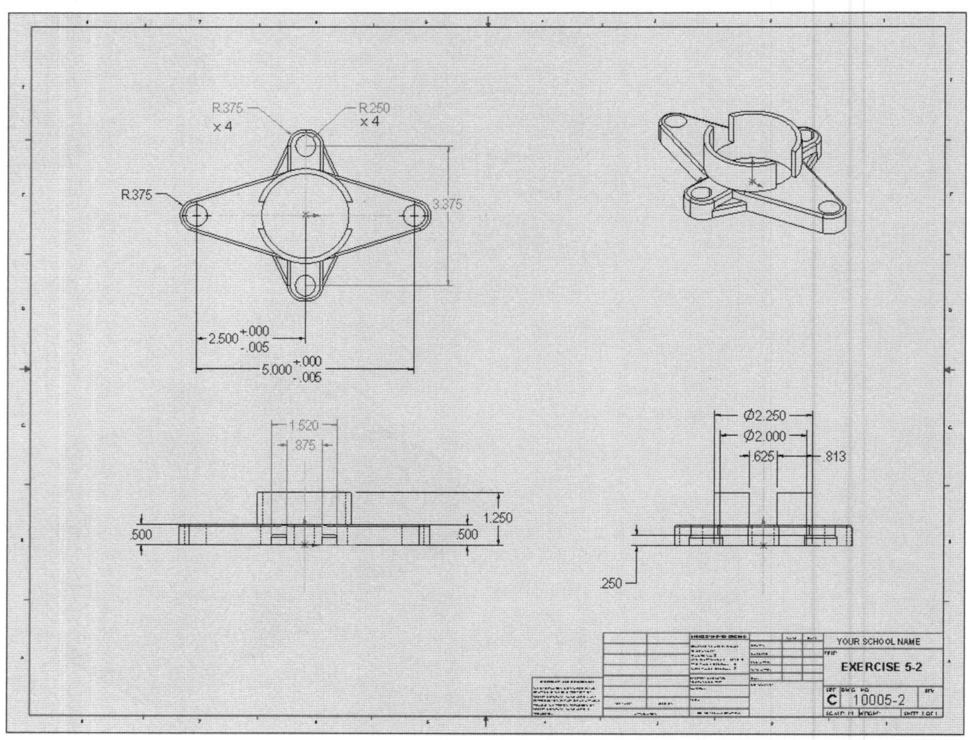

Figure 5-47
Exercise 5-2 Complete

EXERCISE 5-2 **Make a drawing from the model shown in Figure 5-46.**

Step 1. Use the steps described in Project 5-2 to make the drawing as shown in Figure 5-47.

Be sure to use the Drawing Number *10005-2* and choose a C-size drawing format for this drawing.

Step 2. Save your drawing with the file name **Exercise 5-2(your initials)** in at least two places.

Step 3. Print the drawing and turn it in.

EXERCISE 5-3 Make a model (Figure 5-48) from the drawing shown in Figure 5-49.

Step 1. Use the commands and settings from the Projects in this chapter to make the model shown in Figure 5-48. Use the dimensions shown in Figure 5-49 to make the model.

Step 2. Click **Isometric view** so that the model appears as shown in Figure 5-48.

Figure 5-48
Exercise 5-3 Complete

Step 3. Save your drawing (Figure 5-48) with the file name **Exercise 5-3(your initials)** in at least two places.

Step 4. Print the drawing and turn it in.

EXERCISE 5-4 Make a drawing from the model shown in Figure 5-48.

Step 1. Use the steps described in Project 5-2 to make the drawing as shown in Figure 5-49.

Be sure to use the Drawing Number *10005-4* and choose a C-size drawing format for this drawing.

Step 2. Save your drawing with the file name **Exercise 5-4(your initials)** in at least two places.

Step 3. Print the drawing and turn it in.

Figure 5-49
Exercise 5-4 Complete

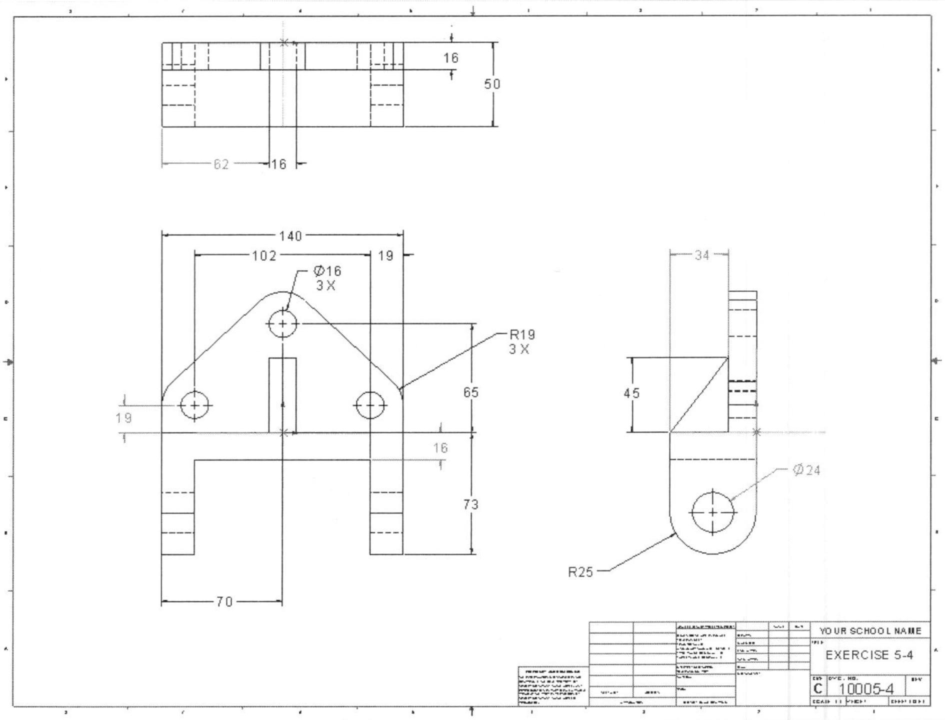

EXERCISE 5-5 **Make a model (Figure 5-50) from the drawing shown in Figure 5-51.**

Step 1. Use the commands and settings from the Projects in this chapter to make the model shown in Figure 5-50. Use the dimensions shown in Figure 5-51 to make the model.

Step 2. Click **Isometric view** so that the model appears as shown in Figure 5-50.

Figure 5-50
Exercise 5-5 Complete

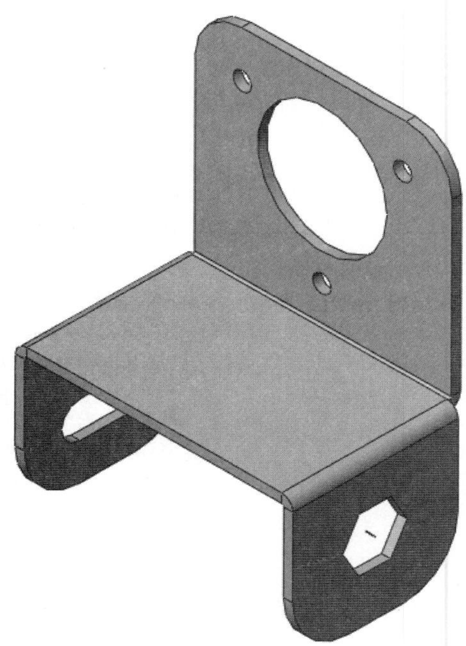

Figure 5-51
Exercise 5-6 Complete

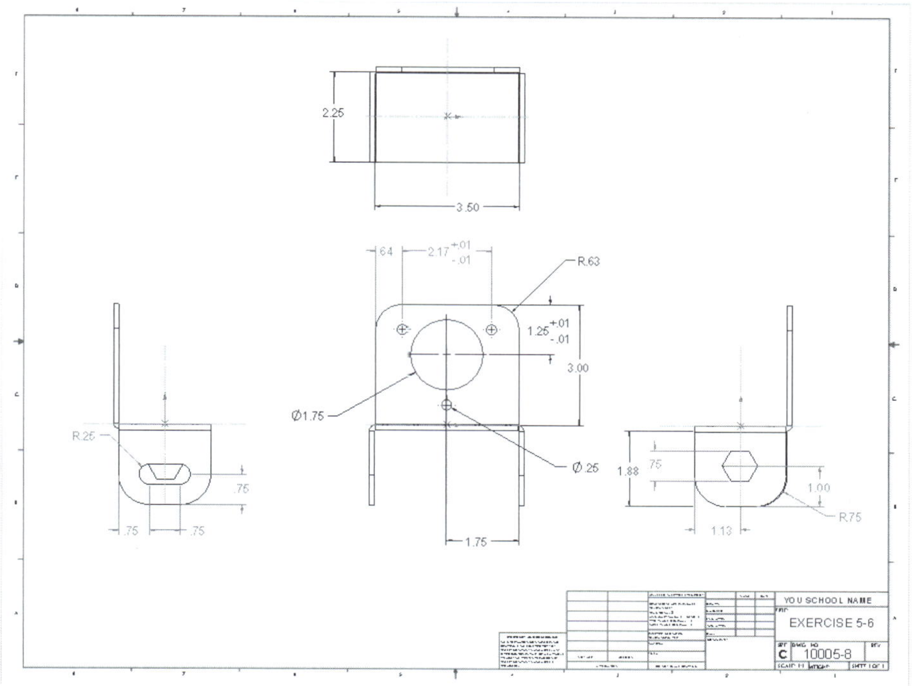

Step 3. Save your drawing (Figure 5-50) with the file name **Exercise 5-5(your initials)** in at least two places.

Step 4. Print the drawing and turn it in.

EXERCISE 5-6 Make a drawing from the model shown in Figure 5-50.

Step 1. Use the steps described in Project 5-2 to make the drawing as shown in Figure 5-51.

Be sure to use the Drawing Number *10005-8* and choose a C-size drawing format for this drawing.

Step 2. Save your drawing with the file name **Exercise 5-6(your initials)** in at least two places.

Step 3. Print the drawing and turn it in.

5 chapterfive

Chapter Summary

This chapter contains the information necessary to make complex models to specific decimal dimensions with tolerances. It also gives you the skills needed to make complex parts with simple and complex holes, to assign materials, and to make drawings from those models. Now you have the skills and information necessary to make complex solid models and drawings.

Review Questions

Circle the most correct answer.

1. Where is the **Hole Wizard** command located?
   a. **Shell** toolbar
   b. **Sketch** toolbar
   c. **Features** toolbar
   d. **Model** menu
2. Which tab on the **Hole Wizard** manager allows you to specify the type of holes?
   a. **Type**
   b. **Position**
   c. **Sketch**
   d. **Model**
3. Where is the **Simple Hole** command for solids located?
   a. **Fillet** Toolbar
   b. **Sketch** Toolbar
   c. **Features** menu
   d. **Model** menu
4. In which category is the **6061 Aluminum** material located?
   a. **Wood**
   b. **Plastic**
   c. **Aluminum**
   d. **Steel**
5. Which of the following is NOT a type of hole in the **Hole Wizard**?
   a. **Counterbore**
   b. **Threaded**
   c. **Countersink**
   d. **Simple Hole**
6. A sketch can be copied from one plane to another.
   a. True
   b. False
7. In a clearance fit, the cylinder is always smaller than the hole.
   a. True
   b. False
8. In an interference fit, the cylinder is always smaller than the hole.
   a. True
   b. False
9. The **Simple Hole** command can make a countersunk hole.
   a. True
   b. False
10. Tolerances cannot be added after a dimension has been added to the drawing.
   a. True
   b. False

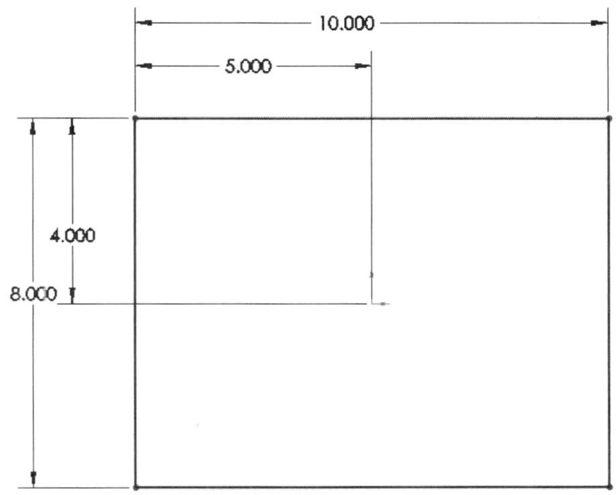

Figure 6-28
Draw a Sketch for an Extruded Boss/Base on the Top Plane and Extrude It 2.00 Inches

Step 6. Make a shell of the extruded model and remove the bottom. Make the walls of the shell 0.625 inch **(D1** under **Parameters)** as shown in Figure 6-29.

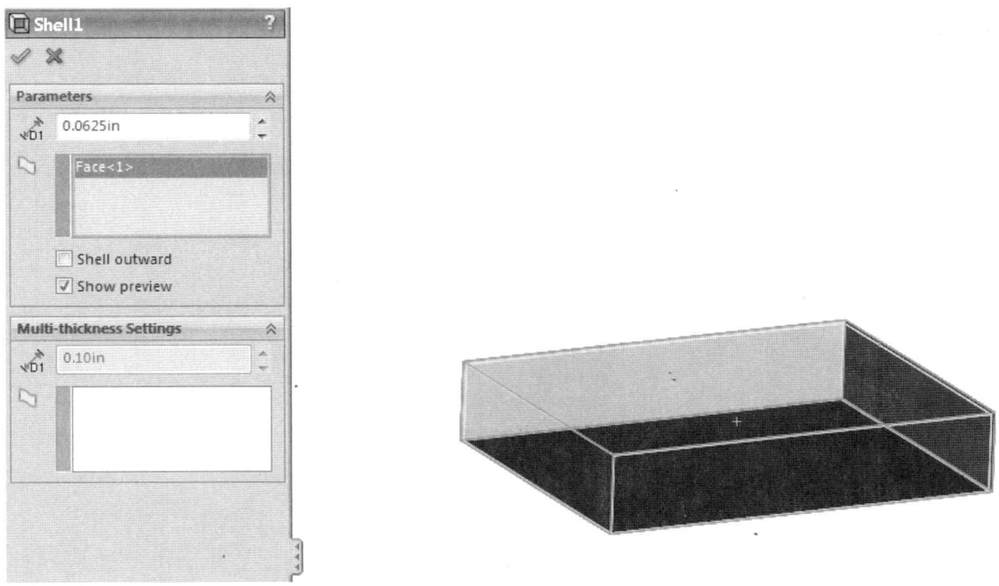

Figure 6-29
Make a Shell of the Extruded Model and Remove the Bottom. Make the Walls of the Shell 0.625 Inch (D1 Under Parameters)

Step 7. Change the **Display Style** to **wireframe** as shown in Figure 6-30.

Step 8. Use **Rip** and select all vertical edges. Make the .125 inch setting as shown in Figure 6-30.

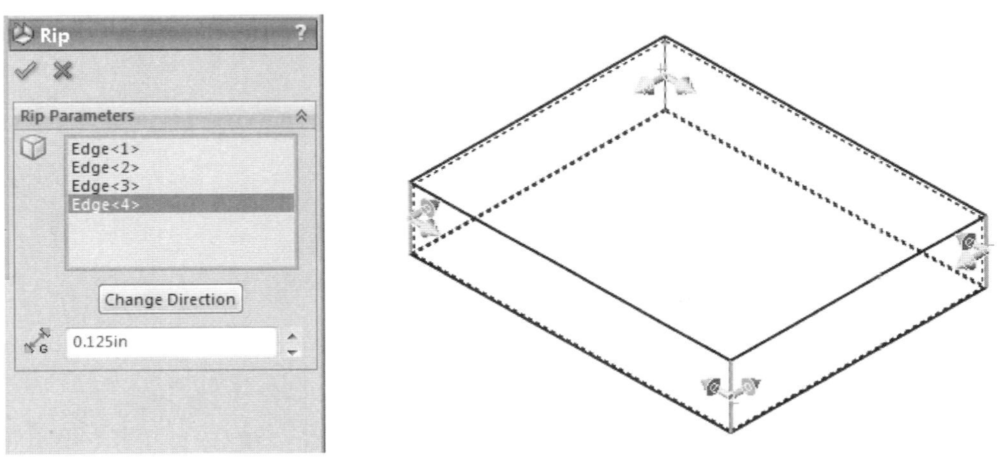

Figure 6-30
Use Rip by Selecting All Vertical Edges to Make a .125 Gap Between Vertical Edges. Make the .125 Inch Setting. (Change Display Style to Wireframe to Select Edges)

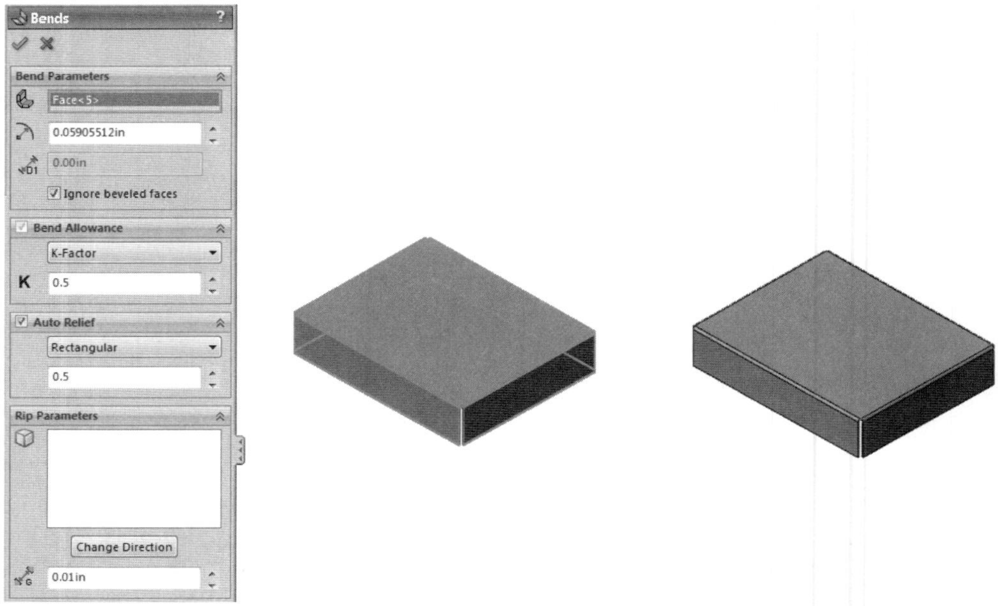

Figure 6-31
Use Insert Bends to Convert the Model to Sheet Metal by Clicking the Top Face and All Outside Vertical Faces

Step 9. Use **Insert Bends** to convert the model to sheet metal by clicking the top face and all outside vertical faces as shown in Figure 6-31.

Step 10. Use **Edge-Flange** to make four 1.00 inch flanges as shown in Figure 6-32.

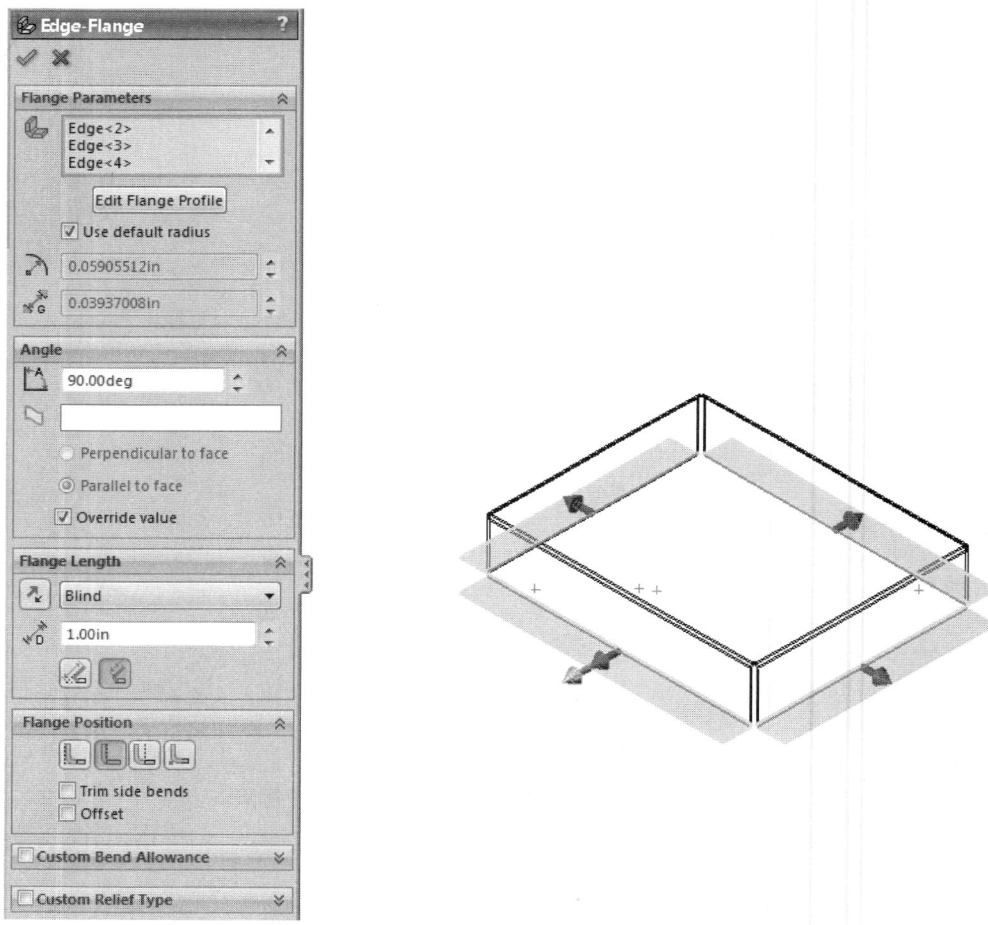

Figure 6-32
Use Edge-Flange to Make Four 1.00 Inch Flanges as Shown

Step 11. Use **Fillet** from the **Features** toolbar to make .500 inch fillets on eight edges as shown in Figure 6-33.

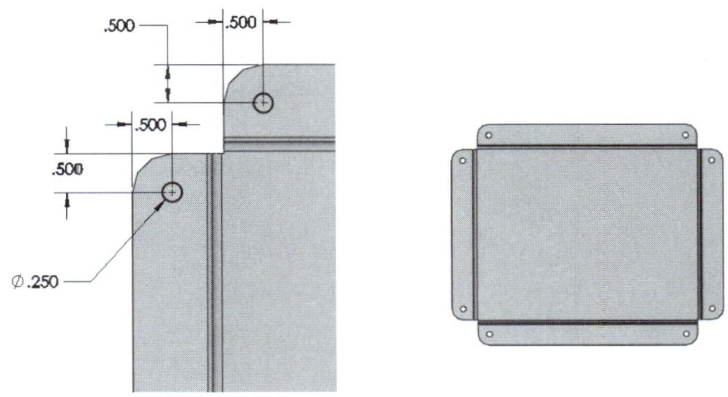

Figure 6-33
Make a .500 Inch Fillet on 8 Edges of the Edge Flanges. Draw a Sketch for an Extrude Cut on All 8 Corners of Edge Flanges, Dimension It and Make the Extruded Cut

Step 12. Draw a sketch for an **Extruded Cut** on all eight corners of the edge flanges and make the **Extruded Cut** as shown in Figure 6-33.

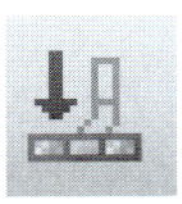

Step 13. Use **Unfold** to flatten the sheet metal part, click the top face as the **Fixed** face, and then click **Collect All Bends** as shown in Figure 6-34.

Figure 6-34
Unfold the Sheet Metal Part, Click: the Top Face as the Fixed Face, and Then Click: Collect All Bends

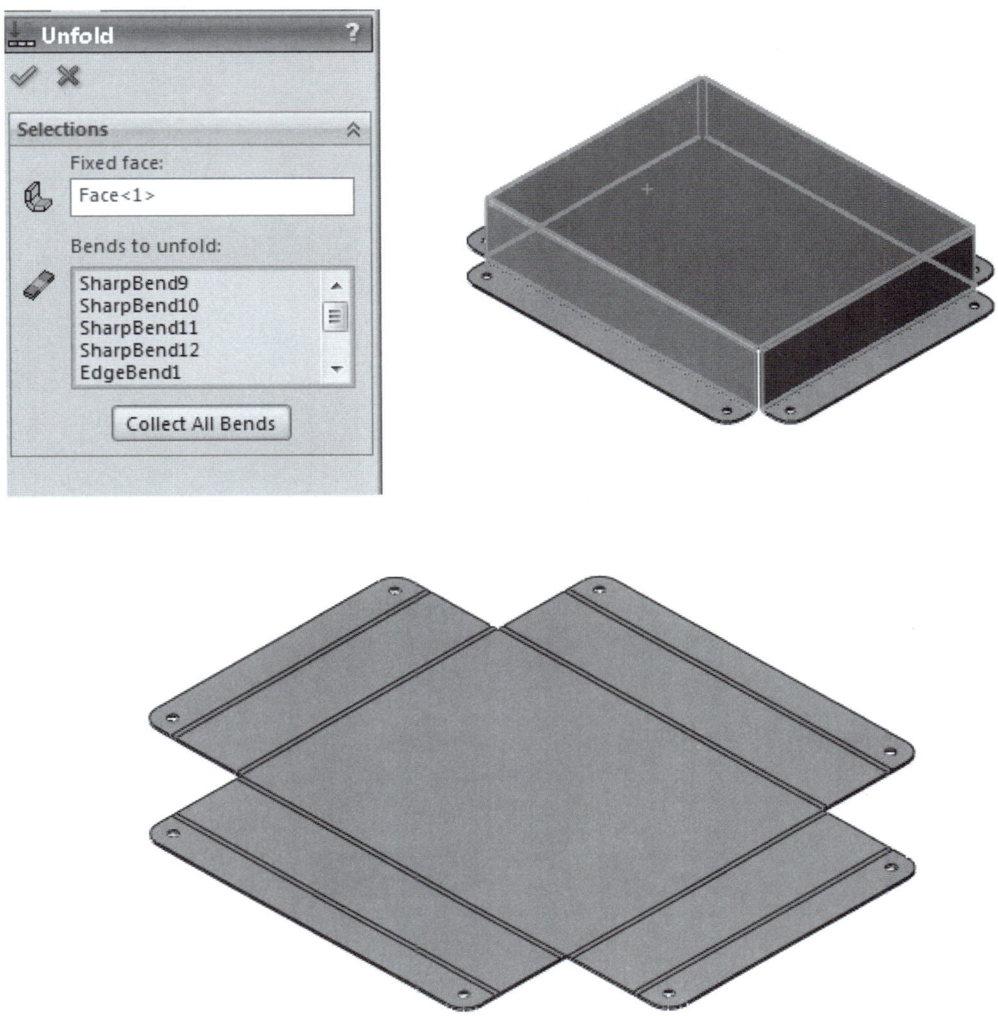

Step 14. Draw a sketch for an **Extruded Cut** using **Straight Slot** and make the cut as shown in Figure 6-35.

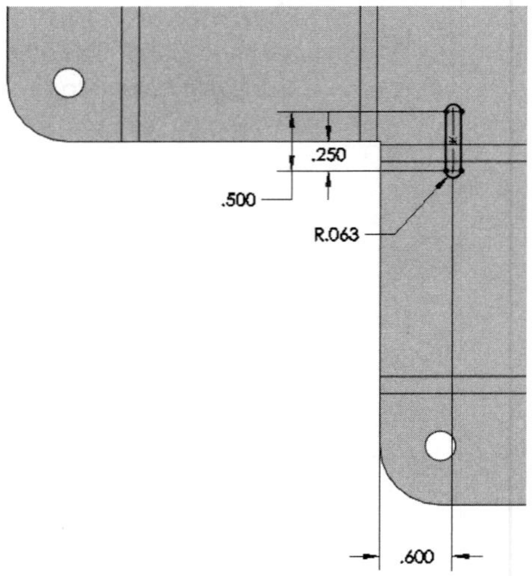

Figure 6-35
Make a Sketch for an Extruded Cut Using Straight Slot, Dimension It, and Make the Cut

Step 15. Use **Linear Pattern** to pattern the **Extruded Cut** 18 instances in **Direction1** and 2 instances in **Direction2** as shown in Figure 6-36.

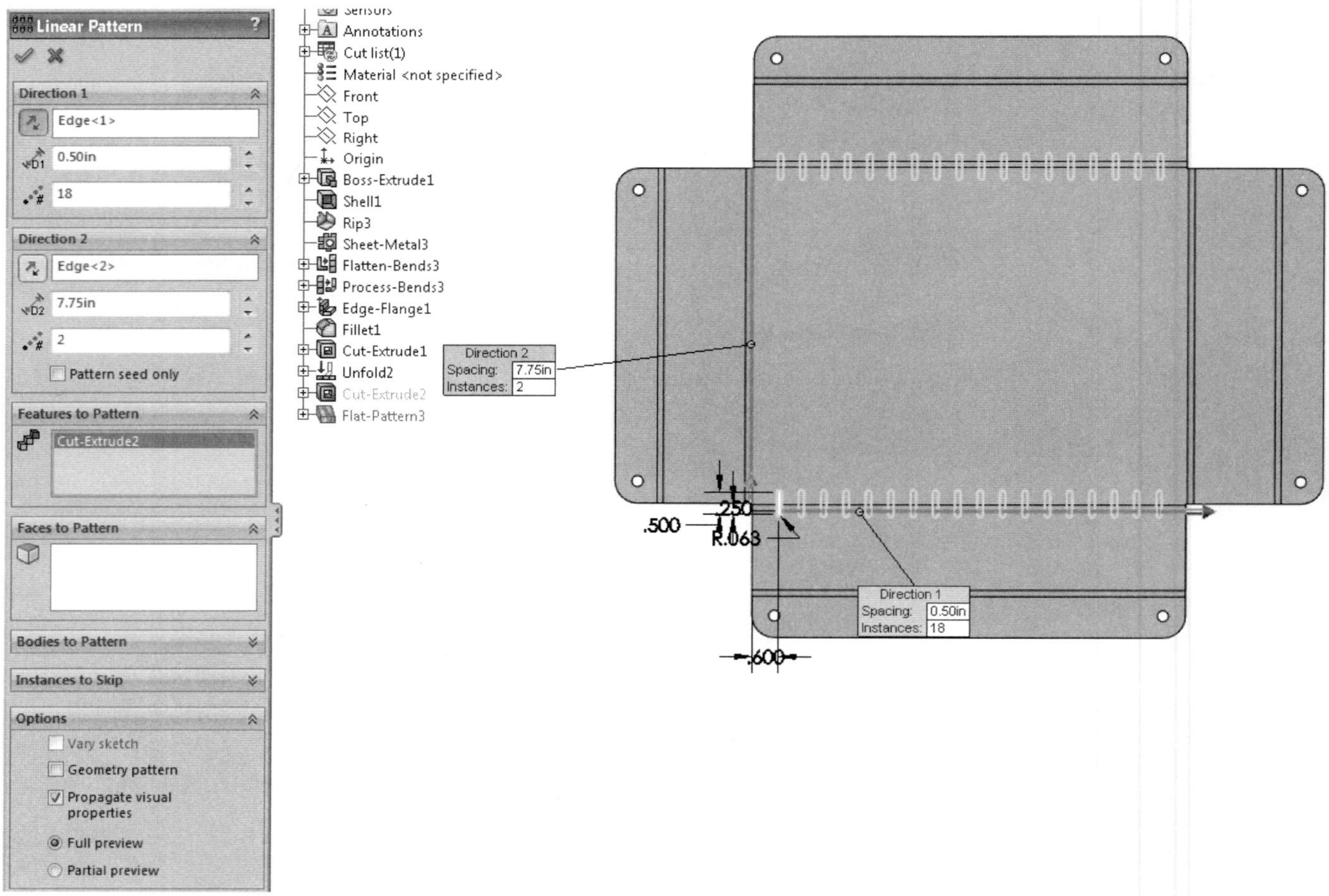

Figure 6-36
Use Linear Pattern to Pattern the Extruded Cut 18 Instances in Direction1 and 2 Instances in Direction2

Step 16. Use **Fold** to fold the sheet metal part, click the top face **(Face<1>)** as the **Fixed face,** and then click **Collect All Bends** as shown in Figure 6-37.

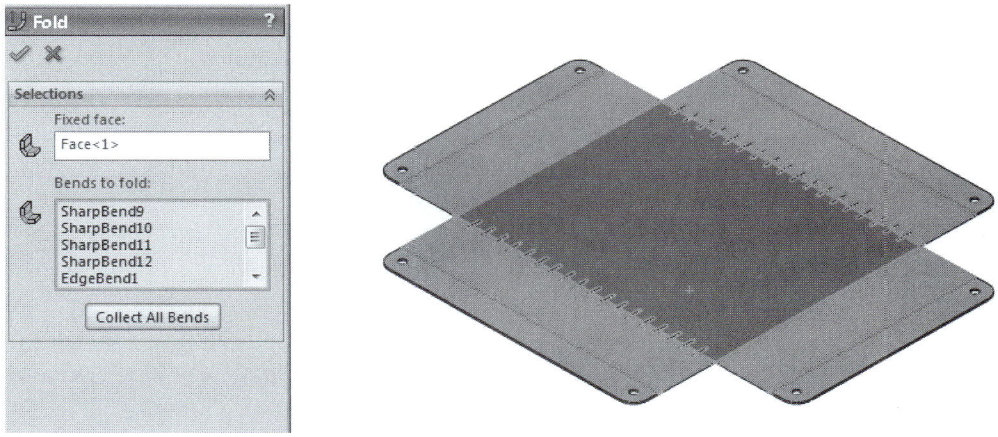

Figure 6-37
Use Fold to Fold the Part with Face 1 (the Top Face) as the Fixed Face and Click: Collect All Bends

Step 17. Click **Isometric view** so the model appears as shown in Figure 6-38.

Step 18. Save your drawing (Figure 6-38) with the file name **Project 6-3(your initials)** in at least two places.

Step 19. Print the drawing and turn it in.

Figure 6-38
Project 6-3 Complete

Project 6-4: Making a Sheet Metal Part Using the Base-Flange, Miter-Flange, Jog, and Hem Commands (Figure 6-39)

Step 1. Click **New....**

Step 2. Click **Part** and click **OK.**

Step 3. Click **Options** on the menu bar, select the **Document Properties** tab, and click **Units** in the list to the left.

Step 4. Make sure the **Units** are correctly set [notice that **IPS** (inch, pound, second) is checked], then click **OK.**

Step 5. Draw a sketch for a **Base Flange** on the top plane and make the base flange with 18 gauge aluminum as shown in Figure 6-40.

Figure 6-39
Project 6-4 Complete

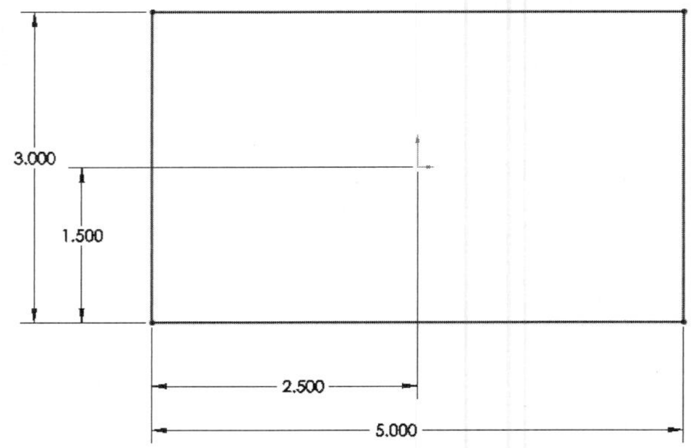

Figure 6-40
Draw a Sketch for a Base Flange and Make the Base Flange with 18 Gauge Steel

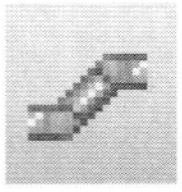

Step 6. Select the top of the base flange as a sketch plane and draw a line and dimension it from the front edge. Click **Jog** and make the settings as shown in Figure 6-41.

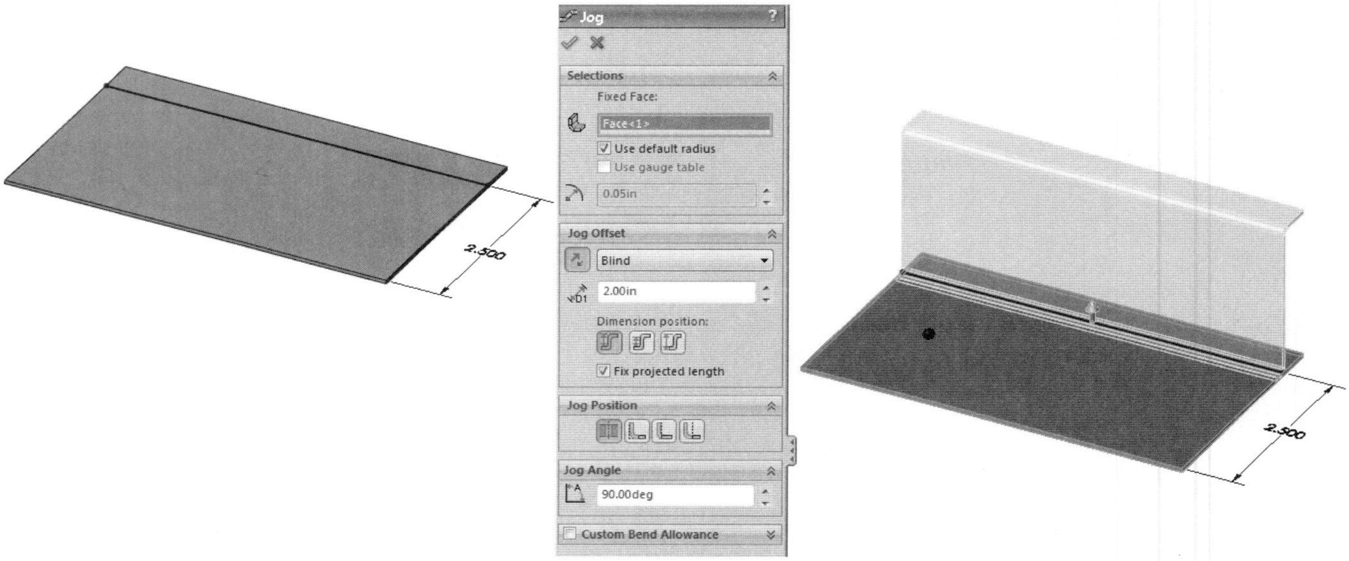

Figure 6-41
Click the Top of the Base Flange as a Sketch Plane. Draw a Line and Dimension It from the Front Edge. Click: Jog. Make the Settings and Click: OK

Step 7. Select the front face of the base flange and draw a .250 inch line. Click **Miter-Flange,** select the additional edges, and make the setting of **.031** in the **Gap distance** box as shown in Figure 6-42.

Step 8. Use the **Hem** command to add hems to four edges as shown in Figure 6-43.

Step 9. Click **Isometric view** so the model appears as shown in Figure 6-44.

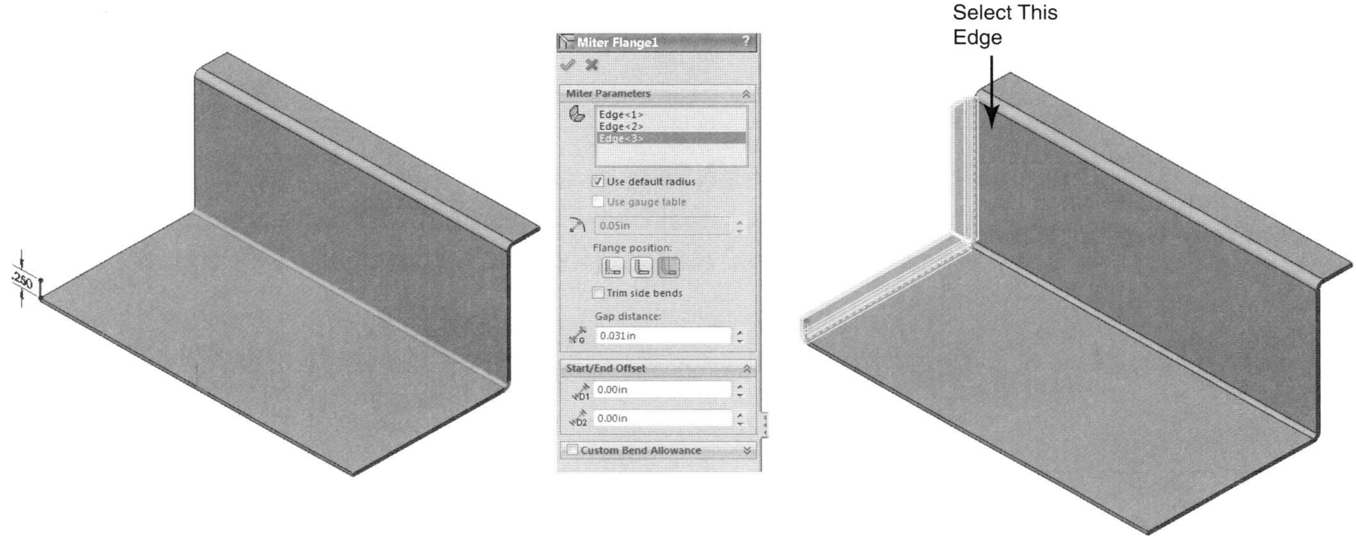

Figure 6-42
Select the Front Face of the Base Flange and Draw a .250 Inch Line. Click: Miter-Flange, and Select the Additional Edge as Shown and Make the Setting of .031 Inch in the Gap Distance Box, Then Click: OK

Figure 6-43
Add Hems to 4 Edges as Shown

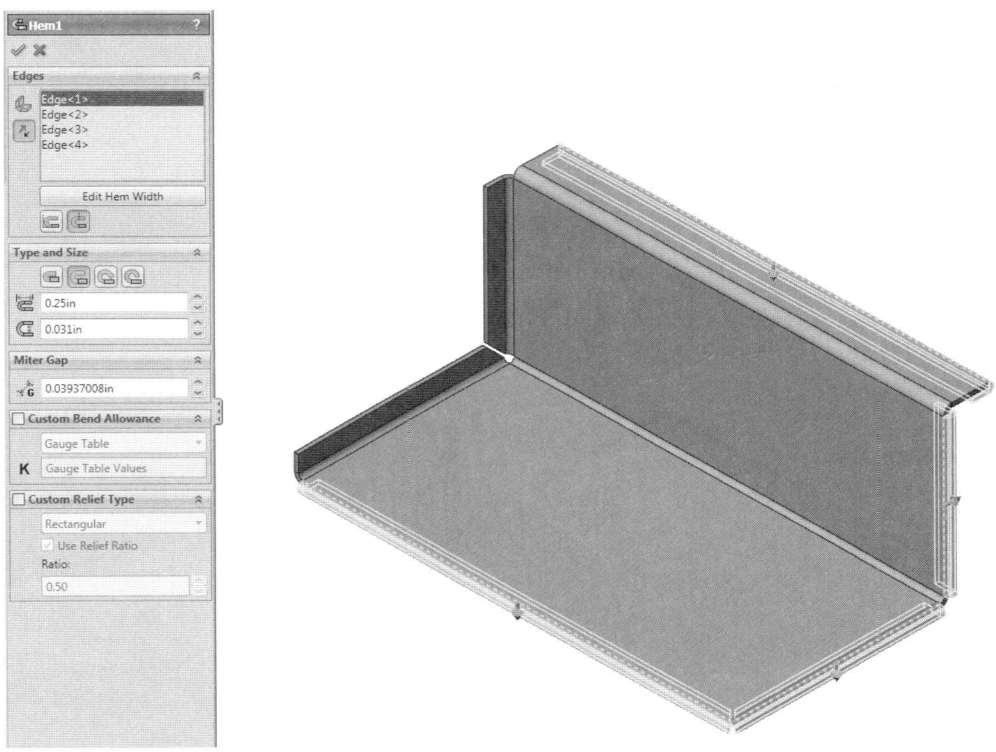

Figure 6-44
Project 6-4 Complete

Step 10. Save your drawing (Figure 6-44) with the file name **Project 6-4 (your initials)** in at least two places.

Step 11. Print the drawing and turn it in.

Project 6-5: **Making a Sheet Metal Part Using the Base-Flange, Miter-Flange, Vent, Extrude Cut, and Revolve Boss/Base Commands (Figure 6-45)**

Step 1. Click **New....**

Step 2. Click **Part** and click **OK.**

Step 3. Click **Options** on the menu bar, select the **Document Properties** tab, and click **Units** in the list to the left.

Step 4. Make sure the **Units** are correctly set [notice that **IPS** (inch, pound, second) is checked], then click **OK.**

Step 5. Draw a sketch for a **Base Flange** on the top plane and make the base flange with 18 gauge aluminum as shown in Figure 6-46.

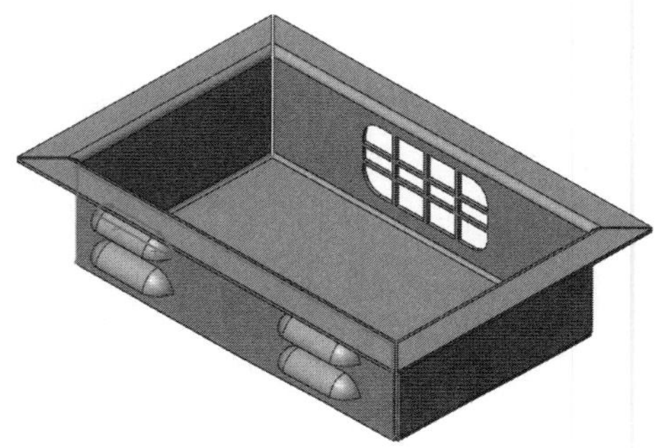

Figure 6-45
Project 6-5 Complete

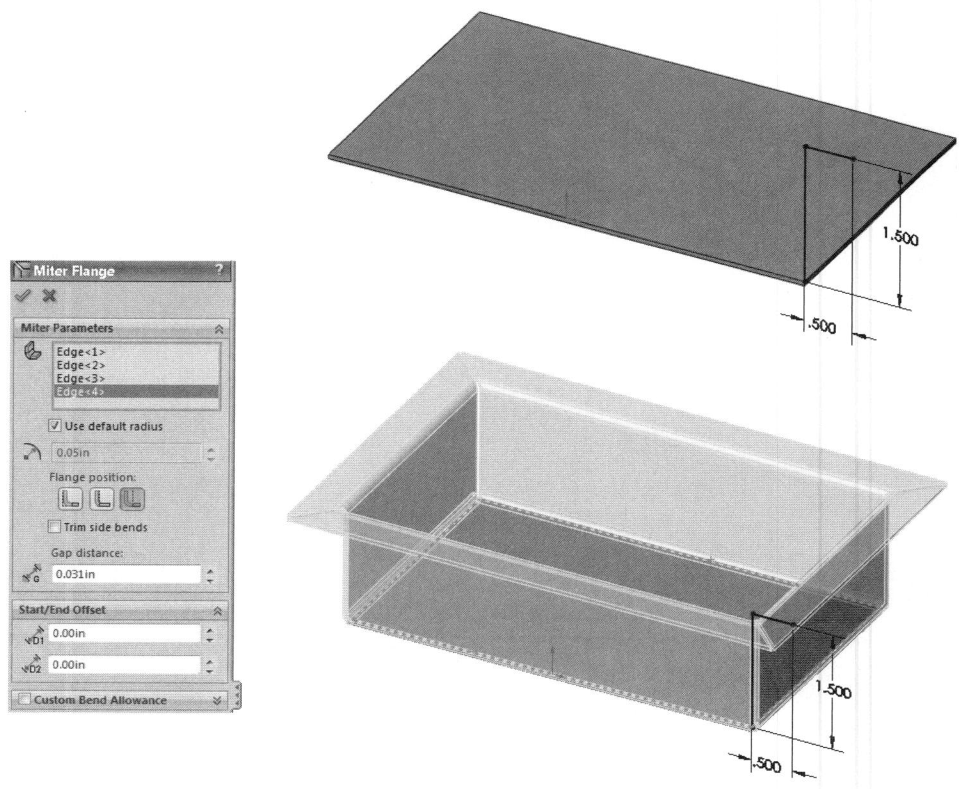

Figure 6-46
Draw a 3.000 Inches × 5.000 Inches Rectangle for a Base Flange on the Top Plane. Make the Base Flange. Click the Front Face of the Base Flange. Make the Sketch for Miter Flanges. Click: Miter-Flange and Click: Three Other Edges. Click: OK

Step 6. Click the front face of the base flange and make the sketch for miter flanges as shown in Figure 6-46.

Step 7. Click **Miter-Flange** and click three other edges as shown in Figure 6-46.

Step 8. Draw and dimension a sketch for a vent on the front plane as shown in Figure 6-47.

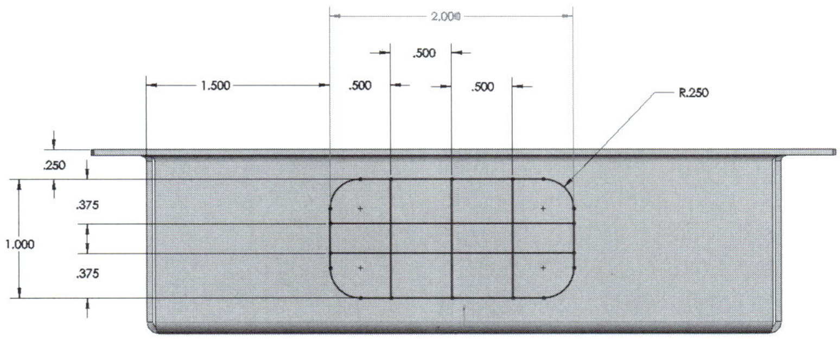

Figure 6-47
Draw and Dimension a Sketch for a Vent on the Front Plane

Step 9. Click **Vent** and select the sketch for the vent boundary, select the inside lines of the sketch for the ribs, then select the back face as the face to place the vent as shown in Figure 6-48.

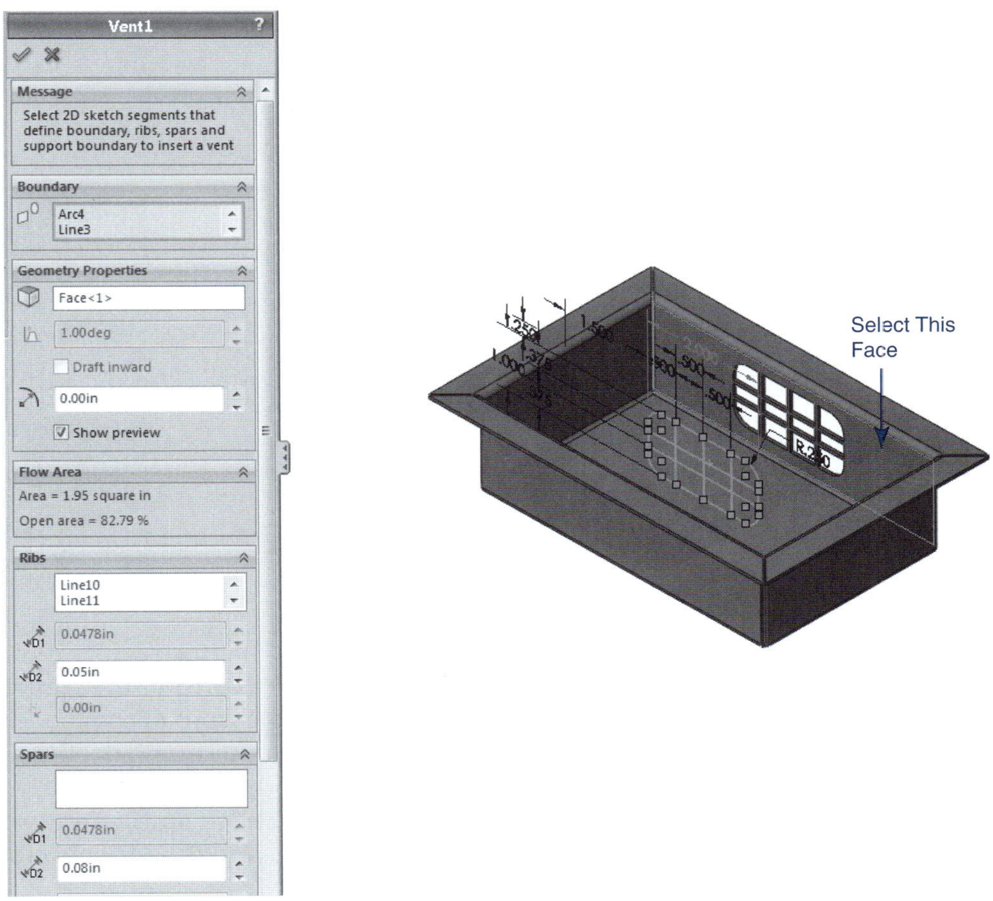

Figure 6-48
Click: Vent. Click the Sketch for the Vent Boundary. Click the Inside Lines of the Sketch for the Ribs. Click the Back Face as the Face to Place the Vent

Step 10. Make a sketch for an **Extruded Cut** and make a **Linear Pattern** of 2 instances in **Direction D1** with 3.00 inches spacing and 2 instances in **Direction D2** with .50 inch spacing and make the cut **Up to next** as shown in Figure 6-49.

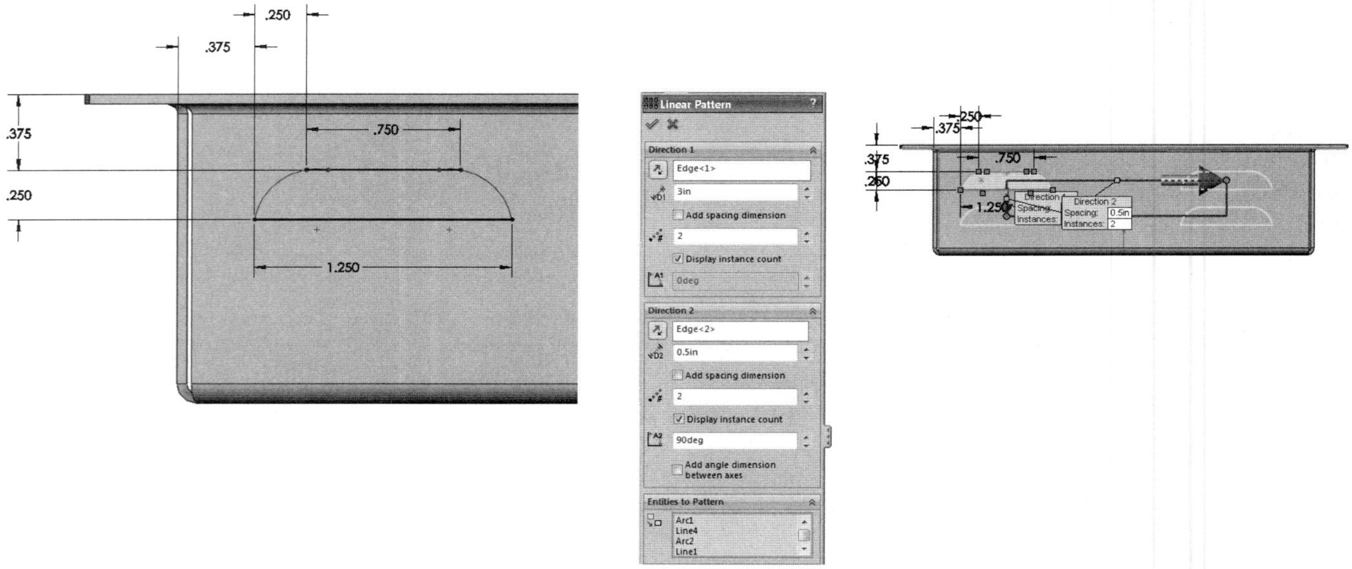

Figure 6-49
Make a Sketch for an Extruded Cut and Make a Linear Pattern of 2 Instances in Direction D1 with 3.00 Inches Spacing and 2 Instances in Direction D2 with .50 Inch Spacing and Make the Cut

Step 11. Make a sketch for a **Revolve Boss/Base** on the front face of the model and revolve it 90 degrees as shown in Figure 6-50.

Figure 6-50
Make a Sketch for a Revolve Boss/Base on the Front Face of the Model and Revolve It 90°

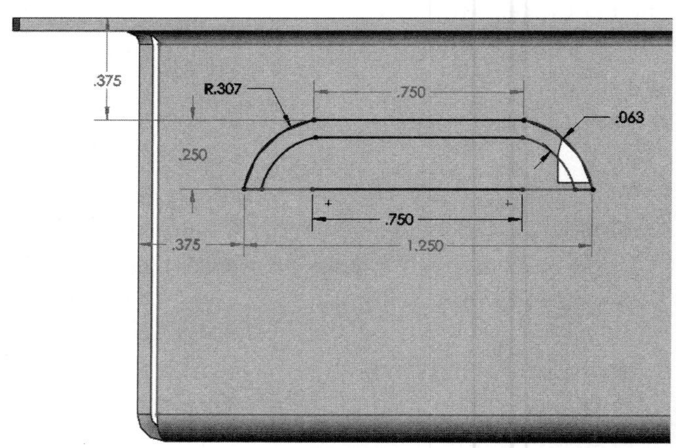

Step 12. Make a **Linear Pattern** of the **Revolved Boss/Base** with a horizontal edge on the front of the model as **Direction1, 3.00 inches** as **D1,** and **2** as the number of instances as shown in Figure 6-51. Make another **Linear Pattern** of the first linear pattern with the vertical edge on the front of the model as **Direction1, .50 inch** as **D1,** and **2** as the number of instances as shown in Figure 6-52.

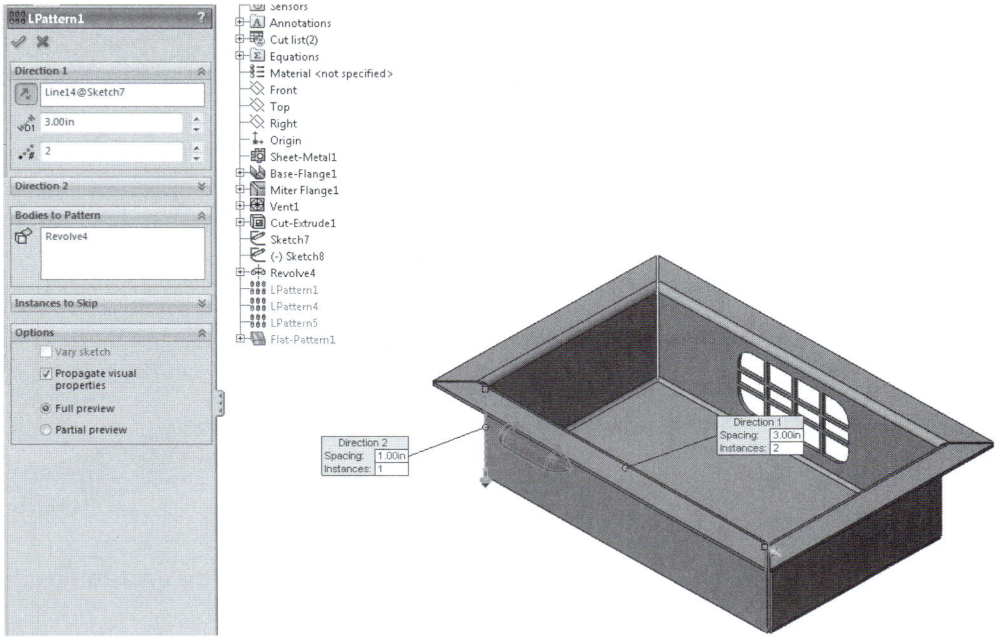

Figure 6-51
Make a Linear Pattern of the Revolved Boss/Base with a Horizontal Edge on the Front of the Model as Direction 1, and 3.00 Inches as D1 and 2 as the Number of Instances and Another Linear Pattern of 2 Instances in the Direction of the Vertical Edge with .50 Inch Spacing

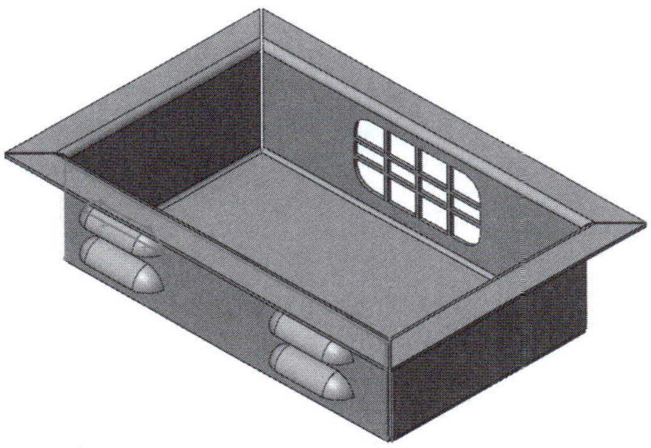

Figure 6-52
Project 6-5 Complete

Step 13. Click **Isometric view** so the model appears as shown in Figure 6-52.

Step 14. Save your drawing (Figure 6-52) with the file name **Project 6-5(your initials)** in at least two places.

Step 15. Print the drawing and turn it in.

Exercises

EXERCISE 6-1 Make a utility tool.

Step 1. Make a sketch for a **Base Flange** on the top plane and extrude it .125 inch as shown in Figure 6-53.

Figure 6-53
Make a Sketch for a Base Flange on the Top Plane and Extrude It .125 Inch

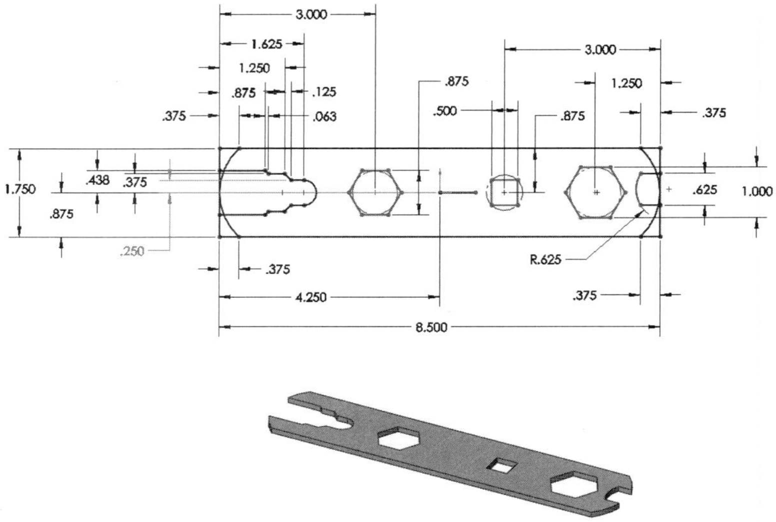

Step 2. Make a sketch for a **Jog** and make the **Jog** as shown in Figure 6-54.

Step 3. Save the drawing (Figure 6-55) with the file name **Exercise 6-1(your initials)** in at least two places.

Step 4. Print the drawing and turn it in.

Figure 6-54
Make a Sketch for a Jog and Make the Jog

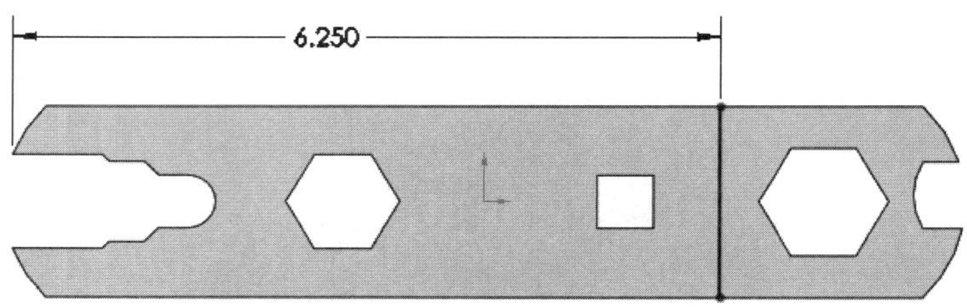

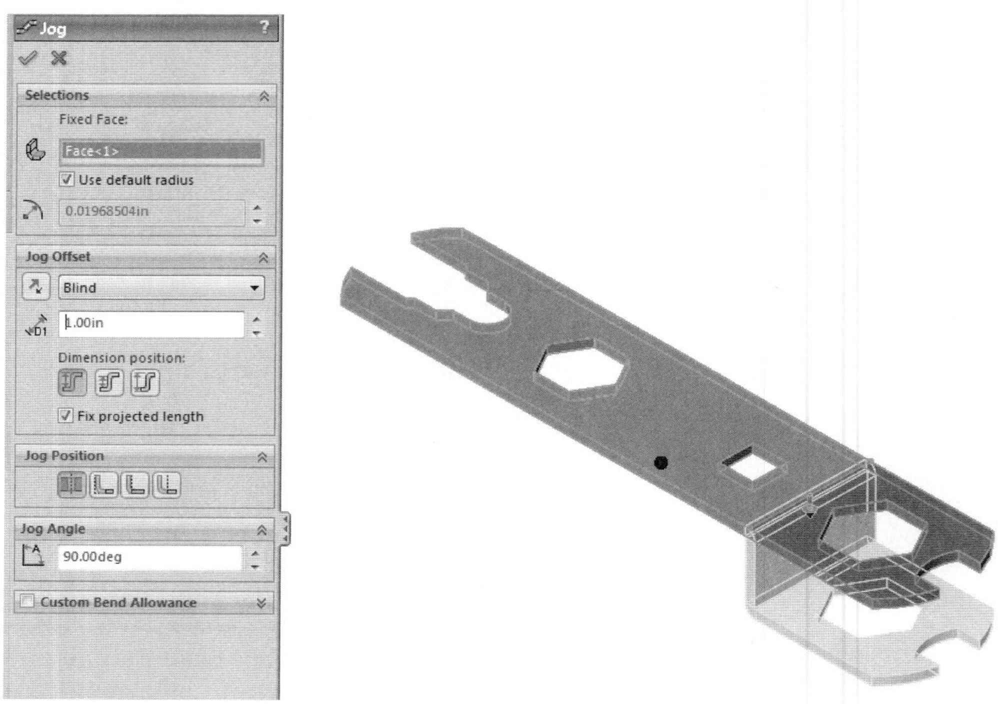

Figure 6-55
Exercise 6-1 Complete

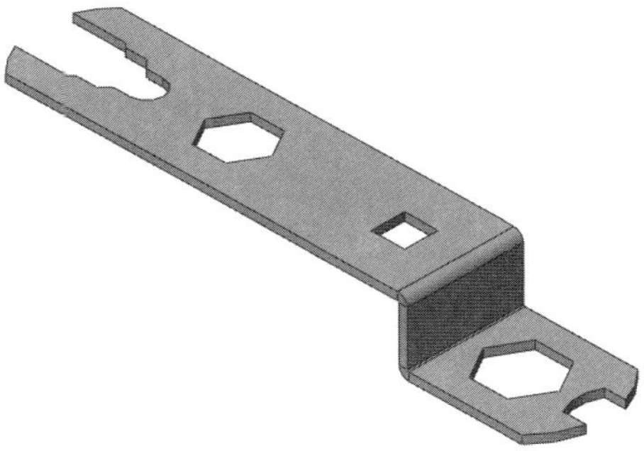

EXERCISE 6-2 **Make five parts for a vent that will be assembled in Chapter 7.**

Exercise 6-2a:

Step 1. Draw a sketch for a Base Flange on the top plane and extrude it 4.00 inches as shown in Figure 6-56.

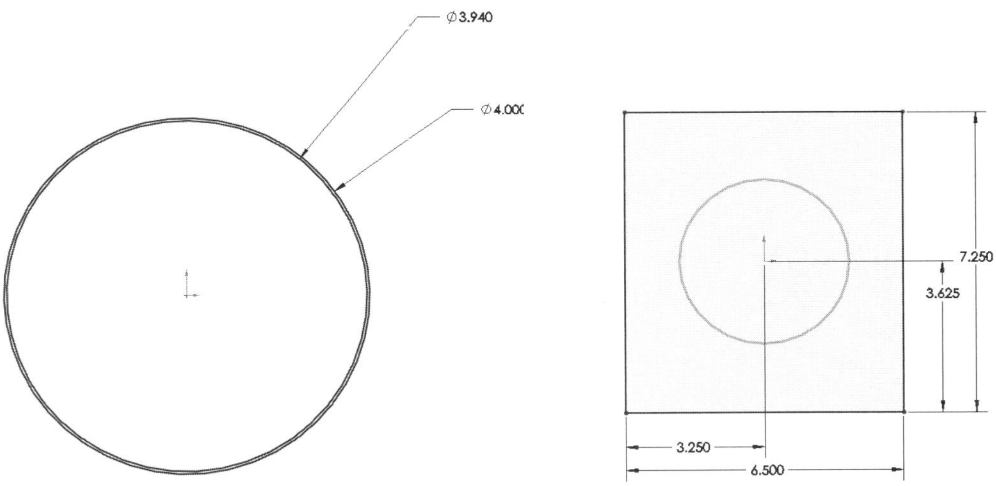

Figure 6-56
Draw a Circle for a Base Flange and Extrude It 4.00 Inches. Draw a 6.500 × 7.250 Rectangle for an Extruded Boss/Base and Extrude It .048 Inch. Draw a 4.00 Inches circle for an Extruded Cut and Make the Cut

Step 2. Draw a 6.500 × 7.250 rectangle for an Extruded Boss/Base on the top plane and extrude it .048 inch as shown in Figure 6-56.

Step 3. Draw a 4.00 inches circle for an **Extruded Cut** as shown in Figure 6-56 and make the cut.

Step 4. Draw a sketch for a vent using Circle, Line, Linear Pattern, and Trim and use that sketch for the vent as shown in Figure 6-57.

Step 5. Save the drawing (Figure 6-57) with the file name **Exercise 6-2a(your initials)** in at least two places.

Step 6. Print the drawing and turn it in.

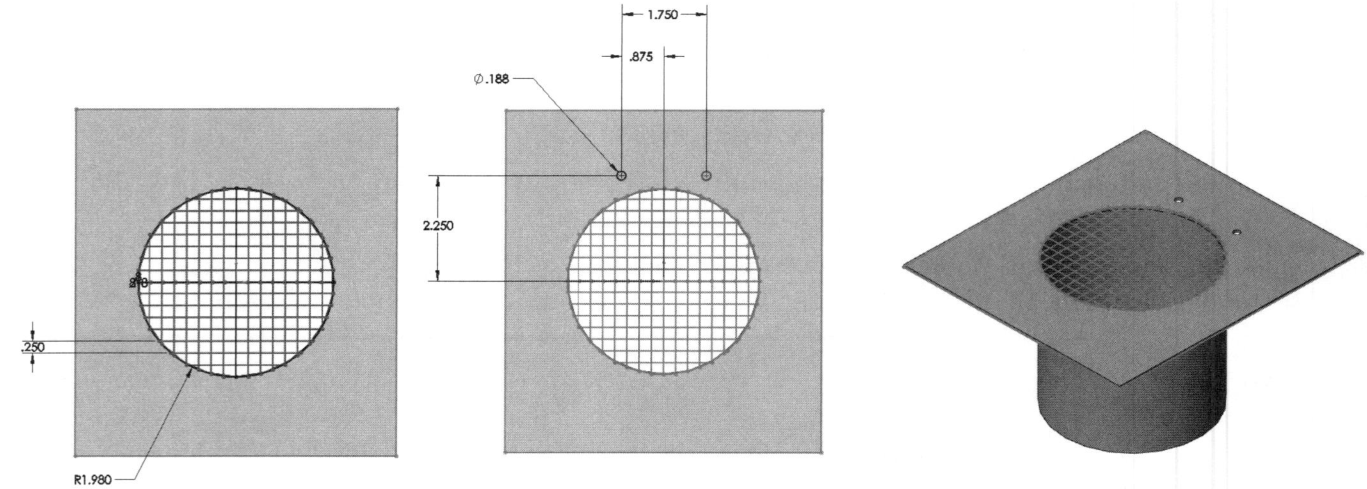

Figure 6-57
Draw a Sketch for a Vent and Make the Vent. Draw a Sketch for an Extruded Cut and Make the Cut. Exercise 6-2a Complete.

Exercise 6-2b:

Step 1. Draw a sketch for a **Base Flange** on the Front Plane as shown in Figure 6-58 and extrude it 7.00 inches with gauge 26 aluminum as shown in Figure 6-59.

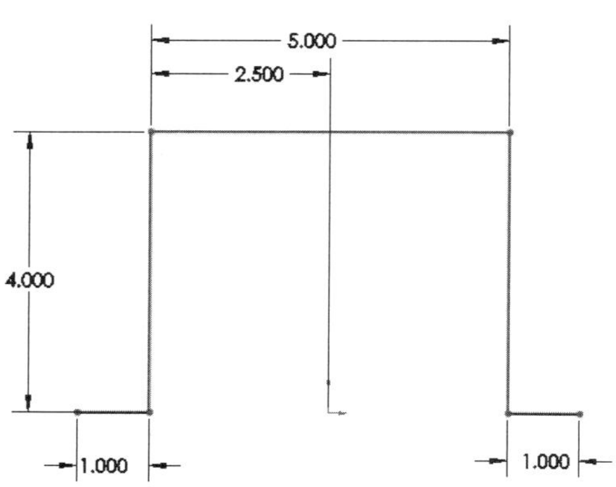

Figure 6-58
Make a Sketch for a Base Flange

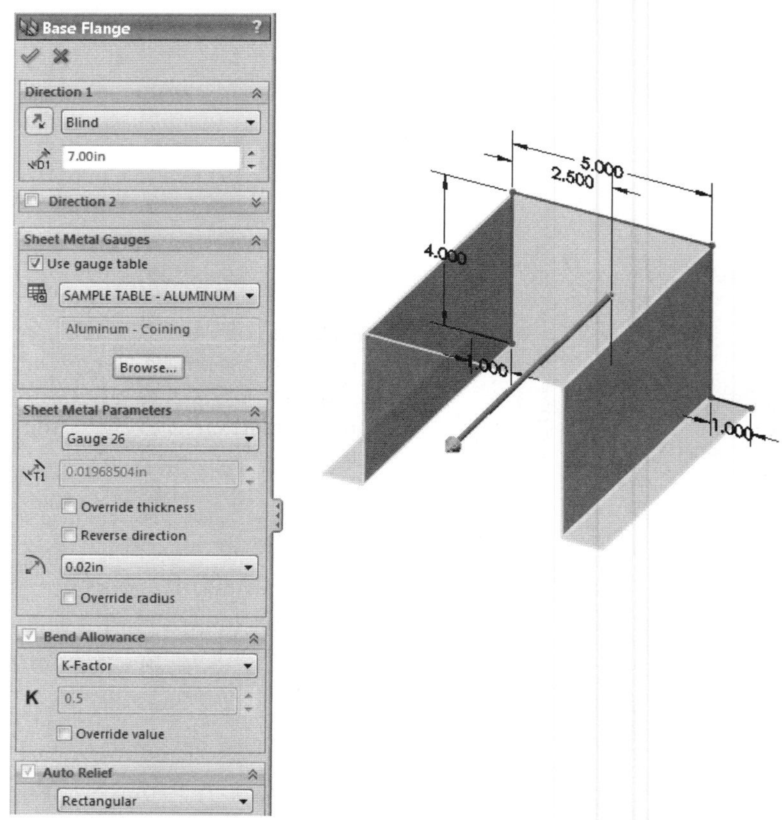

Figure 6-59
Extrude the Base Flange 7.00 Inches with Gauge 26 Aluminum

Step 2. Draw a sketch for an **Extruded Cut** and make the cut as shown in Figure 6-60.

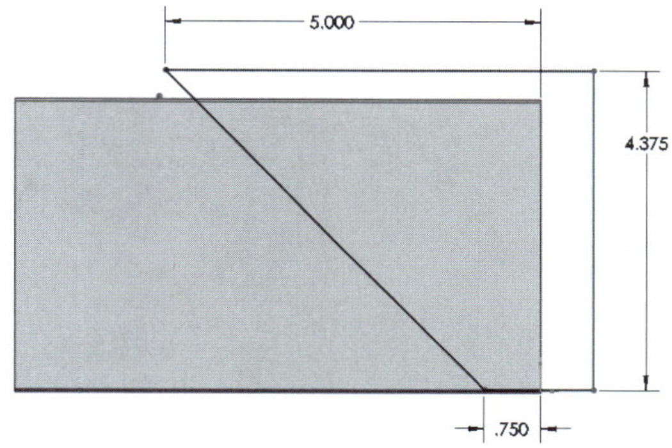

Figure 6-60
Draw a Sketch for an Extruded Cut and Make the Cut

Step 3. Draw a sketch for an **Extruded Cut** and make the cut as shown in Figure 6-61.

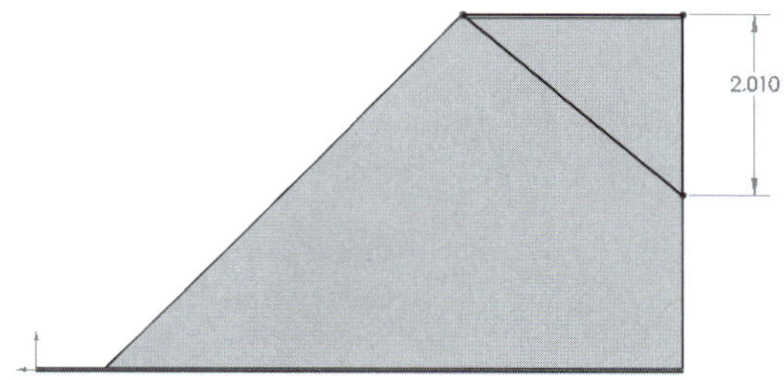

Figure 6-61
Draw a Sketch for an Extruded Cut and Make the Cut

Step 4. Use **Edge-Flange** to close the opening as shown in Figure 6-62.

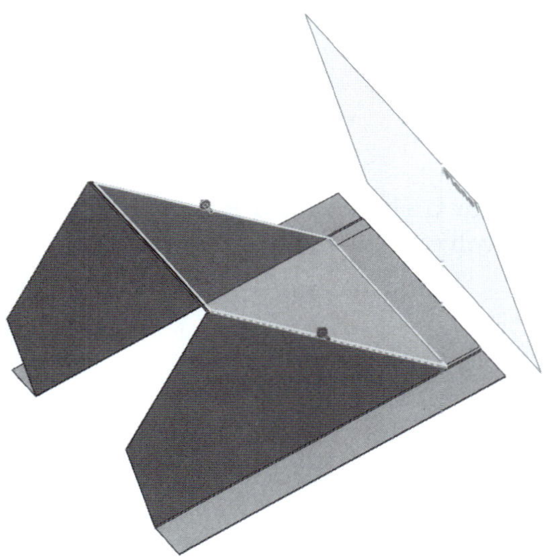

Figure 6-62
Use Edge-Flange to Close the Openirg

Step 5. Use **Hem** to make a .250 inch hem on two edges as shown in Figure 6-63.

Step 6. Save the drawing (Figure 6-64) with the file name **Exercise 6-2b(your initials)** in at least two places.

Step 7. Print the drawing and turn it in.

Figure 6-63
Use Hem to Make a .250 Inch Hem on Two Edges

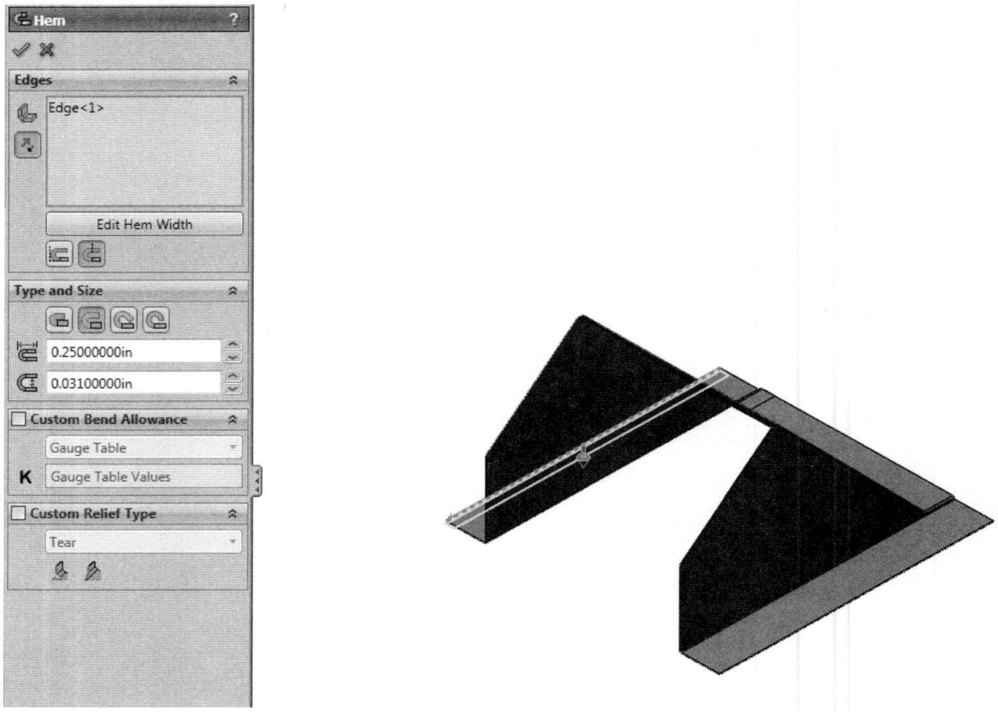

Figure 6-64
Exercise 6-2b Complete

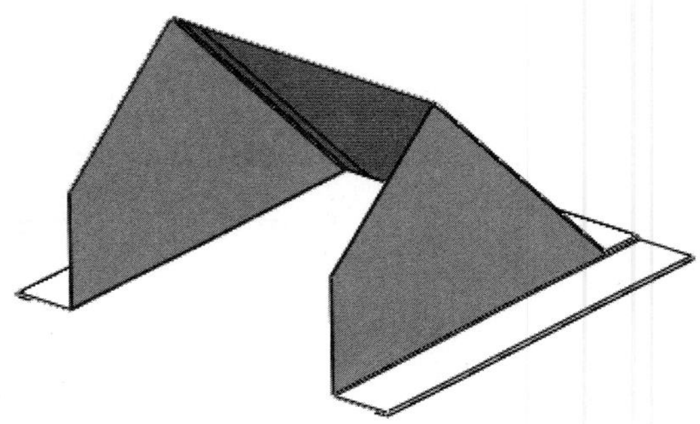

Exercise 6-2c:

Step 1. Draw a sketch for a **Revolved Boss/Base** on the front plane and revolve it as shown in Figure 6-65.

Figure 6-65
Draw a Sketch for a Revolved Boss/Base on the Front Plane and Revolve It

Step 2. Draw a sketch for an **Extruded Boss/Base** and extrude it .048 inch as shown in Figure 6-66.

Step 3. Draw a sketch for an **Extruded Cut** as shown in Figure 6-67 and make the cut.

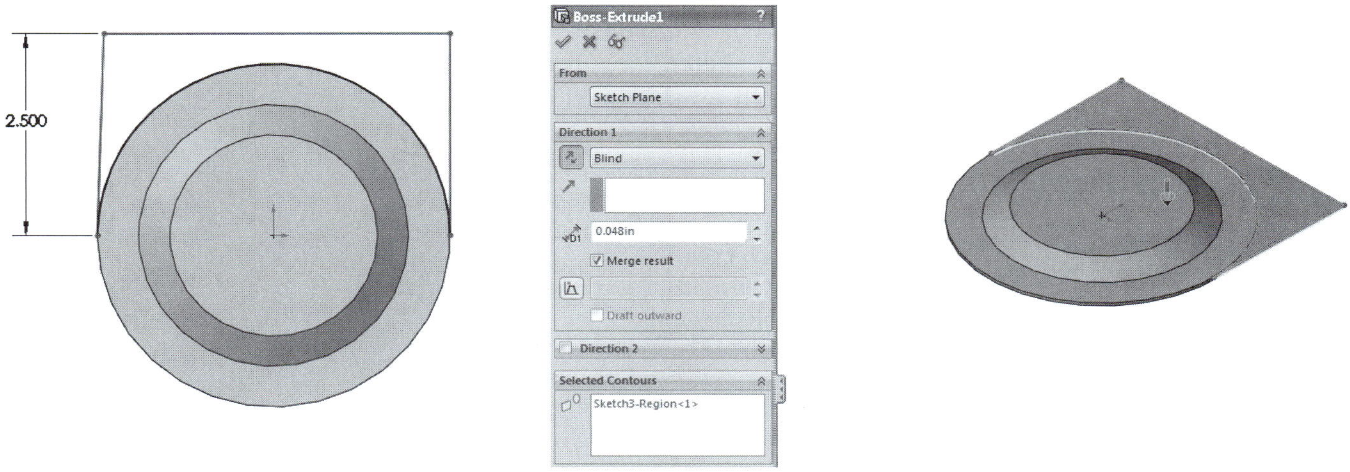

Figure 6-66
Draw a Sketch for an Extruded Boss/Base and Extrude It .048 Inch

Figure 6-67
Draw a Sketch for an Extruded Cut and Make the Cut.
Exercise 6-2c Complete

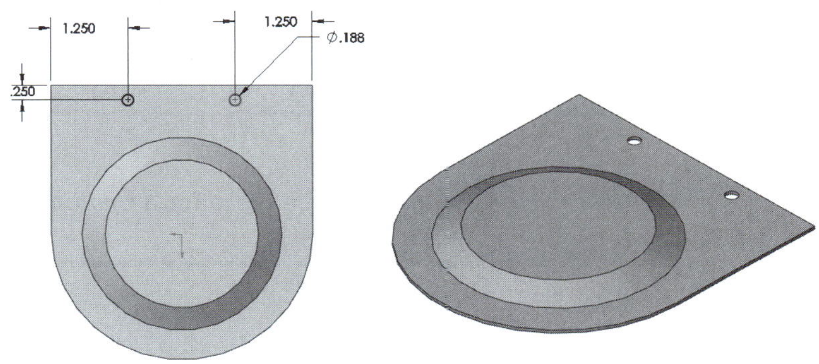

Step 4. Save the drawing (Figure 6-67) with the file name **Exercise 6-2c(your initials)** in at least two places.

Step 5. Print the drawing and turn it in.

Exercise 6-2d:

Step 1. Draw a sketch for a **Revolved Boss/Base** on the front plane and revolve it as shown in Figure 6-68.

Figure 6-68
Draw a Sketch for a Revolved Boss/Base on the Front Plane and Revolve It
Exercise 6-2d Complete

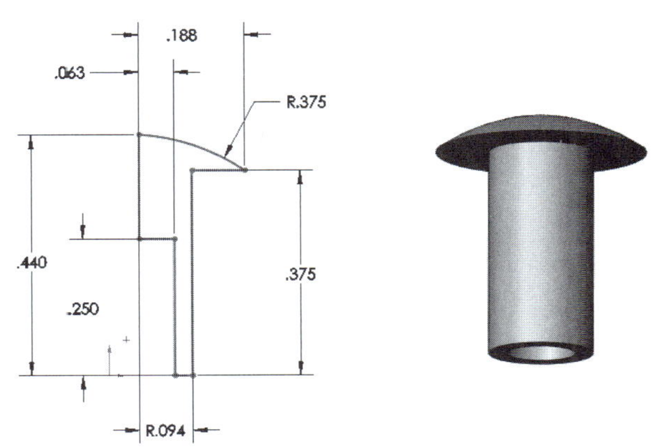

Step 2. Save the drawing (Figure 6-68) with the file name **Exercise 6-2d(your initials)** in at least two places.

Step 3. Print the drawing and turn it in.

Exercise 6-2e:

Step 1. Draw a sketch for a **Revolved Boss/Base** on the front plane and revolve it as shown in Figure 6-69.

Figure 6-69
Draw a Sketch for a Revolved Boss/Base on the Front Plane and Revolve It. Exercise 6-2e Complete

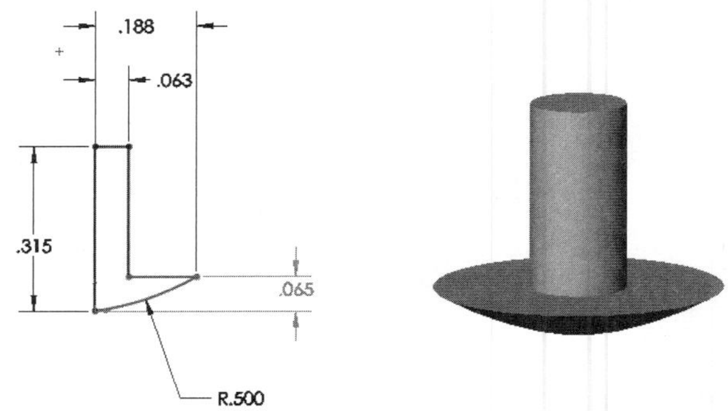

Step 2. Save the drawing (Figure 6-69) with the file name **Exercise 6-2e(your initials)** in at least two places.

Step 3. Print the drawing and turn it in.

EXERCISE 6-3 Make three parts for a mailbox that will be assembled in Chapter 7.

Exercise 6-3a:

Step 1. Draw a sketch for a **Base Flange** on the top plane and extrude it to 26 gauge.

Step 2. Make two **Edge Flanges** .750 inch long as shown in Figure 6-70.

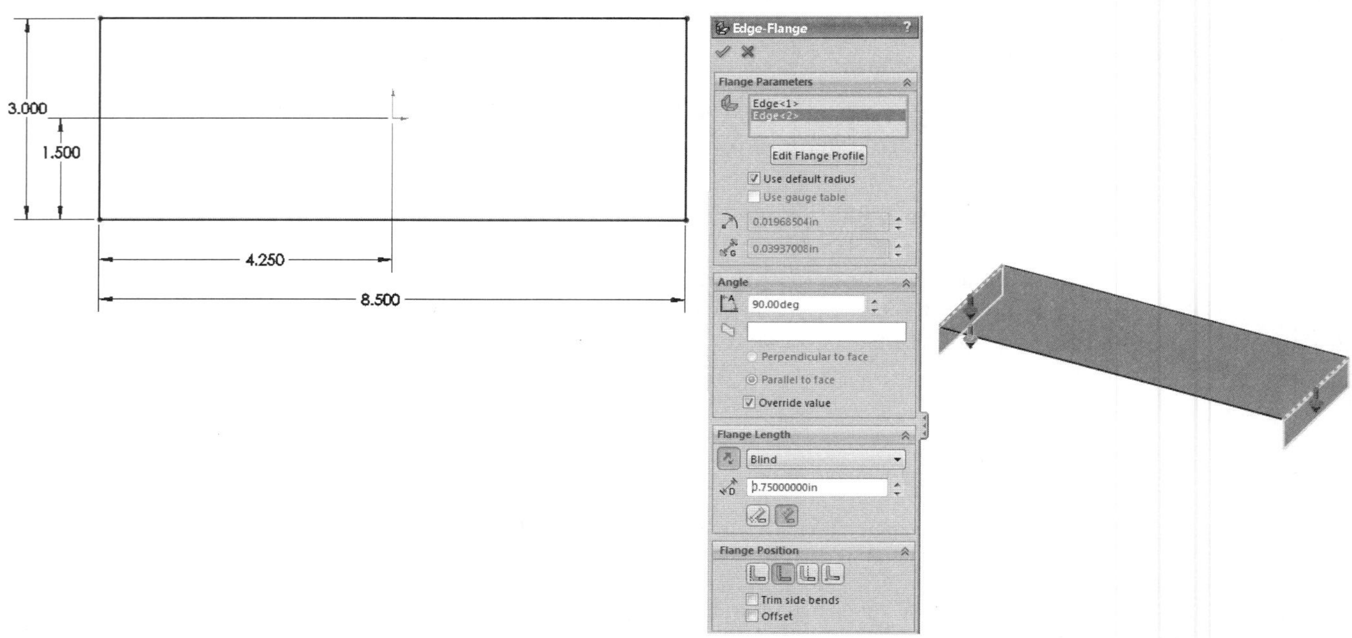

Figure 6-70
Make a Sketch for a Base Flange on the Top Plane and Extrude It to 26 Gauge. Make 2 Edge Flanges .750 Inch Long

Step 3. Draw sketches for a modification to both of the edge flanges using **3 Point Arc** and **Convert Entities** and extrude them using **Base-Flange** as shown in Figure 6-71.

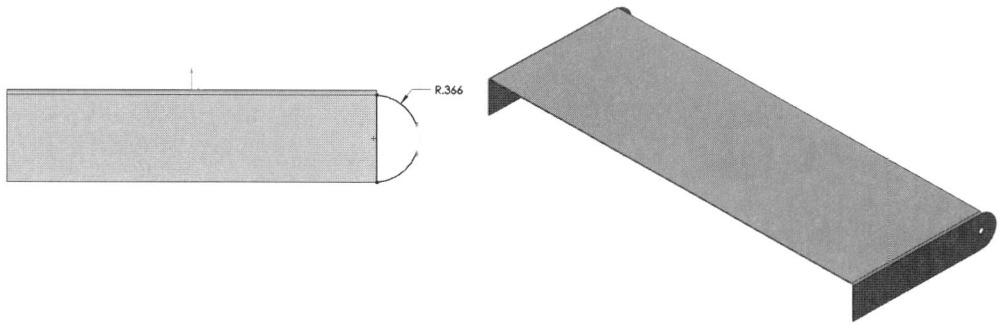

Figure 6-71
Draw a Sketch for a Modification to the Edge Flange Using 3 Point Arc and Convert Entities and Extrude It Using Base-Flange. Make a .125 Inch Cut Hole in the Edge Flanges Through All

Step 4. Draw a sketch of a .125 inch circle in the center of the arc for an **Extruded Cut** and make the cut **Through All.**

Step 5. Make an open hem of 45 degrees and .250 inch radius as shown in Figure 6-72.

Figure 6-72
Make a Hem

Step 6. Make a 45 degree **Edge Flange** 1.000 inch long as shown in Figure 6-73.

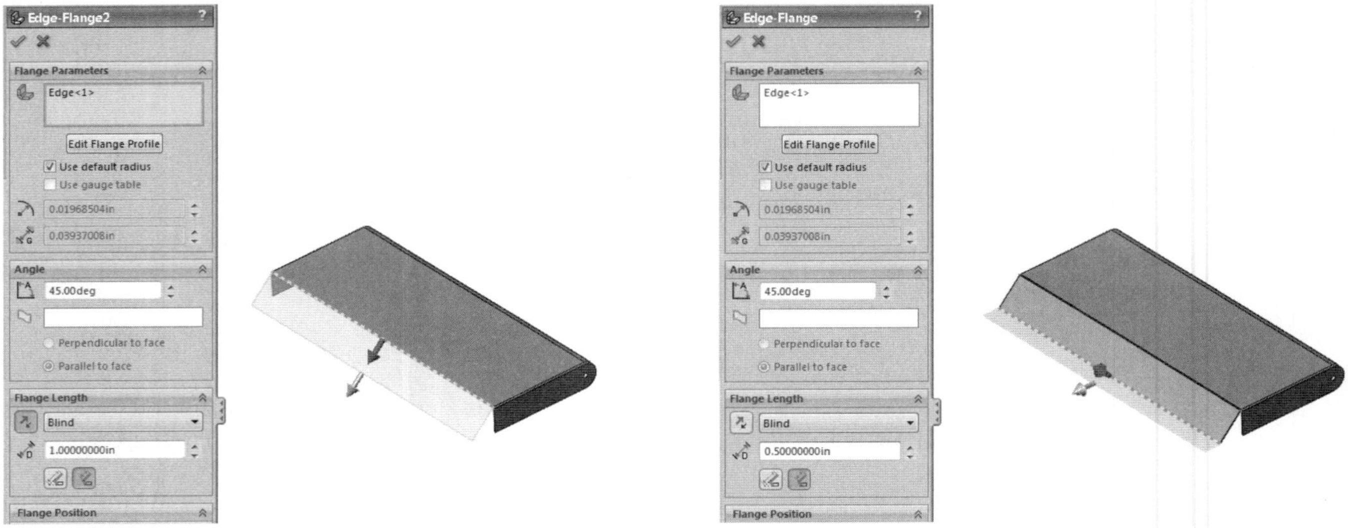

Figure 6-73
Make a 45 Degree Edge Flange 1.000 Inch Long.
Make a 45 Degree Edge Flange .500 Inch Long

Step 7. Make a 45 degree **Edge Flange** .500 inch long as shown in Figure 6-73.

Step 8. Save the drawing (Figure 6-74) with the file name **Exercise 6-3a(your initials)** in at least two places.

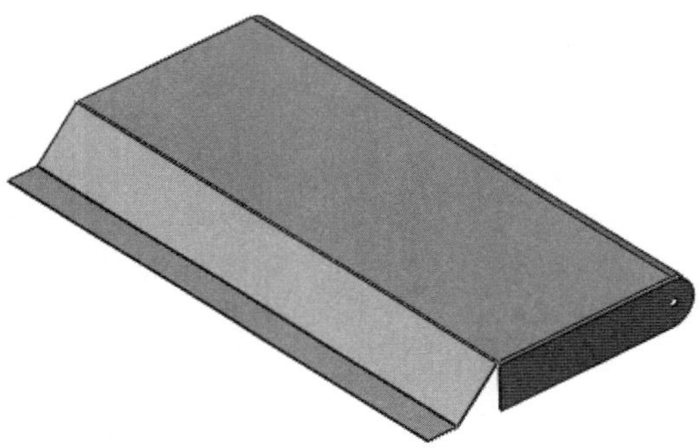

Figure 6-74
Exercise 6-3a Complete

Step 9. Print the drawing and turn it in.

Exercise 6-3b:

Step 1. Draw a sketch for a **Base Flange** on the front plane and extrude it to 28 gauge.

Step 2. Make three **Edge Flanges** 3.500 inches long as shown in Figure 6-75.

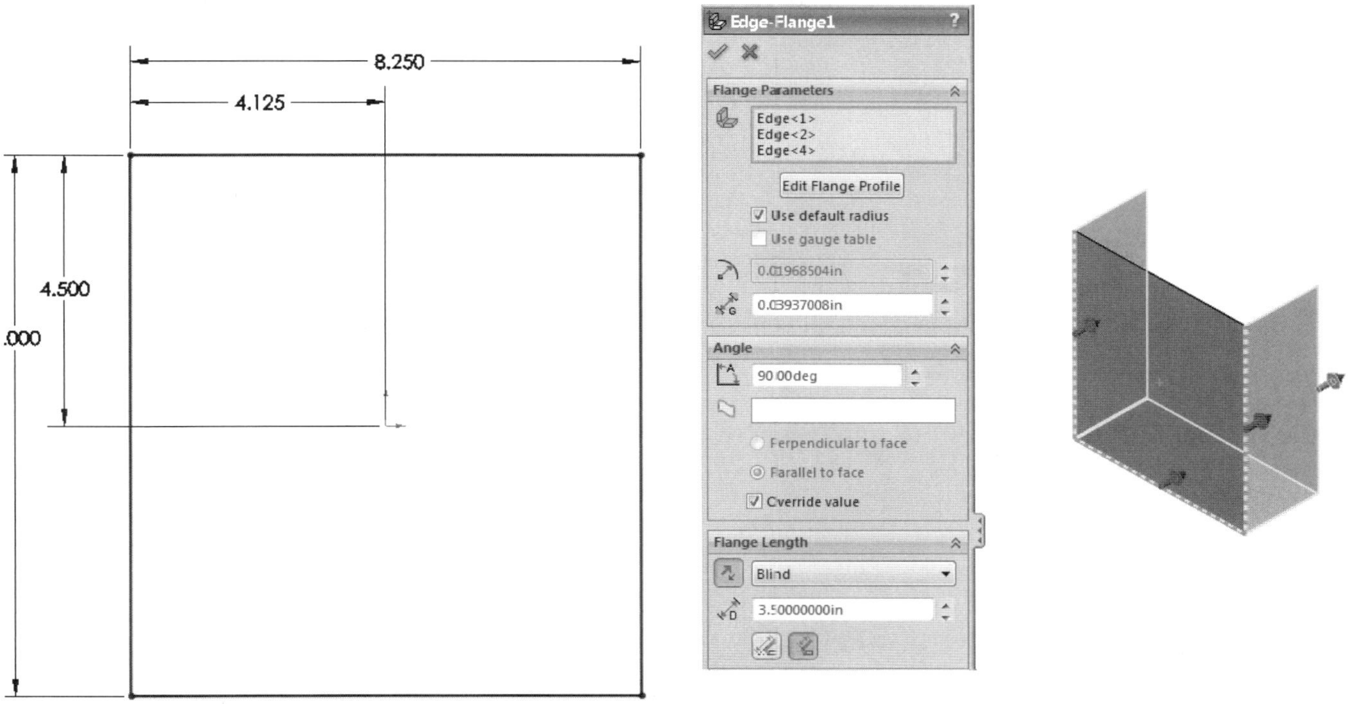

Figure 6-75
Draw a Sketch for a Base Flange on the Front Plane and Extrude It to 28 Gauge. Make 3 Edge Flanges 3.500 Inches Long

Step 3. Draw a sketch to modify the edge flange and extrude it using **Base-Flange** as shown in Figure 6-76.

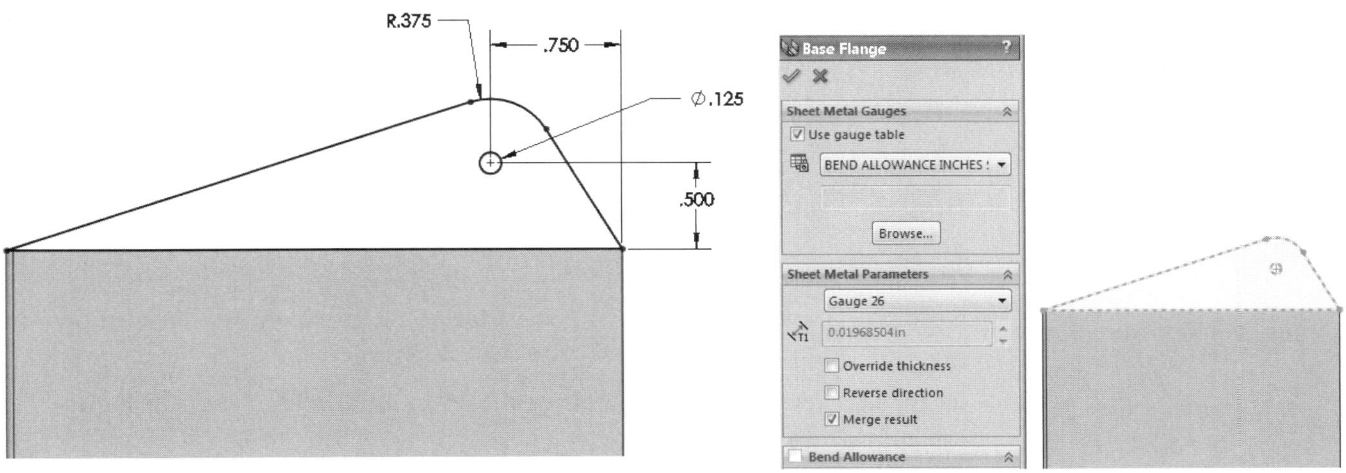

Figure 6-76
Draw a Sketch to Modify the Edge Flange and Extrude It Using Base-Flange

Step 4. Mirror the flange modification as shown in Figure 6-77.

Step 5. Save the drawing (Figure 6-77) with the file name **Exercise 6-3b(your initials)** in at least two places.

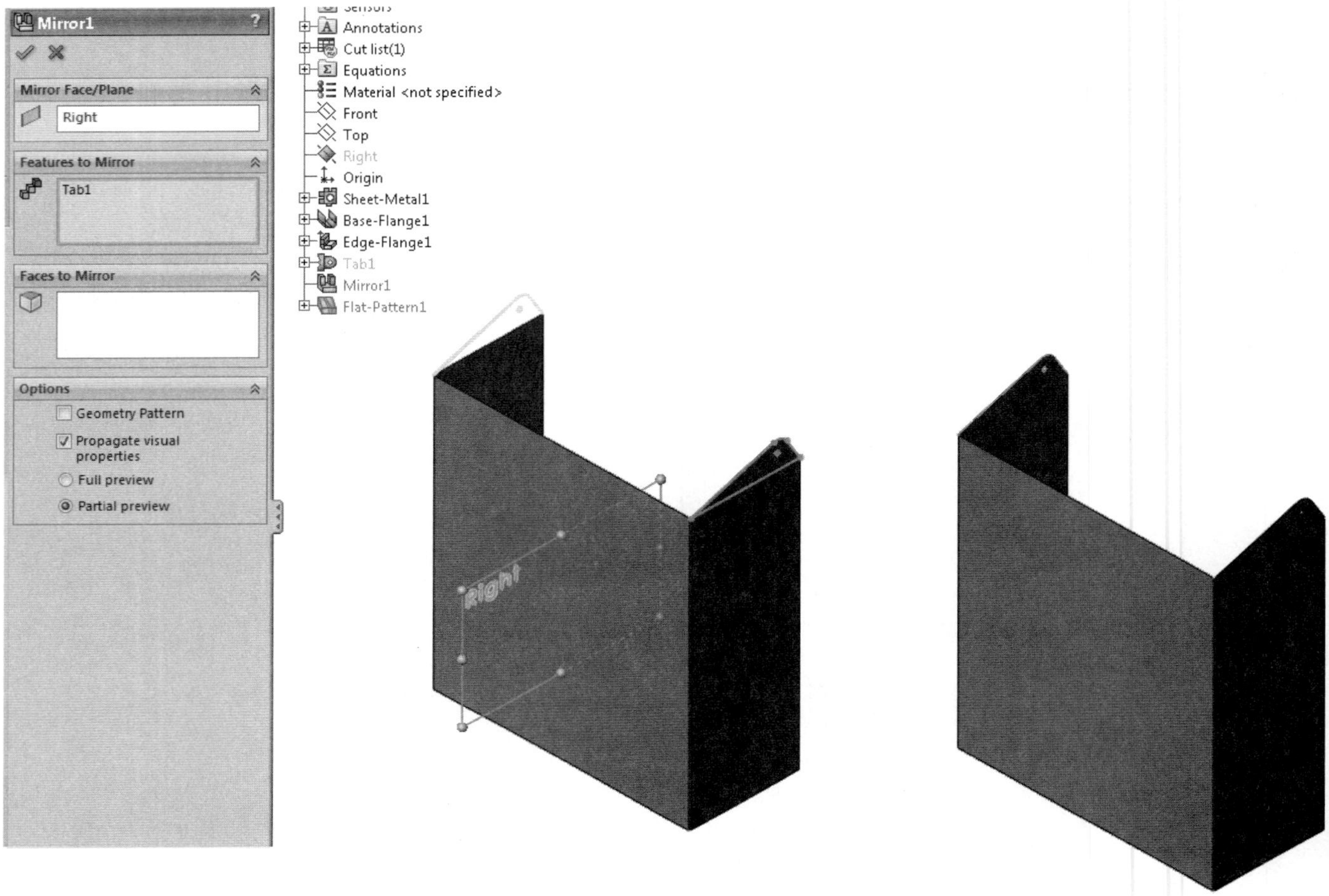

Figure 6-77
Mirror the Flange Modification.
Exercise 6-3b Complete

Step 6. Print the drawing and turn it in.

Exercise 6-3c:

Step 1. Draw a sketch for a **Base Flange** on the front plane as shown in Figure 6-78 and extrude it to 28 gauge.

Step 2. Make three **Edge Flanges** .375 inch long as shown in Figure 6-78.

Step 3. Save the drawing (Figure 6-79) with the file name **Exercise 6-3c(your initials)** in at least two places.

Step 4. Print the drawing and turn it in.

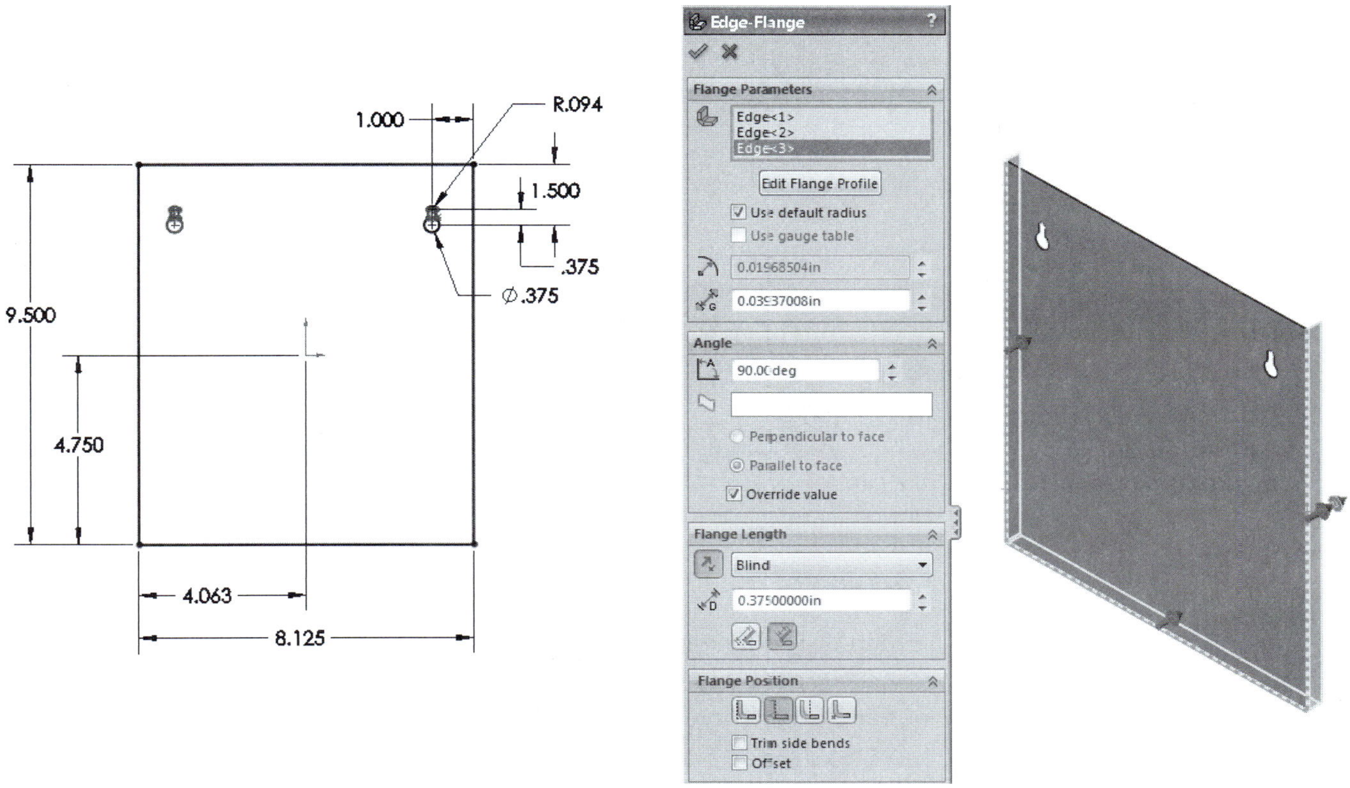

Figure 6-78
Make a Sketch for a Base Flange on the Front Plane (Mirror the Slot at the Top of the Flange and Trim It) and Extrude It to 28 Gauge. Make 3 Edge Flanges .375 Inch Long

Figure 6-79
Exercise 6-3c Complete

6 chapter six

Chapter Summary

This chapter contains additional information necessary to build solid models of sheet metal parts. Now you have the skills and information necessary to produce more complex solid models of sheet metal parts.

Review Questions

Circle the most correct answer.

1. On which toolbar is the **Jog** command located?
   a. **Edit**
   b. **Sketch**
   c. **Features**
   d. **Sheet Metal**

2. Which of the following must be checked in the **Base Flange Manager** to get the setting of 28 gauge?
   a. **Use gauge table**
   b. **Thickness**
   c. **Override thickness**
   d. **Edit Flange Profile**

3. Which of the following must be clicked to make a fillet on a solid?
   a. Two edges
   b. A circular feature
   c. A single edge
   d. A single point

4. Which of these is the same thickness as 18 gauge?
   a. .12500
   b. .03937
   c. .06125
   d. .03125

5. Which area in the **Miter Flange Manager** is used to make a miter flange that is shorter than the edge on which it is made?
   a. **Custom Relief Type**
   b. **Custom Bend Allowance**
   c. **Miter Parameters**
   d. **Start/End Offset**

6. The **Rib** command must have several lines in the sketch.
   a. True
   b. False

7. The **Jog** command must have several lines in the sketch.
   a. True
   b. False

8. A **Miter Flange** can have more than one line in the sketch.
   a. True
   b. False

9. The **Rip** command can be used to make a .125 gap between vertical edges.
   a. True
   b. False

10. Occasionally the **Display Style** must be changed to wireframe to select edges more easily.
   a. True
   b. False

6 chapter six

Matching

Place the correct letter on the line to the right of the numbered item.

1. A feature that must be selected to unfold a sheet metal part _____
2. A command that allows you to make a supporting feature of a part _____
3. A command that makes a vertical segment in a horizontal base flange _____
4. A command that can be used to change a solid to a sheet metal part _____
5. A command that makes a bend on the edge of a sheet metal part _____

a. **Rib**

b. **Jog**

c. **Miter-Flange**

d. **Insert Bends**

e. **Fixed Face**

General Questions

Answer the following questions.

1. In what situations would you use **Rip**?
2. When would you want to use the **Convert to Sheet Metal** command?
3. When would the **Jog** command be more useful than simply using a **Base-Flange**?
4. Why are sheet metal gauges needed?
5. Why are the **Fold** and **Unfold** commands necessary?

7 chapterseven

Sheet Metal Design II

CHAPTER OBJECTIVES

After completing this chapter, you will be able to:

- Make a sheet metal part using the **Base Flange** command.
- Make an assembly using **Mates.**
- Add a sketched bend to a sheet metal part using the **Sketched Bend** command.
- Use **Extruded Cut** to make a cut with a draft angle.
- Make a **Bill of Materials.**
- Add circular and circular split balloons and edit text within those balloons.
- Use **Linear Pattern** to copy features.
- Correctly answer questions regarding SolidWorks sheet metal commands used in this chapter.

The Methods for Making Assemblies and Drawings with Bills of Materials

The major methods used in making assemblies and drawings with bills of materials are as follows.

Inserting components. Before you begin inserting components you can shorten the browsing time by opening all of the parts that are in the assembly as shown in Figure 7-1. After you have all the parts open, choose the largest or most important one and click **Make Assembly from Part.** If you have a subassembly, click **Make Assembly from Assembly.**

Figure 7-1
Open All Parts. Click: Make Assembly from Part. Click: Insert Components. Click: OK to Insert the First Part

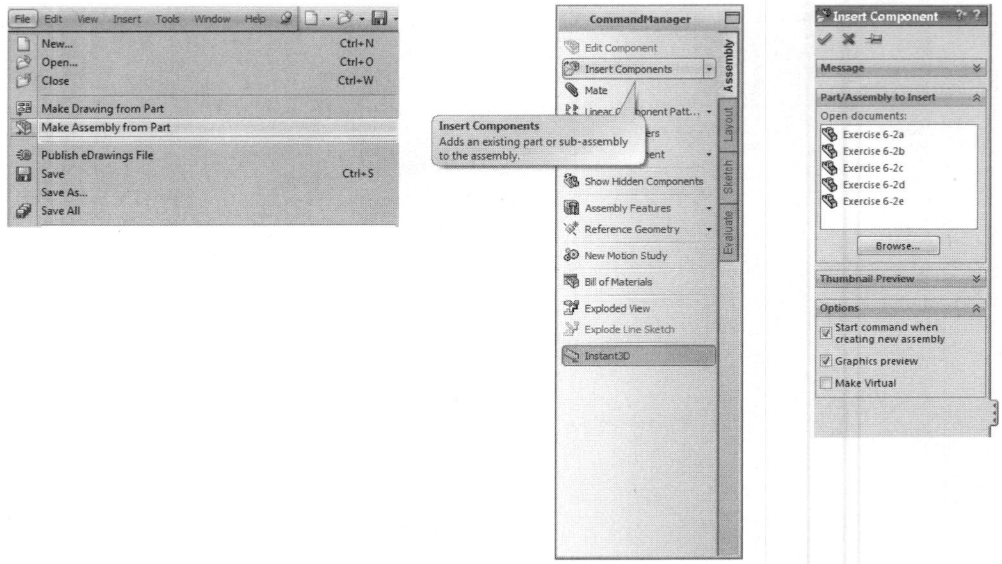

After you have clicked **Make Assembly from Part,** click **Insert Components** and click **OK** to insert the first part. After that click **Insert Components** and click the next part in the **Insert Components Manager** or click **Browse** and locate the next part.

Applying mates. When you have all of the parts of a small assembly or some of the parts of a larger assembly, use mates to assemble the parts. The ones used in this chapter are:

Concentric (Figure 7-2). To use the **Concentric** mate, click **Mate,** then click the cylindrical surface of one part. Next, click the cylindrical surface of the mating part and, if necessary, click the **Concentric** icon in the **Mate** manager.

Figure 7-2
Concentric Mate

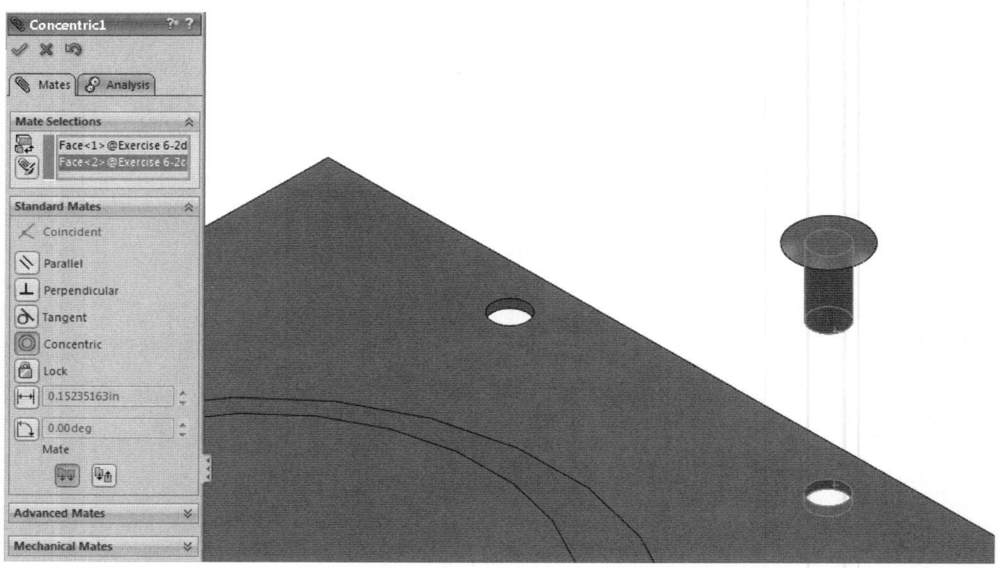

Coincident (Figure 7-3). To use the **Coincident** mate, click **Mate,** then click the flat surface of one part. Next, click the flat surface of the mating part and, if necessary, click the **Coincident** icon in the **Mate** manager.

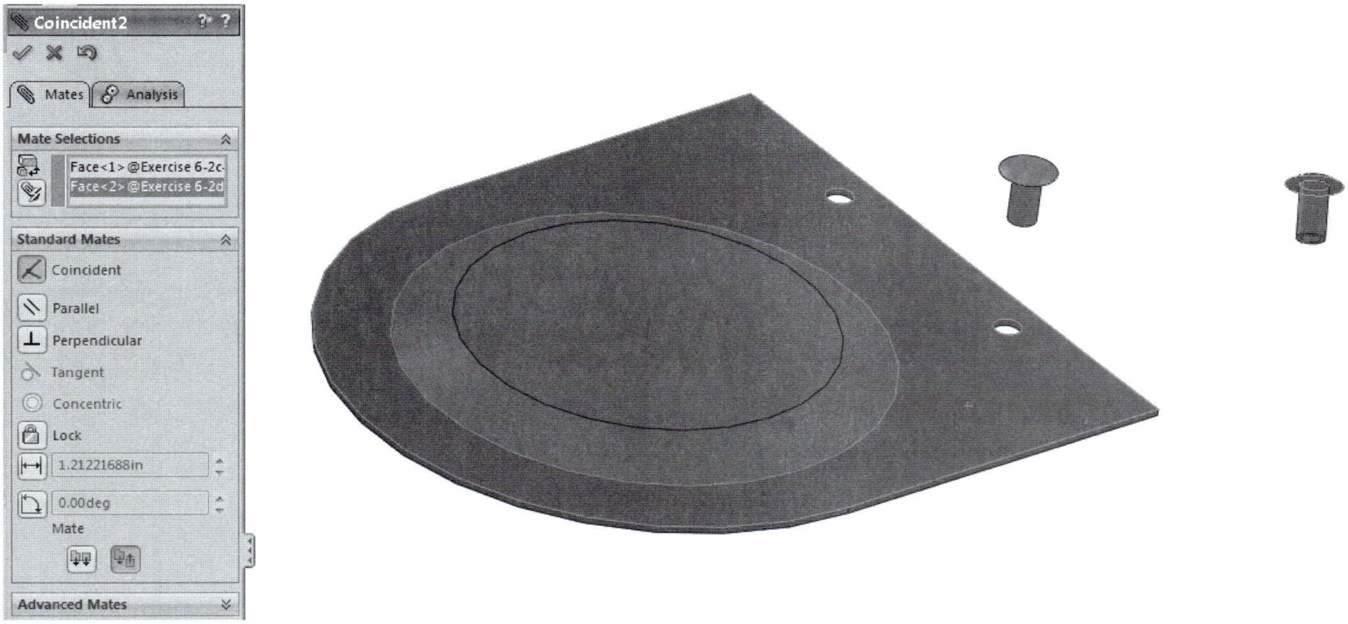

Figure 7-3
Apply Coincident Mate

Parallel (Figure 7-4). To use the **Parallel** mate, click **Mate,** then click an edge of one part. Next, click an edge of the mating part and, if necessary, click the **Parallel** icon in the **Mate** manager.

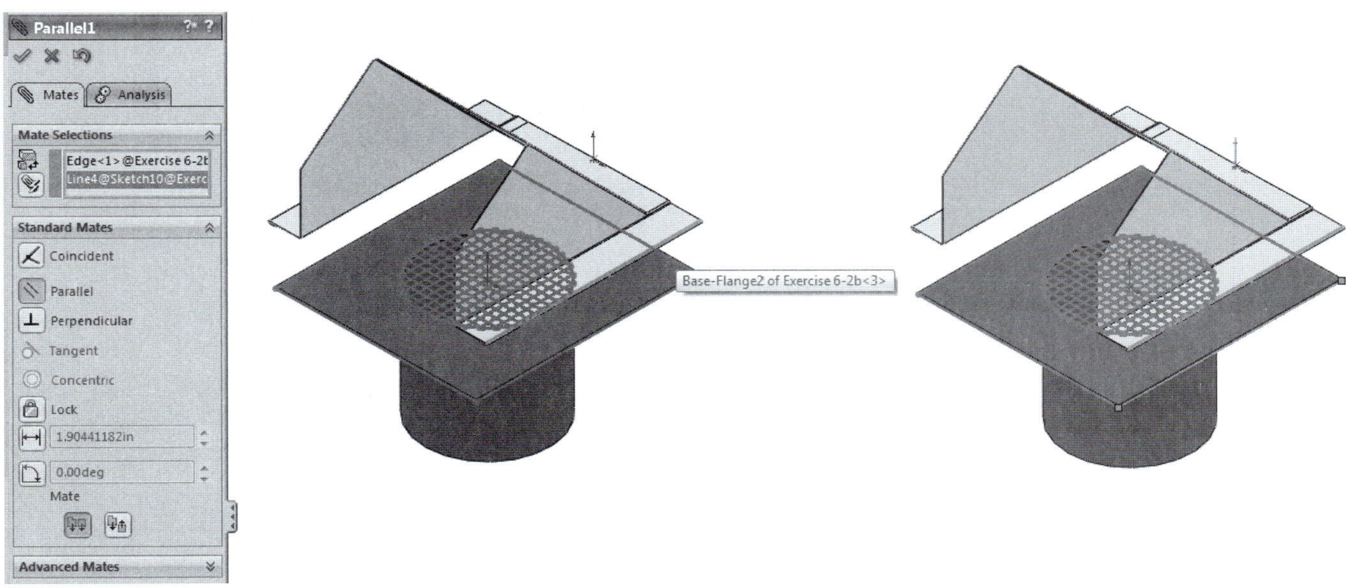

Figure 7-4
Apply Parallel Mate

Bill of Materials (Figure 7-5). When you have made the assembly, click **Make Drawing from Assembly** and select a drawing format for your drawing. After the format appears, click a view from the **View** palette that you want to use for the drawing and drag it onto the format. Right-click in the field of the format and click the word **Tables.** Then, click **Bill of Materials...** and click inside the view. Then, click **OK** and the **Bill of Materials** appears so that you can place it on the drawing. When you need to change the text, double-click the text, highlight it, and type the change. If you need to

Figure 7-5
Accessing Bill of Materials

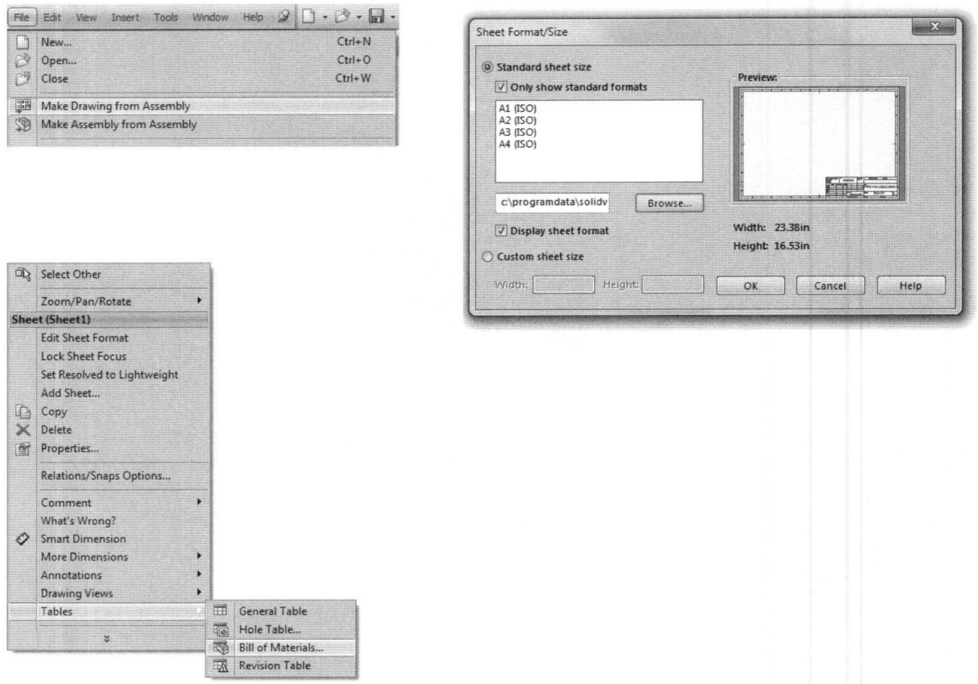

change the font, double-click the text and click the font name as shown in Figure 7-6. If you need to change the alignment, double-click the text and click the **Alignment** icon.

Figure 7-6
Changing the Font and Alignment in the Bill of Materials

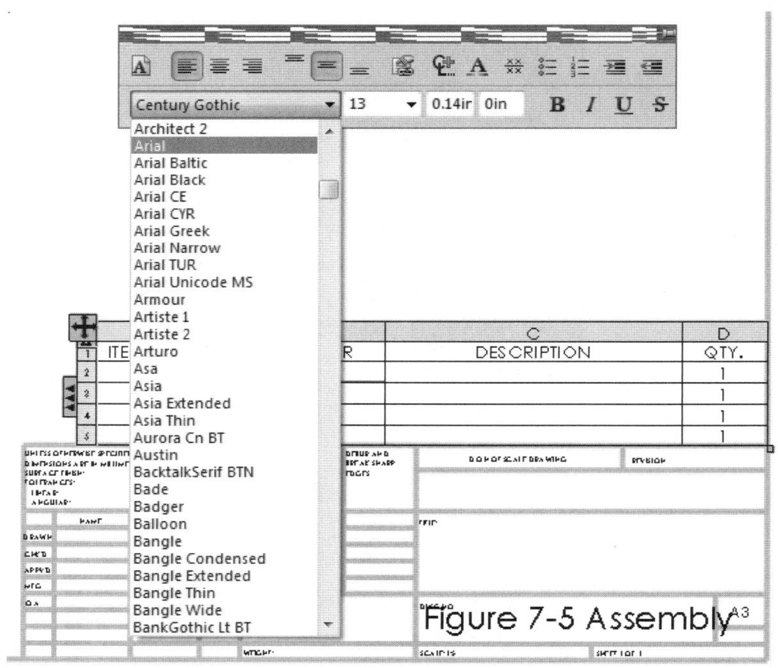

Adding Circular and Circular Split Line Balloons and changing text in them (Figure 7-7). When you have made the Bill of Materials, click **Balloon** from the **Annotation** tab on the **Command Manager.** The **Circular** balloon is the default, so if only one character is needed in each balloon all you have to do is click on the part

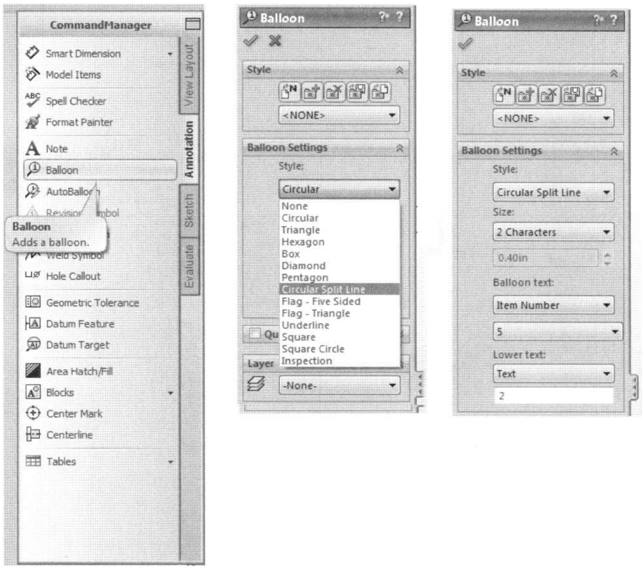

Figure 7-7
Adding Circular and Circular Split Line Balloons and Changing Text in the Balloons

and place the balloon. If you need to show the item number and its quantity, you will need to click **Circular Split Line.** To change the text in a **Circular** balloon, double-click the text and change it. To change the text in a **Circular Split Line,** double-click the text, change the item number, and click **OK** to exit from the command. To change the quantity after you have changed the item number, click on the item number ONLY ONCE and change the text under **Lower text:** in the **Balloon** manager.

Project 7-1: **Making a Sheet Metal Assembly of a Street Sign Using the Base Flange, Extrude, Insert Components, and Mate Commands (Figures 7-8 Through 7-27)**

Step 1. Click **New....**

Figure 7-8
Project 7-1a Complete

Step 2. Click **Part** and click **OK.**

Step 3. Click **Options** on the menu bar, select the **Document Properties** tab, and click **Units** in the list to the left.

Step 4. Make sure the **Units** are correctly set [notice that **IPS** (inch, pound, second) is checked], then click **OK.**

Step 5. Draw a sketch for a **Base Flange** on the front plane and extrude it to 12 gauge using the steel gauge table as shown in Figure 7-9.

Step 6. Change the **Base Flange** color to green using options as shown in Figure 7-10.

Step 7. Save your drawing (Figure 7-8) with the file name **Project 7-1b (your initials)** in at least two places.

Figure 7-9
Make a Sketch for a Base Flange on the Front Plane and Extrude the Base Flange to 12 Gauge

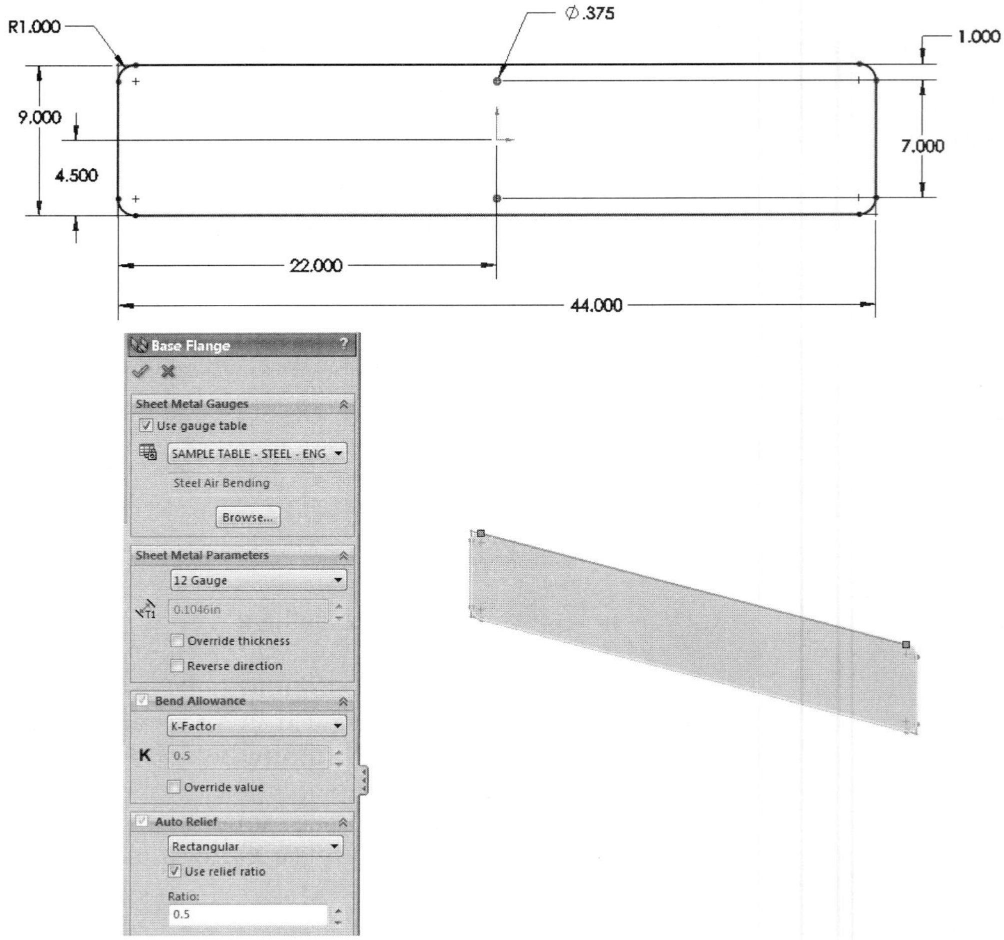

Figure 7-10
Change the Base Flange Color to Green Using Options, Project 7-1b Complete

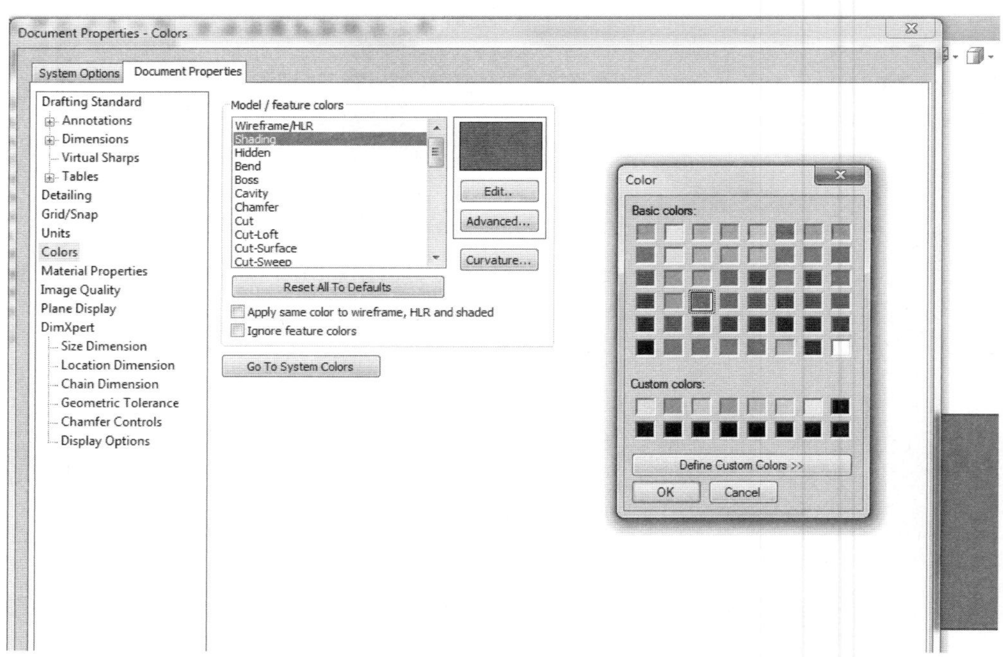

Step 8. Click **New....**

Step 9. Click **Part** and click **OK.**

Step 10. Click **Options** on the menu bar, select the **Document Properties** tab, and click **Units** in the list to the left.

Step 11. Make sure the **Units** are correctly set [notice that **IPS** (inch, pound, second) is checked], then click **OK.**

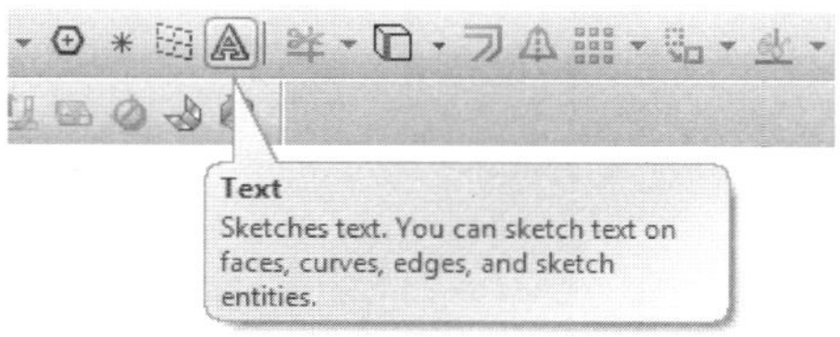

Step 12. Use **Text** with the **Arial** font 4.00 inches high to type the street name and the letters **DR** 2.00 inches high as shown in Figure 7-11.

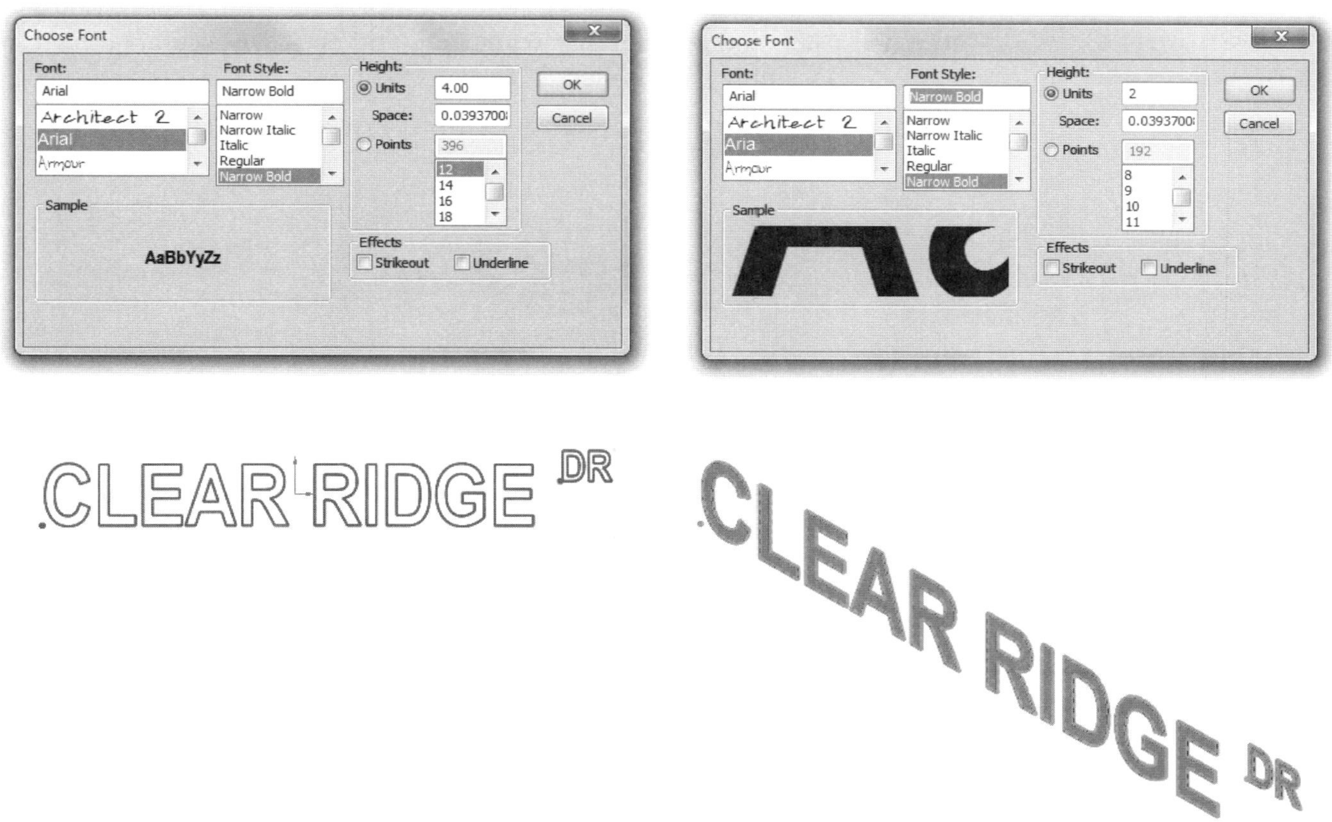

Figure 7-11
Use Text with the Arial Font 4.00 Inches High to Type the Street Name and the Letters "DR" 2.00 Inches High. Extrude the Text

Step 13. Use **Extrude** to extrude the text .10 inch.

Step 14. Right-click **Edit Material** and select **PET Plastic** to change the color to white as shown in Figure 7-12.

Figure 7-12
Right Click: Edit Material and Select PET Plastic. Project 7-1c Complete

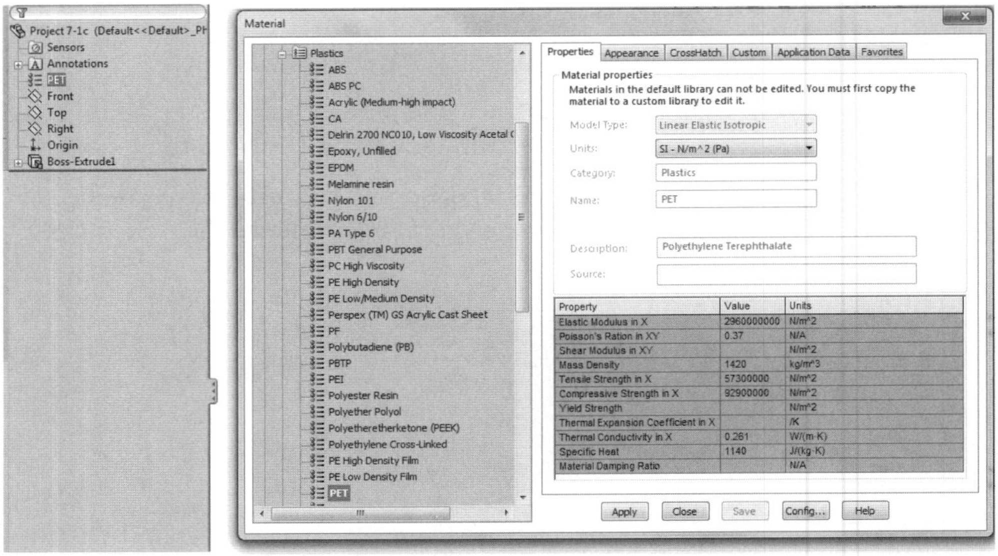

Step 15. Save your drawing with the file name **Project 7-1c(your initials)** in at least two places.

Step 16. Click **Make Assembly from Part** and click **OK** to start the assembly with Project 7-1c.

Step 17. Use **Insert Components** to insert Project 7-1b as shown in Figure 7-13.

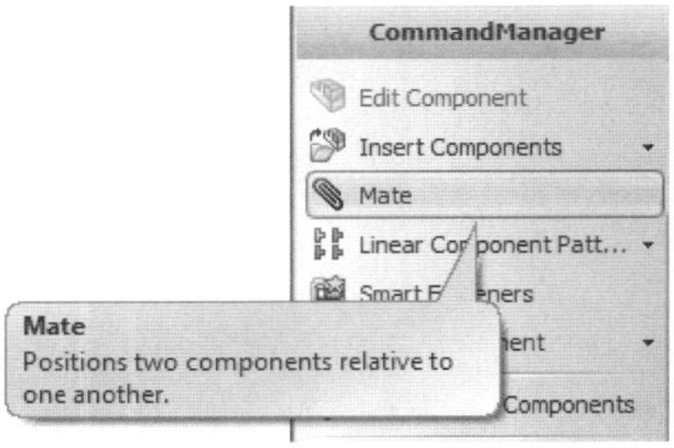

Figure 7-13
Use Insert Components to Insert Part, Project 7-1b

Step 18. Use **Mate** to apply the **Parallel** mate where needed by clicking on one edge of the lettering and one edge of the sign as shown in Figure 7-14.

Figure 7-14
Apply the Parallel Mate Where Needed.
Project 7-1a Complete

Step 19. Save your assembly (Figure 7-14) with the file name **Project 7-1a(your initials)** in at least two places.

Step 20. Print the drawing and turn it in.

Step 21. Start a new drawing on the top plane with inches as the unit of measure as shown in Figure 7-15.

Figure 7-15
Project 7-1c Complete

Step 22. Draw a sketch for a **Base Flange** on the top plane and extrude it 10 feet as shown in Figure 7-16.

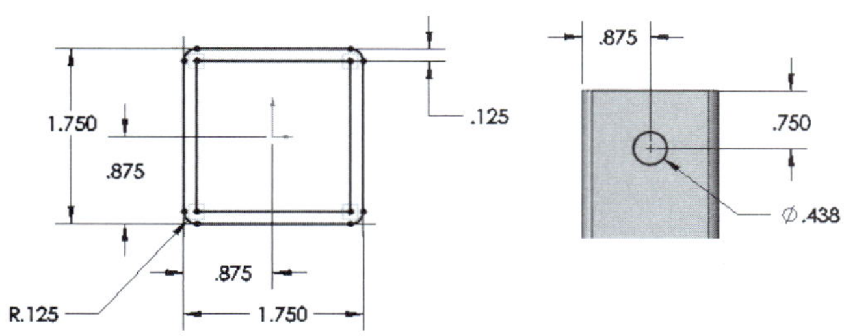

Figure 7-16
Draw a Sketch for a Base Flange on the Top Plane and Extrude It 10 Feet. Draw a Hole for an Extruded Cut as Shown on the Right and Make the Cut Through All

Step 23. Draw a sketch for an **Extruded Cut** using **Circle** and make the cut **Through All** as shown in Figure 7-16.

Step 24. Use **Linear Pattern** to pattern the **Extruded Cut** 120 instances in **Direction1** with spacing of 1.00 inch as shown in Figure 7-17.

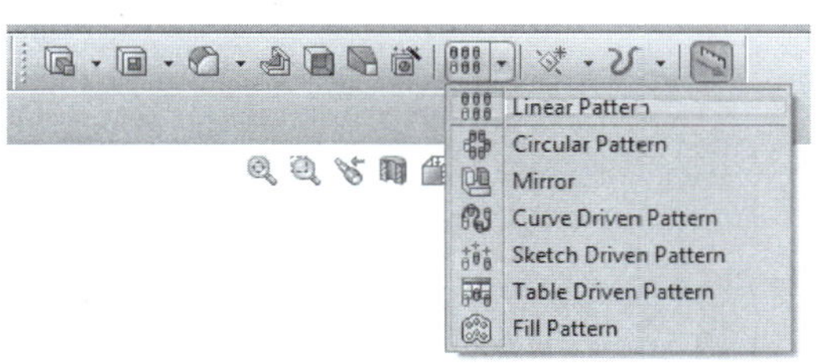

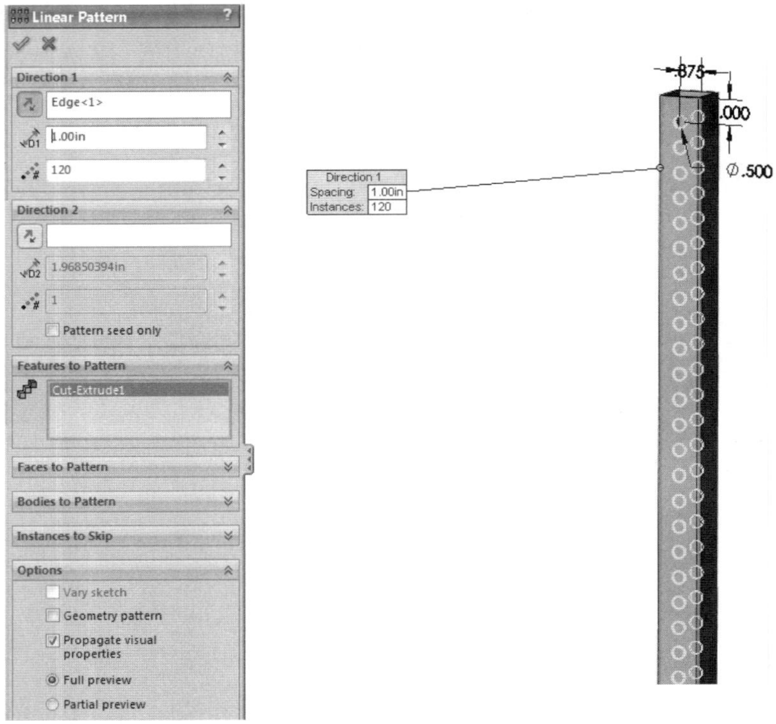

Figure 7-17
Make a Linear Pattern of the Extruded Cut with **Direction1** Spacing of 1.00 Inch and 120 Instances

Figure 7-18
Project 7-1d Complete

Step 25. Save your drawing (Figure 7-18) with the file name **Project 7-1d (your initials)** in at least two places on your computer.

Step 26. Start a new drawing on the top plane with inches as the unit of measure.

Step 27. Draw a sketch for a **Revolved Boss/Base** and revolve it as shown in Figure 7-19.

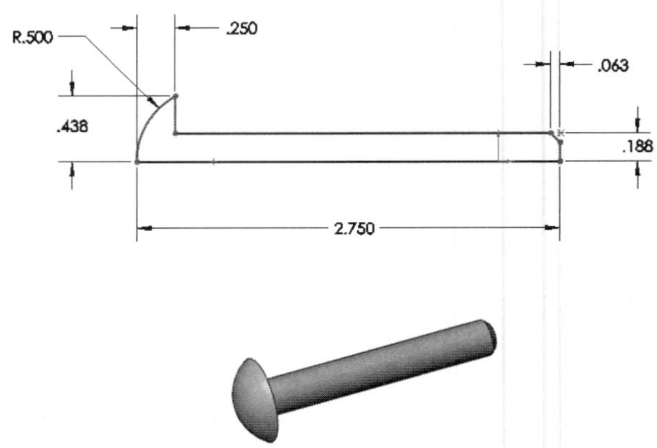

Figure 7-19
Draw a Sketch for a Revolved Boss/Base on the Front Plane and Revolve It

Step 28. Make a plane .26 inch from the bottom of the screw head as shown in Figure 7-20.

Step 29. Draw a sketch for an **Extruded Cut** on **Plane1** and make the cut .188 inch deep as shown in Figure 7-20.

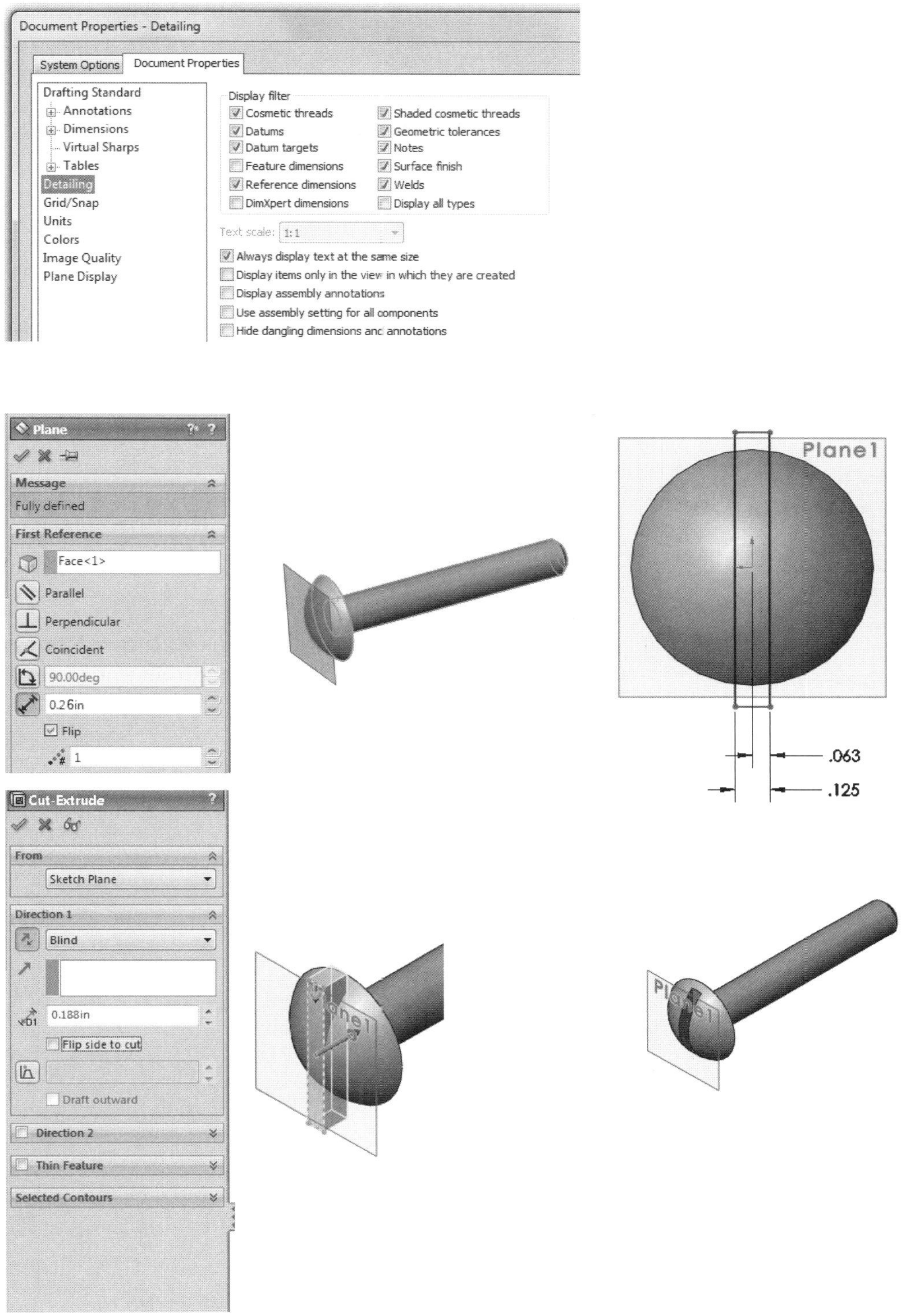

Figure 7-20
Make a Plane .26 Inch from the Bottom of the Screw Head. Draw a Sketch for an Extruded Cut and Make the Cut .188 Inch Deep

Step 30. Select **Options, Document Properties** tab, **Detailing Item,** and **Check Cosmetic Thread.**

Step 31. Click **Cosmetic Thread** and click the bottom edge of the screw so that cosmetic threads appear as shown in Figure 7-21.

Figure 7-21
Apply Cosmetic Thread .375-20 Machine Screw Complete

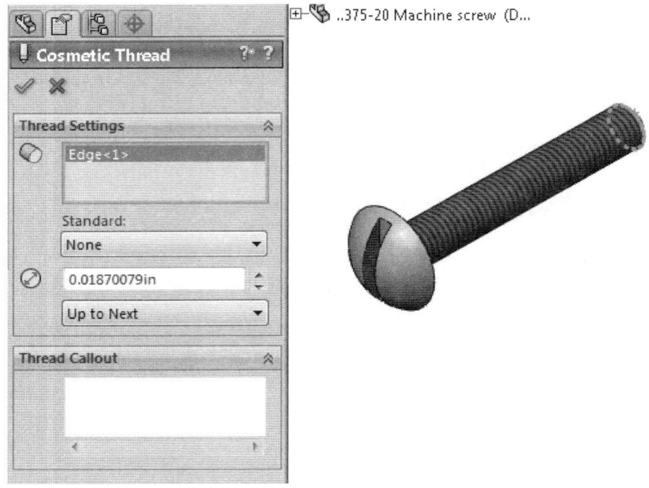

Step 32. Save your drawing with the file name **.375-20 Machine Screw (your initials)** in at least two places on your computer.

Step 33. Start a new drawing on the top plane with inches as the unit of measure.

Step 34. Draw a sketch for an **Extruded Boss/Base** on the front plane and extrude it .07 inch as shown in Figure 7-22.

Figure 7-22
Make a Sketch for an Extruded Boss/Base on the Front Plane and Extrude It .07 Inch.
.375 Lock Washer Complete

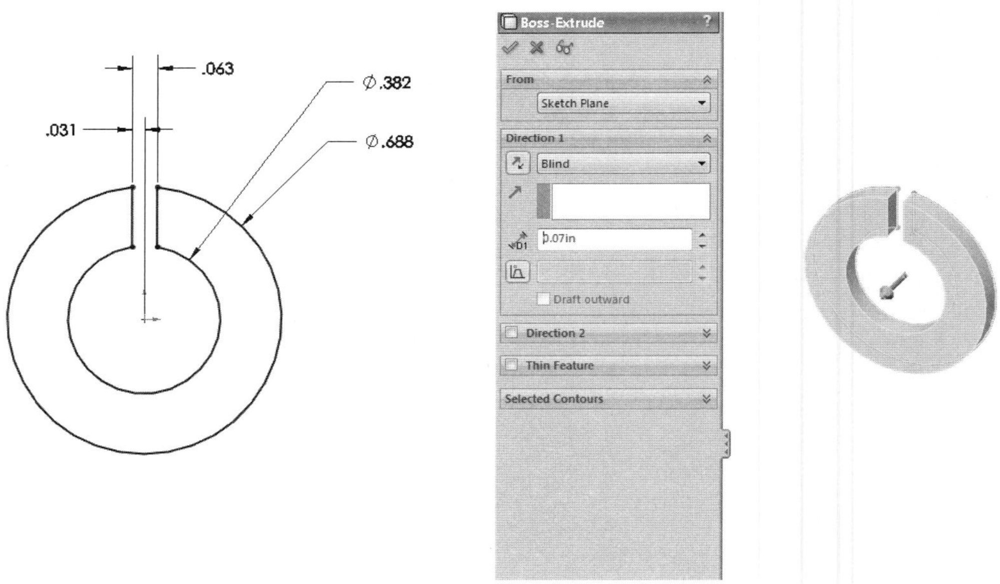

Step 35. Save your drawing with the file name **.375 Lock Washer(your initials)** in at least two places.

Step 36. Start a new drawing on the top plane with inches as the unit of measure.

Step 37. Draw a sketch (the hexagon) for an **Extruded Boss/Base** on the front plane and extrude it .25 inch as shown in Figure 7-23.

Step 38. Select the front of the extrusion as the sketch plane and draw a .250 circle for an **Extruded Cut** as shown on the right side of Figure 7-23.

Figure 7-23
Draw a Sketch for an Extruded Boss/Base On the Front Plane and Extrude It .25 Inch. Select the Front of the Extrusion as the Sketch Plane and Draw a .250 Circle for an Extruded Cut

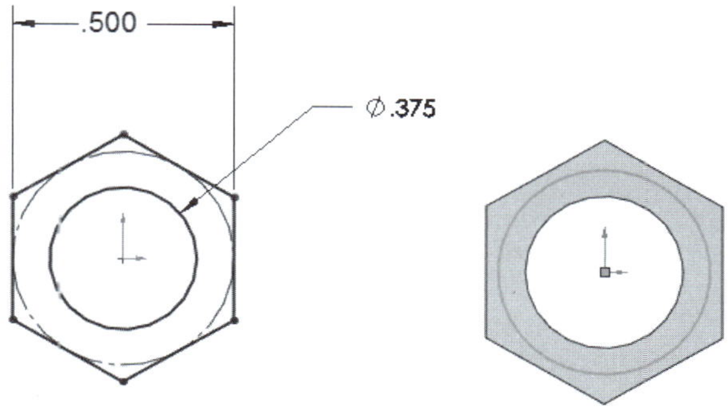

Step 39. Make an **Extruded Cut** on the front side of the extrusion with .30 inch as **Direction1** and a 60 degree draft angle as shown in Figure 7-24.

Figure 7-24
Make an Extruded Cut on the Front Side, Repeat That Cut for the Back Side. Add Cosmetic Threads. .375-20 Nut Complete

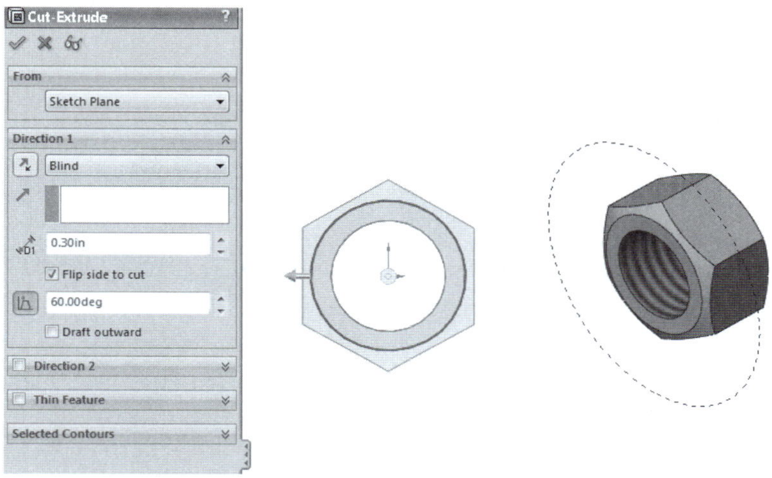

Step 40. Repeat Step 37 for the back side of the nut.

Step 41. Apply **Cosmetic Threads** to the nut.

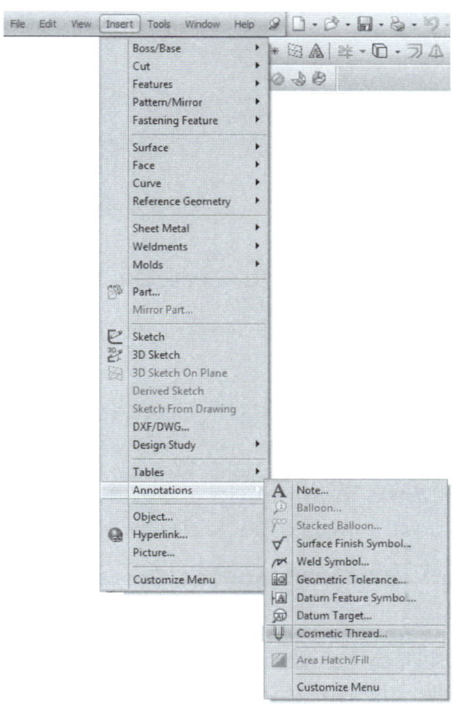

Step 42. Save your drawing with the file name **.375-20 Nut(your initials)** in at least two places.

Step 43. Open **Project 7-1c, Project 7-1d, .375-20 Machine Screw, .375-20 Nut,** and **.375 Lock Washer,** click **Make Assembly from Assembly,** and click **OK** to start the assembly with **Project 7-1d.**

Step 44. Use **Insert Components** to insert **Project 7-1c** as shown in Figure 7-25.

Figure 7-25
Select Make Assembly from Assembly. Click: Insert Components (the Post and the Sign Assembly), and Apply Concentric and Coincident Mates

Step 45. Use **Mate** to apply the **Concentric** mate by clicking the top hole of the post and the top hole of the sign as shown in Figure 7-25.

Step 46. Use **Mate** to apply the **Coincident** mate by clicking the front of the post and the back of the sign as shown in Figure 7-25.

Step 47. Insert **.375-20 Machine Screw, .375-20 Nut,** and **.375 Lock Washer.**

Step 48. Apply **Concentric** and **Coincident** mates to all three components as shown in Figure 7-26.

Figure 7-26
Insert .375-20 Machine Screw, .375 Lock Washer and .375-20 Nut. Apply Concentric and Coincident Mates to All Three Components.
Make a Linear Pattern of 2 Instances and 6.00 Inches in **Direction1**

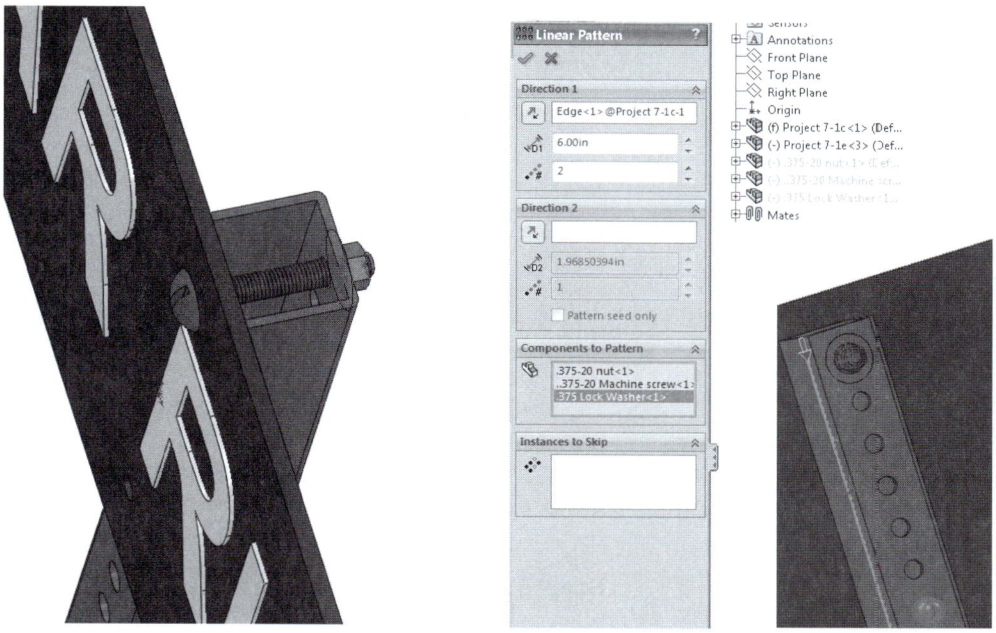

Step 49. Make a **Linear Pattern** of 2 instances and 6.00 inches in **Direction1.**

Step 50. Save your assembly (Figure 7-27) with the file name **Project 7-1 (your initials)** in at least two places.

Step 51. Print the drawing and turn it in.

Figure 7-27
Project 7-1 Complete

Project 7-2: **Making a Sheet Metal Assembly of a Tiller Using the Base Flange, Extrude, Extrude Cut, Sketched Bend, Linear Pattern, Insert Components, and Mate Commands (Figures 7-28 Through 7-57)**

Step 1. Click **New....**

Step 2. Click **Part** and click **OK.**

Step 3. Click **Options** on the menu bar, select the **Document Properties** tab, and click **Units** in the list to the left.

Figure 7-28
Project 7-2 Complete

Step 4. Make sure the **Units** are correctly set [notice that **IPS** (inch, pound, second) is checked], then click **OK.**

Step 5. Draw a 1.25 inches circle on the top plane and extrude it as shown in Figure 7-29.

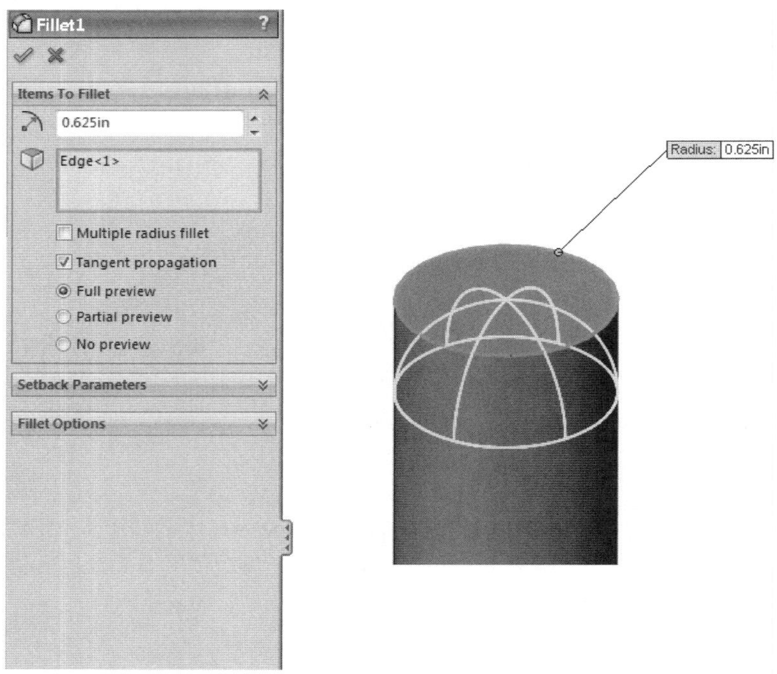

Figure 7-29
Draw a 1.25 Inches Circle on the Top Plane and Extrude It 54 Inches. Make a .625 Radius Fillet on the Top of the Extrusion

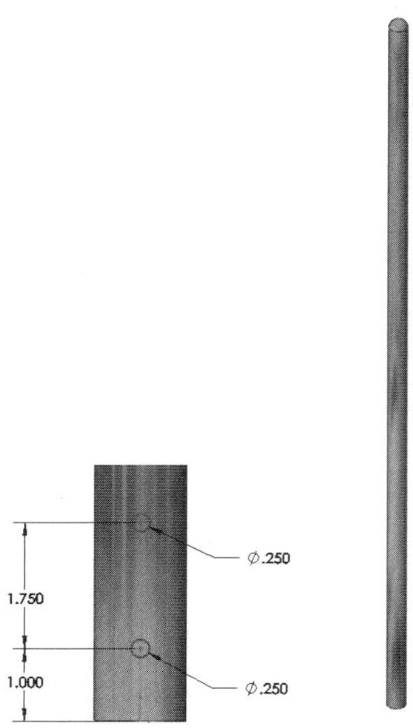

Figure 7-30
Make a Sketch for an Extruded Cut on the Right Plane and Make the Cut Through All. Apply Oak Material Using Edit Material Project 7-2a Complete

Step 6. Make a .625 inch radius **Fillet** on the top of the extrusion as shown in Figure 7-29.

Step 7. Draw a sketch for an **Extruded Cut** on the right plane and make the cut **Through All** as shown in Figure 7-30.

Step 8. Apply **Oak** material using **Edit Material** as shown in Figure 7-30.

Step 9. Save your drawing (Figure 7-30) with the file name **Project 7-2a(your initials)** in at least two places.

Step 10. Click **New....**

Step 11. Start a new drawing on the top plane with inches as the unit of measure.

Step 12. Draw a sketch for a **Base Flange** on the top plane and trim it so that only a semicircle remains, then extrude it 3.00 inches using 10 gauge steel as shown in Figure 7-31.

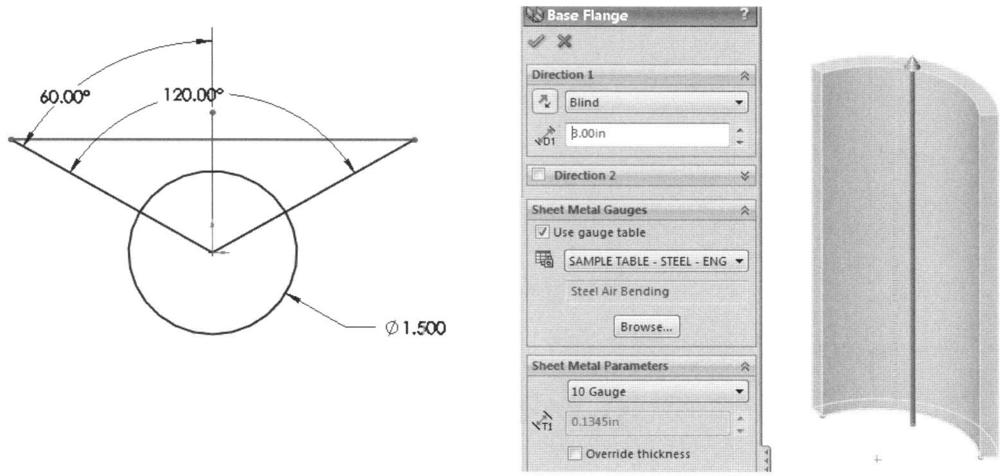

Figure 7-31
Draw a Sketch on the Top Plane and Trim It so Only a Semicircle Remains. Extrude the Base Flange 3.00 Inches Using 10 Gauge Steel

Step 13. Draw a sketch for a **Base Flange** on the right plane and extrude it 1.50 inches from the mid plane as shown in Figure 7-32.

Step 14. Use **Convert Entities** to draw a sketch for an **Extruded Boss/Base** and extrude it **Up to Next** as shown in Figure 7-32.

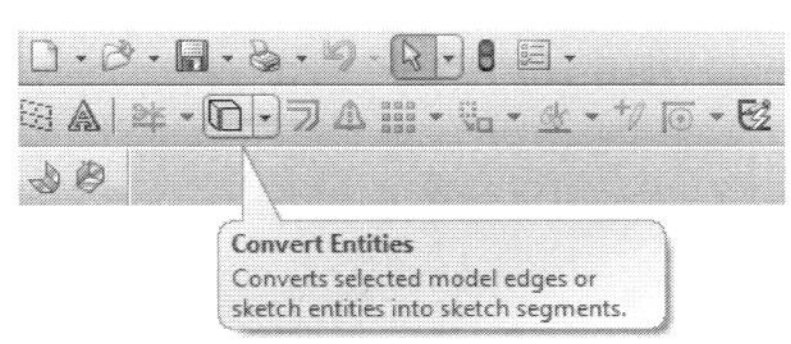

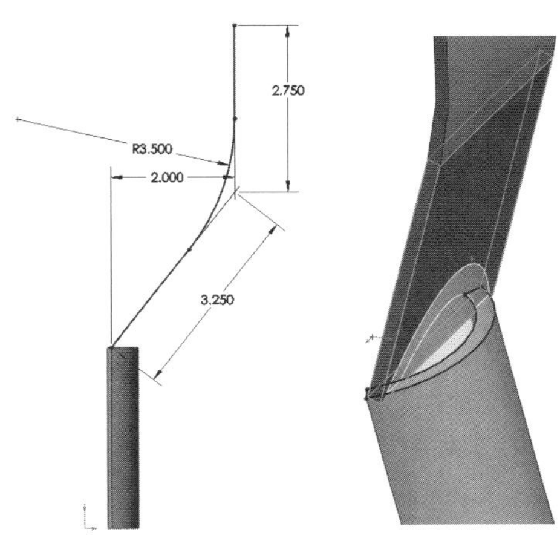

Figure 7-32
Make a Sketch for a Base Flange on the Right Plane and Extrude It 1.50 Inches from the Mid Plane. Use Convert Entities to Make a Sketch for an Extrude Boss/Base and Extrude It Up to Next

Step 15. Draw a sketch for an **Extruded Cut** and make the cut **Through All** as shown in Figure 7-33.

Step 16. Draw a sketch for an **Extruded Cut** on the front plane and make the cut **Through All** as shown in Figure 7-34.

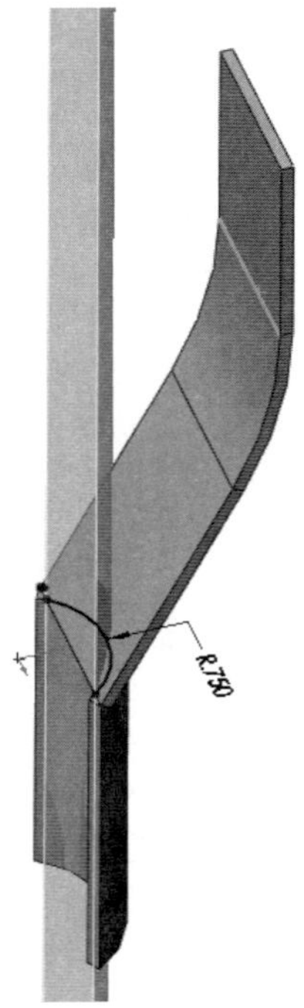

Figure 7-33
Draw a Sketch for an Extruded Cut on the Surface Shown and Make the Cut Through All

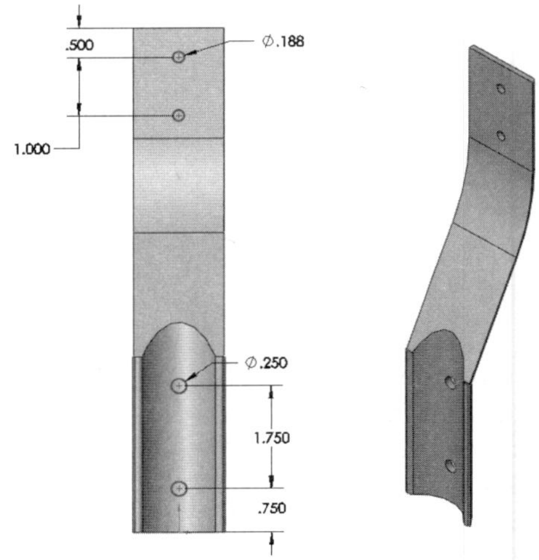

Figure 7-34
Make a Sketch for an Extruded Cut on the Front Plane and Make the Cut Through All. Project 7-2b Complete

Step 17. Save your drawing (Figure 7-34) with the file name **Project 7-2b (your initials)** in at least two places.

Step 18. Click **New....**

Step 19. Start a new drawing on the top plane with inches as the unit of measure.

Step 20. Draw a sketch for a **Base Flange** on the right plane and extrude it to 18 gauge steel as shown in Figure 7-35.

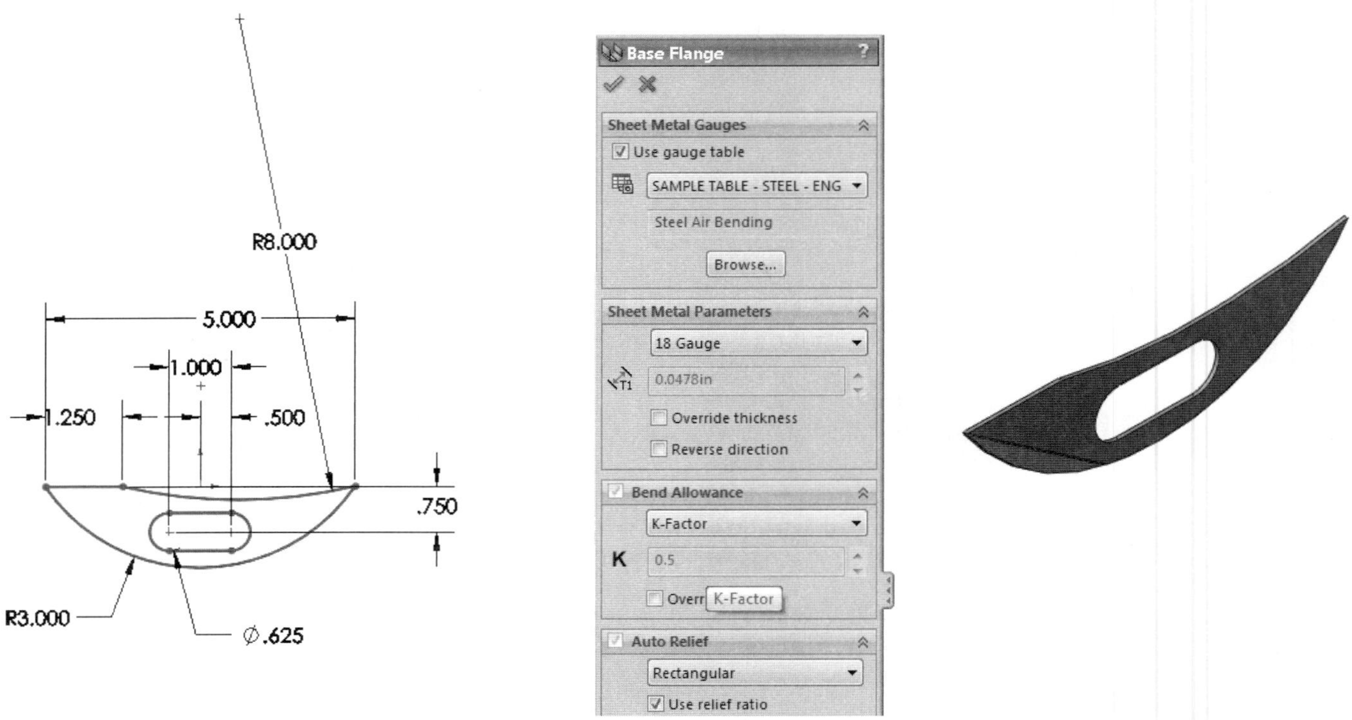

Figure 7-35
Make a Sketch for a Base Flange on the Right Plane and Extrude the Base Flange to 18 Gauge Steel

Step 21. Make a sketch on the front face of the base flange for a **Sketched Bend** and make the bend as shown in Figure 7-36.

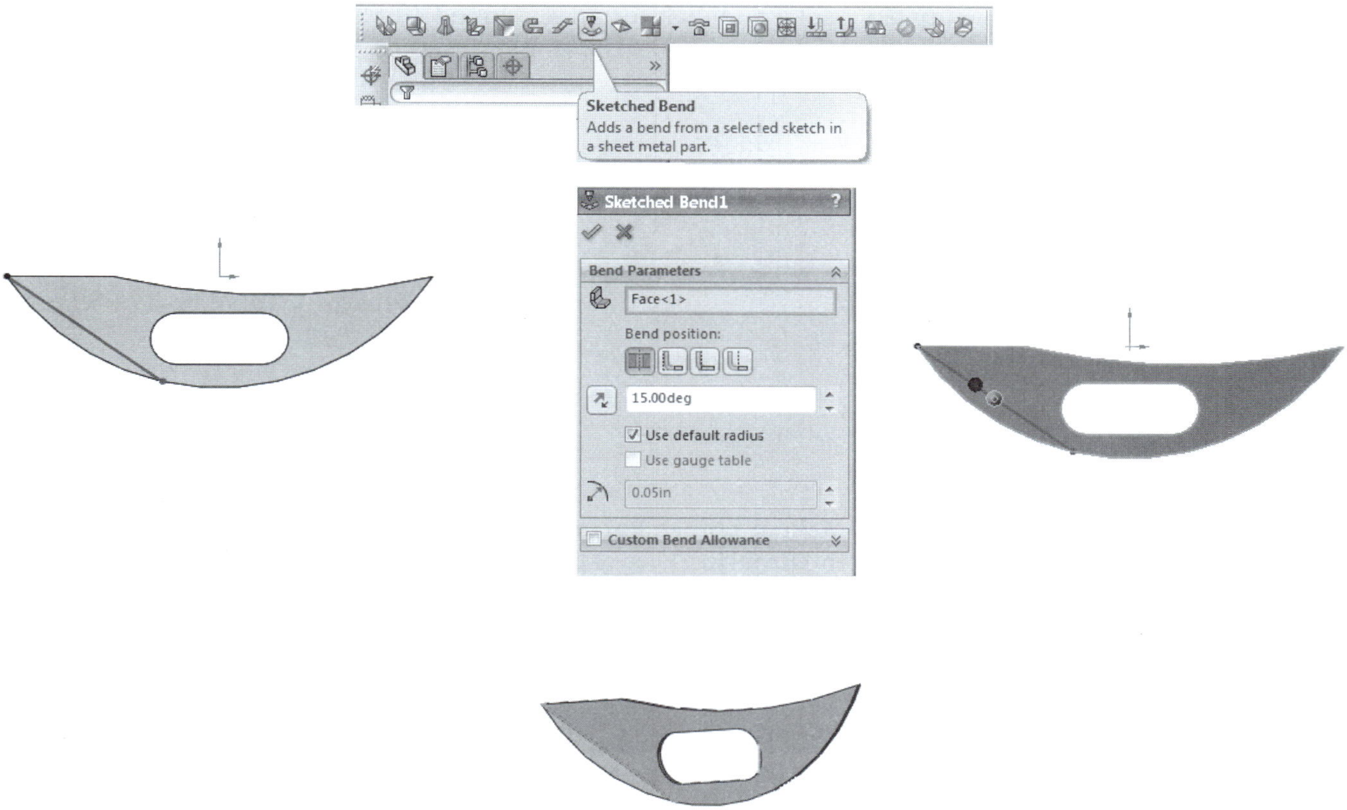

Figure 7-36
Make a Sketch on the Front Face of the Base Flange for a Sketched Bend and Make the Bend. Project 7-2c Complete

Step 22. Save your drawing (Figure 7-36) with the file name **Project 7-2c (your initials)** in at least two places.

Step 23. Click **New....**

Step 24. Start a new drawing on the top plane with inches as the unit of measure.

Step 25. Draw a sketch for an **Extruded Boss/Base Thin Feature** and extrude it 14.5 inches using 18 gauge steel (.0478 inch) as shown in Figure 7-37.

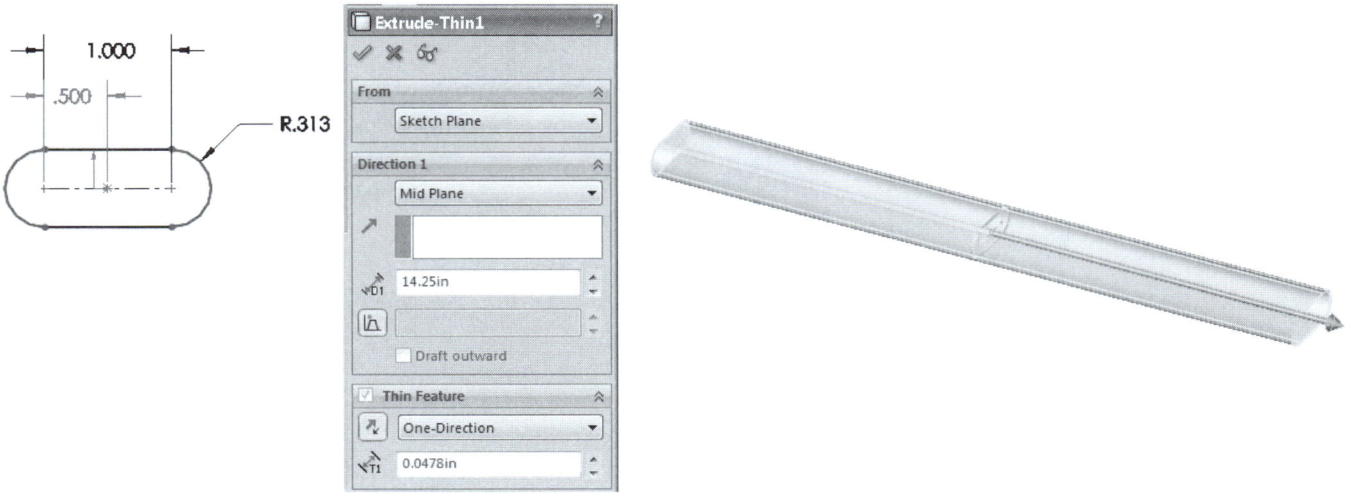

Figure 7-37
Draw a Sketch for an Extruded Boss/Base Thin Feature on the and Extrude It 14.25 Inches to 18 Gauge Steel

Step 26. Make a sketch for an **Extruded Cut** on the top of the extrusion and make the cut .048 inch deep as shown in Figure 7-38.

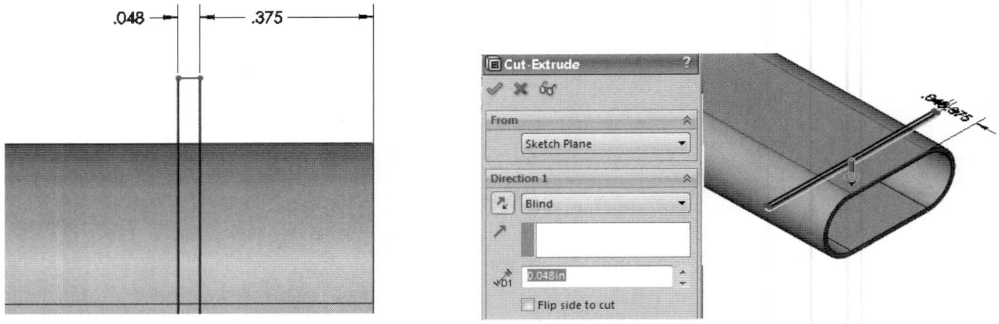

Figure 7-38
Make a Sketch for an Extruded Cut on the Top of the Extrusion and Make the Cut .048 Inch Deep

Step 27. Make a **Linear Pattern** of 19 instances with spacing of .75 inch in **Direction1** as shown in Figure 7-39.

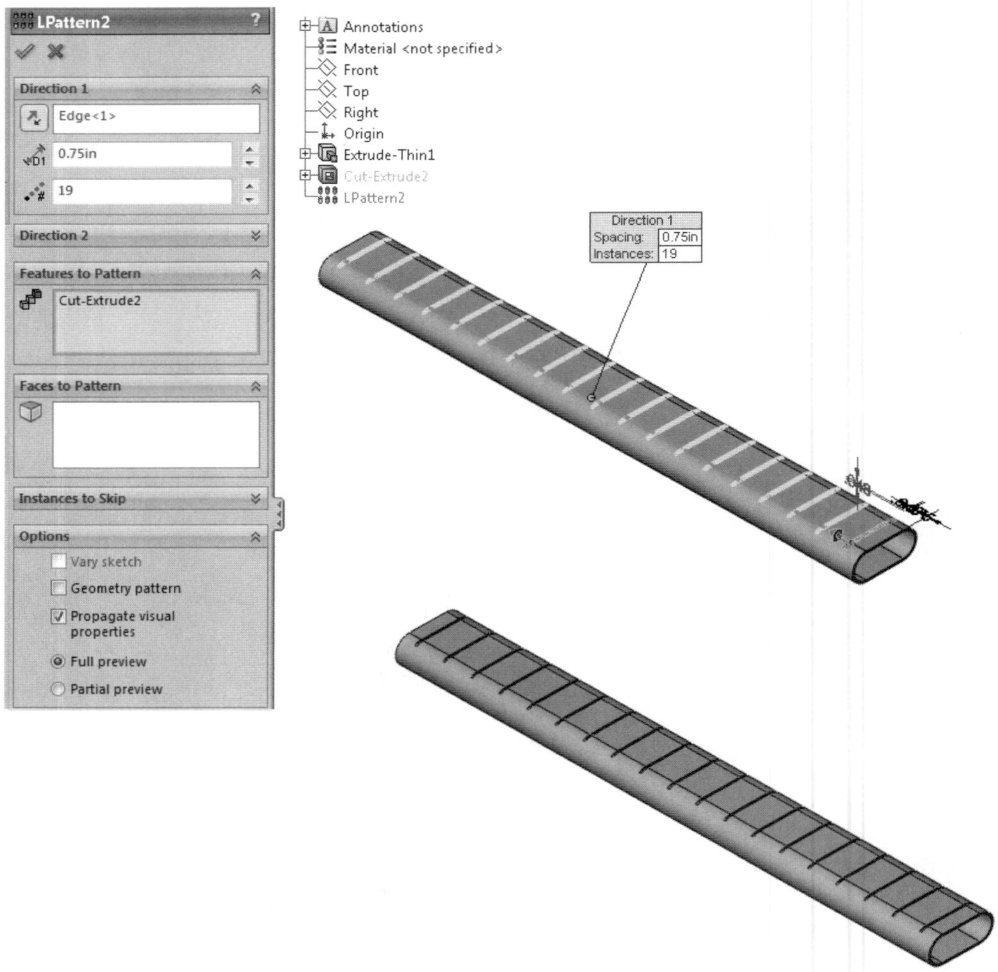

Figure 7-39
Make a Linear Pattern of 19 Instances with Spacing of .75 Inch in **Direction1**. Project 7-2d Complete

Step 28. Save your drawing (Figure 7-39) with the file name **Project 7-2d (your initials)** in at least two places.

Step 29. Open **Project 7-2c.**

Step 30. Draw a sketch for a **Base Flange** on the surface of **Project 7-2c** as shown in Figure 7-40.

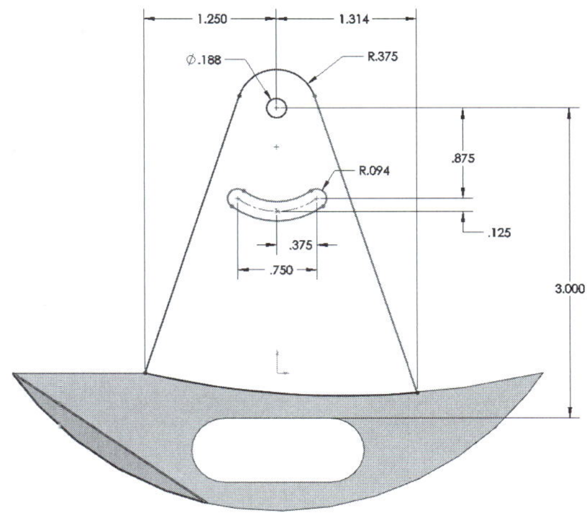

Figure 7-40
Draw a Sketch for a Base Flange on the Surface of Project 7-2c

Step 31. Extrude the base flange to 18 gauge steel as shown in Figure 7-41.

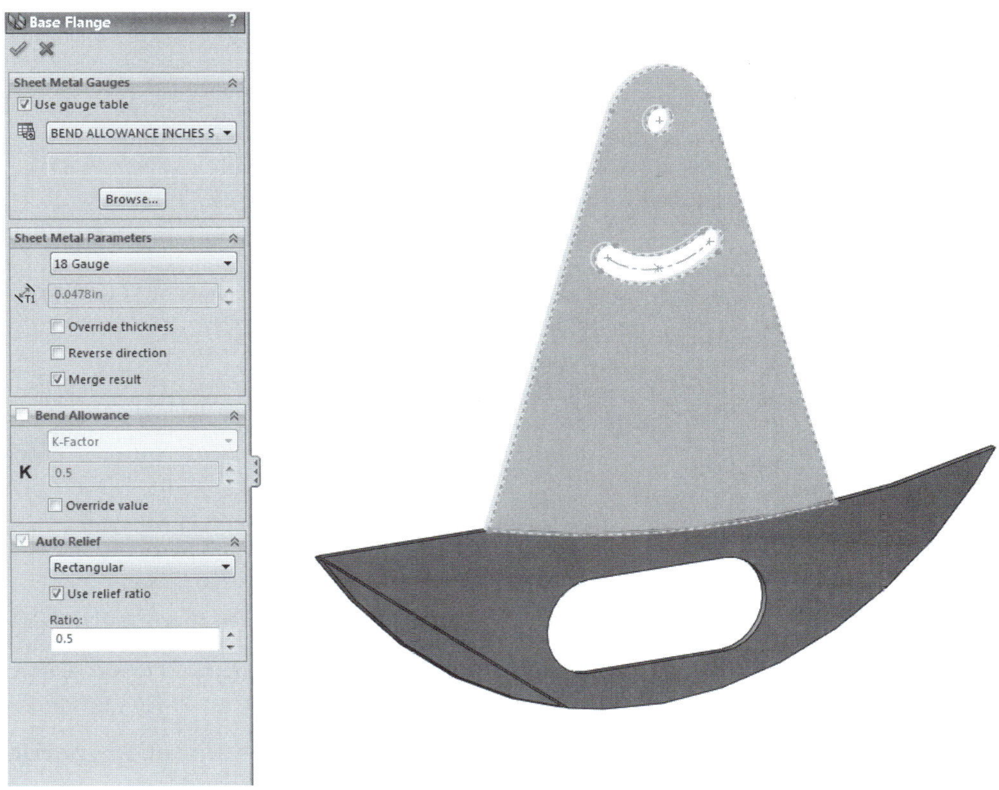

Figure 7-41
Extrude the Base Flange to 18 Gauge Steel

Step 32. Make .25 inch **Fillets** on two edges as shown in Figure 7-42.

Step 33. Save your drawing (Figure 7-42) with the file name **Project 7-2e (your initials)** in at least two places.

Step 34. Start a new drawing on the top plane with inches as the unit of measure.

Figure 7-42
Make .25 Inch Fillets on Two Edges.
Project 7-2e Complete

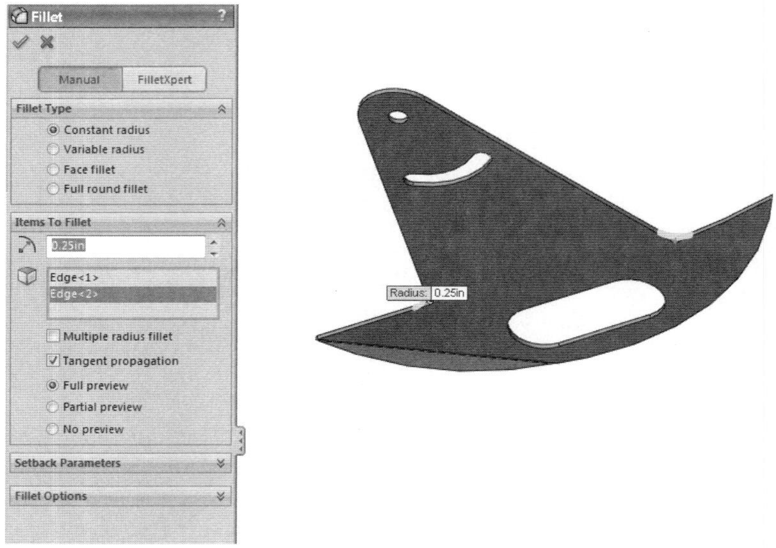

Step 35. Draw a sketch for a **Revolved Boss/Base** and revolve it as shown in Figure 7-43.

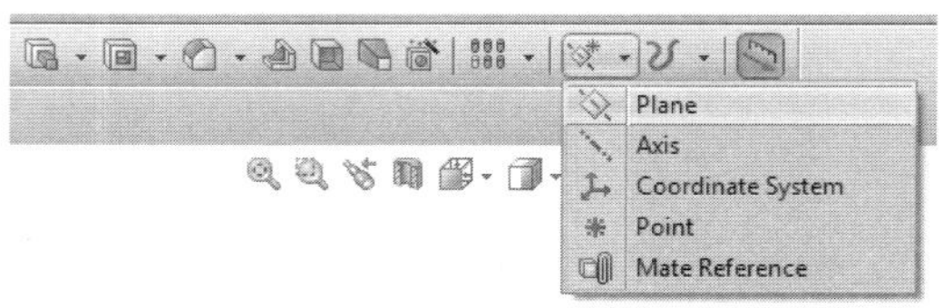

Figure 7-43
Draw a Sketch for a Revolved Boss/Base on the Front Plane and Revolve It. Add an Extruded Cut for the Screw Head. Apply Cosmetic Threads. .250-20 × 2.000 Lg Machine Screw Complete

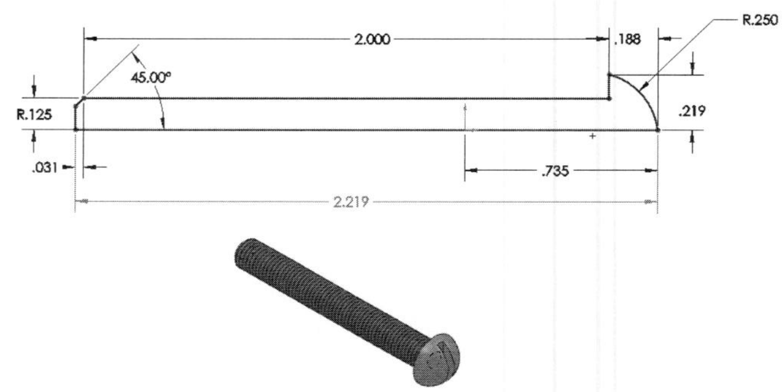

Step 36. Make a plane .188 inch from the bottom of the screw head.

Step 37. Draw a sketch for an **Extruded Cut** on **Plane1** and make the cut .125 inch deep as shown in Figure 7-43.

Step 38. Select **Options, Document Properties** tab, **Detailing Item,** and check **Cosmetic Thread.**

Step 39. Click **Cosmetic Thread** and click the bottom edge of the screw so that cosmetic threads appear as shown in Figure 7-43.

Step 40. Save your drawing with the file name **.250-20 × 2.000 Lg Machine Screw(your initials)** in at least two places.

Step 41. Start a new drawing on the top plane with inches as the unit of measure.

Step 42. Draw a sketch for a **Revolved Boss/Base** and revolve it as shown in Figure 7-44.

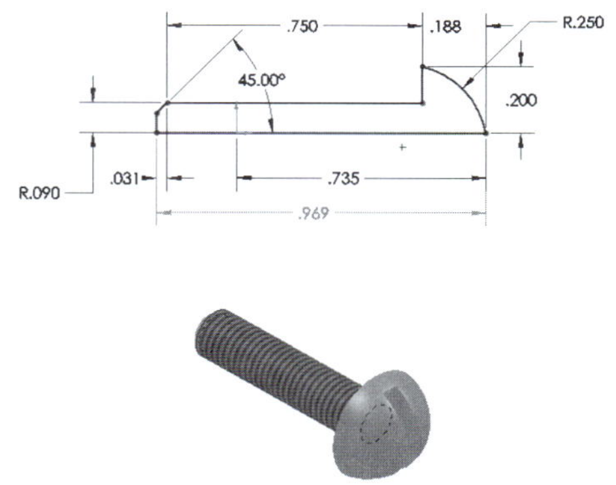

Figure 7-44
Draw a Sketch for a Revolved Boss/Base on the Front Plane and Revolve It. Add an Extruded Cut for the Screw Head. Apply Cosmetic Threads. .188-20 × .750 Lg Machine Screw Complete

Step 43. Make a plane .188 inch from the bottom of the screw head.

Step 44. Draw a sketch for an **Extruded Cut** on **Plane1** and make the cut .125 inch deep as shown in Figure 7-44.

Step 45. Select **Options, Document Properties** tab, **Detailing Item,** and check **Cosmetic Thread.**

Step 46. Click **Cosmetic Thread** and click the bottom edge of the screw so that cosmetic threads appear as shown in Figure 7-44.

Step 47. Save your drawing with the file name **.188-20 × .750 Lg Machine Screw(your initials)** in at least two places.

Step 48. Draw a sketch for an **Extruded Boss/Base** on the front plane and extrude it .05 inch as shown in Figure 7-45.

Step 49. Save your drawing with the file name **.250 Lock Washer(your initials)** in at least two places on your computer.

Step 50. Draw a sketch for an **Extruded Boss/Base** on the front plane and extrude it .05 inch as shown in Figure 7-46.

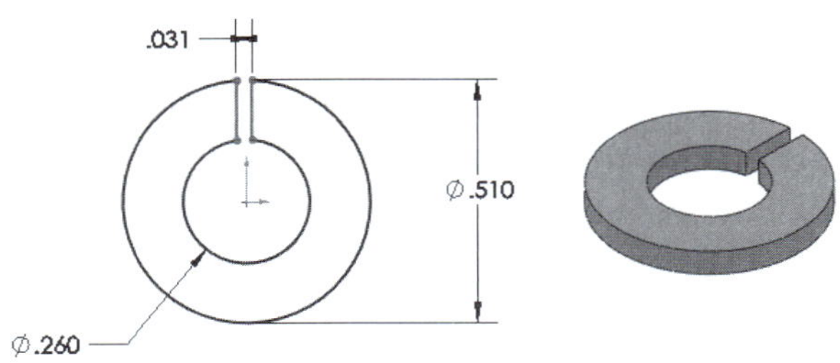

Figure 7-45
Make a Sketch for an Extruded Boss/Base on the Front Plane and Extrude It .05 Inch. .250 Lock Washer Complete

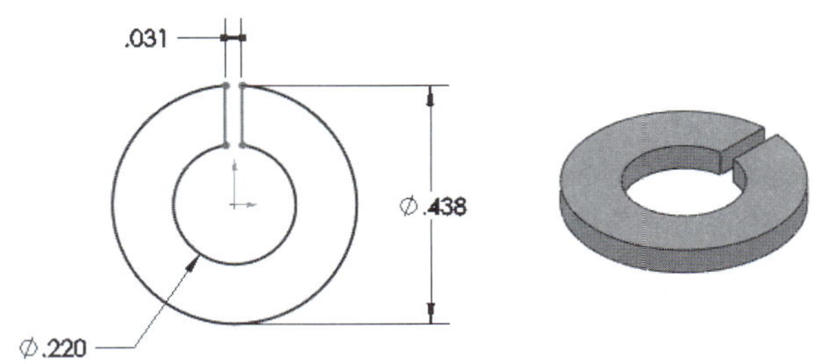

Figure 7-46
Make a Sketch for an Extruded Boss/Base on the Front Plane and Extrude It .05 Inch. .188 Lock Washer Complete

Step 51. Save your drawing with the file name **.188 Lock Washer(your initials)** in at least two places.

Step 52. Start a new drawing on the top plane with inches as the unit of measure.

Step 53. Draw a sketch (the hexagon) for an **Extruded Boss/Base** on the front plane and extrude it .30 inch from the mid plane.

Step 54. Select the front of the extrusion as the sketch plane and draw a .438 circle for an **Extruded Cut** as shown on the left side of Figure 7-47.

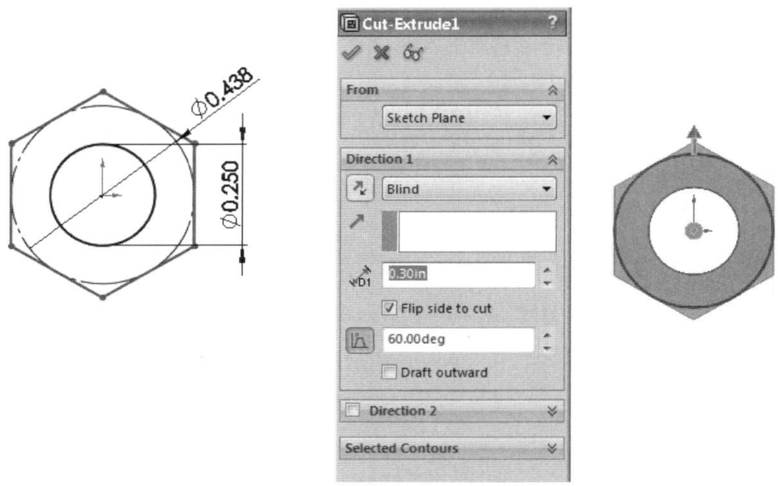

Figure 7-47
Draw a Sketch for an Extruded Boss/Base on the Front Plane and Extrude It .30 Inch from the Mid Plane. Make an Extruded Cut Using a .438 Diameter Circle, .30 Inch Deep with a Draft Angle of 60 Degrees

Step 55. Make an **Extruded Cut** on the front side of the extrusion with .30 inch as **Direction1** and a 60 degree draft angle as shown in Figure 7-47.

Step 56. **Mirror** the **Extruded Cut** using the top plane as the **Mirror Plane** as shown in Figure 7-48.

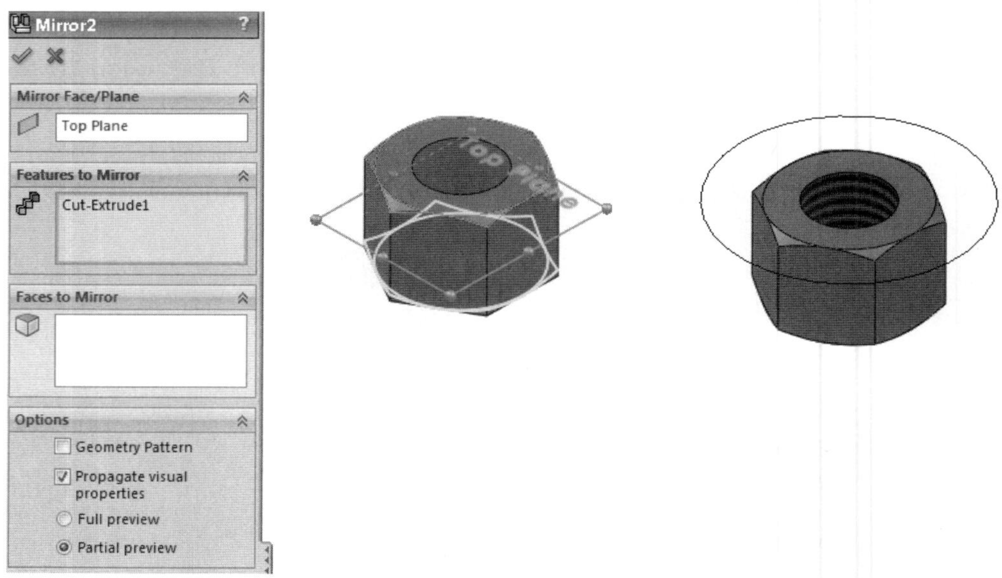

Figure 7-48
Mirror the Extruded Cut Using the Top Plane as the Mirror Plane. Apply Cosmetic Threads. .250-20 Nut Complete

Step 57. Apply **Cosmetic Threads** to the nut.

Step 58. Save your drawing with the file name **.250-20 Nut(your initials)** in at least two places.

Step 59. Start a new drawing on the top plane with inches as the unit of measure.

Step 60. Draw a sketch (the hexagon) for an **Extruded Boss/Base** on the front plane and extrude it .30 inch from the mid plane.

Step 61. Select the front of the extrusion as the sketch plane and draw a .329 circle for an **Extruded Cut** as shown on the left side of Figure 7-49.

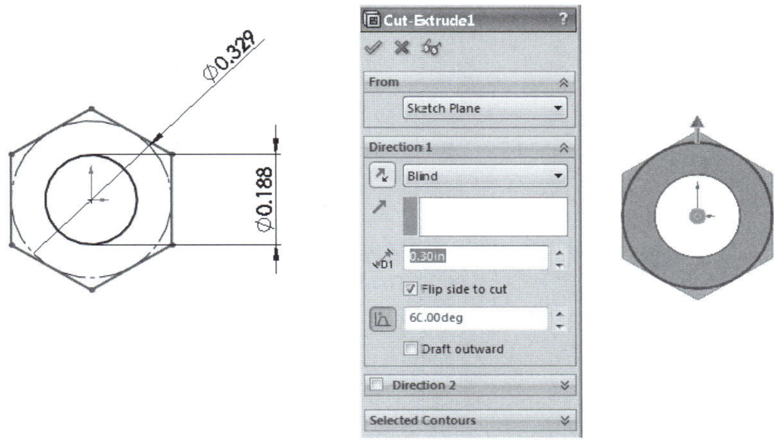

Figure 7-49
Draw a Sketch for an Extruded Boss/Base on the Front Plane and Extrude It .30 Inch from the Mid Plane Make an Extruded Cut Using a .329 Diameter Circle, .30 Inch Deep with a Draft Angle of 60 degrees

Step 62. Make an **Extruded Cut** on the front side of the extrusion with .30 inch as **Direction1** and a 60 degree draft angle as shown in Figure 7-49.

Step 63. **Mirror** the **Extruded Cut** using the top plane as the **Mirror Plane** as shown in Figure 7-50.

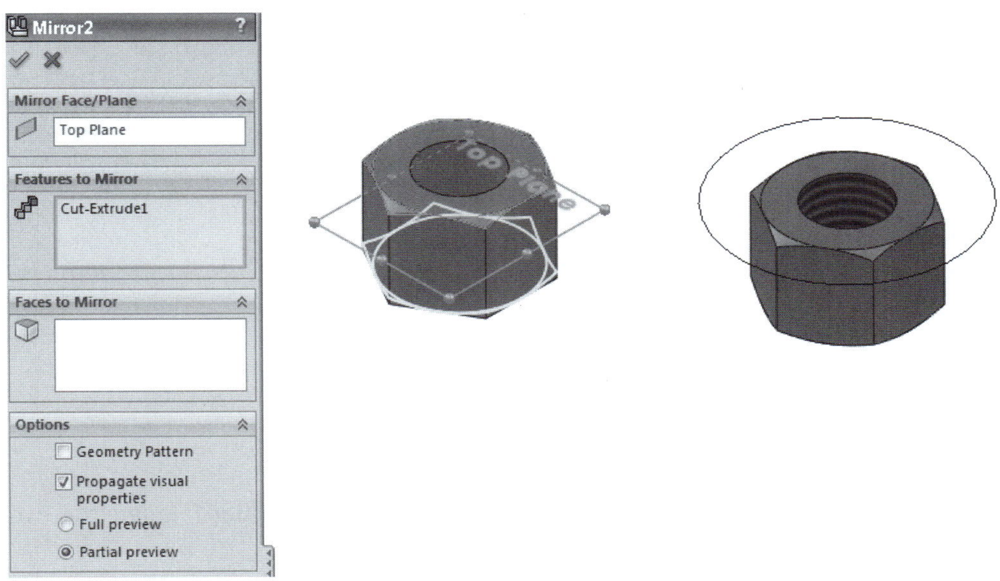

Figure 7-50
Mirror the Extruded Cut Using the Top Plane as the Mirror Plane. Apply Cosmetic Threads. .188-20 Nut Complete

Step 64. Apply **Cosmetic Threads** to the nut.

Step 65. Save your drawing with the file name **.188-20 Nut(your initials)** in at least two places.

Step 66. Open **Project 7-2d.**

Step 67. Insert **Project 7-2e** twice and **Project 7-2c** as shown in Figure 7-51.

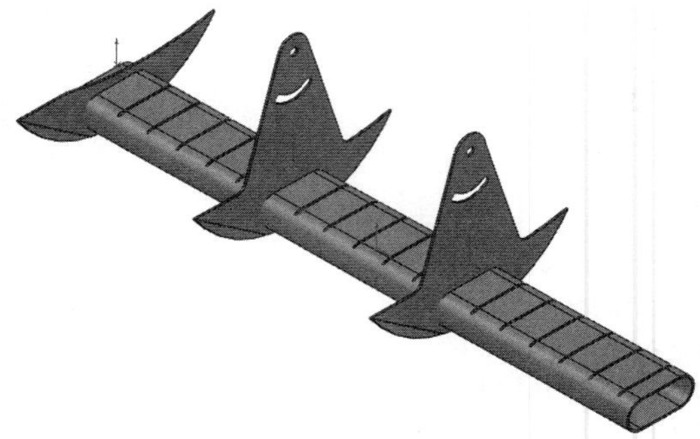

Figure 7-51
Insert Project 7-2e Twice and Project 7-2c and Apply Concentric and Coincident Mates

Step 68. Use **Linear Pattern** to pattern **Project 7-2c** 19 instances in **Direction1** with .75 inch spacing as shown in Figure 7-52.

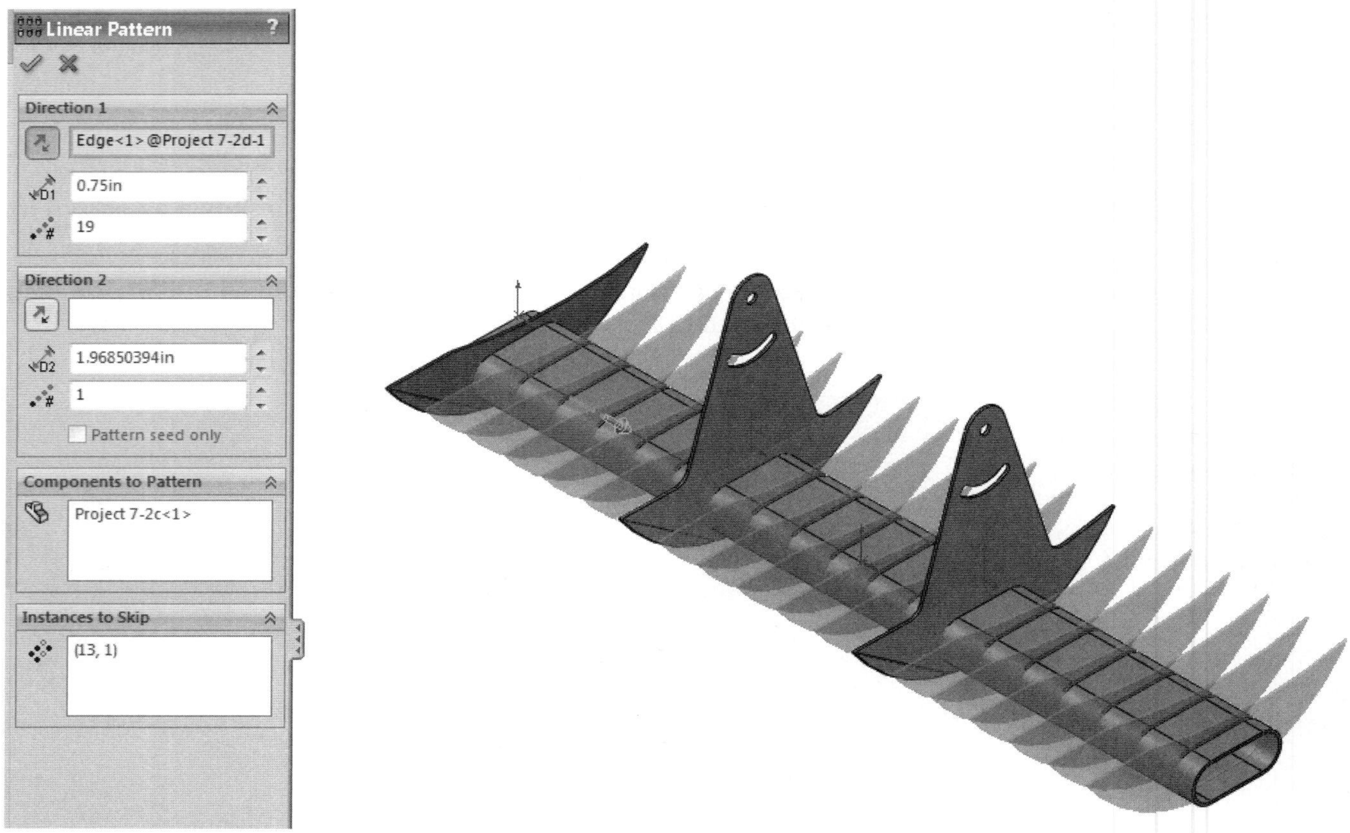

Figure 7-52
Make a Linear Pattern of 19 Instances and .75 Inch Spacing in **Direction1**. Delete the Two Project 7-2c's in Spaces Where Project 7-2e's Are

Step 69. Delete the two **Project 7-2c**'s in spaces where **Project 7-2e**'s are located.

Step 70. Insert **Project 7-2b** twice and apply **Concentric** and **Coincident** mates as shown in Figure 7-53.

Figure 7-53
Insert Project 7-2b Twice and Apply Concentric and Coincident Mates

Step 71. Insert **Project 7-2a**(the oak handle).

Step 72. Apply a concentric mate to one of the .250 inch holes and a tangent mate between the handle and one **Project 7-2b** as shown in Figure 7-54.

Step 73. Insert **.188-20 × .750 Lg Machine Screw, .188 Lock Washer,** and **.188-20 Nut** as shown in Figure 7-55.

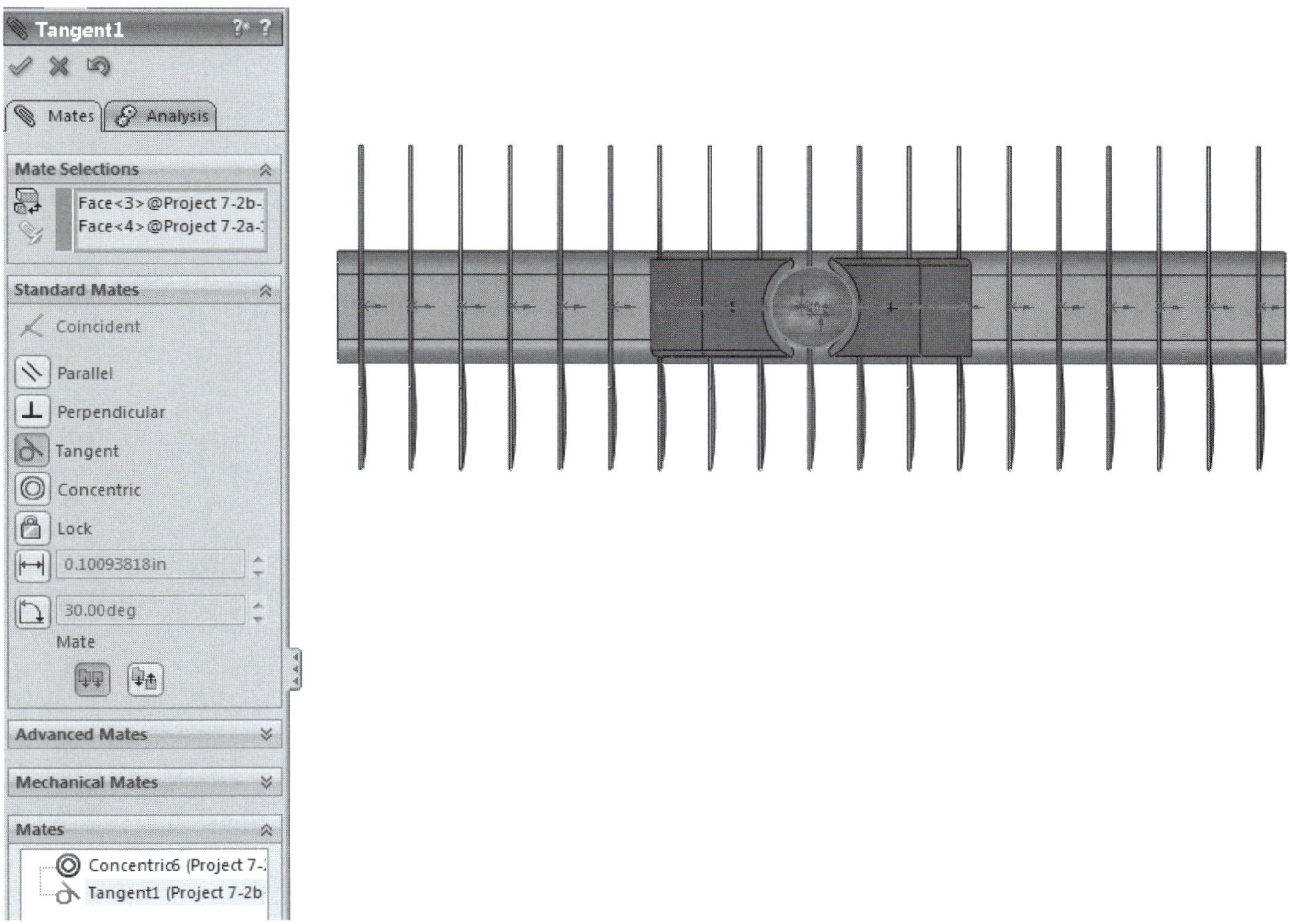

Figure 7-54
Insert Project 7-2a (the Handle) and Apply a Concentric Mate to One of the .250 Inch Holes and a Tangent Mate Between the Handle and One Project 7-2b

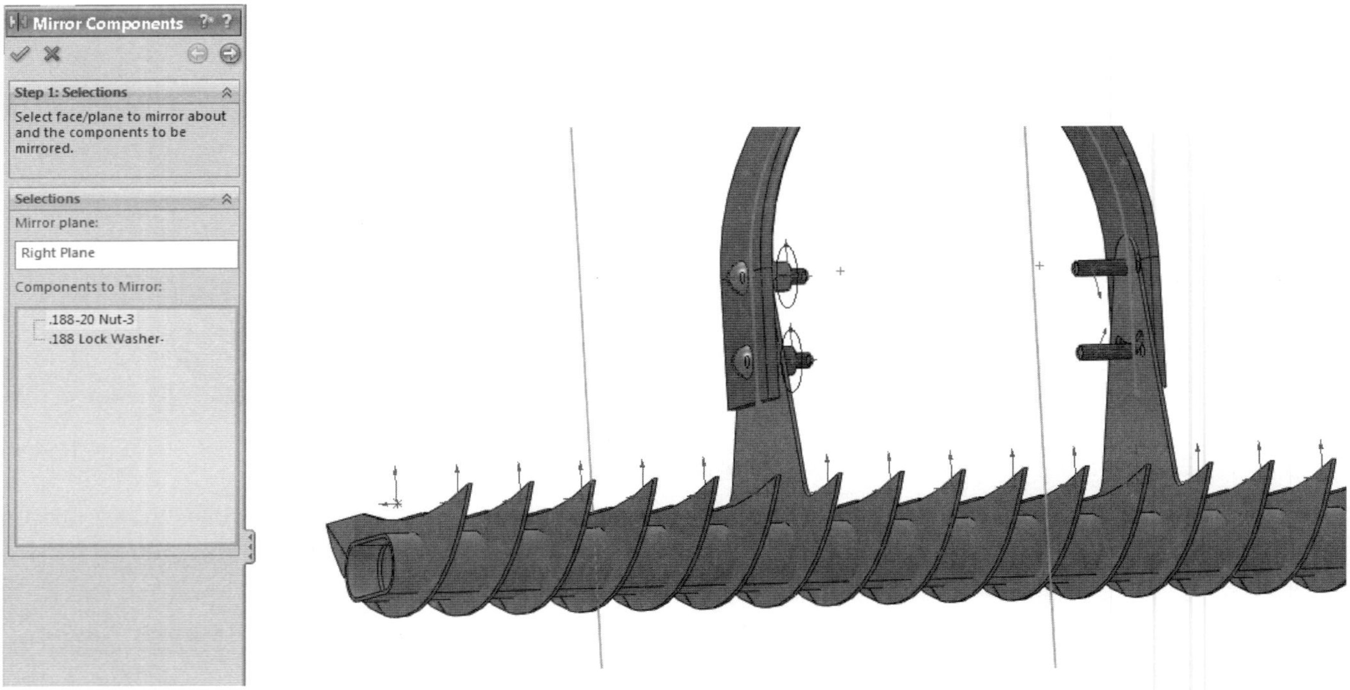

Figure 7-55
Insert .188-20 × .750 Lg Machine Screw, .188 Lock Washer and .188-20 Nut and Apply Concentric and Coincident Mates. Mirror the Inserted Components Using the Right Plane as the Mirror Plane

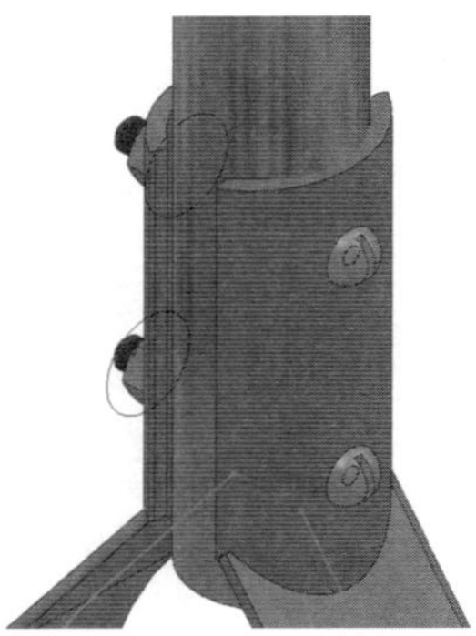

Figure 7-56
Insert .250-20 × 2.000 Lg Machine Screw, .250 Lock Washer and .250-20 Nut and Apply Concentric and Coincident Mates

Step 74. Apply **Concentric** and **Coincident** mates to the fasteners and **Projects 7-2b** and **7-2e** as shown in Figure 7-55.

Step 75. **Mirror** the inserted fasteners using the right plane as the **Mirror Plane** as shown in Figure 7-55.

Step 76. Insert **.250-20 × 2.000 Lg Machine Screw, .250 Lock Washer,** and **.250-20 Nut** as shown in Figure 7-56.

Step 77. Apply **Concentric** and **Coincident** mates to the fasteners and **Project 7-2a** as shown in Figure 7-56.

Step 78. Save your assembly (Figure 7-57) with the file name **Project 7-2 (your initials)** in at least two places.

Step 79. Print the drawing and turn it in.

Figure 7-57
Project 7-2 Top Assembly Complete

Exercises

EXERCISE 7-1 Make a vent assembly.

Step 1. Open Exercise 6-2a and select **Make Assembly from Part.**

Step 2. Insert **Exercises 6-2b, 6-2c, 6-2d,** and **6-2e.**

Step 3. Apply **Concentric** and **Coincident** mates to **Exercises 6-2a, b, c,** and **d** as shown in Figure 7-58.

Figure 7-58
Open Exercise 6-2a and Select Make Assembly from Part. Insert Exercises 6-2b, c, d and e. Apply Concentric and Coincident Mates

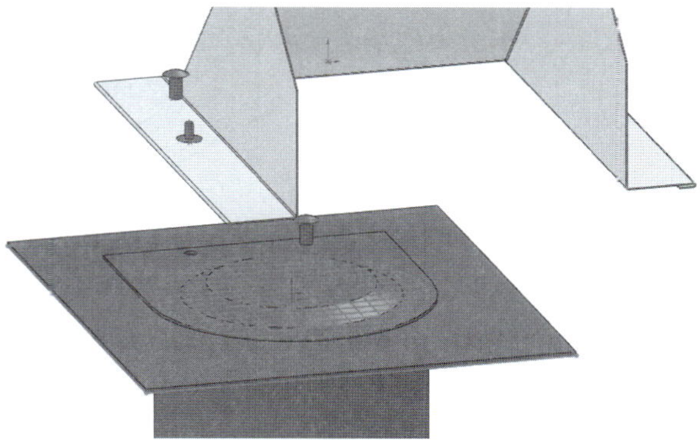

Step 4. Apply **Coincident** and **Parallel** mates to **Exercises 6-2a** and **e** as shown in Figure 7-59.

Figure 7-59
Apply Coincident and Parallel Mates. Exercise 7-1 Complete

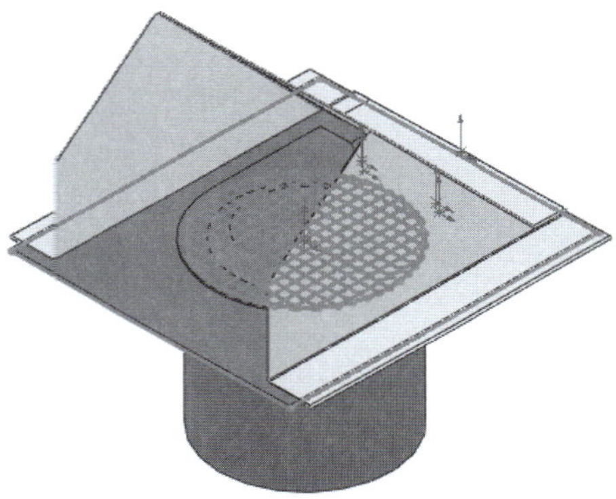

Step 5. Save the drawing (Figure 7-59) with the file name **Exercise 7-1 (your initials)** in at least two places.

Step 6. Print the drawing and turn it in.

EXERCISE 7-2 Make a drawing with a Bill of Materials for a vent assembly.

Step 1. Open **Exercise 7-1,** select a **C-Size Format,** and drag the isometric view onto the format from the **View** palette as shown in Figure 7-60.

Step 2. Right-click in the field of the format and click the word **Tables.** Then, click **Bill of Materials...** and click inside the view. Then, click **OK** and the **Bill of Materials** appears so that you can place it on the drawing as shown in Figure 7-61.

Figure 7-60
Drag Isometric View onto the Sheet Format

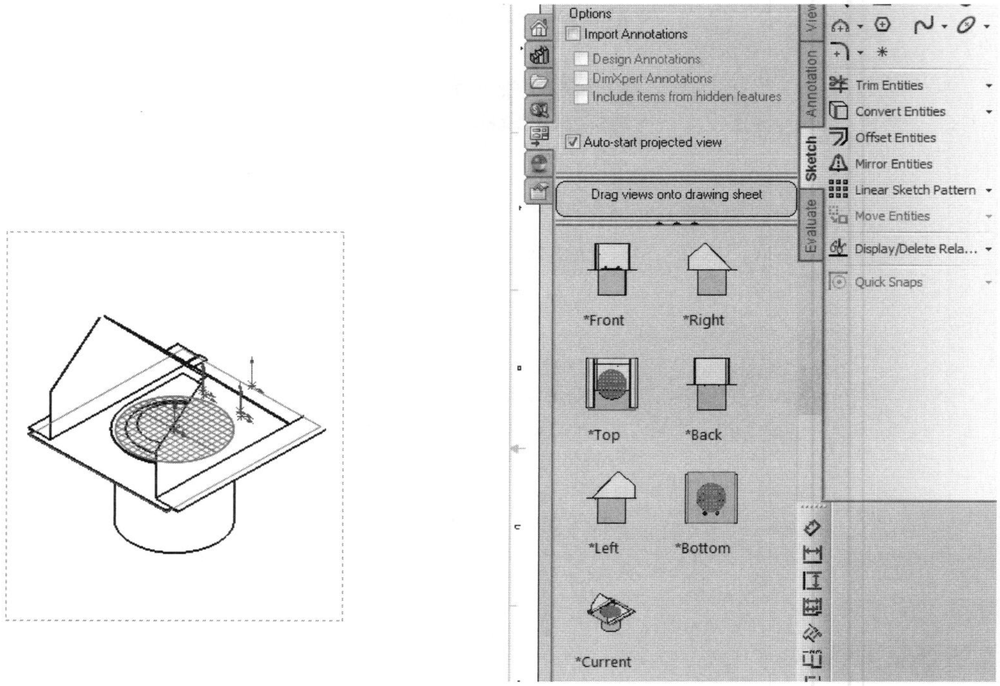

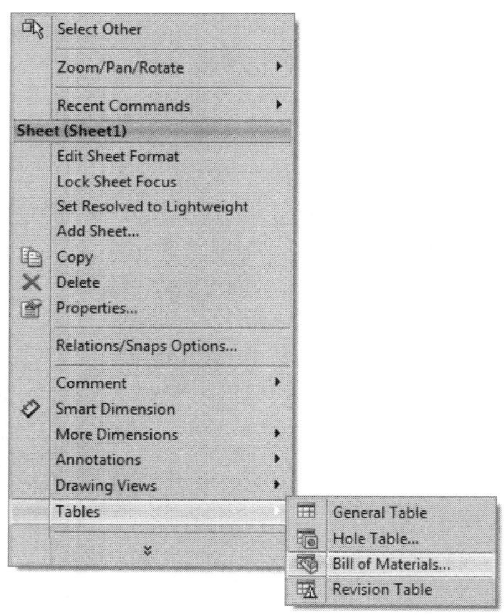

Figure 7-61
Make a Bill of Materials, Add Balloons, and Edit Sheet Format.
Exercise 7-2 Complete

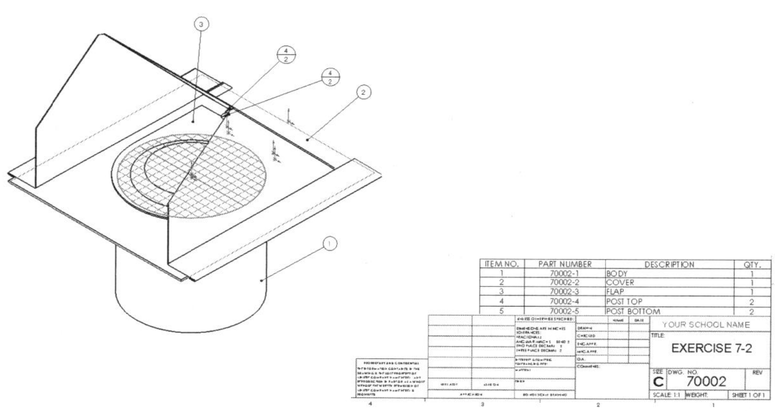

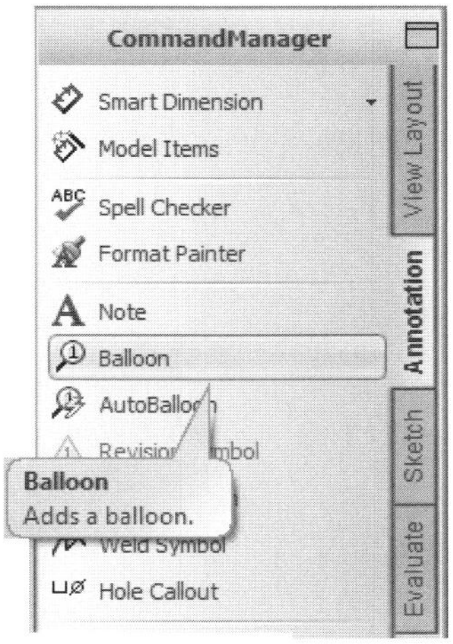

Step 3. Click **Balloon** from the **Annotation** tab on the **Command Manager.** The **Circular** balloon is the default, so if only one character is needed in each balloon all you have to do is click on the part and place the balloon. If you need to show the item number and its quantity, you will need to click **Circular Split Line.** To change the text in a **Circular** balloon, double-click the text and change it.

To change the text in a **Circular Split Line,** double-click the text, change the item number, and click **OK** to exit from the command. To change the quantity after you have changed the item number, click on the item number ONLY ONCE and change the text under **Lower text:** in the **Balloon** manager as shown in Figure 7-61.

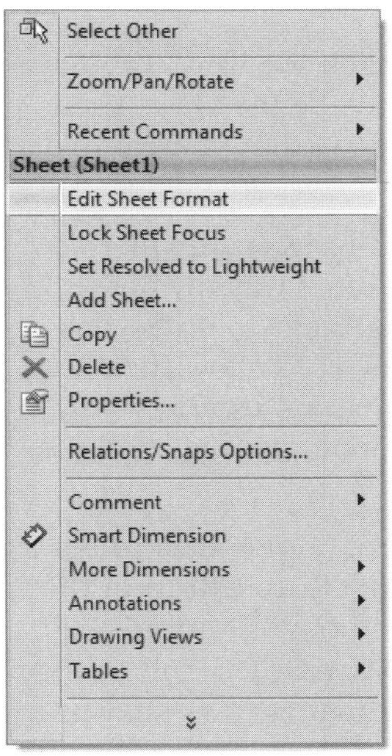

Step 4. Right-click in the drawing, click **Edit Sheet Format,** and make the changes as shown in Figure 7-61.

Step 5. Right-click in the drawing and click **Edit Sheet.**

Step 6. Save the drawing (Figure 7-61) with the name **Exercise 7-2(your initials)** in at least two places.

Step 7. Print the drawing and turn it in.

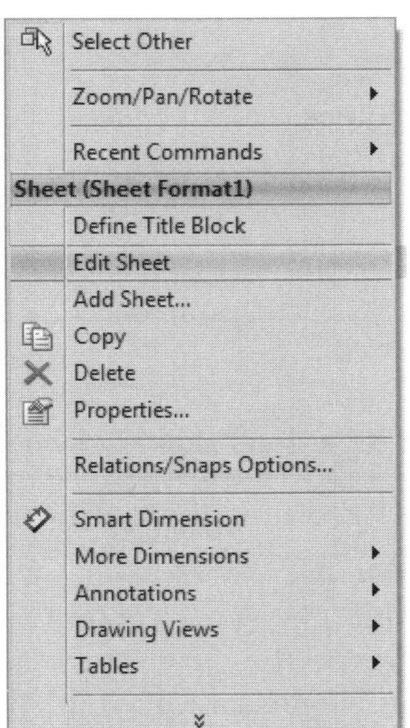

EXERCISE 7-3 Make a mail box assembly.

Step 1. Open **Exercise 6-3a** and select **Make Assembly from Part.**

Step 2. Insert **Exercises 6-3b** and **6-3c.**

Step 3. Apply **Concentric, Coincident,** and **Parallel** mates to **Exercises 6-3a, b,** and **d** as shown in Figure 7-62.

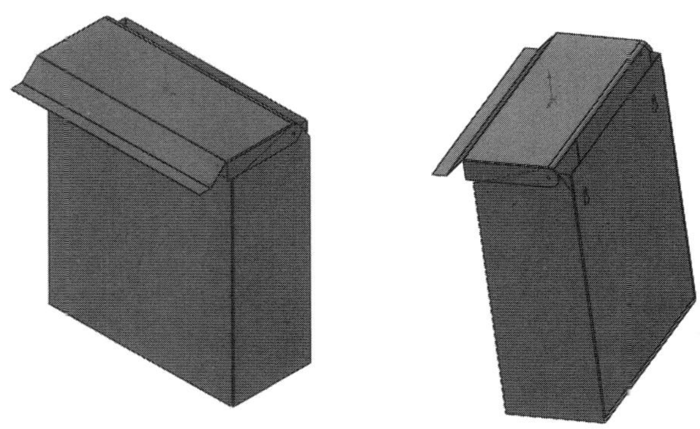

Figure 7-62
Open Exercise 6-3a and Select Make Assembly from Part. Insert Exercises 6-3b and c. Apply Coincident, Concentric and Parallel Mates.
Exercise 7-3 Complete

Step 4. Save the drawing (Figure 7-62) with the file name **Exercise 7-3 (your initials)** in at least two places.

Step 5. Print the drawing and turn it in.

EXERCISE 7-4 Make a drawing with a Bill of Materials for a mail box assembly.

Step 1. Open **Exercise 7-3,** select a **B-Size Format,** and drag the isometric view onto the format from the **View** palette as shown in Figure 7-63.

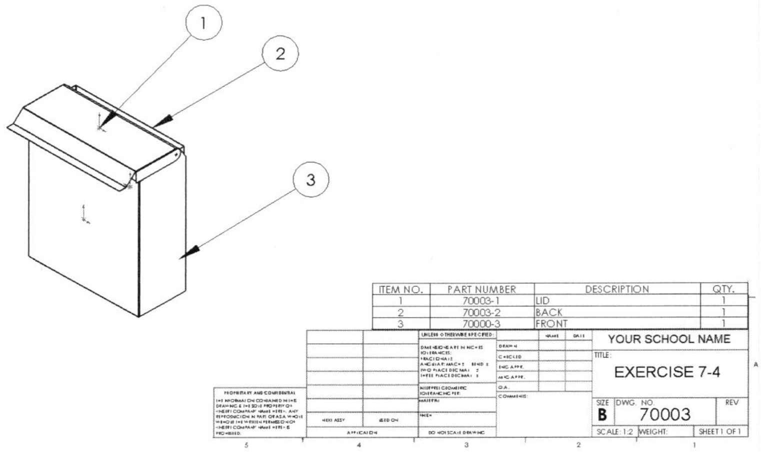

Figure 7-63
Make a Bill of Materials, Add Balloons, and Edit Sheet Format. Exercise 7-4 Complete

Step 2. Right-click in the field of the format and click the word **Tables.** Then, click **Bill of Materials...** and click inside the view. Then, click **OK** and the **Bill of Materials** appears so that you can place it on the drawing as shown in Figure 7-63.

Step 3. Click **Balloon** from the **Annotation** tab on the **Command Manager.** The **Circular** balloon is the default, so if only one character is needed in each balloon, all you have to do is click on the part and place the balloon. If you need to show the item number and its quantity, you will need to click **Circular Split Line.** To change the text in a **Circular** balloon, double-click the text and change it.

To change the text in a **Circular Split Line,** double-click the text and change the item number and click **OK** to exit from the command. To change the quantity after you have changed the item number, click on the item number ONLY ONCE and change the text under **Lower text:** in the **Balloon** manager as shown in Figure 7-61.

Step 4. Right-click in the drawing, click **Edit Sheet Format,** and make the changes as shown in Figure 7-63.

Step 5. Right-click in the drawing and click **Edit Sheet.**

Step 6. Save the drawing (Figure 7-63) with the file name **Exercise 7-4 (your initials)** in at least two places.

Step 7. Print the drawing and turn it in.

EXERCISE 7-5 Make a drawing with a Bill of Materials for a street sign assembly.

Step 1. Open **Project 7-1,** select an **A-Size Format,** and drag the isometric view onto the format from the **View** palette.

Step 2. Right-click in the field of the format and click the word **Tables.** Then, click **Bill of Materials...** and click inside the view. Then, click **OK** and the **Bill of Materials** appears so you can place it on the drawing.

Step 3. Click **Balloon** from the **Annotation** tab on the **Command Manager.** The **Circular** balloon is the default, so if only one character is needed in each balloon all you have to do is click on the part and place the balloon. If you need to show the item number and its quantity, you will need to click **Circular Split Line.** For items that do not show (3 and 5), use the **Circular Split Line** balloon and click any open space so that a question mark appears. After you have the balloon with the question mark, move them as shown in Figure 7-64. To change the text in a **Circular** balloon, double-click the text and change it.

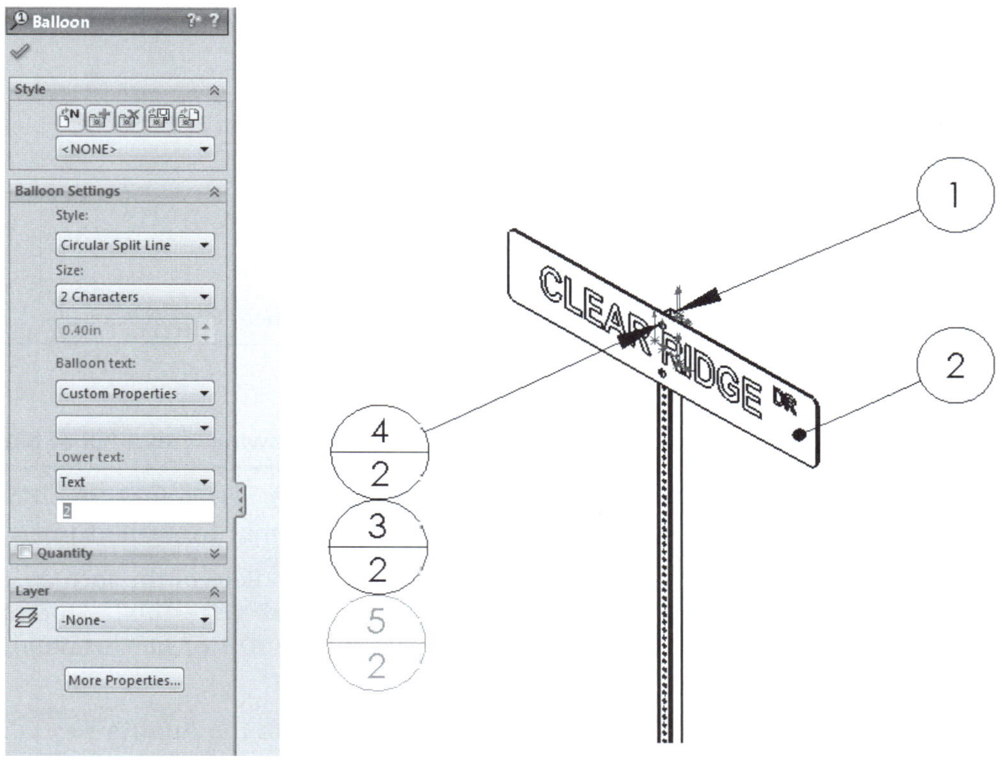

Figure 7-64
Add Balloons. For Items that Do Not Show (3 and 5), Use the Circular Split Line and Click: Any Open Space so a Question Mark Appears. Change that Item Number and Use Lower Text to Change the Quantity

To change the text in a **Circular Split Line,** double-click the text, change the item number, and click **OK** to exit from the command. To change the quantity after you have changed the item number, click on the item number ONLY ONCE and change the text under **Lower text:** in the **Balloon** manager as shown in Figure 7-64.

Step 4. Right-click in the drawing, click **Edit Sheet Format,** and make the changes as shown in Figure 7-65.

Step 5. Right-click in the drawing and click **Edit Sheet.**

Step 6. Save the drawing (Figure 7-65) with the file name **Exercise 7-5 (your initials)** in at least two places.

Step 7. Print the drawing and turn it in.

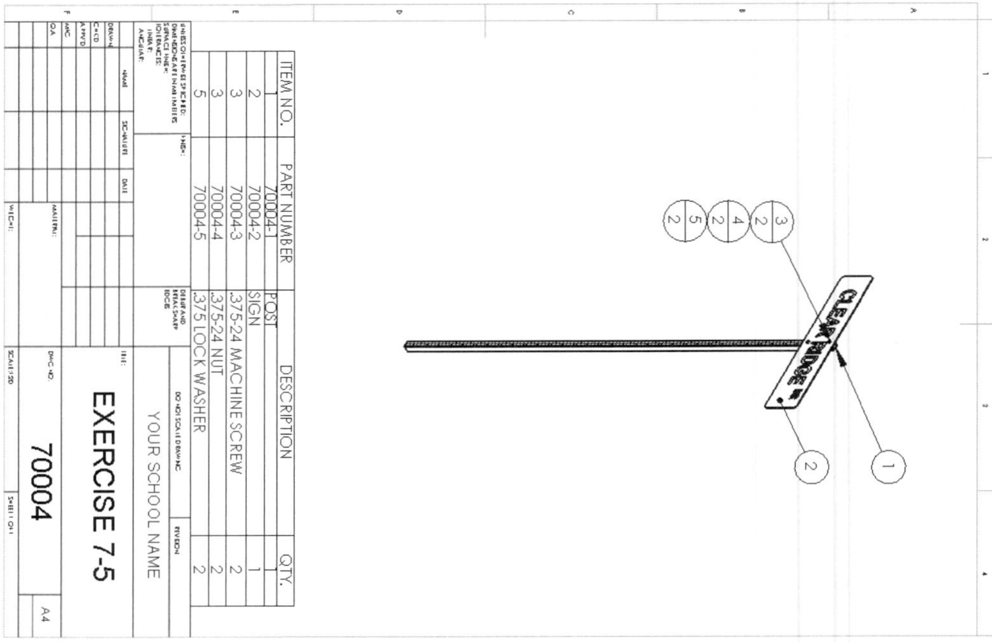

Figure 7-65
Make a Bill of Materials. Edit Sheet Format. Exercise 7-5 Complete

EXERCISE 7-6 **Make a drawing with a Bill of Materials for a tiller assembly.**

Step 1. Open **Project 7-2,** select an **A-Size Format,** and drag the isometric view onto the format from the **View** palette.

Step 2. Right-click in the field of the format and click the word **Tables.** Then, click **Bill of Materials...** and click inside the view. Then, click **OK** and the **Bill of Materials** appears so that you can place it on the drawing.

Step 3. Click **Balloon** from the **Annotation** tab on the **Command Manager.** The **Circular** balloon is the default, so if only one character is needed in each balloon all you have to do is click on the part and place the balloon. If you need to show the item number and its quantity, you will need to click **Circular Split Line.** For items that do not show, use the **Circular Split Line** balloon and click any open space so that a question mark appears. After you have the balloon with the question mark, move them as shown in Figure 7-66. To change the text in a **Circular** balloon, double-click the text and change it.

To change the text in a **Circular Split Line,** double-click the text, change the item number, and click **OK** to exit from the command. To change the quantity after you have changed the item number, click on the item number ONLY ONCE and change the text under **Lower text:** in the **Balloon** manager as shown in Figure 7-66.

Step 4. Right-click in the drawing, click **Edit Sheet Format,** and make the changes as shown in Figure 7-66.

Step 5. Right-click in the drawing and click **Edit Sheet.**

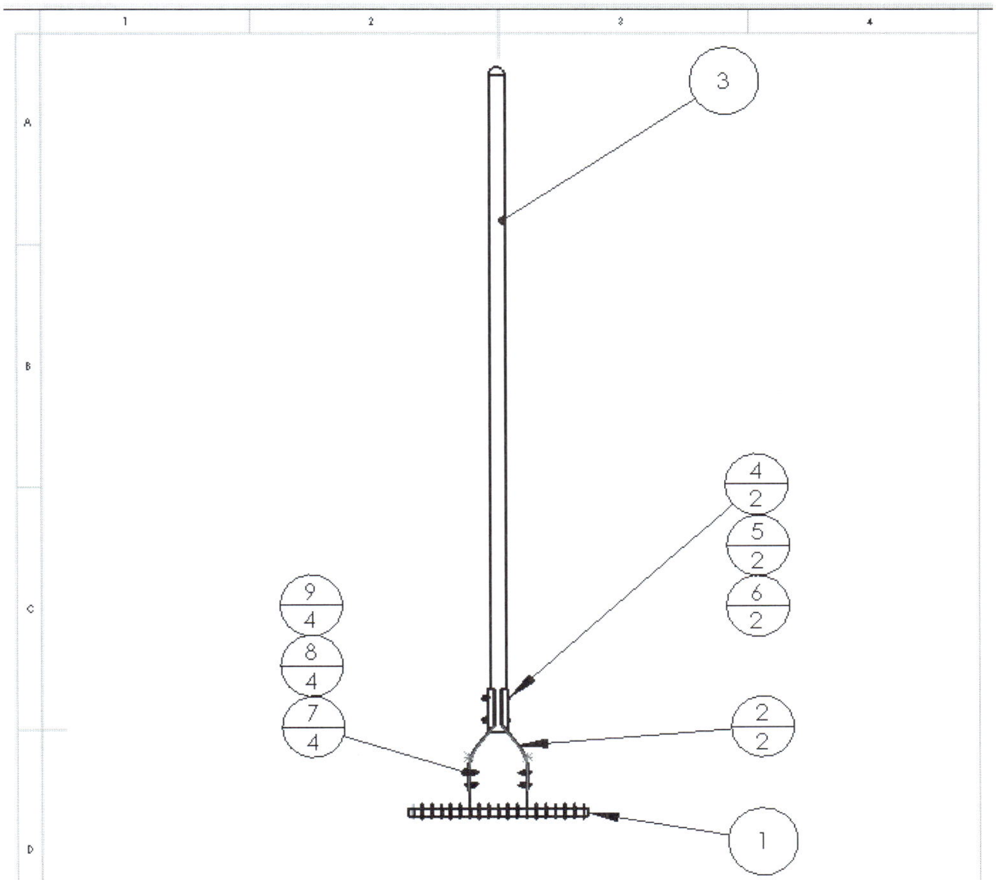

ITEM NO.	PART NUMBER	DESCRIPTION	QTY.
1	70006	BLADE SUBASSEMBLY	1
2	70005-1	POST BRACKET	2
3	70005-2	POST	1
4	70005-3	.25-20 X 2.00 LG MACHINE SCREW	2
5	70005-4	.25-20 Nut	2
6	70005-5	.25 LOCK WASHER	2
7	70005-6	.188-20 X .750 LG MACHINE SCREW	4
8	70005-7	.188-20 NUT	4
9	70005-8	.188 LOCK WASHER	4

YOUR SCHOOL NAME

TITLE: EXERCISE 7-6

DWG NO. 70005 A4

SCALE: 1:20 SHEET 1 OF 1

Figure 7-66
Make a Bill of Materials. Add Balloons, and Edit Sheet Format.
Exercise 7-6 Complete

Step 6. Save the drawing (Figure 7-66) with the file name **Exercise 7-6 (your initials)** in at least two places.

Step 7. Print the drawing and turn it in.

7 seven

Chapter Summary

This chapter contains additional information necessary to build solid models of sheet metal parts. It also contains information on making assemblies and drawings with bills of materials. Now you have the skills and information necessary to produce more complex solid models of sheet metal parts, assemblies, and drawings with bills of materials.

Review Questions

Circle the most correct answer.

1. Which of the following can shorten the time needed to make an assembly?
   a. Do not open any part
   b. Open only subassemblies
   c. Open all the parts of the assembly
   d. Open only the major part
2. Which of the following must be clicked to apply a **Concentric** mate?
   a. Cylindrical surfaces
   b. Edges
   c. Flat surfaces
   d. Perpendicular surfaces
3. Which of the following must be clicked to apply a **Coincident** mate?
   a. Cylindrical surfaces
   b. Edges
   c. Flat surfaces
   d. Perpendicular surfaces
4. Which of the following must be clicked to apply a **Parallel** mate?
   a. Cylindrical surfaces
   b. Edges
   c. Flat surfaces
   d. Perpendicular surfaces
5. Which item on the right-click menu that appears after the view has been placed in the drawing contains the **Bill of Materials**?
   a. **Annotations**
   b. **Drawing Views**
   c. **Tables**
   d. **Edit Sheet Format**
6. The drawing view must be clicked to make a **Bill of Materials**.
   a. True
   b. False
7. **Circular Split Line** balloons contain only one character.
   a. True
   b. False
8. Circular balloons show the quantity of the item.
   a. True
   b. False
9. The **Sketched Bend** command can be used to make a bend.
   a. True
   b. False
10. **Convert Entities** cannot be used to make a sketch.
    a. True
    b. False

7 chapter seven

Matching

Place the correct letter on the line to the right of the numbered item.

1. A feature that can be used to make a cut at an angle _____
2. A command that allows you to make threads on a cylinder _____
3. A command that is used to assemble parts _____
4. A heading that must be clicked to allow the **Cosmetic Thread** box to be checked _____
5. A command that makes a bend on the edge of a sheet metal part _____

a. **Cosmetic Thread**

b. **Draft Angle**

c. Detailing item

d. **Sketched Bend**

e. **Mates**

General Questions

Answer the following questions.

1. In what situations would you use **Sketched Bend**?
2. When would you want to use a **Bill of Materials**?
3. Why are assemblies needed?
4. Why are **Circular Split Line** balloons needed?
5. Why are the mates needed?

chapter eight

8 Making a Chassis, a Front Panel, a Rear Panel, and a Cover for a Sheet Metal Assembly

CHAPTER OBJECTIVES

After completing this chapter, you will be able to:

- Make a chassis to specific 3-place decimal-inch measurements for a sheet metal assembly using the **Base Flange/Tab** command and add countersunk holes using the **Hole Wizard.**
- Make a front panel to specific 3-place decimal-inch measurements for a sheet metal assembly using **Rectangle, Straight Slot,** and add countersunk and counterbored holes using the **Hole Wizard.**
- Make a rear panel to specific 3-place decimal-inch measurements for a sheet metal assembly using **Rectangle** and add countersunk and clearance holes using the **Hole Wizard** and **Extrude Cut.**
- Make a cover to specific 3-place decimal-inch measurements for a sheet metal assembly using **Rectangle,** and add countersunk and clearance holes using the **Hole Wizard** and **Extrude Cut.**
- Make drawings from the sheet metal parts and correctly space the 2-place and 3-place decimal dimensions and annotations that are inserted or added.
- Correctly answer questions regarding SolidWorks sheet metal and drawing commands used in this chapter.

Project 8-1: Making the Chassis

Figure 8-1 shows the completed chassis.

Begin the chassis as described next:

Step 1. Click **New....**

Step 2. Click **Part** and click **OK.**

Step 3. Click **Options** on the menu bar, select the **Document Properties** tab, and click **Units** in the list to the left.

Figure 8-1
Chassis Complete

Step 4. Make sure the **Units** are correctly set as shown in Figure 8-2 [notice that **IPS** (inch, pound, second) is checked], then click **OK.**

Figure 8-2
Set Units to Inches

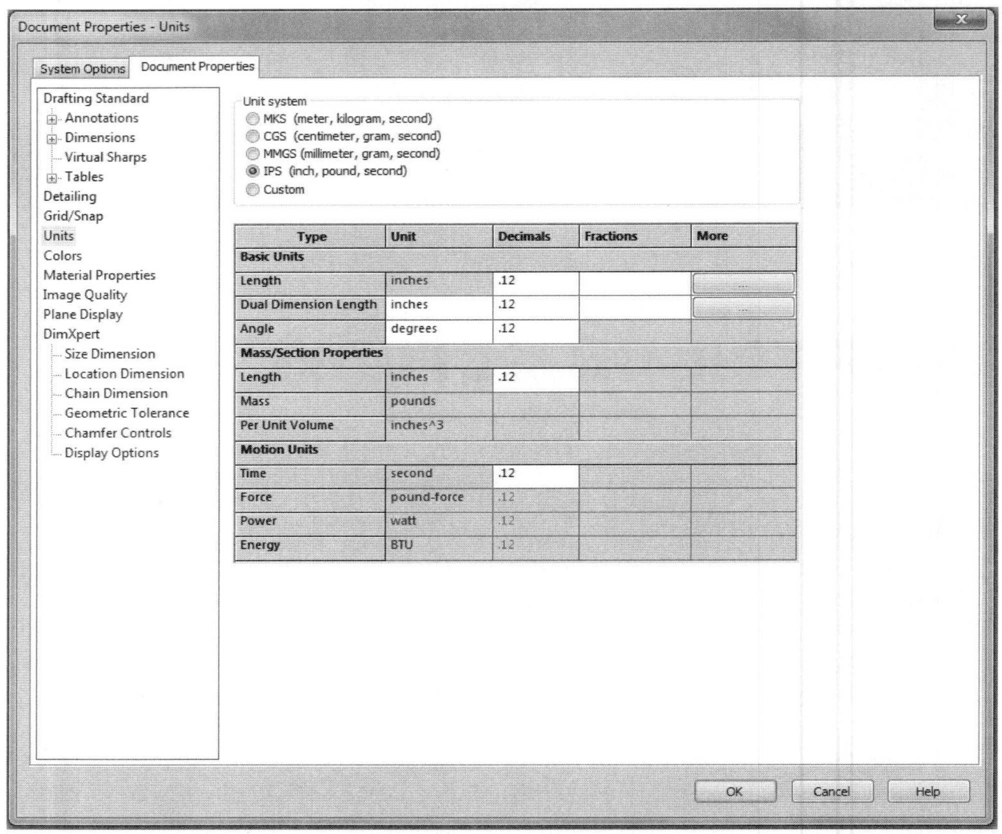

Step 5. Use the **Line, Add Relation**, and **Smart Dimension** commands to draw Figure 8-3 on the front plane.

Step 6. Click **Exit Sketch** and make the settings as shown in Figure 8-4. Make sure to type **6.75** in the **D1** text box to make the chassis 6.75 inches long and **0.09** in **T1** under **Sheet Metal Parameters** so that the metal thickness is .09 inch.

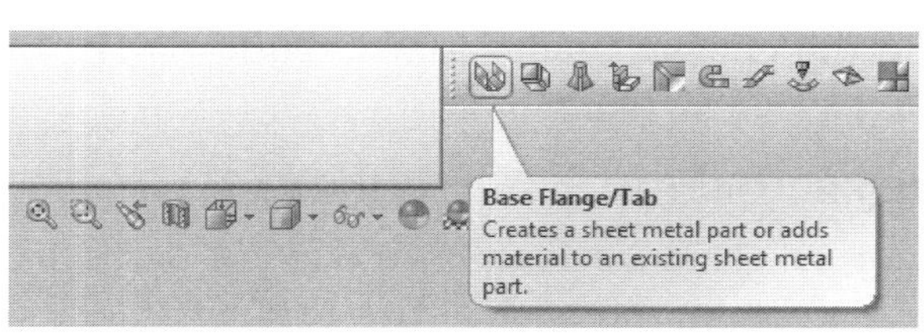

Figure 8-3
Make a Sketch for a Base Flange

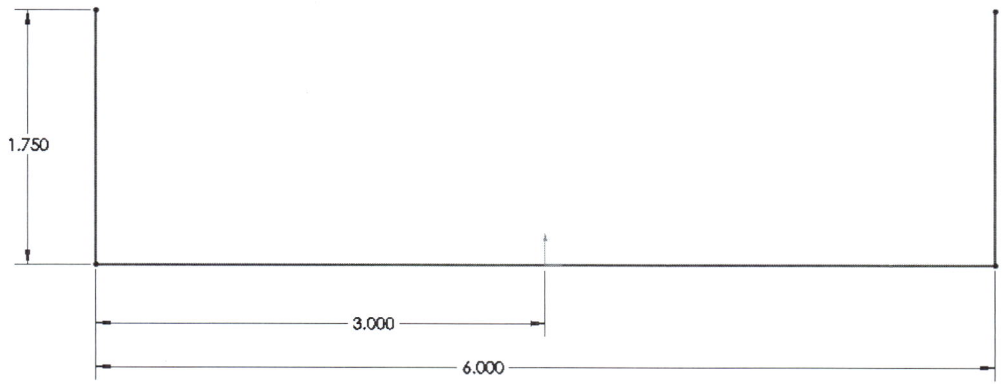

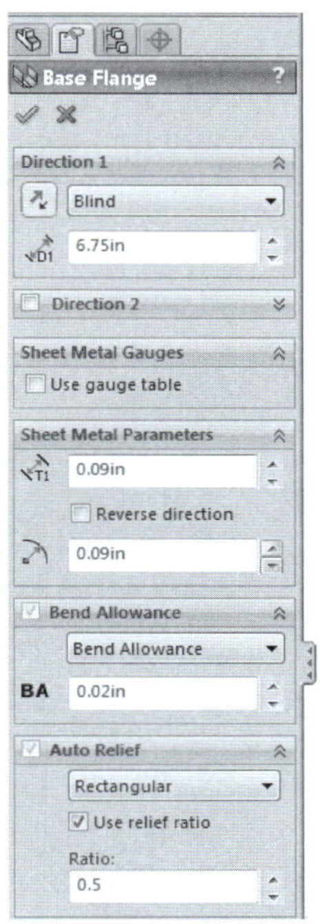

Figure 8-4
Extrude the Base Flange 6.75 Inches in **Direction1** with .09 Inch Thickness, T1

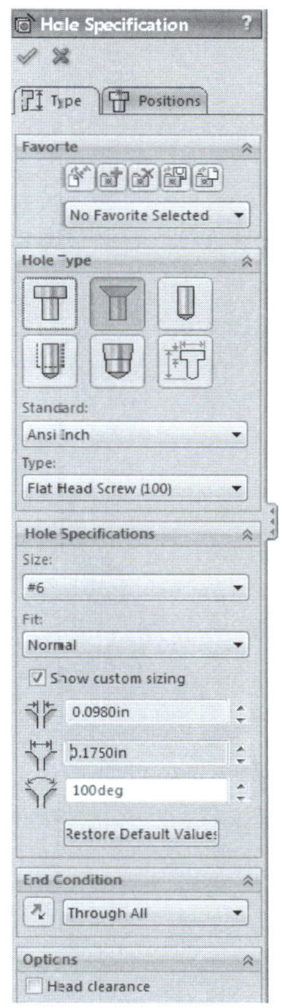

Figure 8-5
Make Settings for a Countersunk Hole Using Hole Wizard

Step 7. Click **OK** (the check mark under the words **Base Flange**).

Step 8. Click **Hole Wizard** and make the settings shown in Figure 8-5 under the **Type** tab. (You are making countersunk holes on the far side. To be safe, turn the model over by selecting the bottom view from the **Heads-up View** toolbar and put the holes on the bottom so that the countersunk holes will be on the correct side.)

Step 9. Click the **Positions** tab and click the location of all nine holes shown in Figure 8-6. (The location is not critical because you will dimension the holes in Step 10.)

Figure 8-6
Position 9 Holes and Dimension Their Locations

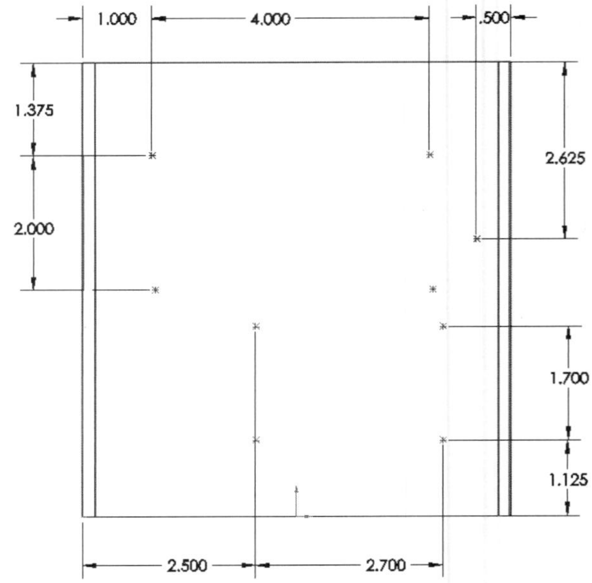

Step 10. Use **Smart Dimension** to dimension the holes as shown in Figure 8-6.

Step 11. Click **OK** and rotate the model so that you can see the countersink. The holes should appear as shown in Figure 8-7.

Figure 8-7
Rotate the Model

Step 12. Click the outside plane of one of the 1.75 inches sides and click **Sketch.**

Step 13. Click **Exit Sketch.**

Step 14. Click **Hole Wizard** and make the settings shown in Figure 8-5 under the **Type tab.**

Step 15. Click the **Positions** tab and click the location of all eight holes shown in Figure 8-8. (The location is not critical because you will dimension the holes in Step 16.)

Step 16. Use **Smart Dimension** to dimension the holes as shown in Figure 8-8.

Figure 8-8
Position 8 Holes and Dimension Their Locations

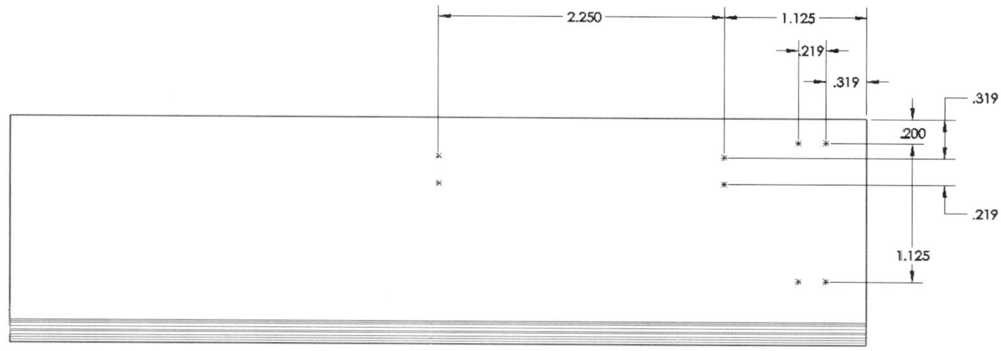

Step 17. Click **OK** and rotate the model to the isometric view.

Step 18. Click **Plane** (from **Reference Geometry**) and click **Front Plane** in the **Feature Manager**. Make the settings as shown in Figure 8-9 and click **OK.**

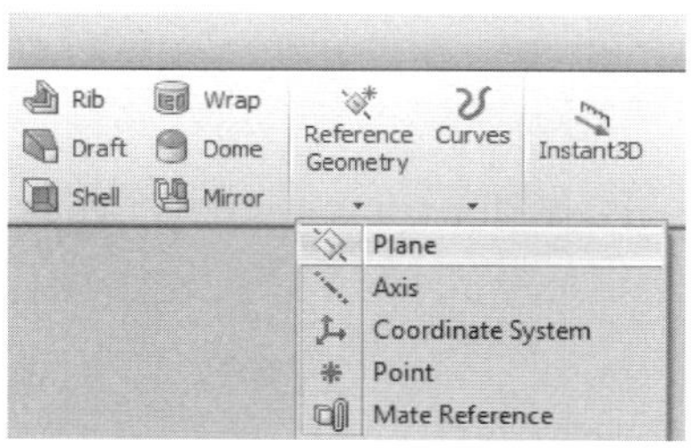

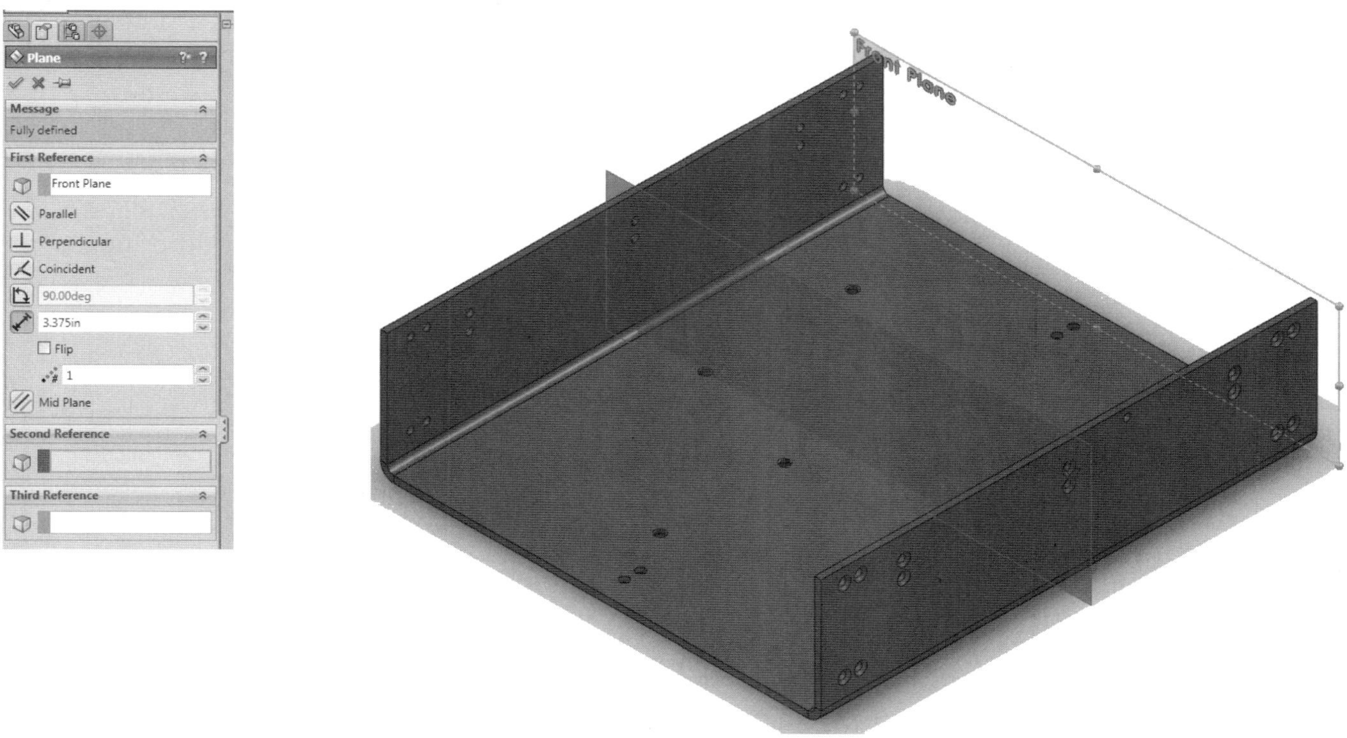

Figure 8-9
Make a Plane 3.375 Inches from the Front Plane

Step 19. Click **Mirror**, click **Features to Mirror** in the **Mirror Manager,** and click **CSK for #6 Flat Head Machine Screw (100) 3** in the **Feature Manager** (Figure 8-10).

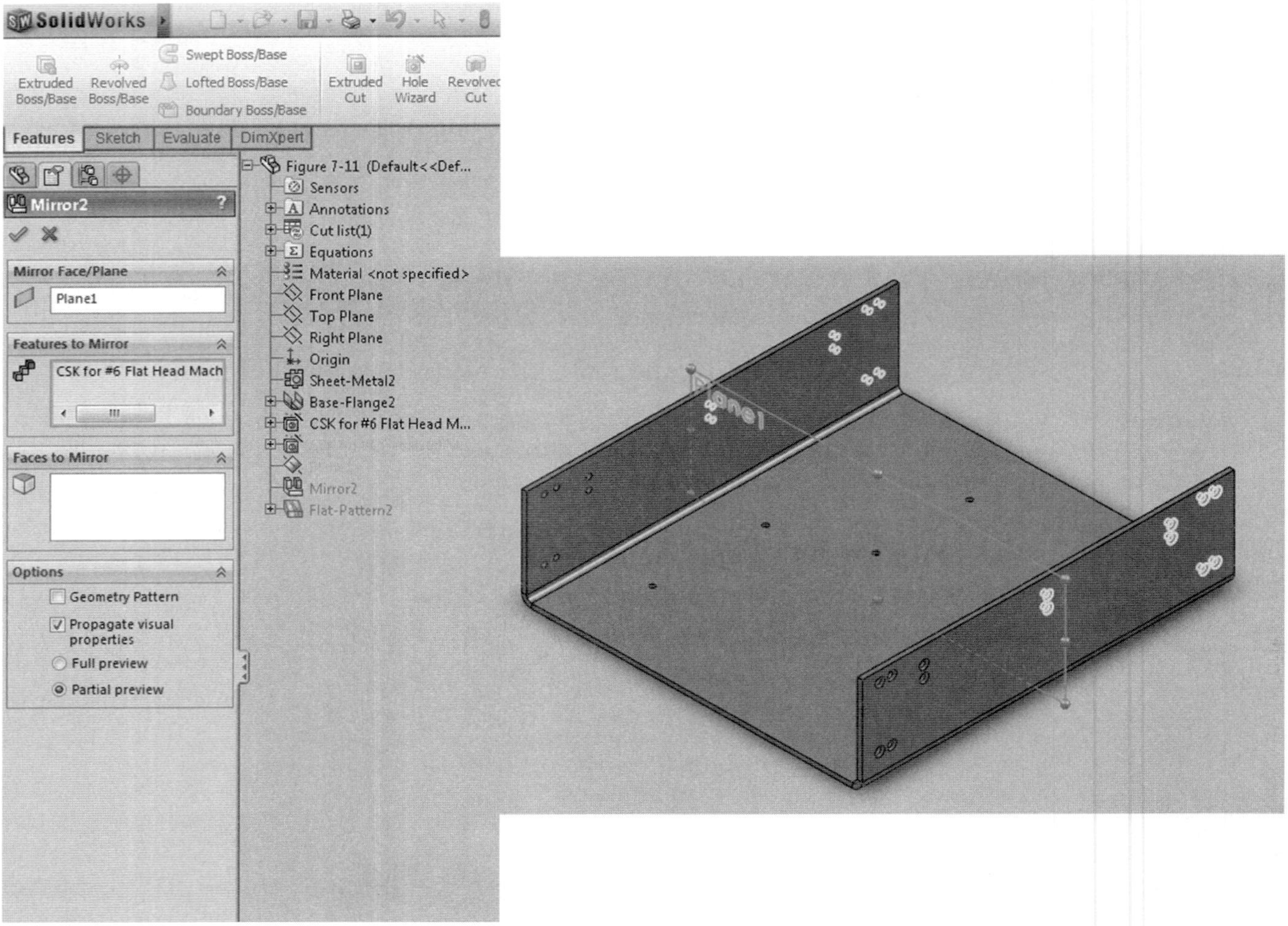

Figure 8-10
Mirror 8 Holes Using Plane1 as the Mirror Plane

Step 20. Save the part as **Chassis** in at least two places and print the model (Figure 8-11).

Figure 8-11
Chassis Complete

Project 8-2: **Making the Front Panel**

Figure 8-12 shows the completed front panel.

Begin the front panel as described next:

Step 1. Click **New....**

Step 2. Click **Part** and click **OK.**

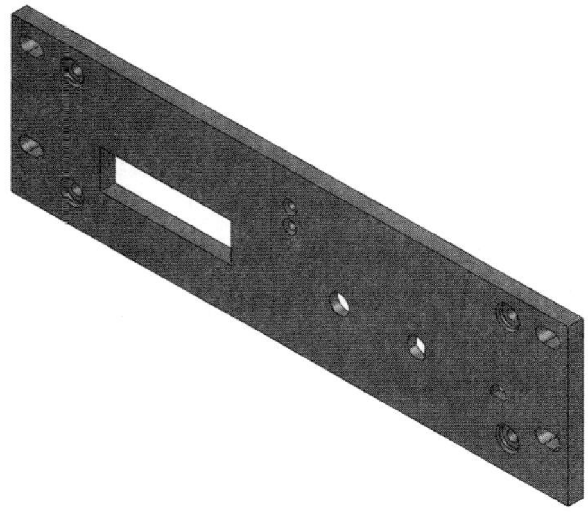

Figure 8-12
Front Panel Complete

Step 3. Click **Front Plane** as the sketch plane.

Step 4. Click **Options** on the menu bar, select the **Document Properties** tab, and click **Units** in the list to the left.

Step 5. Make sure the **Units** are correctly set as shown in Figure 8-2 [notice that **IPS** (inch, pound, second) is checked], then click **OK**.

Step 6. Use the **Line, Rectangle, Circle, Straight Slot,** and **Smart Dimension** commands to draw Figure 8-13.

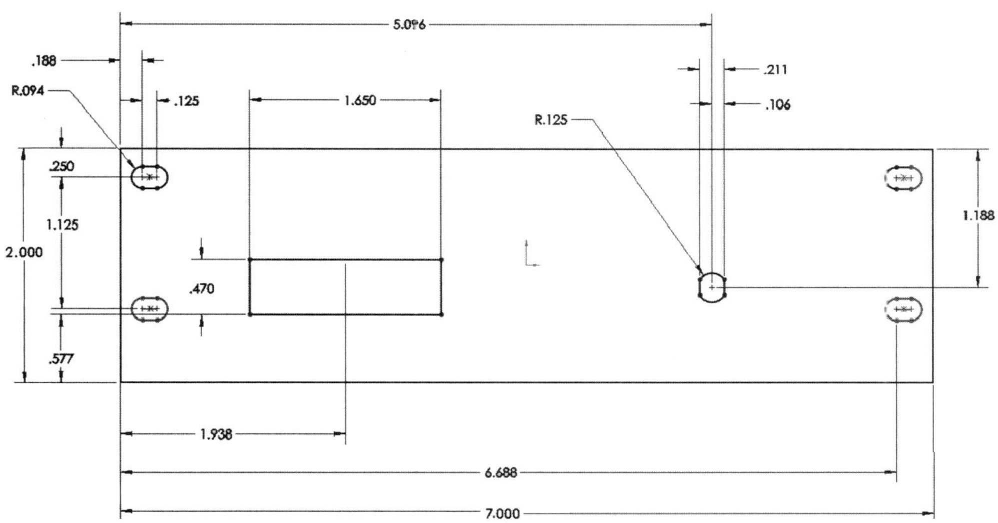

Figure 8-13
Make a Sketch for a Base Flange and Extrude It .190 Inch

Figure 8-14
Make Settings for a Hole Using Hole Wizard

Step 7. Click **Exit Sketch** and **Extrude** the sketch .190.

Step 8. Click **OK.**

Step 9. Click **Hole Wizard** and make the settings shown in Figure 8-14 under the **Type** tab.

Step 10. Click the **Positions** tab and click the location of both holes shown in Figure 8-15.

Step 11. Use **Smart Dimension** to dimension the holes as shown in Figure 8-15 and click **OK.**

Step 12. Click **OK.**

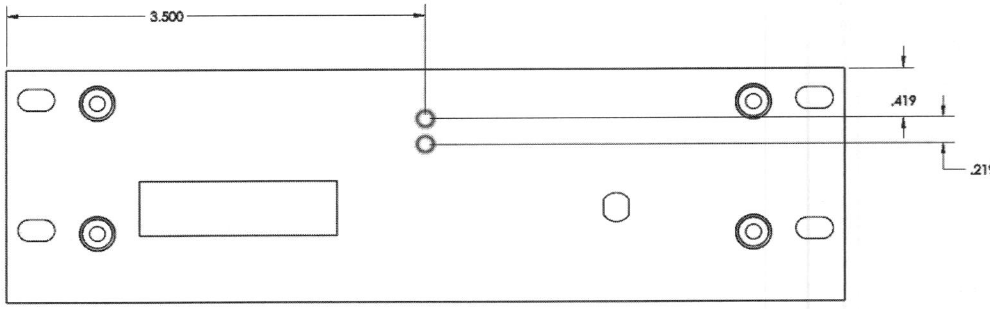

Figure 8-15
Position 2 Holes and Dimension Their Location

Step 13. Click **Hole Wizard** and make the settings for the counterbored hole shown in Figure 8-16 under the **Type** tab.

Step 14. Click the **Positions** tab and click the location of the hole shown in Figure 8-17.

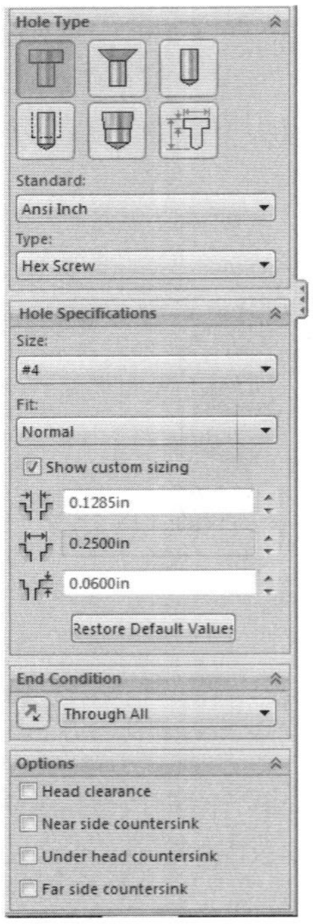

Figure 8-16
Make Settings for a Counterbored Hole Using Hole Wizard

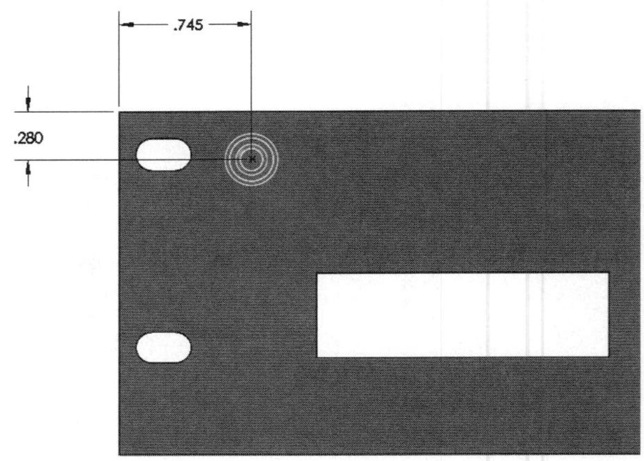

Figure 8-17
Position the Hole and Dimension Its Location

Step 15. Use **Smart Dimension** to dimension the hole as shown in Figure 8-17.

Step 16. Click **OK.**

Step 17. Click **OK.**

Step 18. Click **Linear Pattern** and click a horizontal edge with a spacing of 5.48 inches for **Direction1,** a vertical edge with a spacing of 1.125 inches for **Direction2,** the **CBORE** for **#4 Hex Machine Screw 2** in **Features to Pattern,** and **2** for the number of holes in both directions as shown in Figure 8-18.

Step 19. Click **OK.**

Step 20. Click **Hole Wizard** and make the settings for the counterbored hole shown in Figure 8-19 under the **Type** tab.

Step 21. Click the **Positions** tab and click the location of the hole shown in Figure 8-20.

Step 22. Use **Smart Dimension** to dimension the hole as shown in Figure 8-20.

Step 23. Click **OK.**

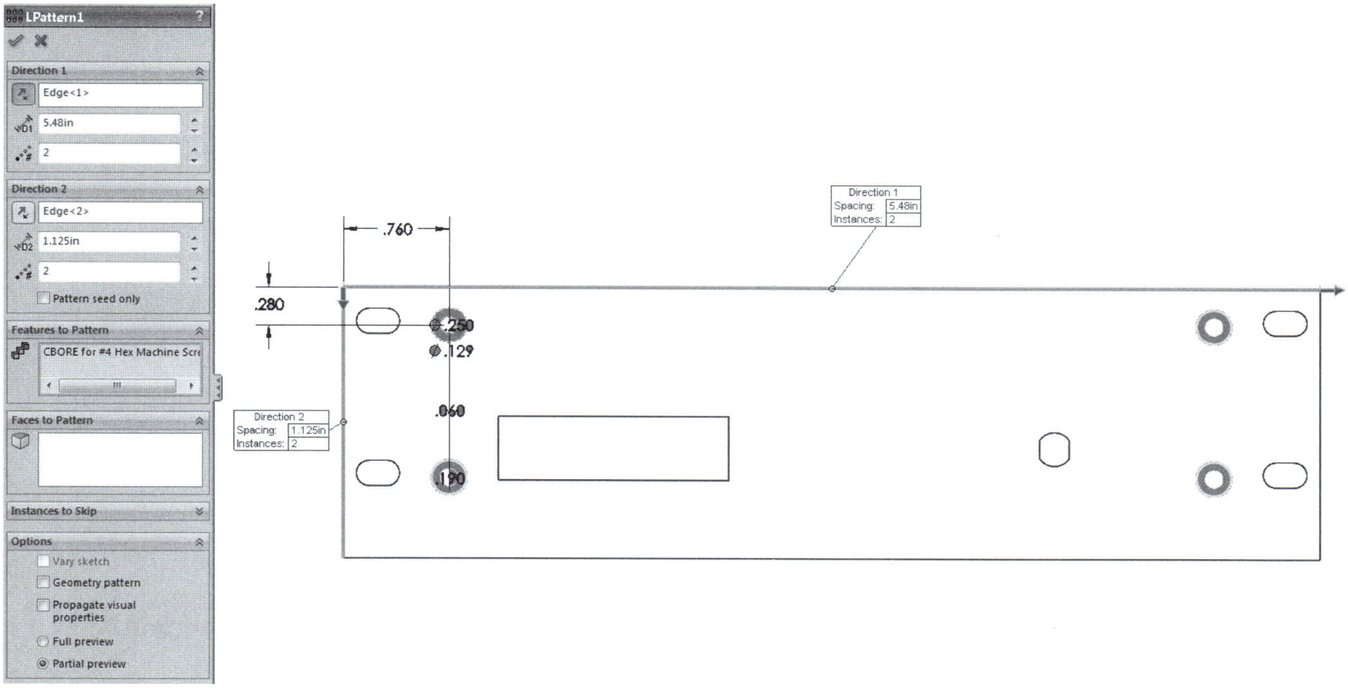

Figure 8-18

Make a Linear Pattern of 2 Instances with 2 Spacing of 5.48 Inches in Direction1 and 1.125 Inches in Direction2

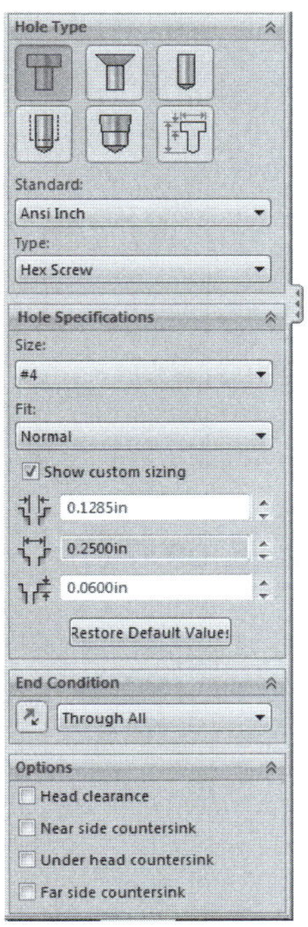

Figure 8-19

Make Settings for a Counterbored Hole Using Hole Wizard

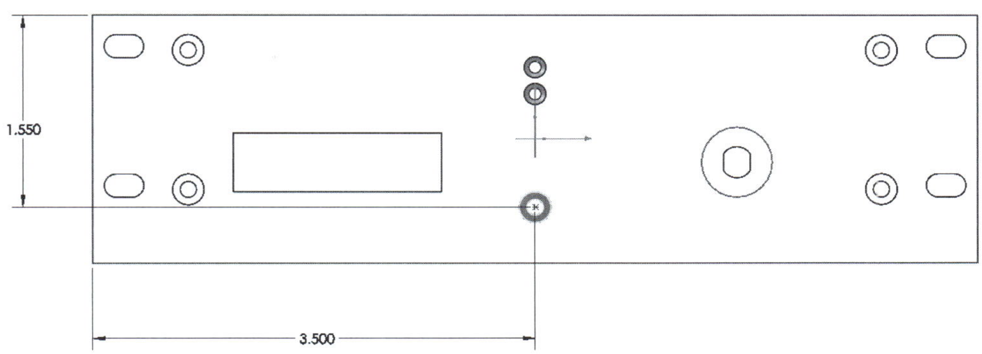

Figure 8-20

Position the Hole and Dimension Its Location

Step 24. Click **OK.**

Step 25. Click the front face of the **Front Panel** as the sketch plane and click **Sketch.**

Step 26. Use the **Line, Circle,** and **Smart Dimension** commands to draw Figure 8-21.

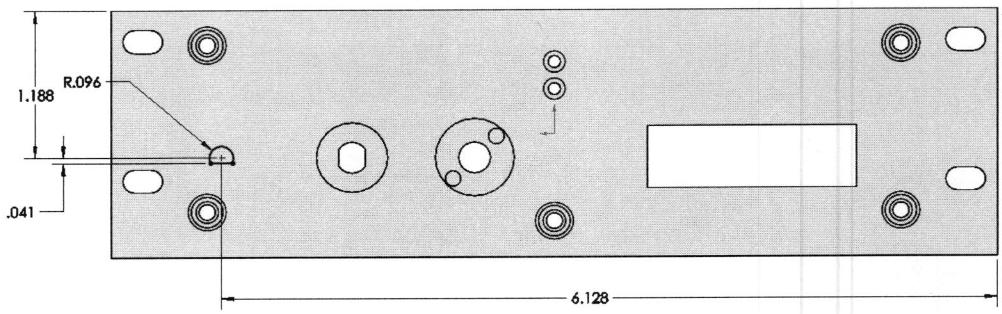

Figure 8-21
Make a Sketch for an Extruded Cut and Make the Cut Through All

Step 27. Click **Exit Sketch** and do an **Extrude Cut Through All.**

Step 28. Click **OK.**

Step 29. Use **View Orientation** to choose the **Back** view.

Step 30. Click the back face of the **Front Panel** as the sketch plane and click **Sketch.**

Step 31. Use the **Circle** and **Smart Dimension** commands to draw Figure 8-22.

Figure 8-22
Make a Sketch for an Extruded Cut and Make the Cut .09 Deep

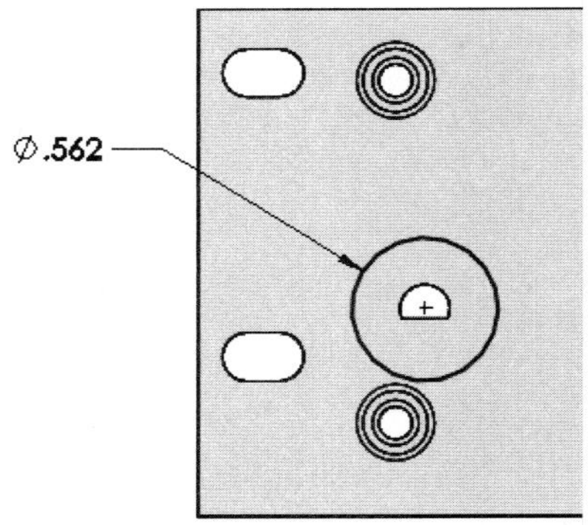

Step 32. Click **Exit Sketch** and do an **Extrude Cut .09** deep.

Step 33. Click **OK.**

Step 34. Click **Hole Wizard** and make the settings for the counterbored hole shown in Figure 8-23 under the **Type** tab.

Step 35. Click the **Positions** tab and click the location of the hole shown in Figure 8-23.

Step 36. Use **Smart Dimension** to dimension the hole as shown in Figure 8-23.

Step 37. Click **OK.**

Step 38. Click **OK.**

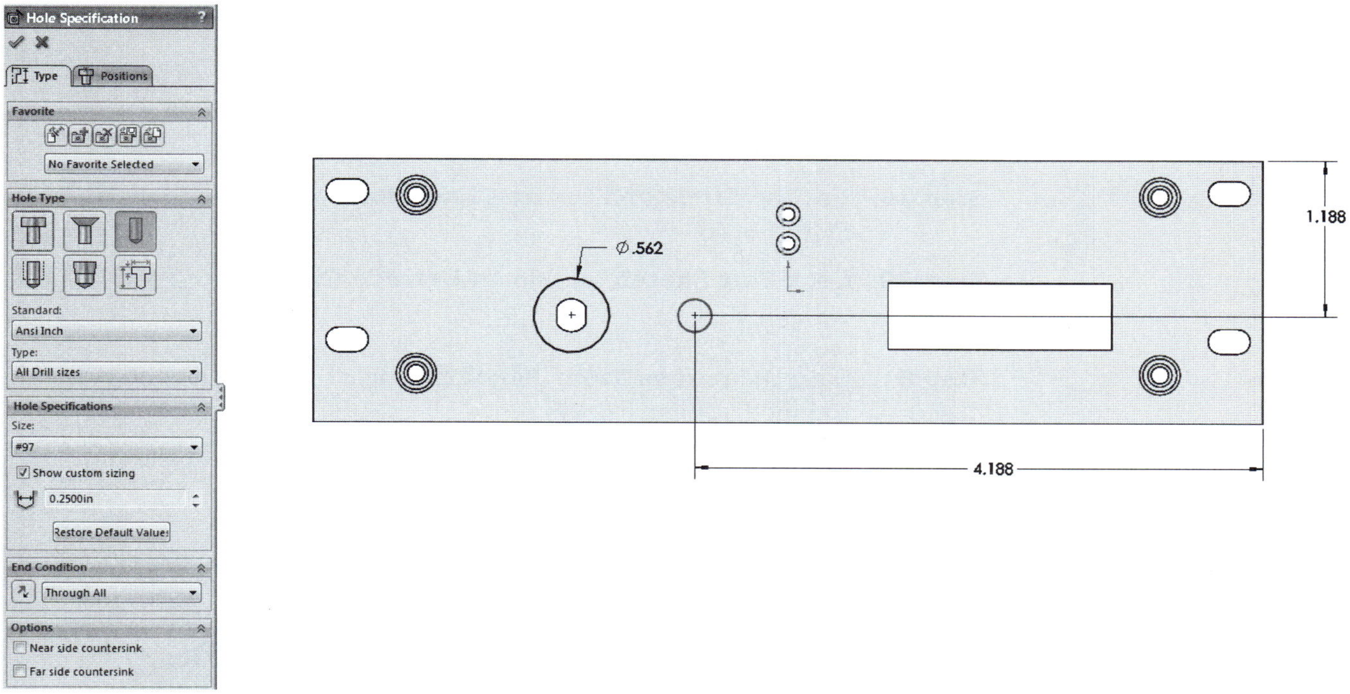

Figure 8-23
Make Settings for a Hole Using Hole Wizard. Position the Hole and Dimension Its Location

Step 39. Click the back face of the **Front Panel** as the sketch plane and click **Sketch.**

Step 40. Use the **Circle** and **Smart Dimension** commands to draw Figure 8-24.

Step 41. Click **Exit Sketch** and do an **Extrude Cut .09** deep.

Step 42. Click **OK.**

Step 43. Click the bottom of the **Extrude Cut** as the sketch plane and click **Sketch.**

Step 44. Use the **Circle** and **Smart Dimension** commands to draw Figure 8-25.

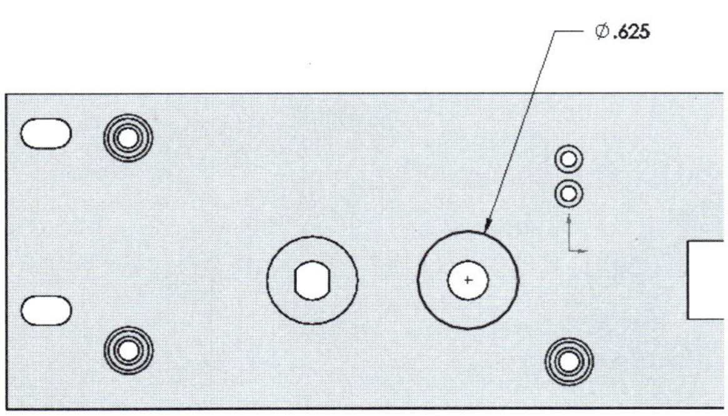

Figure 8-24
Make a Sketch for an Extruded Cut and Make the Sketch .09 Inch Deep

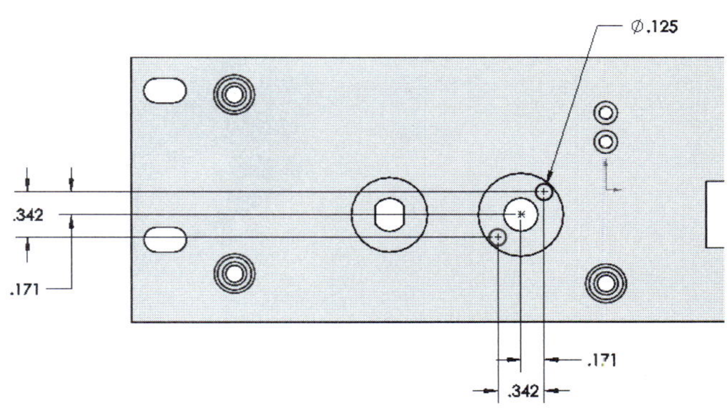

Figure 8-25
Make a Sketch for an Extruded Cut and Make the Cut .09 Deep

Step 45. Click **Exit Sketch and do an Extruded Cut .09 Deep.**

Step 46. Click **OK.**

Step 47. Click the back face of the **Front Panel** as the sketch plane and click **Sketch.**

Step 48. Use the **Circle** and **Smart Dimension** commands to draw Figure 8-26.

Step 49. Click **Exit Sketch** and do an **Extrude Cut .09** deep.

Step 50. Click **OK.**

Step 51. Save the part as **Front Panel** in at least two places and print the model (Figure 8-27).

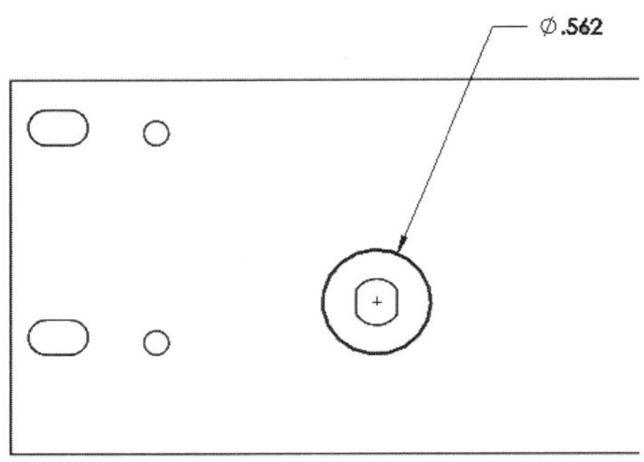

Figure 8-26
Make a Sketch for an Extruded Cut and Make the Cut .09 Inch Deep

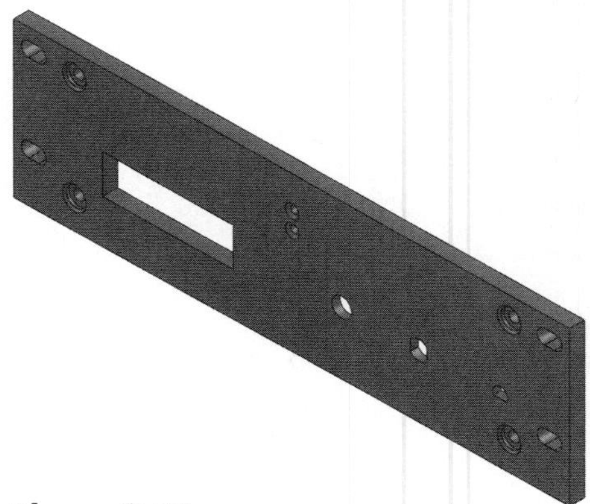

Figure 8-27
Front Panel Complete

Project 8-3: **Making the Rear Panel**

Figure 8-28 shows the completed rear panel.

Begin the rear panel as described next:

Step 1. Click **New....**

Step 2. Click **Part** and click **OK.**

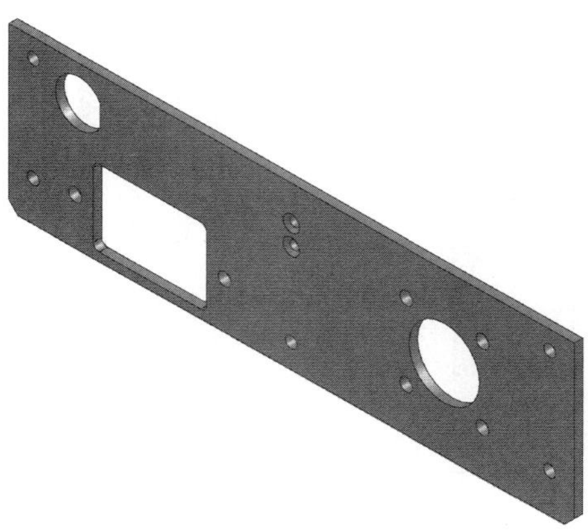

Figure 8-28
Rear Panel Compete

Step 3. Click **Front Plane** as the sketch plane.

Step 4. Click **Options** on the menu bar, select the **Document Properties** tab, and click **Units** in the list to the left.

Step 5. Make sure the **Units** are correctly set as shown in Figure 8-2 [notice that **IPS** (inch, pound, second) is checked], then click **OK.**

Step 6. Use the **Line** and **Smart Dimension** commands to draw Figure 8-29.

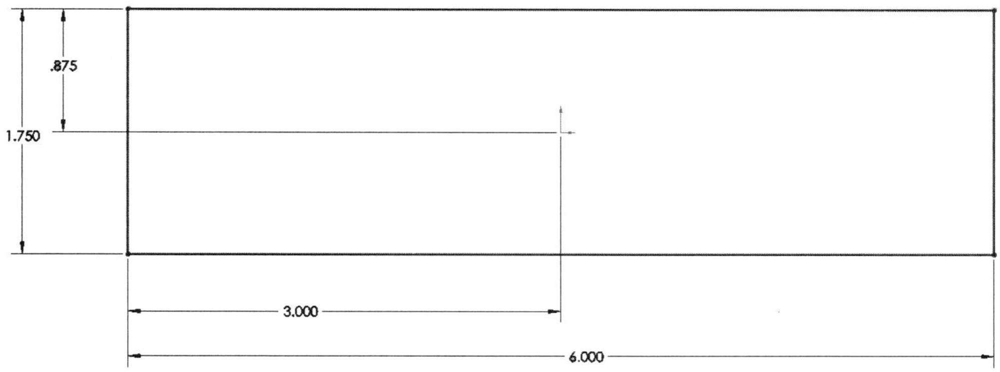

Figure 8-29
Make a Sketch for a Base Flange and Extrude It .190 Inch

Step 7. Click **Exit Sketch** and **Extrude** the sketch .190.

Step 8. Click **OK.**

Step 9. Click the front face of the **Rear Panel** as the sketch plane and click **Sketch.**

Step 10. Use the **Line, Circle, Rectangle,** and **Smart Dimension** commands to draw Figure 8-30.

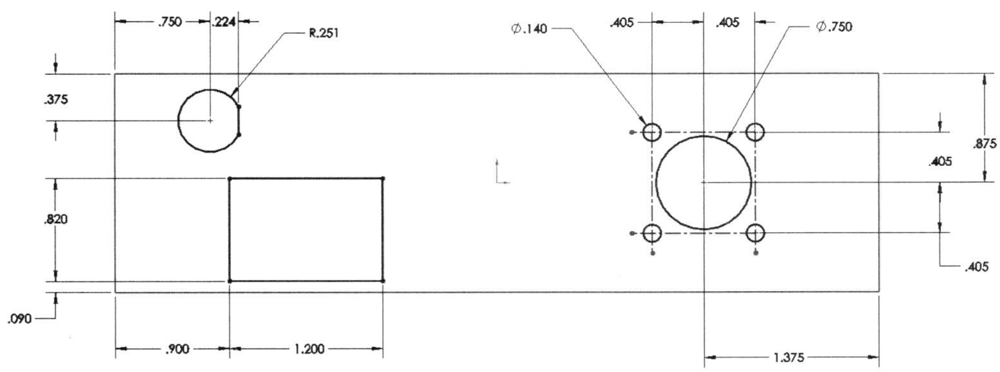

Figure 8-30
Make a Sketch for an Extruded Cut and Make the Sketch Through All

Step 11. Click **Exit Sketch** and do an **Extrude Cut Through All.**

Step 12. Click **OK.**

Step 13. Click the front face of the **Rear Panel** as the sketch plane and click **Sketch.**

Figure 8-31
Make a Sketch for an Extruded Cut and Make the Cut Through All

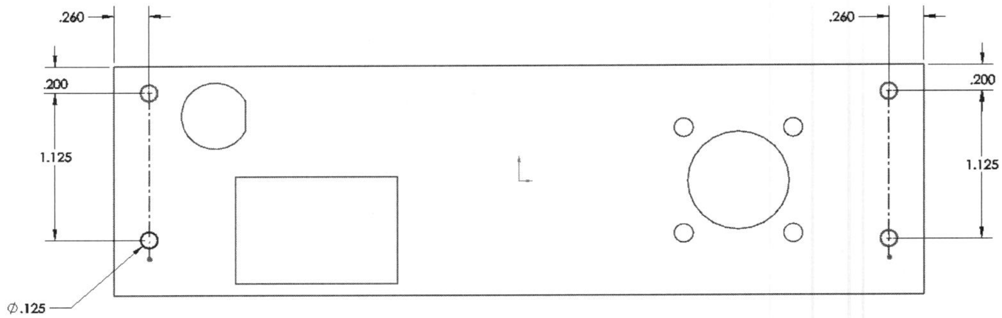

Step 14. Use the **Circle** and **Smart Dimension** commands to draw the remaining holes for Figure 8-31.

Step 15. Click **Exit Sketch** and do an **Extrude Cut Through All.**

Step 16. Click **OK.**

Step 17. Click **Hole Wizard** and make the settings shown in Figure 8-32 under the **Type** tab.

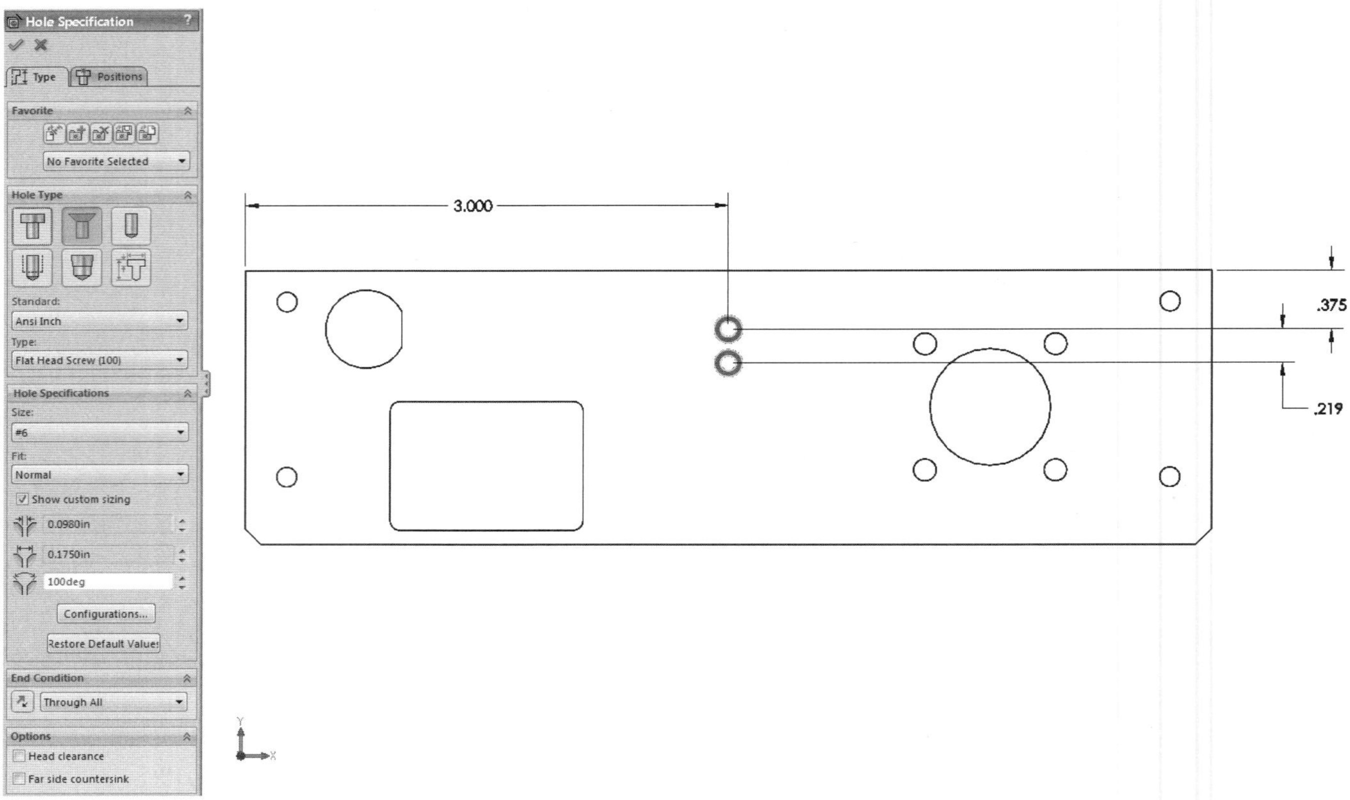

Figure 8-32
Make Settings for a Countersunk Hole Using Hole Wizard. Position the Holes and Dimension Them

Step 18. Click the **Positions** tab and click the location of both dimensioned holes shown in Figure 8-32.

Step 19. Use **Smart Dimension** to dimension the holes as shown in Figure 8-32 and click **OK.**

Step 20. Click **OK. Make two .125 diameter chamfers and four .0625 radius fillets in the rectangular hole** as shown in Figure 8-32.

Step 21. Click the front face of the **Rear Panel** as the sketch plane and click **Sketch.**

Step 22. Use the **Circle** and **Smart Dimension** commands to draw the remaining holes for Figure 8-33.

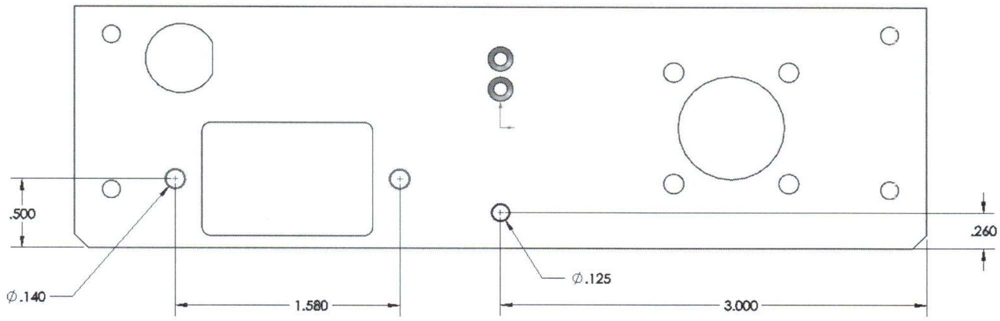

Figure 8-33
Make a Sketch for an Extruded Cut and Make the Cut Through All

Step 23. Click **Exit Sketch** and do an **Extrude Cut Through All.**

Step 24. Click **OK.**

Step 25. Save the part as **Rear Panel** in at least two places and print the model (Figure 8-34).

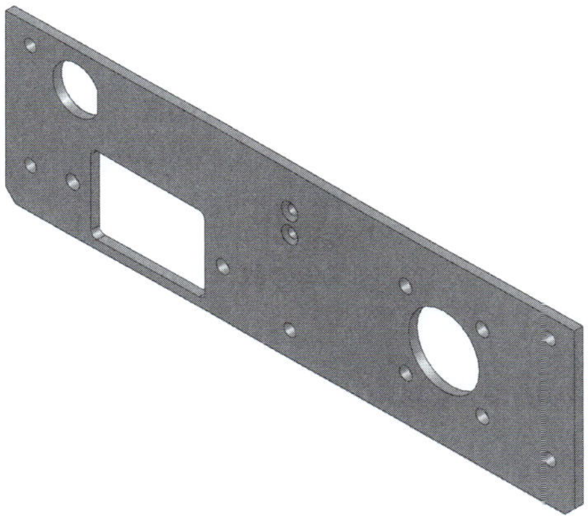

Figure 8-34
Rear Panel Complete

Project 8-4: Making the Chassis Cover

Figure 8-35 shows the completed cover.

Begin the cover as described next:

Step 1. Click **New....**

Step 2. Click **Part** and click **OK.**

Step 3. Click **Top Plane** as the sketch plane.

Figure 8-35
Chassis Cover Complete

Step 4. Click **Options** on the menu bar, select the **Document Properties** tab, and click **Units** in the list to the left.

Step 5. Make sure the **Units** are correctly set as shown in Figure 8-2 [notice that **IPS** (inch, pound, second) is checked], then click **OK.**

Step 6. Use the **Line** and **Smart Dimension** commands to draw Figure 8-36.

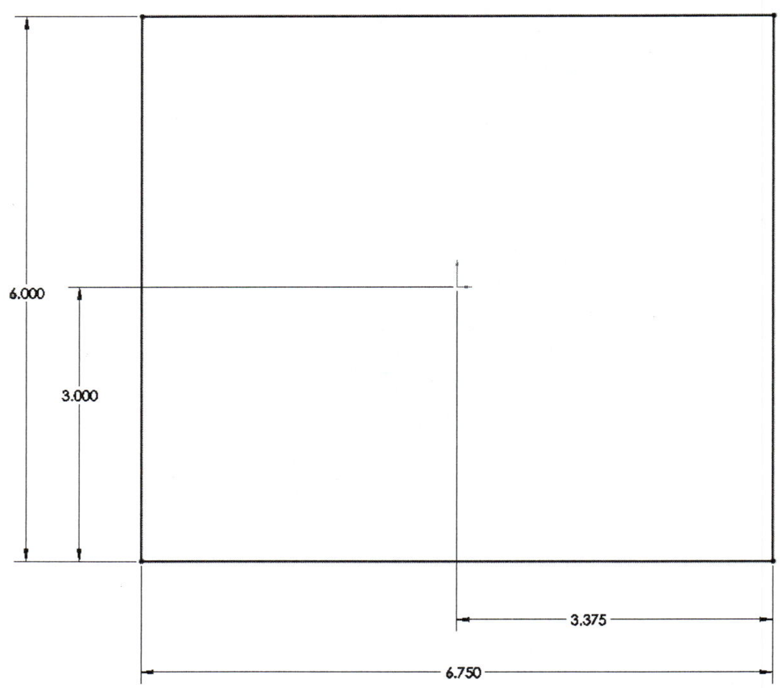

Figure 8-36
Make a Sketch for a Base Flange and Extrude It .190 Inch

Step 7. Click **Exit Sketch** and **Extrude** the sketch .190.

Step 8. Click **OK.**

Step 9. Click **Hole Wizard** and make the settings shown in Figure 8-37 under the **Type** tab.

Step 10. Click the **Positions** tab and click the location of all eight holes shown in Figure 8-37. (The location is not critical because you will dimension the holes in Step 11.)

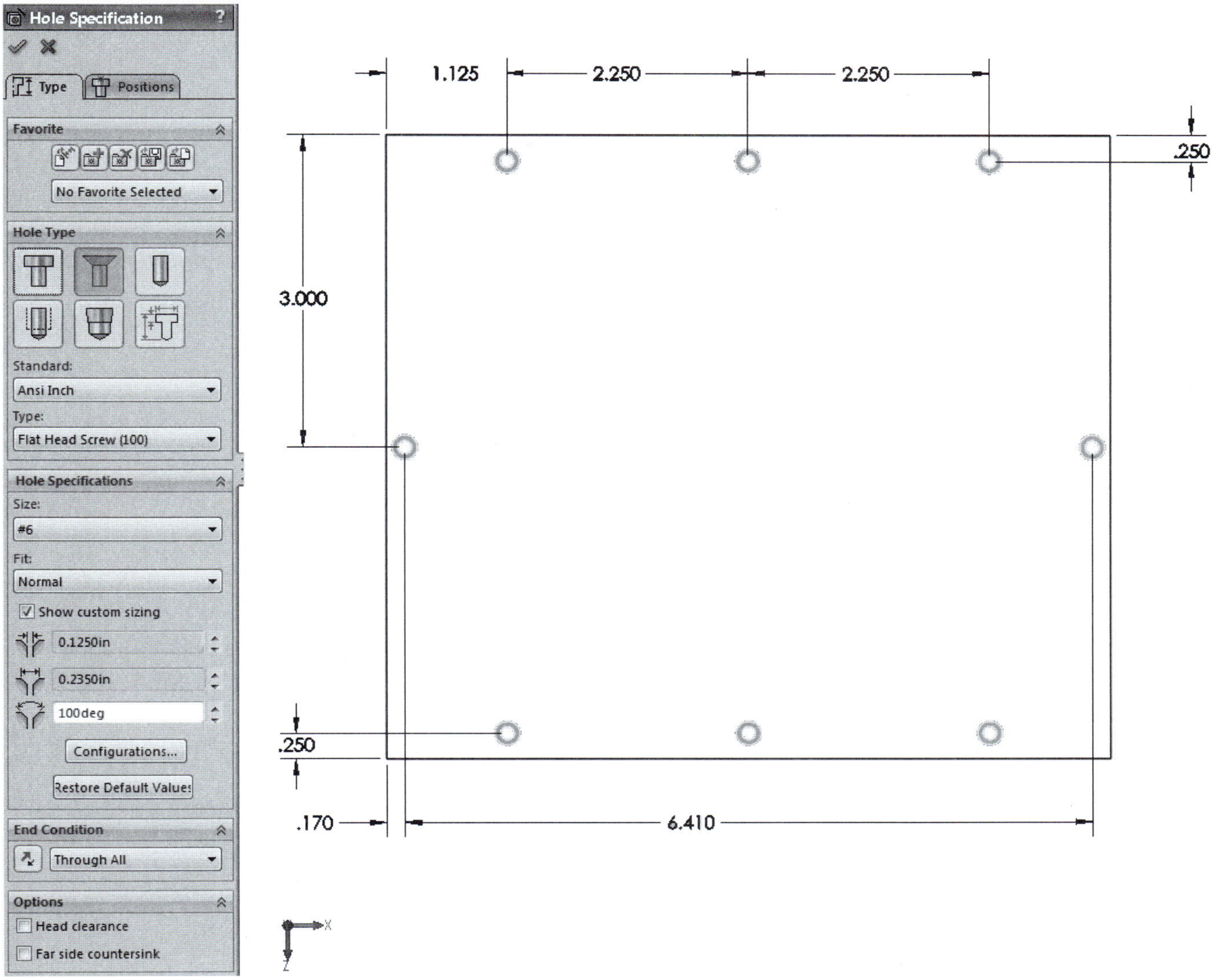

Figure 8-37
Make Settings for a Countersunk Hole Using Hole Wizard. Position the Holes and Dimension Their Location

Step 11. Use **Smart Dimension** to dimension the holes as shown in Figure 8-37.

Step 12. Click **OK** and rotate the model to the isometric view.

Step 13. Save the part as **Chassis Cover** in at least two places and print the model (Figure 8-38).

Figure 8-38
Chassis Cover Complete

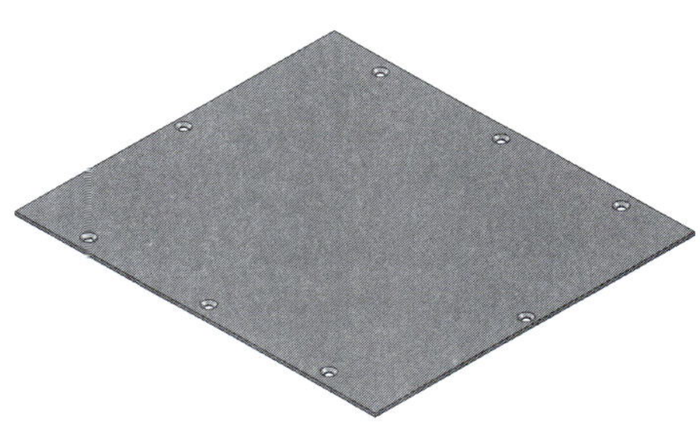

Exercises

EXERCISE 8-1 **Make a drawing from the Chassis.**

Step 1. Open the **Chassis** model.

Step 2. Click the **Make Drawing from Part** command.

Step 3. Using the methods described in Chapter 2, make a drawing that looks similar to Figure 8-39.

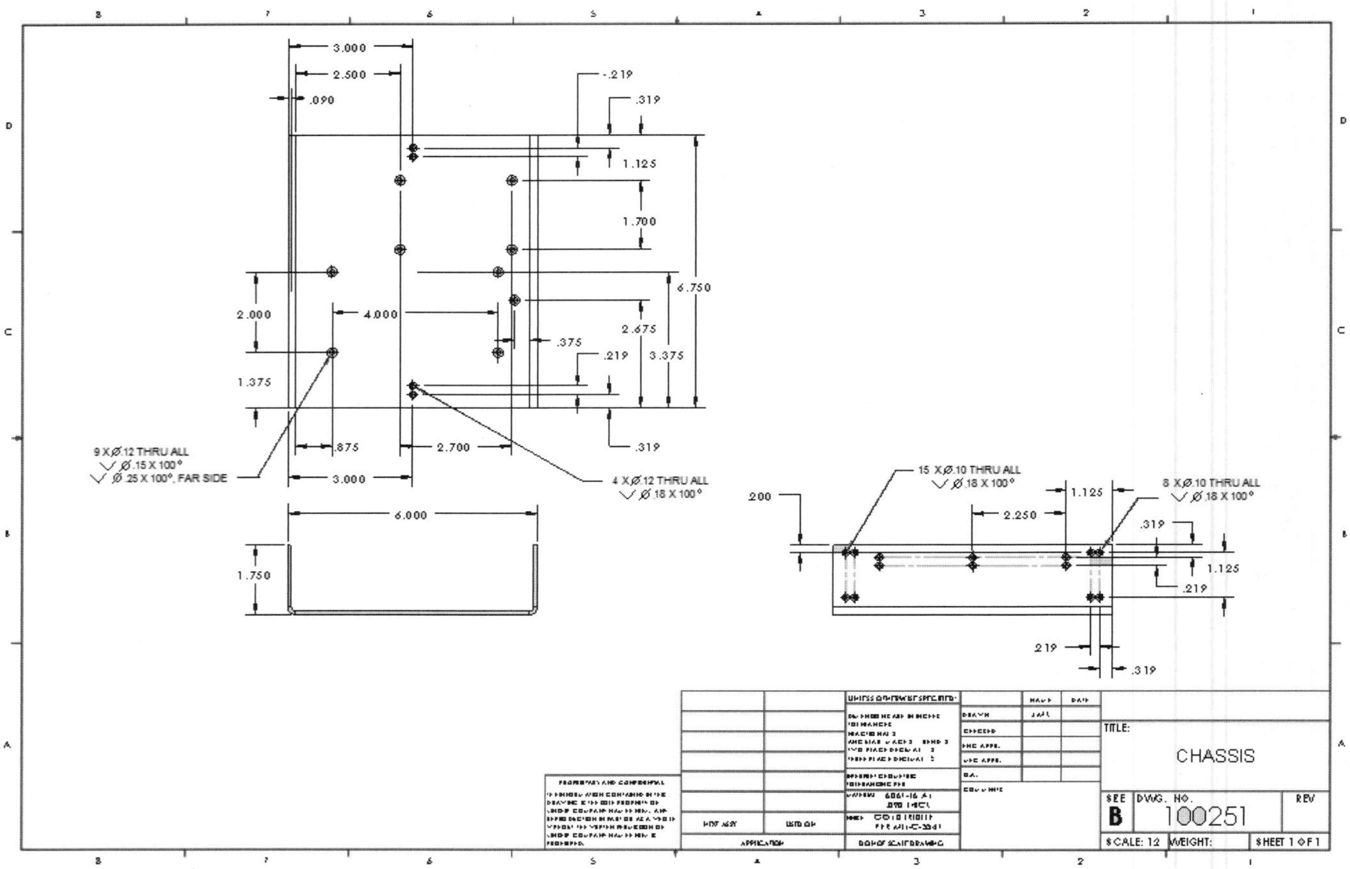

Figure 8-39
Make a Drawing from the Chassis Part

Step 4. Save the drawing as **100251** in at least two places.

Step 5. Print the drawing on a B-Size sheet and turn it in.

EXERCISE 8-2 **Make a drawing from the Front Panel.**

Step 1. Open the **Front Panel** model.

Step 2. Click the **Make Drawing from Part** command.

Step 3. Using the methods described in Chapter 2, make a drawing that looks similar to Figure 8-40.

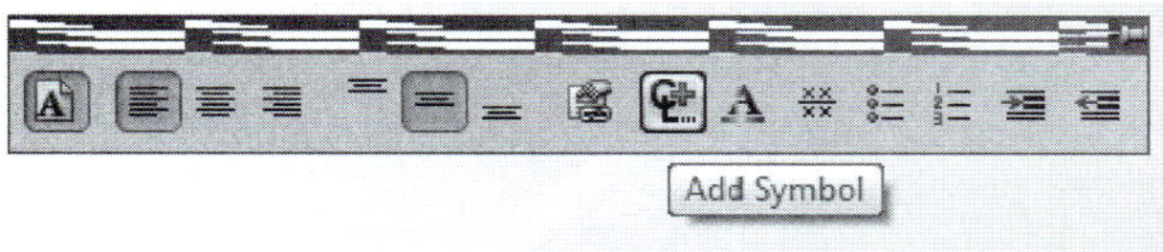

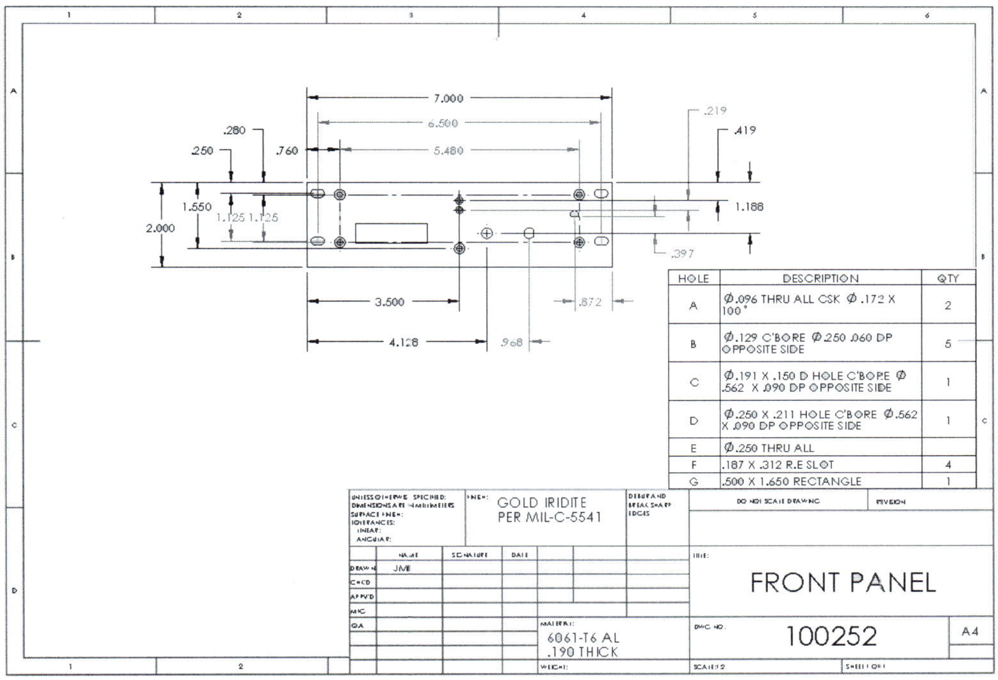

Figure 8-40
Make a Drawing from the Front Panel Part

Step 4. Save the drawing as **100252** in at least two places.

Step 5. Print the drawing on a B-Size sheet and turn it in.

EXERCISE 8-3 Make a drawing from the Rear Panel.

Step 1. Open the **Rear Panel** model.

Step 2. Click the **Make Drawing from Part** command.

Step 3. Using the methods described in Chapter 2, make a drawing that looks similar to Figure 8-41.

Step 4. Save the drawing as **100253** in at least two places.

Step 5. Print the drawing on a B-Size sheet and turn it in.

Figure 8-41
Make a Drawing from the Rear Panel Part

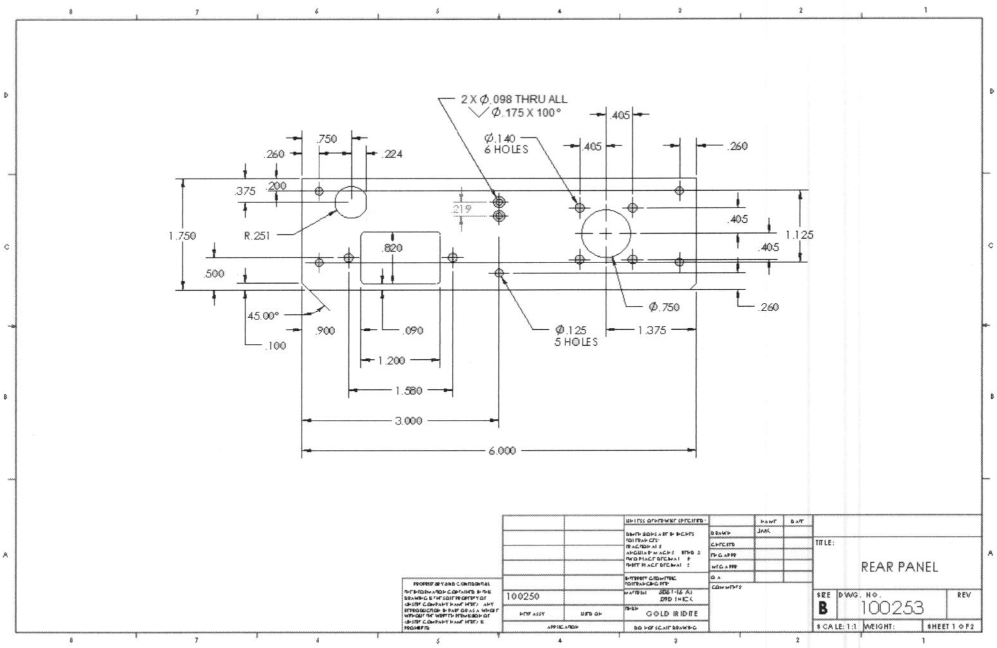

EXERCISE 8-4 Make a drawing from the Chassis Cover.

Step 1. Open the **Chassis Cover** model.

Step 2. Click the **Make Drawing from Part** command.

Step 3. Using the methods described in Chapter 2, make a drawing that looks similar to Figure 8-42.

Step 4. Save the drawing as **100254** in at least two places.

Step 5. Print the drawing on a B-Size sheet and turn it in.

Figure 8-42
Make a Drawing from the Chassis Cover Part

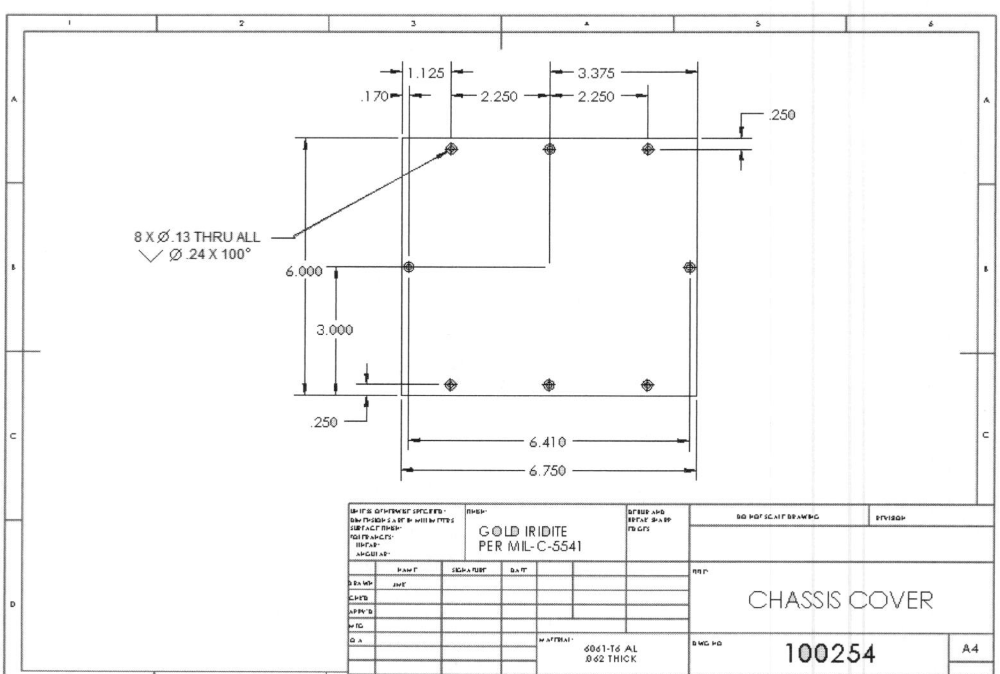

8 chaptereight

Chapter Summary

This chapter contains the information necessary to build solid models of sheet metal parts. In addition you have used previously learned skills to make drawings from those parts. Now you have the skills and information necessary to produce complex solid models of sheet metal parts and make 2-dimensional drawings from them.

Review Questions

Circle the most correct answer.

1. On which menu is the **Base Flange** command located?
   a. **Edit** c. **Insert**
   b. **View** d. **Tools**
2. The **Hole Wizard** allows you to make which of these types of holes from the **Type** tab?
   a. **Through** c. **Countersink**
   b. **Counterbore** d. All of these
3. Which of the following allows you to make a hole without using the **Hole Wizard**?
   a. **Extrude Boss/Base** c. **Extrude Cut**
   b. **Revolve Boss/Base** d. **Chamfer**
4. Which of these commands allows you to make a hole that is a rectangle with a full radius on both ends?
   a. **Straight Slot** c. **Hole Wizard**
   b. **Simple Hole** d. **Extrude Cut**
5. What setting is used for dimensions to appear as inches?
   a. **Dimensions** c. **Units**
   b. **Annotations** d. **Detailing**
6. The **Edit Sheet** command is used to edit the sheet format.
   a. True b. False
7. The **Base Flange** command is used to make the overall shape of a sheet metal part.
   a. True b. False
8. The **Chamfer** command cannot be used in the features mode.
   a. True b. False
9. All counterbored holes must be done with the **Hole Wizard.**
   a. True b. False
10. All countersunk holes have the same angle for the countersink.
    a. True b. False

8 chaptereight

Matching

Place the correct letter on the line to the right of the numbered item.

1. A command that gives you a feature that can be used to mirror holes ______	a. **Hole Wizard**
2. A command that allows you to edit a sheet format ______	b. **Fillet**
3. A command that makes a counterbored hole ______	c. **Edit Sheet Format**
4. A command that can be used to duplicate a series of holes ______	d. **Plane**
5. A command that makes a rounded ______	e. **Linear Pattern Surface**

General Questions

Answer the following questions.

1. In what situations would you use the **Hole Wizard**?
2. When would you want to use the **Extrude Cut** command?
3. What commands are used to import dimensions into a drawing from a model?
4. How can the **Mirror** command be used to make a pattern of many holes?
5. When can **Smart Dimension** be used in making a drawing from a part?

chapter nine

Assemble Nutplates on the Chassis and Front Panel for the Assembly

CHAPTER OBJECTIVES

After completing this chapter, you will be able to:

- Make a nutplate to specific 3-place decimal inch measurement for the chassis and front panel.
- Assemble nutplates on the chassis and front panel.
- Make an **Assembly** from the chassis assembly and front panel assembly.
- Make drawings from the chassis and front panel.
- Correctly space the 2-place and 3-place decimal dimensions and annotations that are inserted or added.
- Correctly answer questions regarding SolidWorks commands used in this chapter.

Project 9-1: Making the Nutplate

Figure 9-1 shows the completed nutplate.

Step 1. Start SolidWorks.

Step 2. Click **New....**

The **New SolidWorks Document** dialog box appears.

Step 3. Click **Part** and click **OK.**

The SolidWorks screen appears.

Step 4. Click **Options** on the menu bar, select the **Document Properties** tab, and click **Units.**

Make sure the **Units** are correctly set to **IPS** (inch, pound, second) and click **OK.**

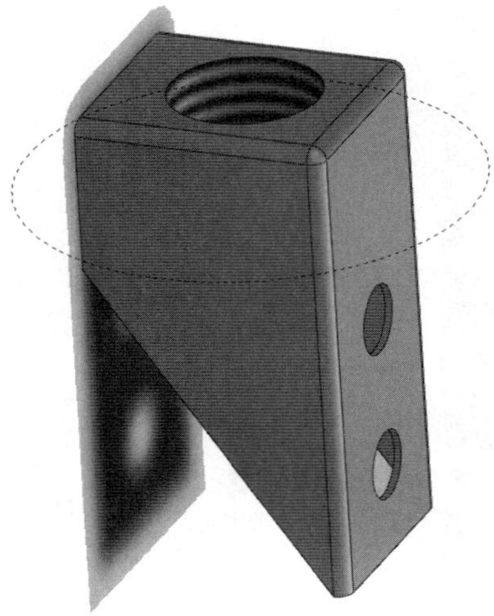

Figure 9-1
The Completed Nutplate

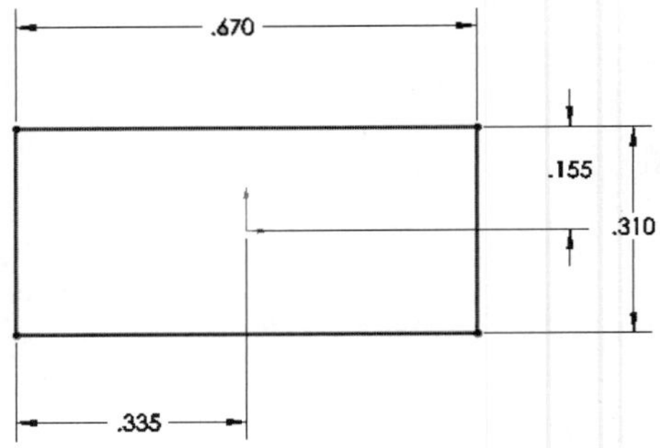

Figure 9-2
Draw and Dimension the Rectangle on the Top Plane and Extrude It .35 Inch

Step 5. Use the **Rectangle** and **Smart Dimension** commands to draw Figure 9-2 on the top plane.

Step 6. Click **Exit Sketch** and **Extrude** the rectangle .35 inch.

Step 7. Use the **Shell** command to make the settings shown in Figure 9-3 and **Shell** the rectangle.

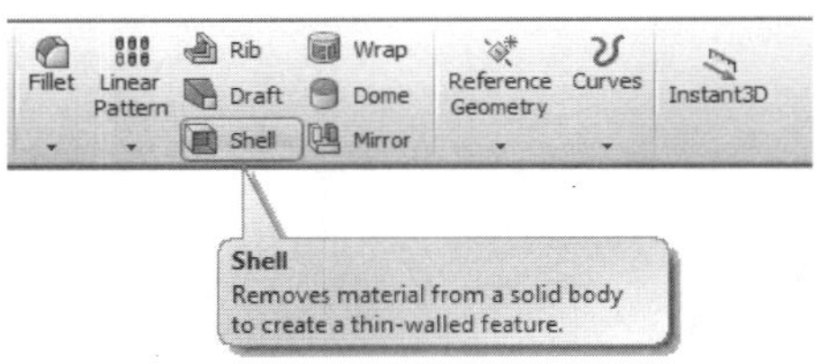

Figure 9-3
Shell .02 Inch the Extruded Rectangle

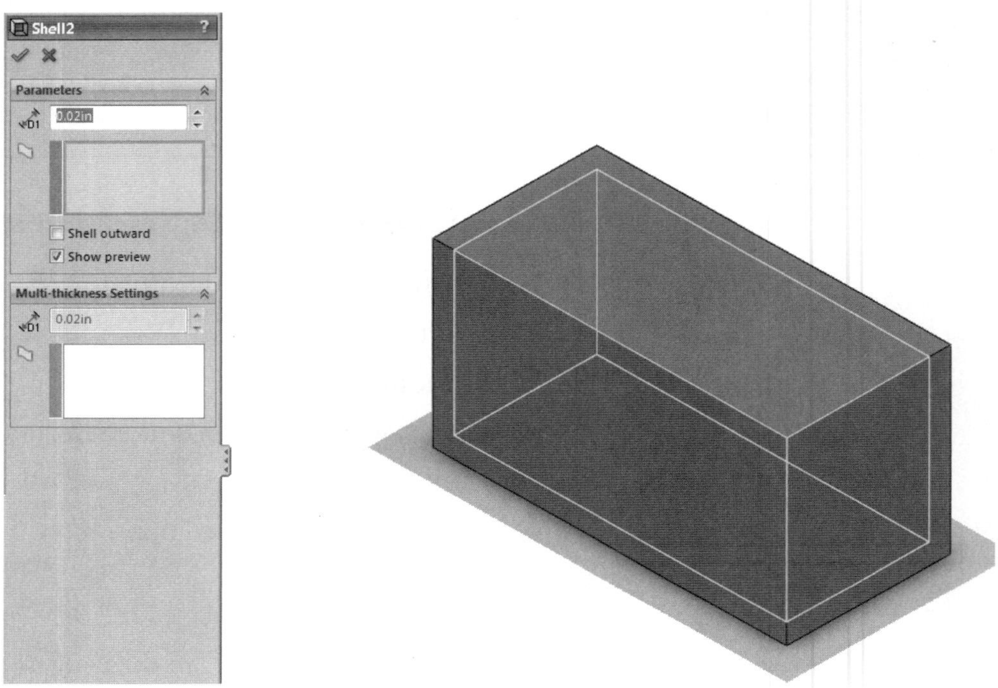

Step 8. Use the **Fillet** command to fillet the edges shown in Figure 9-4 .02 inch. You will have to rotate the view to fillet the hidden vertical edge because if you pick what you think is that edge you will fillet the edge that is inside the shell.

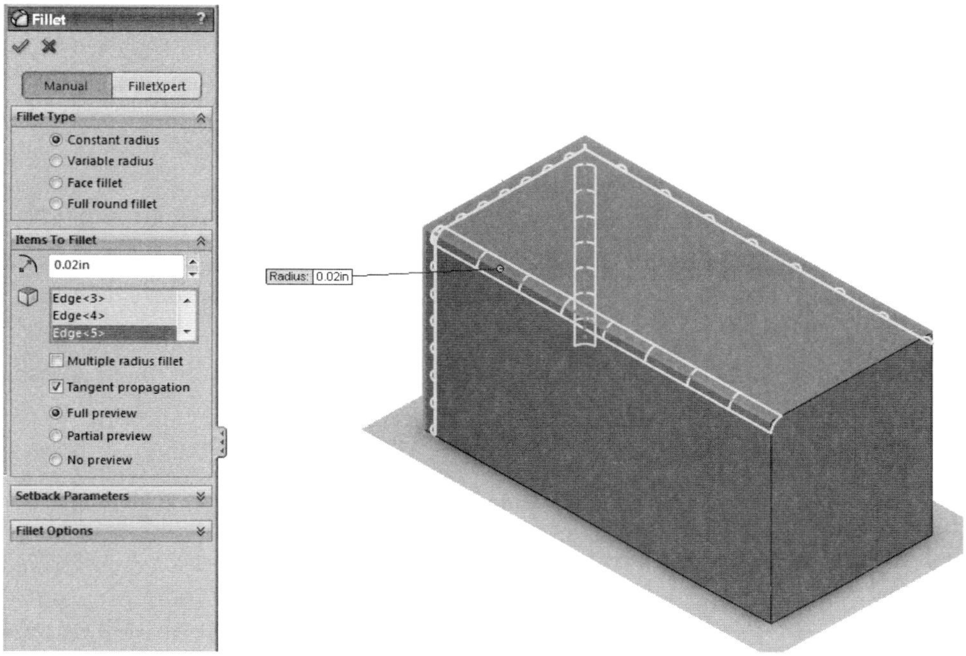

Figure 9-4
Fillet 5 Edges .02 Inch

Step 9. Click the bottom face of the extruded rectangle and click **Sketch.** Then draw and dimension a rectangle on the bottom face of the extruded rectangle as shown in Figure 9-5. The length of the rectangle is not important as long as it extends outside the extruded rectangle.

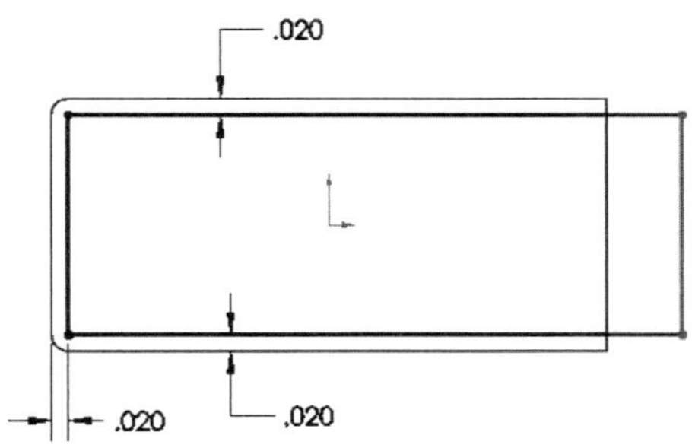

Figure 9-5
Draw a Rectangle on the Bottom Face
That Is .02 Inch from 3 Edges

Step 10. Click **Exit Sketch** and use the **Extrude Cut** command to extrude the rectangle .33 inch as shown in Figure 9-6.

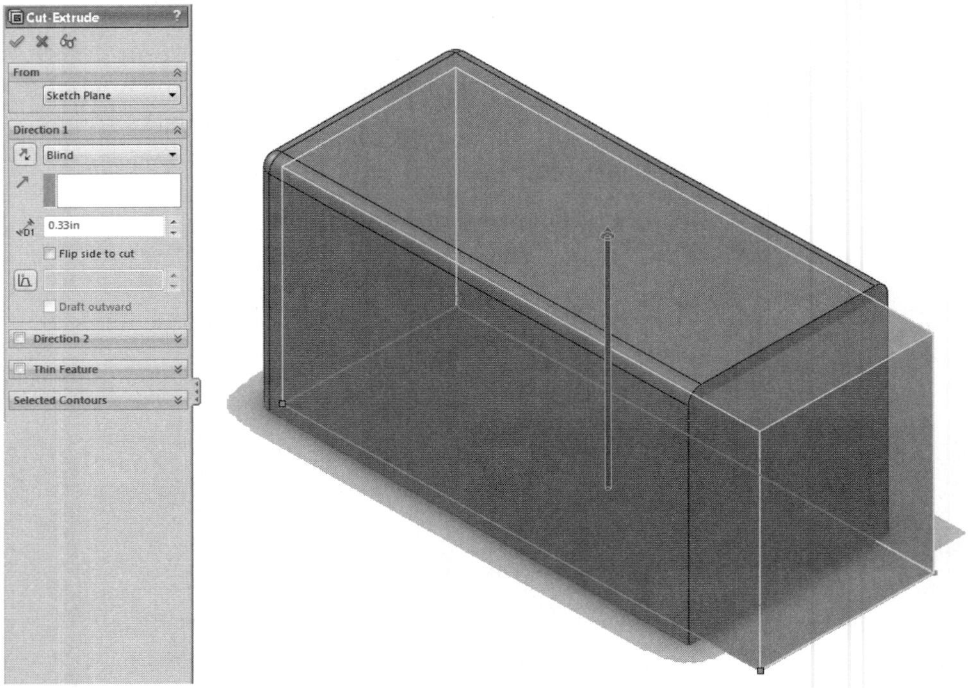

Figure 9-6
Use the Extrude Cut Command to Extrude the Rectangle .33 Inch

Step 11. Draw and dimension the sketch on the front face as shown in Figure 9-7.

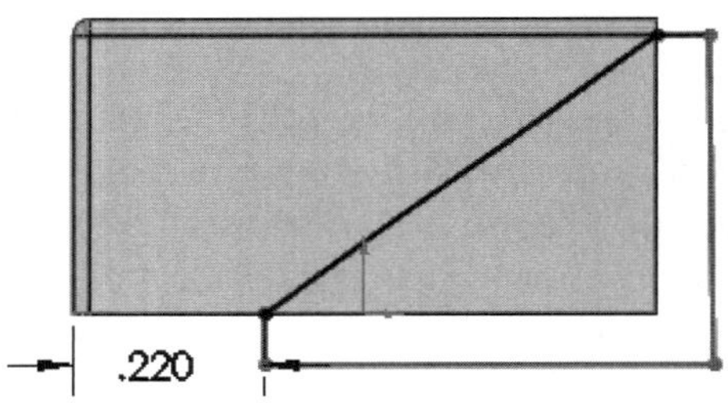

Figure 9-7
Draw a Sketch on the Front Face

Step 12. Click **Exit Sketch** and use the **Extrude Cut** command to extrude the sketch **Through All** as shown in Figure 9-8.

Step 13. Click the top face and use **Hole Wizard** to make the settings on the **Type** tab shown in Figure 9-9.

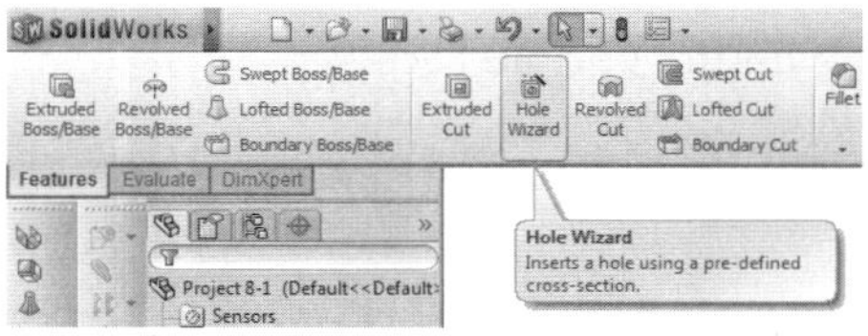

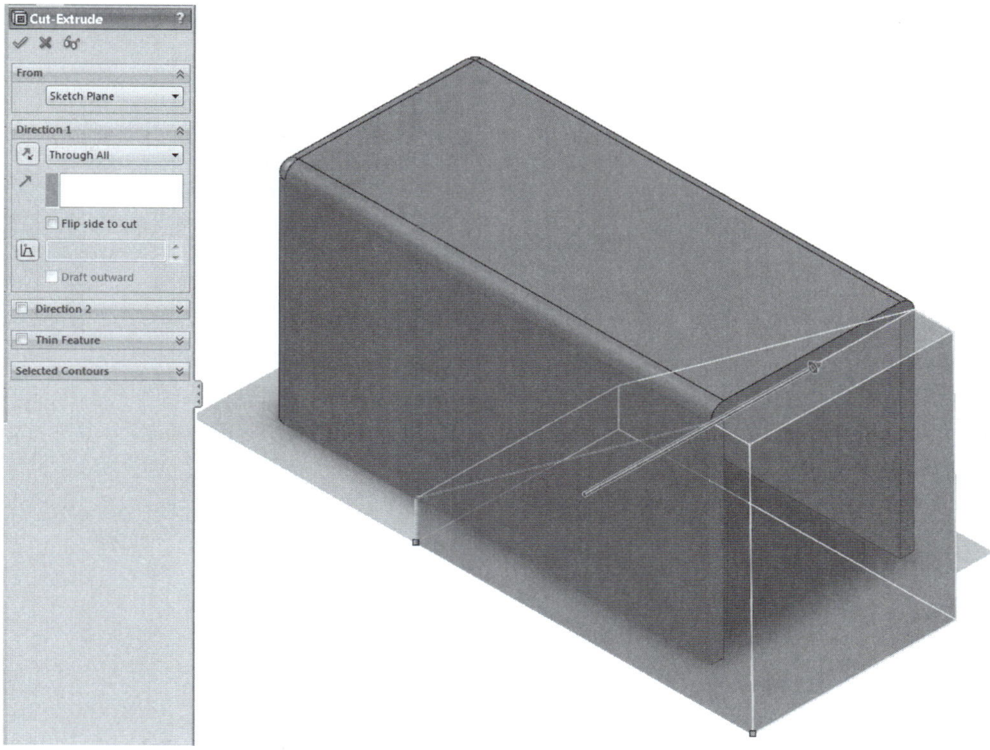

Figure 9-8
Use Extrude Cut to Extrude the Sketch Through All

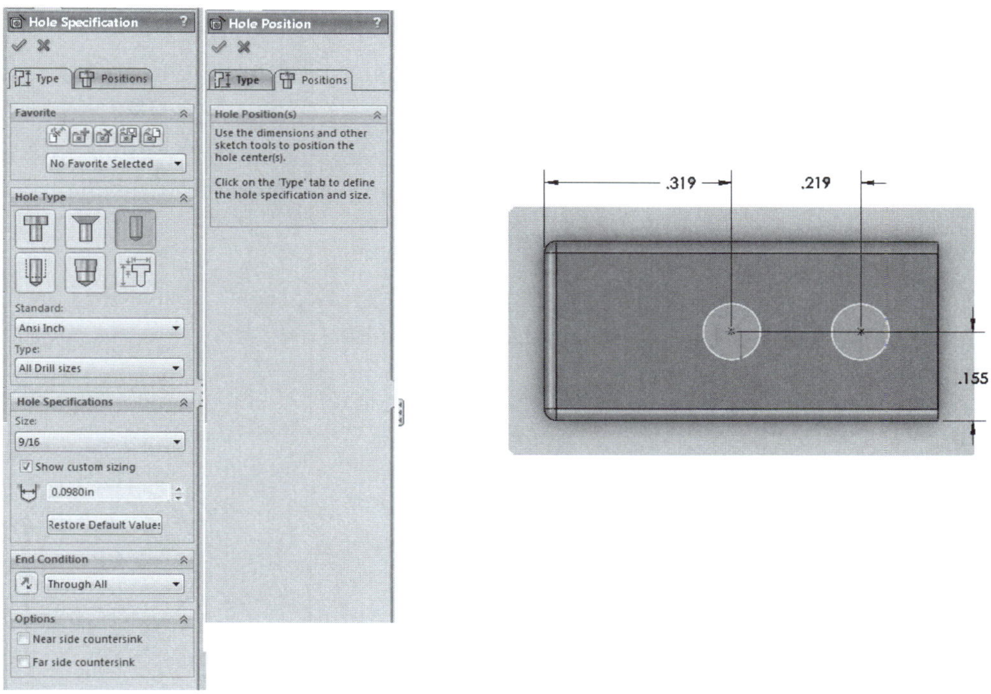

Figure 9-9
Use Hole Wizard to Make Two Holes on the Top Face

Step 14. Click the **Positions** tab, click another hole, and add dimensions as shown in Figure 9-9 and a horizontal relation between the first hole and the second hole. Then click **OK.**

Step 15. Click the left face and use **Hole Wizard** to make the settings on the **Type** tab shown in Figure 9-10.

Figure 9-10
Use Hole Wizard to Make a 9/16 Inches Hole in the Left Face

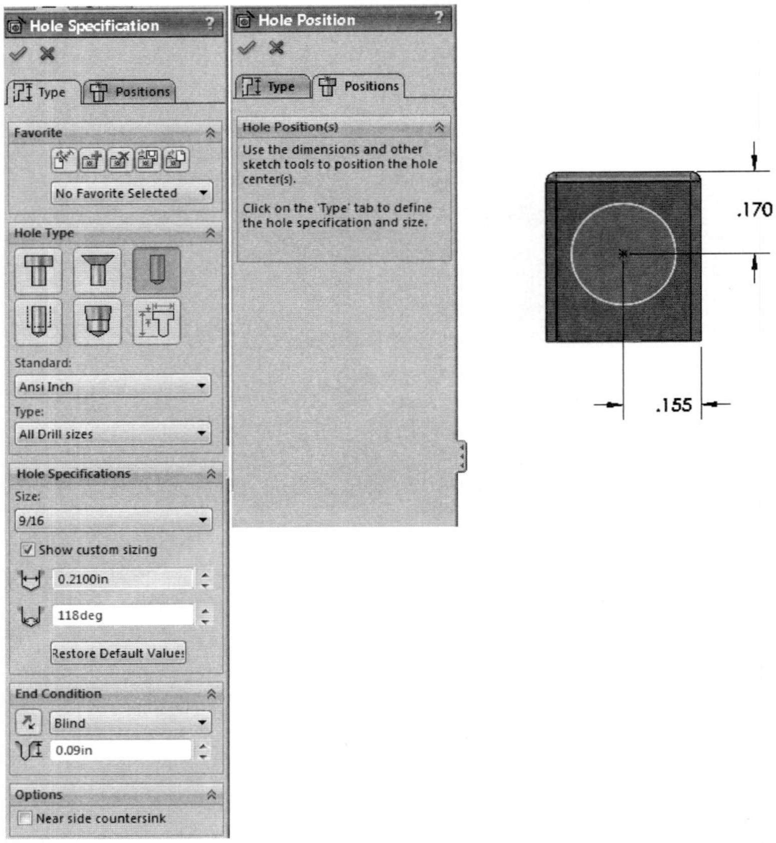

Step 16. Click the **Positions** tab and add dimensions as shown in Figure 9-10. Then click **OK.**

Step 17. Click the inside of the left face and click **Sketch.** Then draw and dimension the circle and hexagon as shown in Figure 9-11.

Figure 9-11
Draw and Dimension a Circle and a Hexagon on the Inside of the Left Face and Extrude the Area .125 Inch

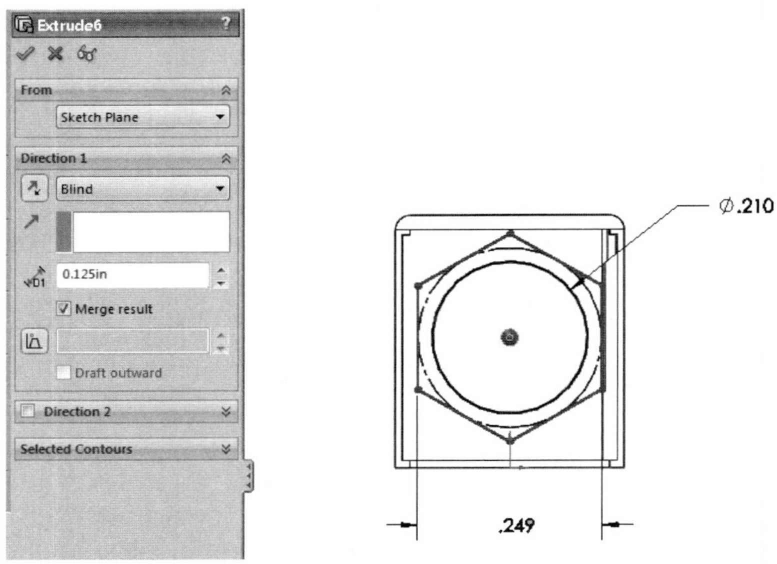

Step 18. Click **Exit Sketch** and **Extrude** the area between the circle and hexagon .125 inch.

Step 19. Use **Edit Material** to apply a finish of **Gold Iridite** (you will have to choose **Pure Gold** in the finish dialog box) as shown in Figure 9-12.

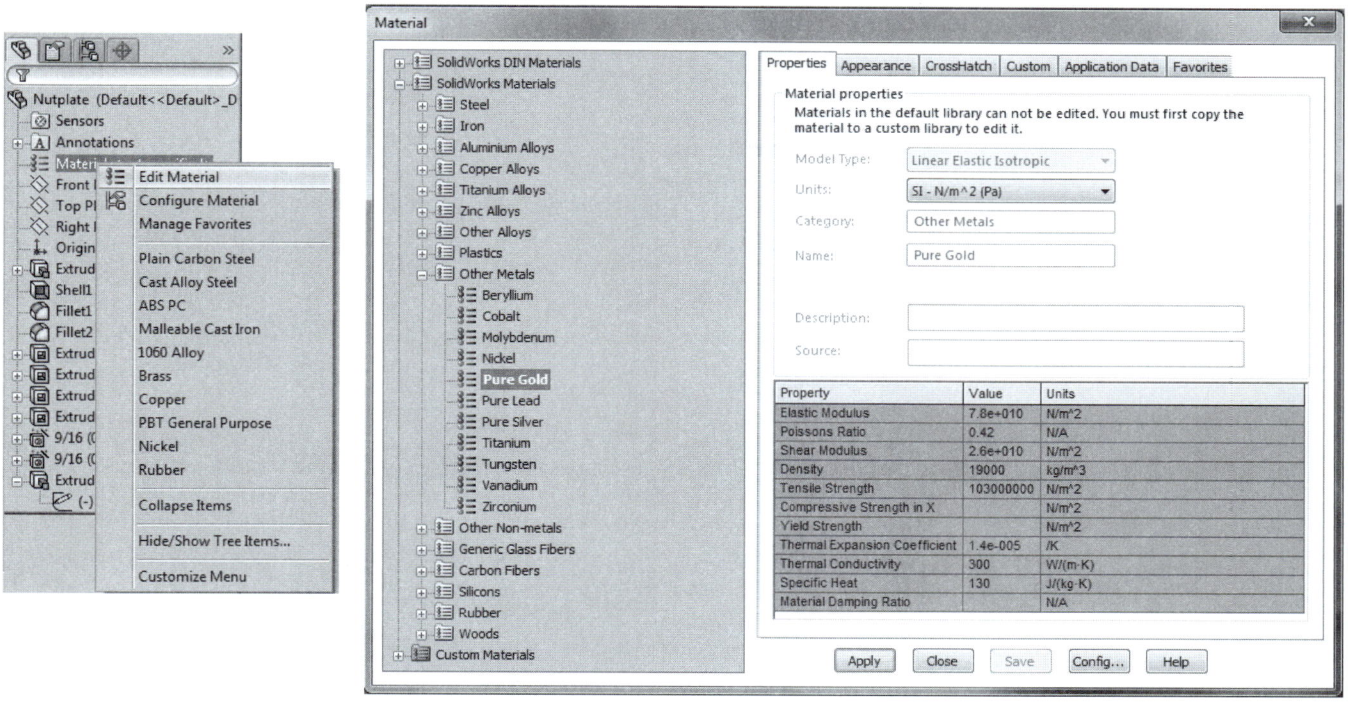

Figure 9-12
Assign the Gold Iridite Finish to the Nutplate Using Edit Material

Step 20. Use the **Options, Document Properties, Detailing** tab to check **Shaded cosmetic threads** as shown in Figure 9-13.

Step 21. Use **Cosmetic Thread...** from the **Insert** menu and click the edge of the 9/16 inches hole to show threads as shown in Figure 9-14.

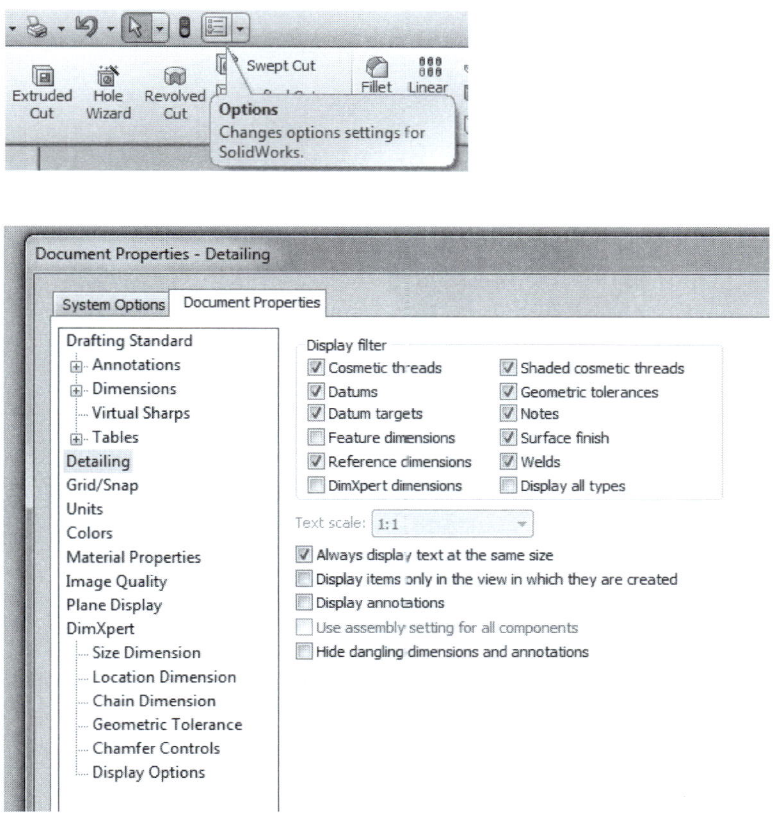

Figure 9-13
Use Options, Document Properties, Detailing Tabs to Check Shaded Cosmetic Threads

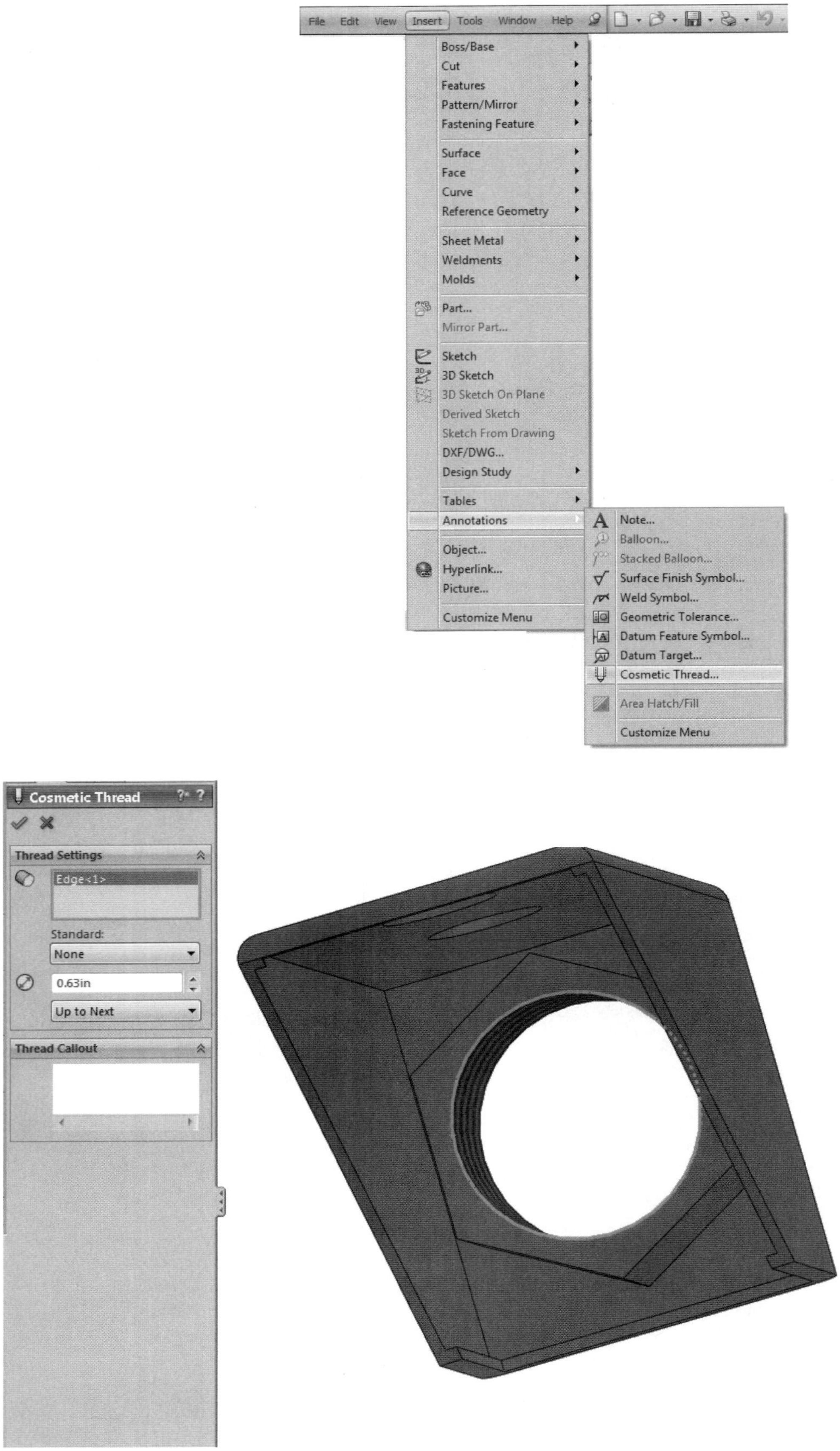

Figure 9-14
Use Cosmetic Thread... from the Insert Menu to Show Threads on the 9/16 Inches Diameter Hole

Step 22. Save your drawing with the file name **Project 9-1(your initials)** in at least two places.

Step 23. Print the drawing and turn it in.

Project 9-2: Assemble Nutplates on the Chassis for the Assembly Using Flat Head Screws

Figure 9-15 shows the chassis with nutplates assembled.

Step 1. Start SolidWorks.

Step 2. Click **Open....**

The **Open SolidWorks Document** dialog box appears.

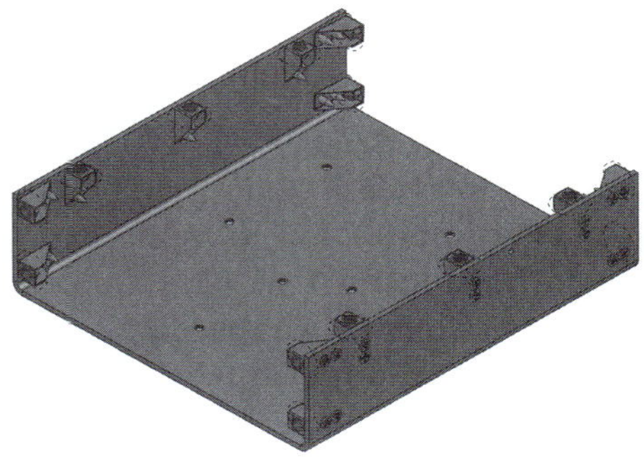

Figure 9-15
Chassis with Nutplates Assembled

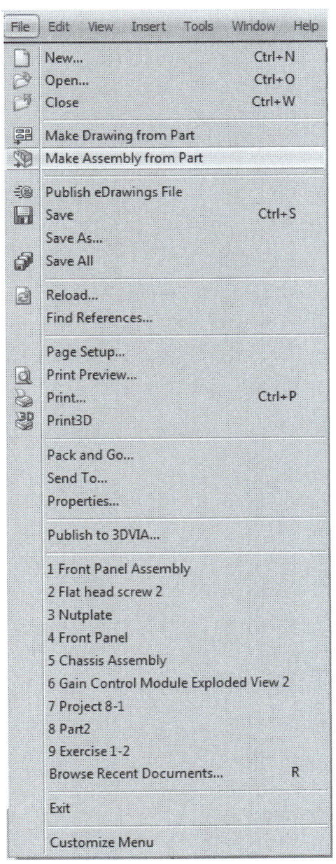

Step 3. Click **Chassis** and click **Open.**

Step 4. Click **Make Assembly from Part.**

The **Begin Assembly Manager** appears.

Step 5. Click **OK** with the word **Chassis** in the **Open documents:** box as shown in Figure 9-16.

Figure 9-16
Begin the Assembly

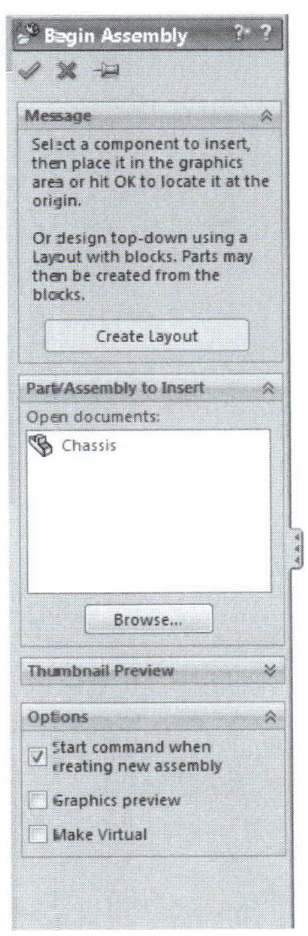

Step 6. Click **Insert Components,** click the **Nutplate** as shown in Figure 9-17, and click **Open.**

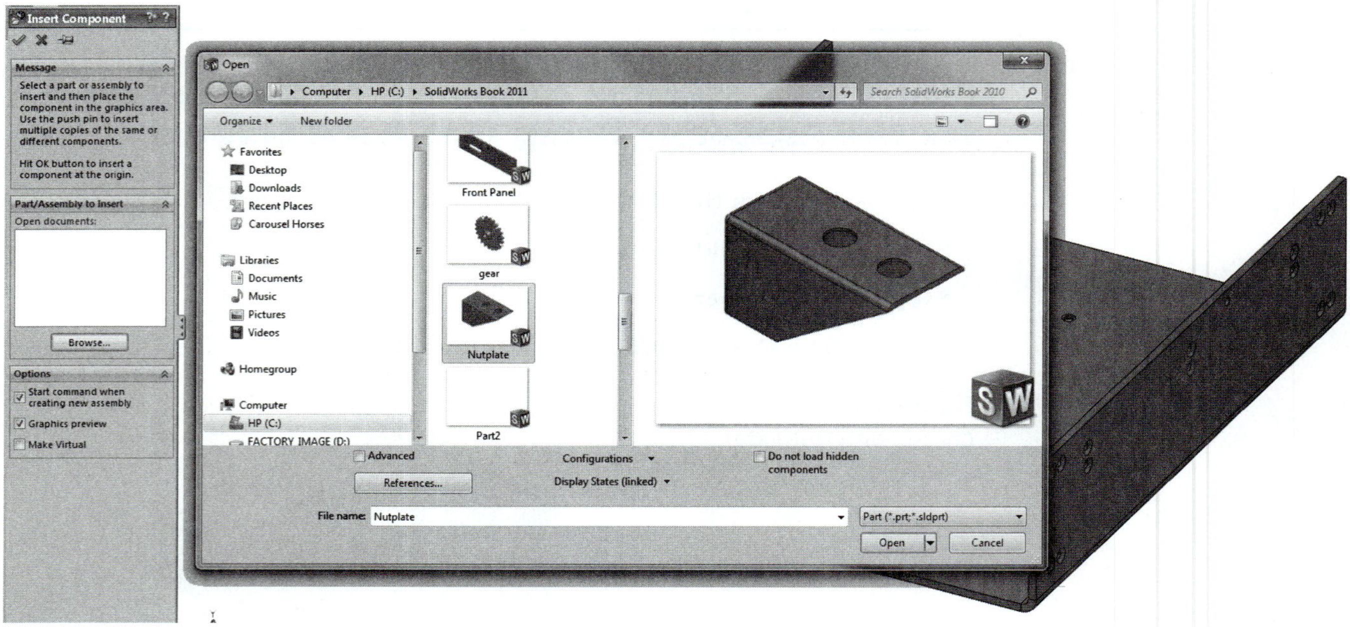

Figure 9-17
Insert the Nutplate

Step 7. Click the **Nutplate,** and right-click and **Rotate** the **Nutplate** as shown in Figure 9-18.

Figure 9-18
Rotate the Nutplate Using Move with Triad

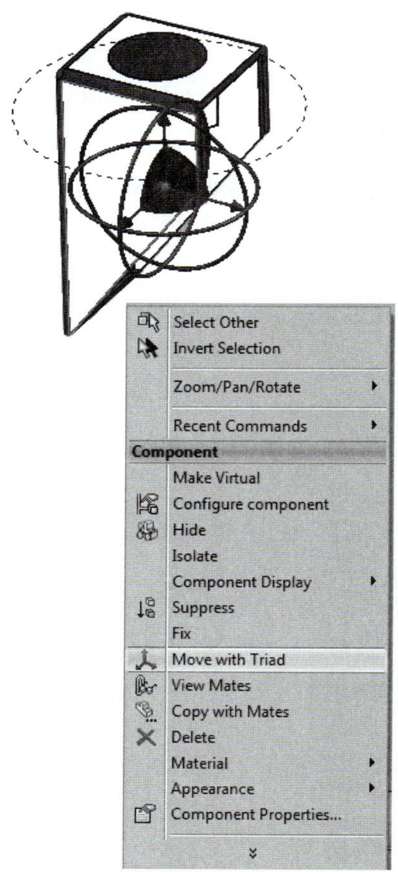

Step 8. Click **Mate,** then the **Concentric** icon as shown in the **Mate Manager.** Then, click the top small hole in the nutplate and the top hole at the right in the chassis as shown in Figure 9-19. Then, click **OK.** With the **Mate Manager** still active, repeat the **Concentric** mate for the bottom hole in the nutplate.

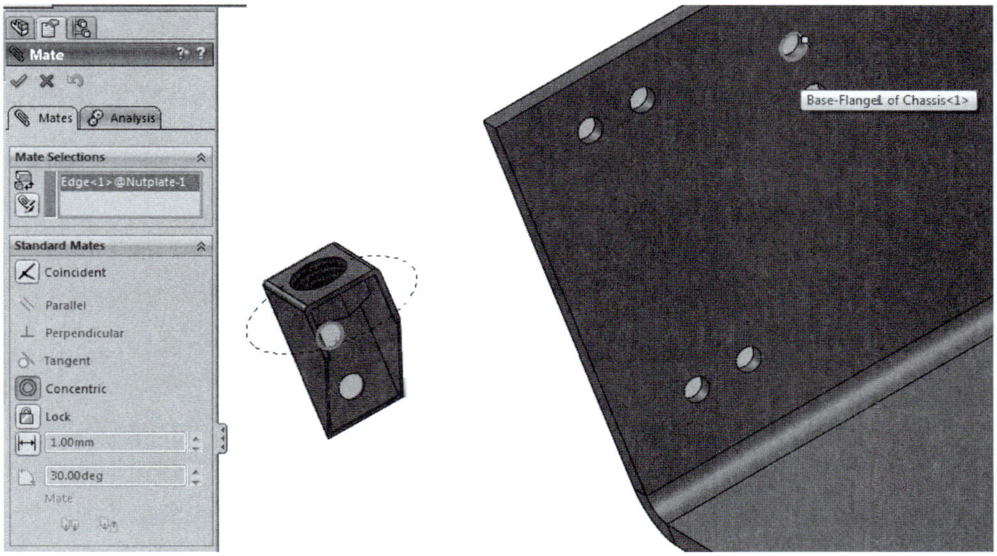

Figure 9-19
Apply Concentric Mate

Step 9. Click **Mate,** then the **Coincident** icon as shown in the **Mate Manager.** Then, click the large vertical right plane of the nutplate as shown in Figure 9-20. Then, click the inside vertical left plane in the chassis as shown in Figure 9-21. Then, click **OK.** Then, click **OK** again to close the **Mate Manager.**

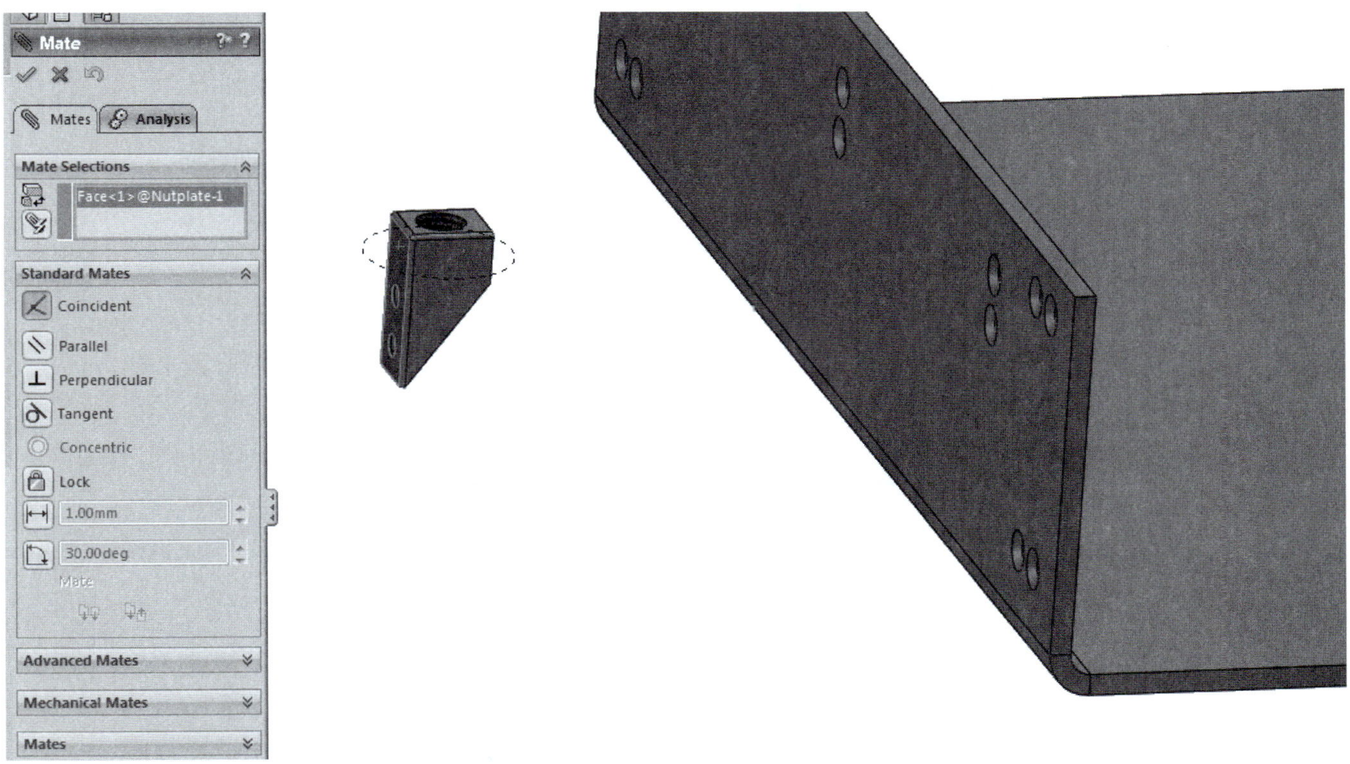

Figure 9-20
Apply the Coincident Mate, First Plane

Figure 9-21
Apply the Coincident Mate, Second Plane

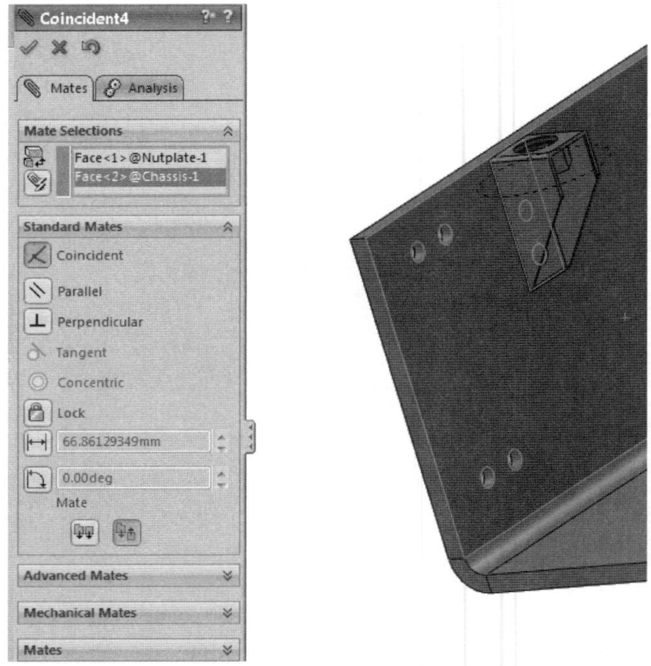

Step 10. Click **Insert Components,** click the flat head screw as shown in Figure 9-22, and click **Open.**

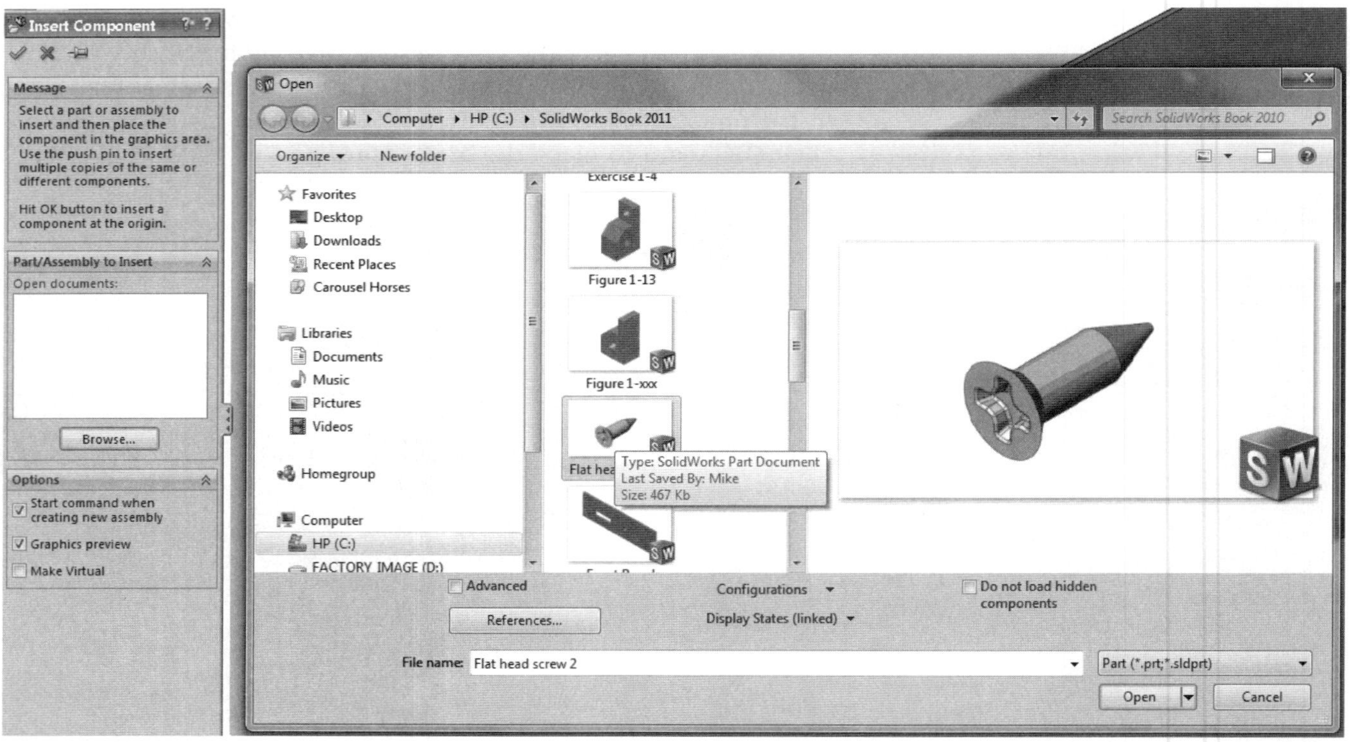

Figure 9-22
Insert the Flat Head Screw

Step 11. Click the flat head screw, and right-click and **Rotate** the flat head screw to the position shown in Figure 9-23.

Step 12. Click **Mate,** then the **Concentric** icon as shown in the **Mate Manager.** Then, click the cylindrical part of the screw and the bottom mounting hole for the nutplate as shown in Figure 9-23.

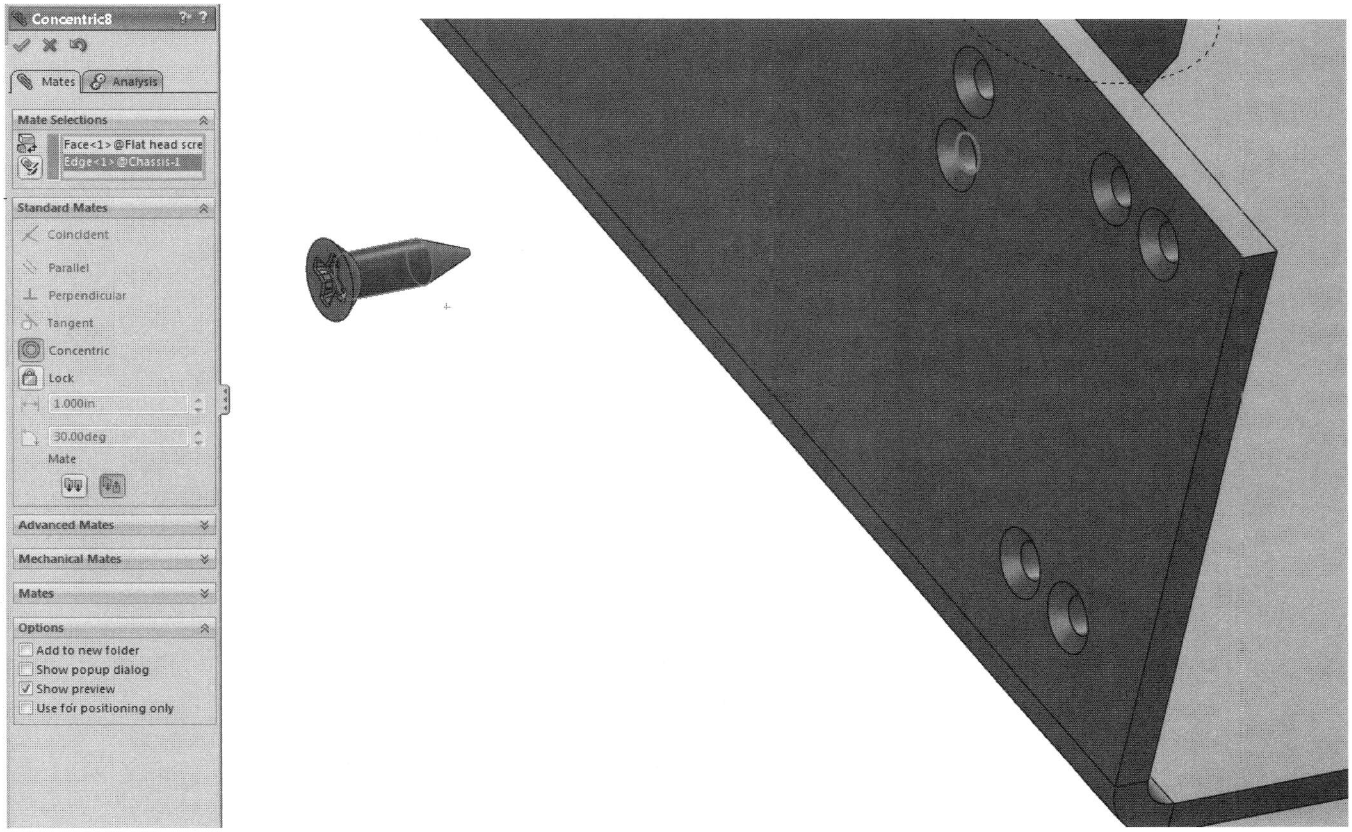

Figure 9-23
Apply the Concentric Mate

Step 13. Click **Mate,** then the **Coincident** icon as shown in the **Mate Manager.** Then, click the outside vertical left plane in the chassis as shown in Figure 9-24. Then, click the flat area on the screw

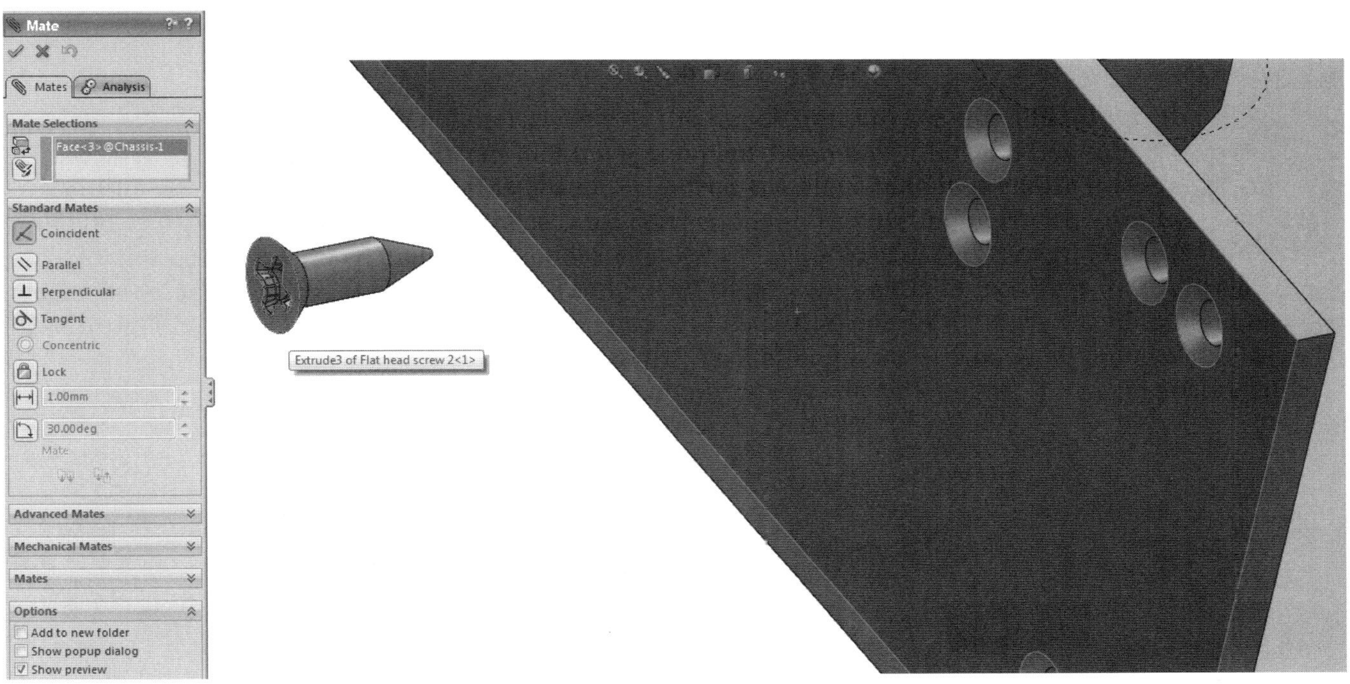

Figure 9-24
Apply the Coincident Mate, First Plane

as shown in Figure 9-25. Then, click **OK.** Then, click **OK** again to close the **Mate Manager.**

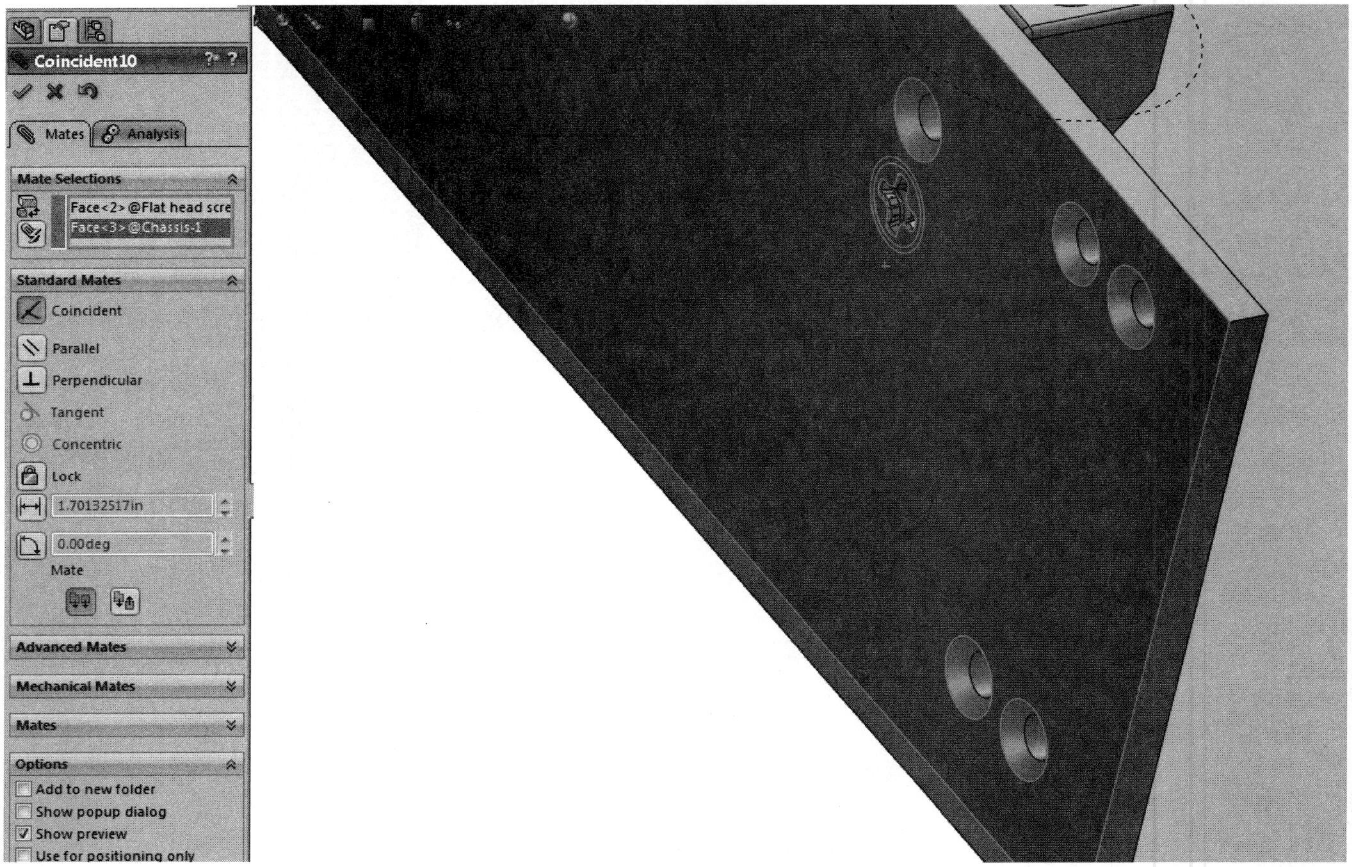

Figure 9-25
Apply the Coincident Mate, Second Plane

Step 14. Click **Linear Pattern.** Then, click a left vertical edge of the chassis and make the settings of **.219** inch in **D1** text box and **2** in the # text box. Then, click the **Components to Pattern** area in the **Linear Pattern Manager** and click the **Flat head screw** in the **Feature Manager Design Tree** as shown in Figure 9-26. Then, click **OK.**

The flat head screw is copied as shown in Figure 9-27.

Step 15. Click **Linear Pattern.** Then, click a left horizontal edge of the chassis and make the settings of **2.250** inches in the **D1** text box and **3** in the # text box. Then, click the **Components to Pattern** area in the **Linear Pattern Manager** and click the nutplate and the linear pattern of the screws in the **Feature Manager Design Tree** as shown in Figure 9-28. Then, click **OK.**

Step 16. Click **Insert Components,** click the **Nutplate,** and click **Open.**

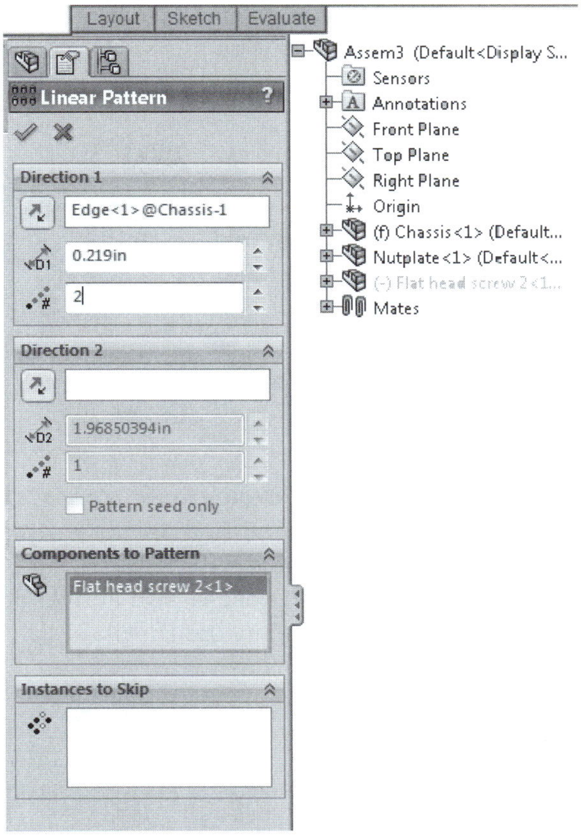

Figure 9-26
Copy the Screw Using Linear Pattern

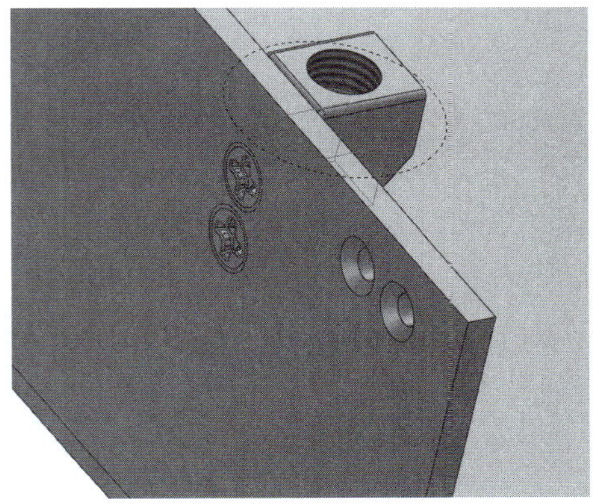

Figure 9-27
Screw Copied

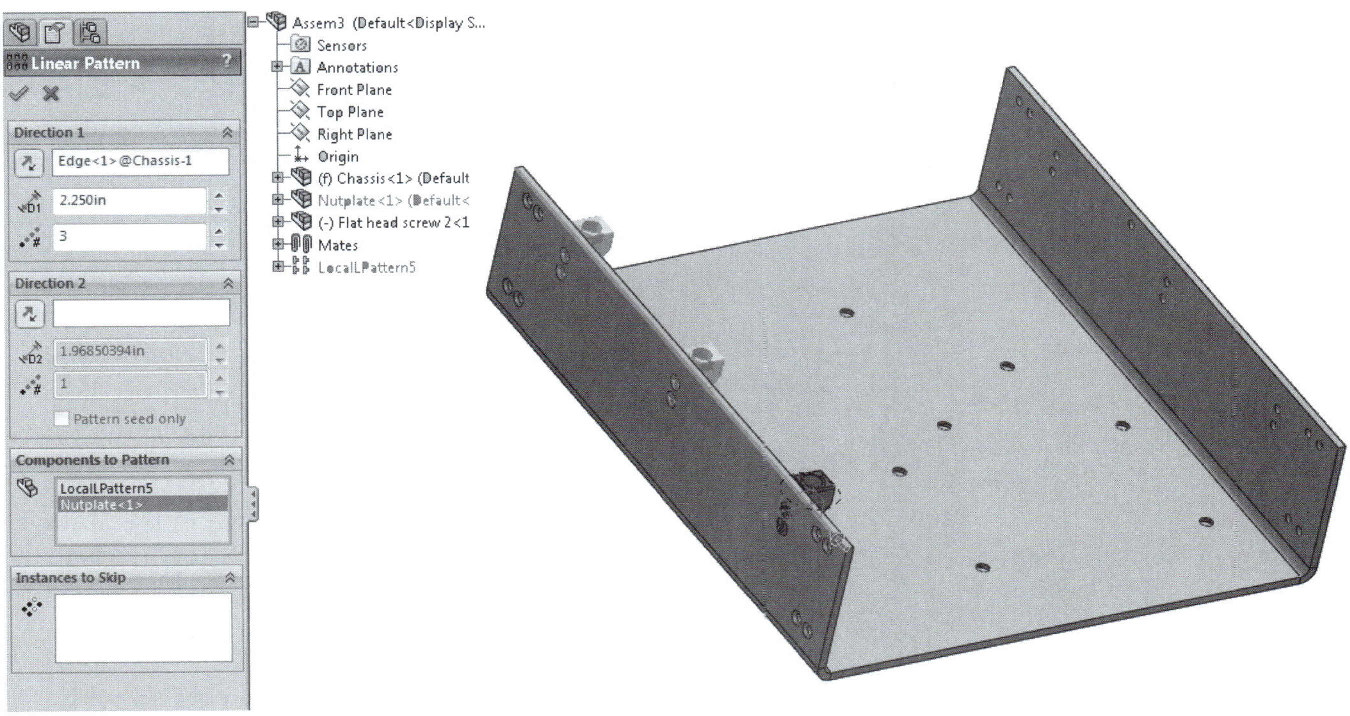

Figure 9-28
Nutplates and Screws Copied

Step 17. Click the **Nutplate,** and right-click and **Rotate** the **Nutplate** using **Move with Triad** as shown in Figure 9-29.

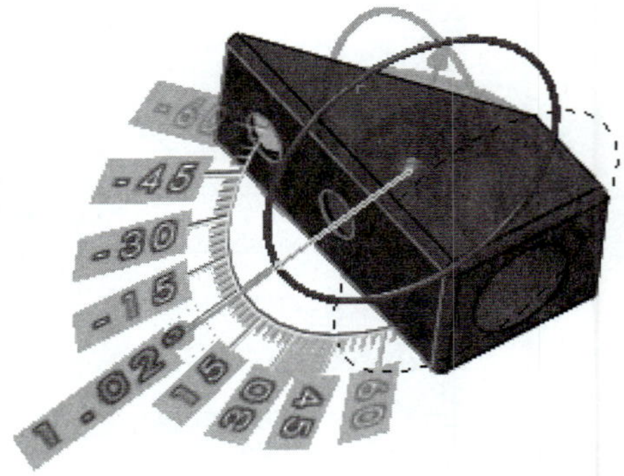

Figure 9-29
Rotate the Nutplate Using Move with Triad

Step 18. Click **Mate,** then the **Concentric** icon in the **Mate Manager.** Then, click the left small hole in the nutplate and the top hole at the left in the chassis as shown in Figure 9-30. Then, click **OK.** With the **Mate Manager** still active, repeat the **Concentric** mate for the right hole in the nutplate.

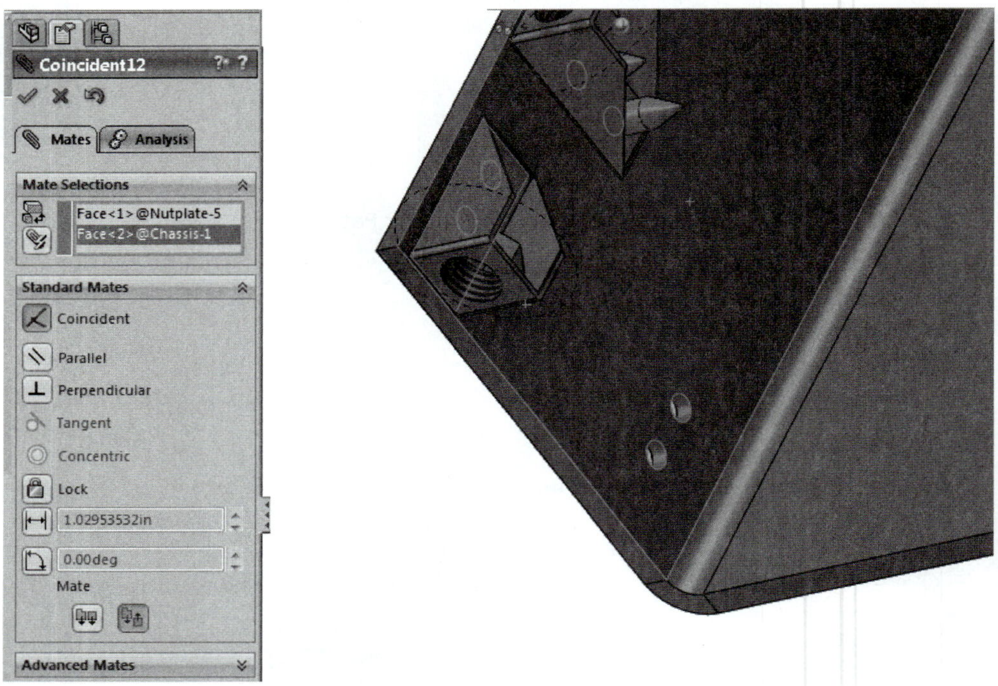

Figure 9-30
Apply the Coincident Mate

Step 19. Click **Mate,** then the **Coincident** icon as shown in the **Mate Manager.** Then, click the large vertical plane of the nutplate as shown in Figure 9-31. Then, click the inside vertical left plane in the chassis as shown in Figure 9-31. Then, click **OK.** Then, click **OK** again to close the **Mate Manager.**

YOU WILL NOW COPY THE TWO HORIZONTAL NUTPLATES TO THE OPPOSITE END OF THE CHASSIS USING THE MIRROR COMPONENTS. START WITH MAKING A PLANE FOR THE **MIRROR COMPONENTS** COMMAND.

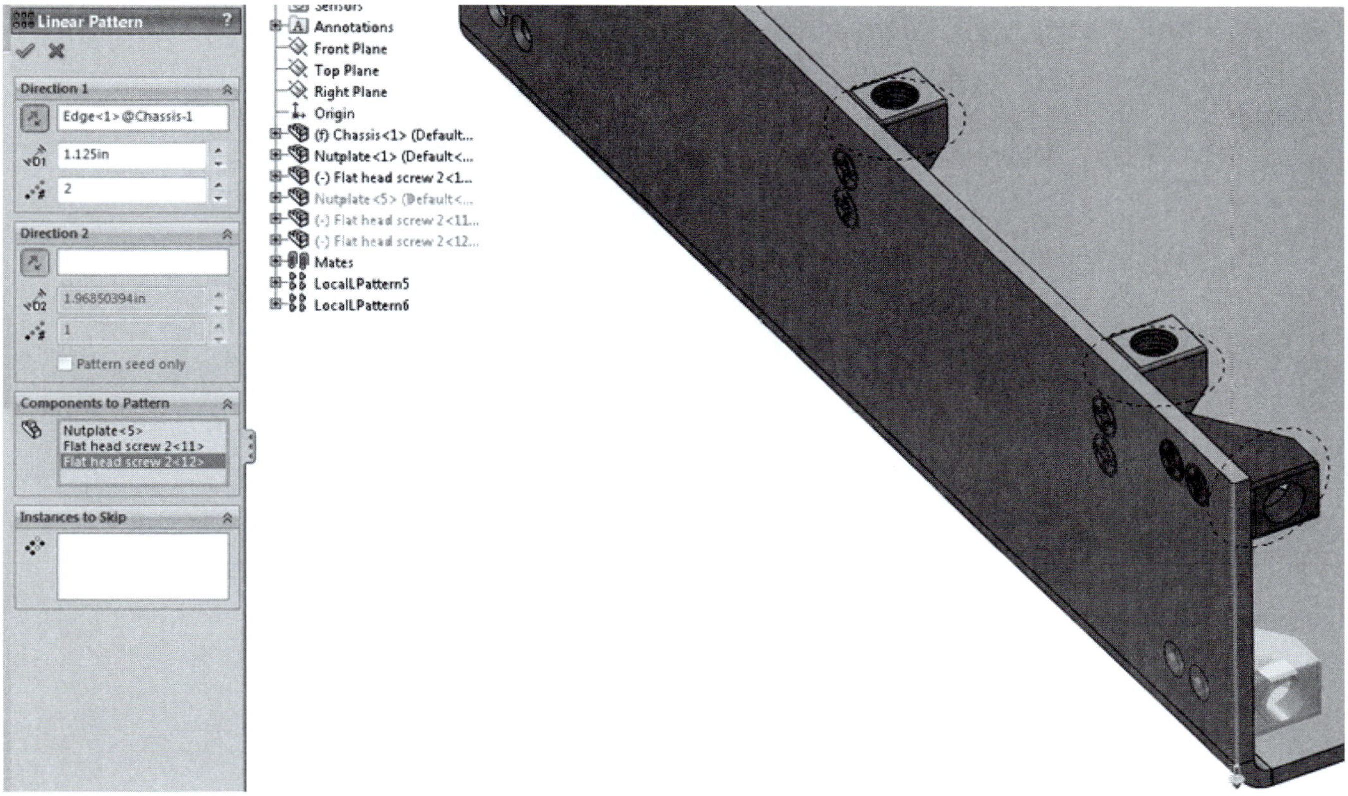

Figure 9-31
Copy the Nutplate and Screws Using Linear Pattern

Step 20. Click **Plane.** Then, click the **Front Plane** in the **Feature Manager Design Tree.** Then, type **3.375** in the **Offset Distance** box as shown in Figure 9-32 and click **OK.**

Step 21. Click **Mirror Components.** Then, with **PLANE1** shown in the **Mirror Components Command Manager,** click **Next** (the arrow pointing right). Then, click the top horizontal **Nutplate,** the two **Flat Head Screws,** and the **Local Pattern** you just made so that they appear in the **Mirror Components Command Manager** as shown in Figure 9-33. Then, click **OK.**

Step 22. Click **Plane.** Then, click the right outside face of the chassis. Then, type **3.000** in the **Offset Distance** box as shown in Figure 9-34 and click **OK.**

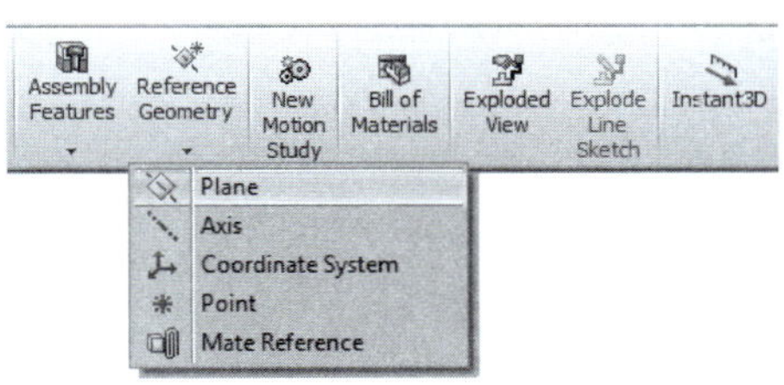

Step 23. Click **Mirror Components.** Then, with **PLANE2** shown in the **Mirror Components Command Manager,** click **Next** (the arrow pointing right). Then, click all of the nutplates, screws, and **Local Patterns** you just made so they appear in the **Mirror Components Command Manager** as shown in Figure 9-35. Then, click **OK.**

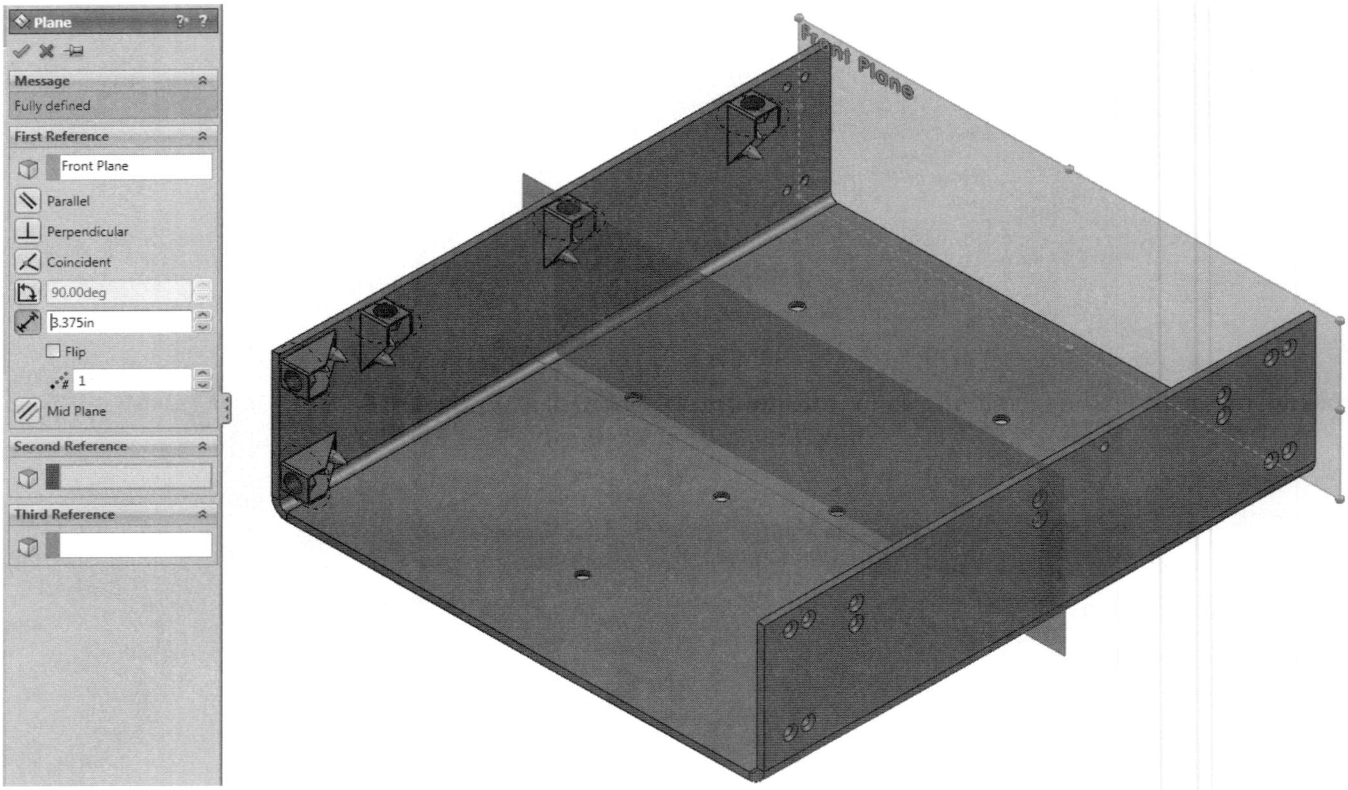

Figure 9-32
Make a Plane Offset 3.375 Inches from the Front Plane

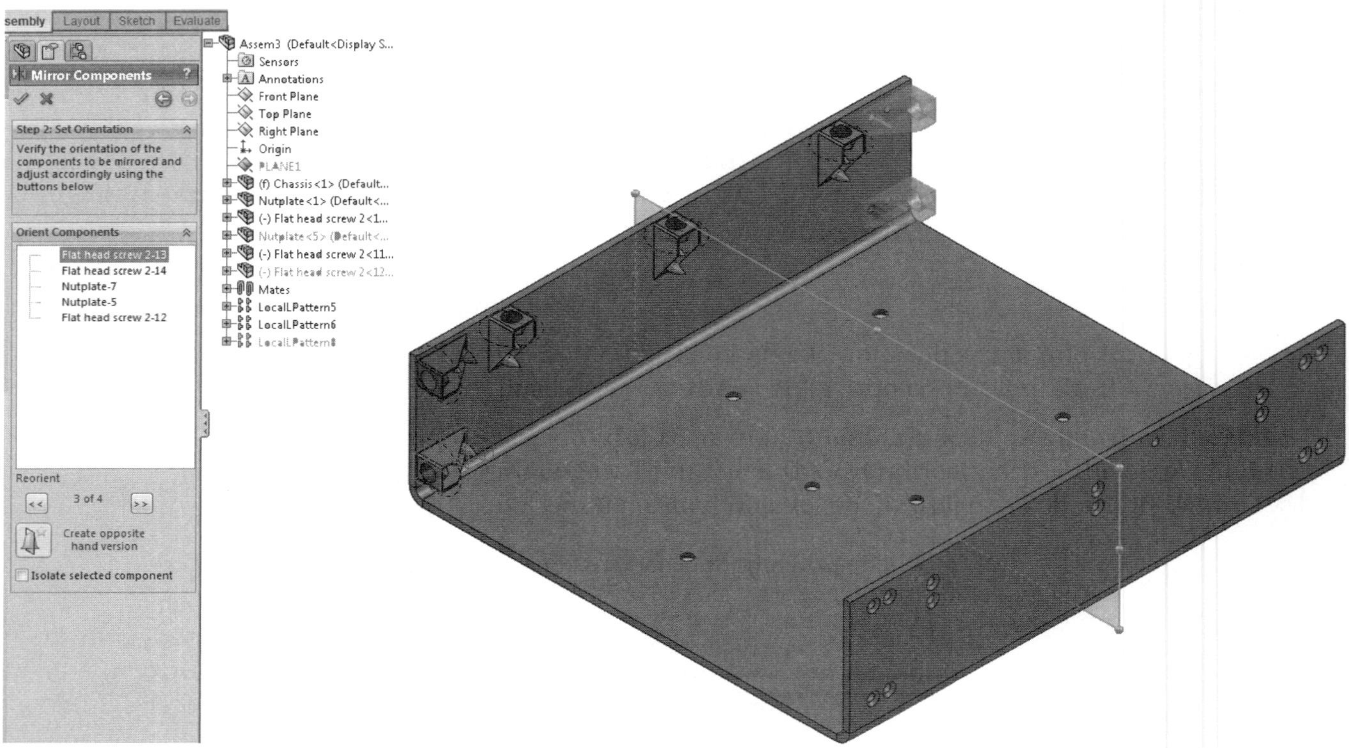

Figure 9-33
Copy Nutplates 1/3 Screws Using the Mirror Components Command and Plane1

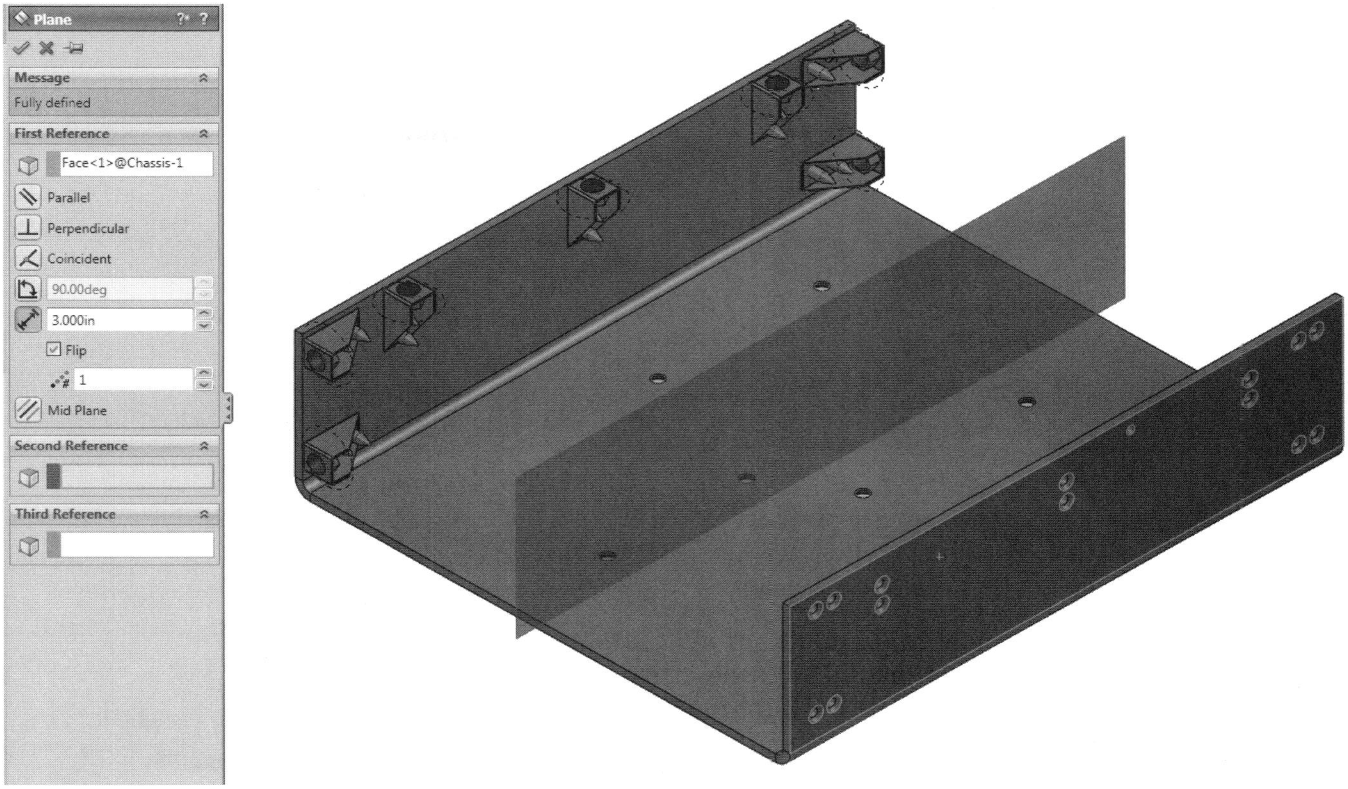

Figure 9-34
Make Plane2 and Offset It 3.000 Inches from the Right Outside Face of the Chassis

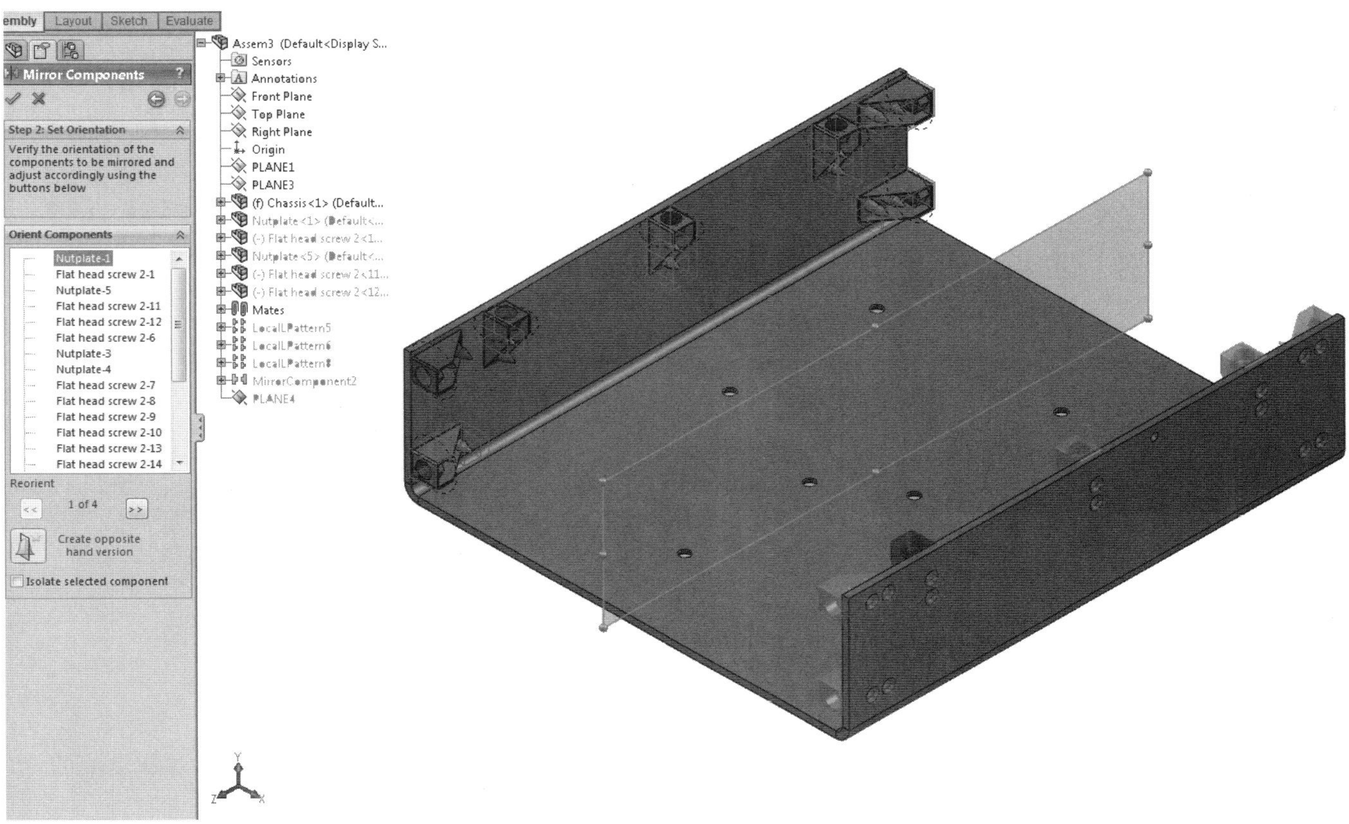

Figure 9-35
Mirror Nutplates and Screws Using Mirror Components and Plane2

All nutplates with screws are copied as shown in Figure 9-36.

Figure 9-36
Chassis with All Nutplates Assembled

Step 24. Save your drawing with the file name **Project 9-2(your initials)** in at least two places.

Step 25. Print the drawing and turn it in.

Project 9-3: Assemble Nutplates on the Front Panel for the Assembly Using Flat Head Screws

Figure 9-37 shows the front panel with nutplates assembled.

Step 1. Start SolidWorks.

Step 2. Click **Open....**

The **Open SolidWorks Document** dialog box appears.

Step 3. Click **Front Panel** and click **Open.**

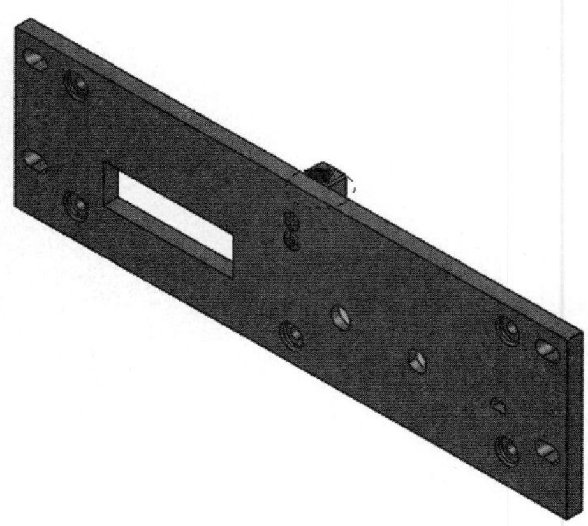

Figure 9-37
Front Panel with Nutplate Assembled

Step 4. Click **Make Assembly from Part.**

The **Begin Assembly Manager** appears.

Step 5. Click **OK** with the word **Front Panel** in the **Open documents:** box as shown in Figure 9-38.

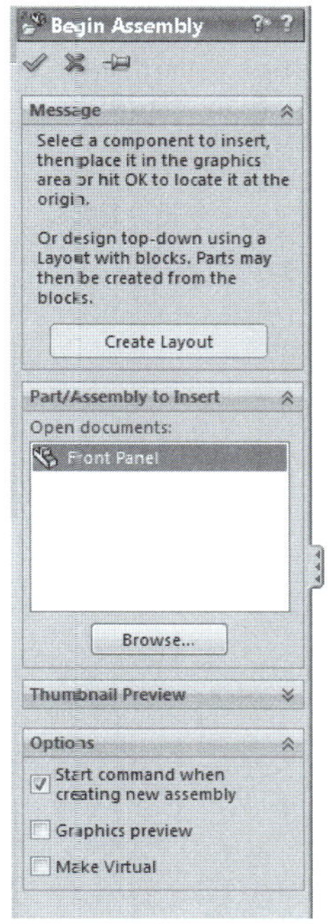

Figure 9-38
Begin the Front Panel Assembly

Step 6. Click **Insert Components** and click the **Nutplate** as shown in Figure 9-39 and click **Open.**

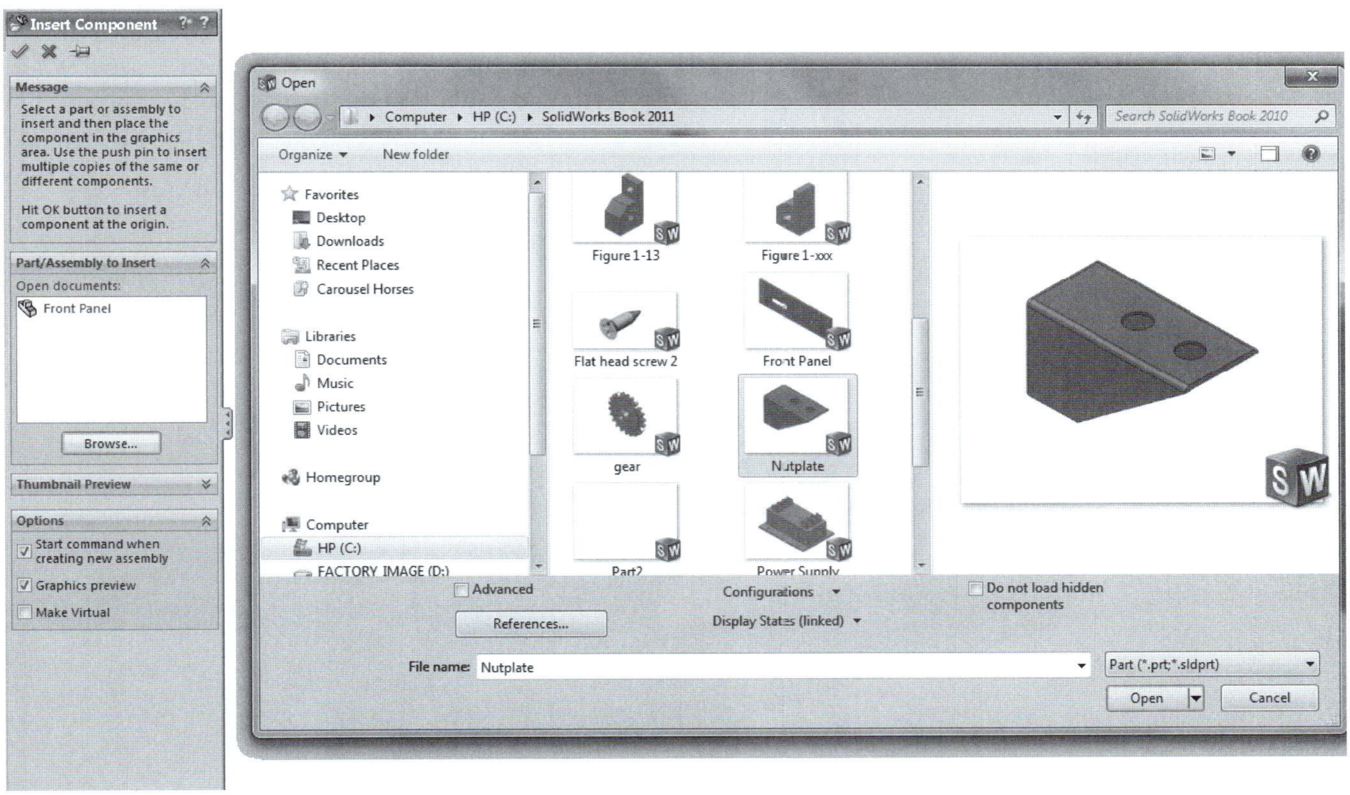

Figure 9-39
Insert the Nutplate

Step 7. Click the **Nutplate,** and right-click and **Rotate** the **Nutplate** using **Move with Triad** as shown in Figure 9-40.

Figure 9-40
Rotate the Nutplate Using Move with Triad

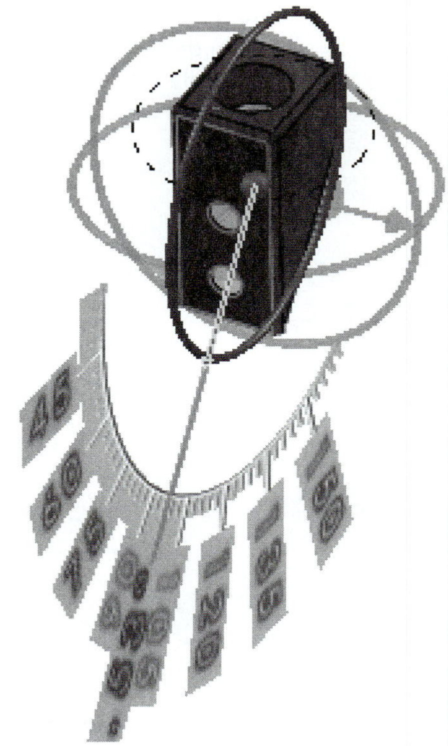

Step 8. Click **Mate,** then the **Concentric** icon as shown in the **Mate Manager.** Then, click the top small hole in the nutplate and the top hole at the right in the chassis as shown in Figure 9-41. Then, click **OK.** With the **Mate Manager** still active, repeat the **Concentric** mate for the bottom hole in the nutplate.

Figure 9-41
Apply Concentric Mate

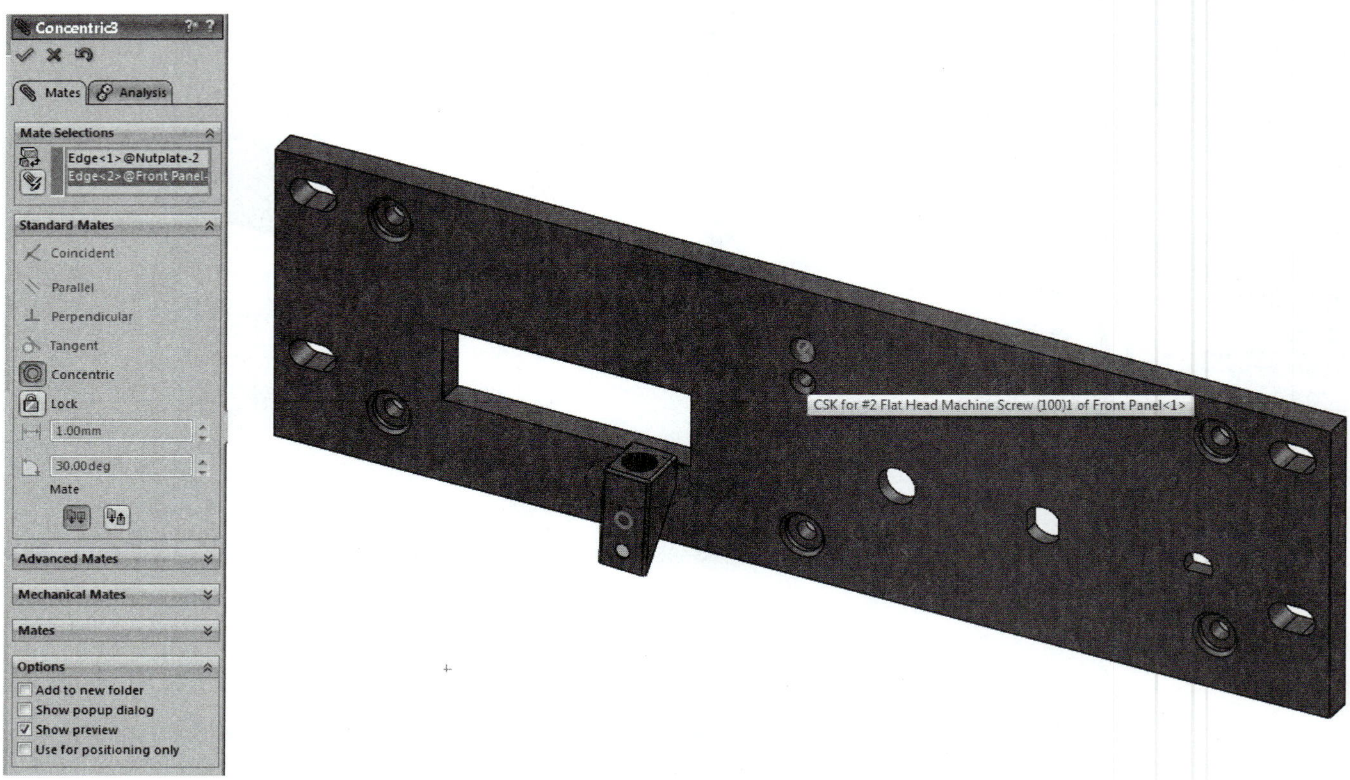

Step 9. Click **Mate,** then the **Coincident** icon as shown in the **Mate Manager.** Then, click the large vertical plane of the nutplate as shown in Figure 9-42. Then, click the inside vertical plane of the front panel as the second plane as shown in Figure 9-42. Then, click **OK.** Then, click **OK** again to close the **Mate Manager.**

Figure 9-42
Apply the Coincident Mate, Second Plane

Step 10. Click **Insert Components,** click the flat head screw as shown in Figure 9-43, and click **Open.**

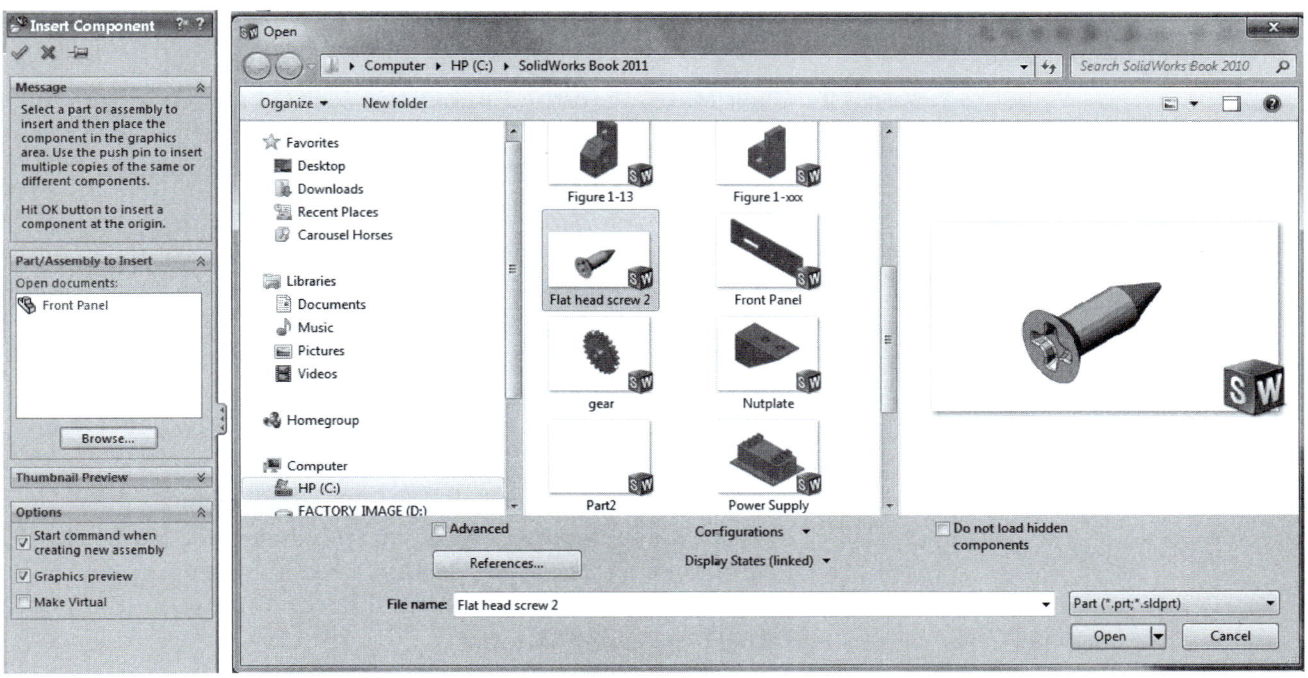

Figure 9-43
Insert the Flat Head Screw

Step 11. Click the flat head screw, and right-click and **Rotate** the flat head screw to the position shown in Figure 9-44 if it is not already in that position.

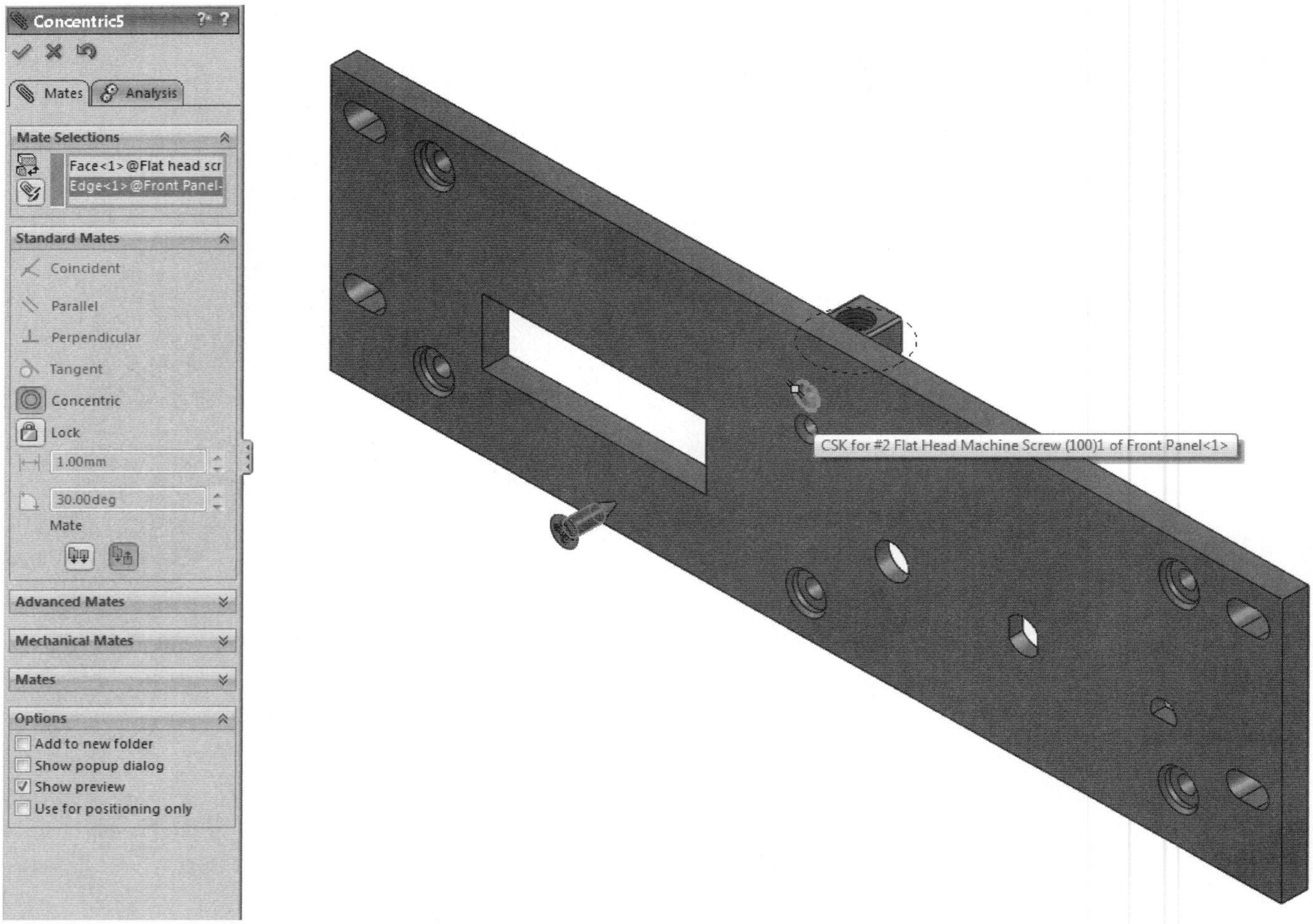

Figure 9-44
Apply the Concentric Mate

Step 12. Click **Mate,** then the **Concentric** icon as shown in the **Mate Manager.** Then, click the cylindrical part of the screw and the top mounting hole for the front panel as shown in Figure 9-44.

Step 13. Click **Mate,** then the **Coincident** icon as shown in the **Mate Manager.** Then, click the flat area on the screw as the first plane as shown in Figure 9-45. Then, click the outside plane of the front panel. Then, click **OK.** Then, click **OK** again to close the **Mate Manager.**

Step 14. Click **Linear Pattern.** Then, click a vertical edge of the front panel and make the settings of **.219** inch in **D1** text box and **2** in the # text box. Then, click the **Components to Pattern** area in the **Linear Pattern Manager** and click the **Flat head screw** in the **Feature Manager Design Tree** as shown in Figure 9-46. Then, click **Reverse Direction** if needed (the icon just above **D1**). Then, click **OK.**

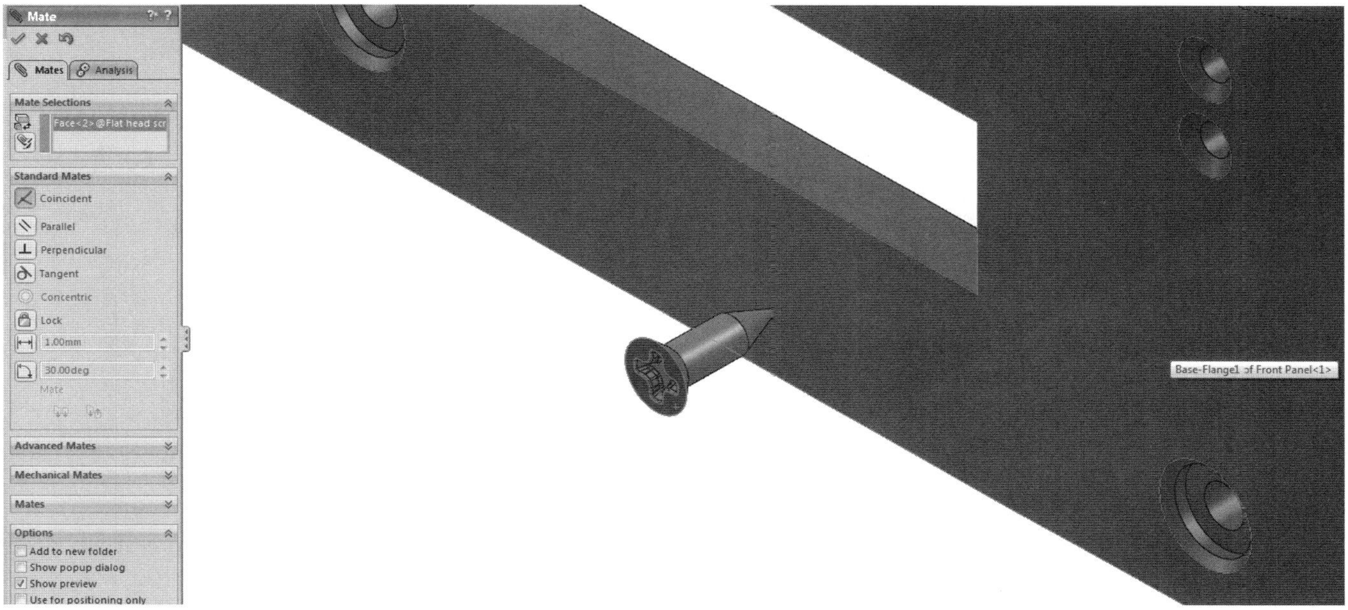

Figure 9-45
Apply the Coincident Mate First Plane

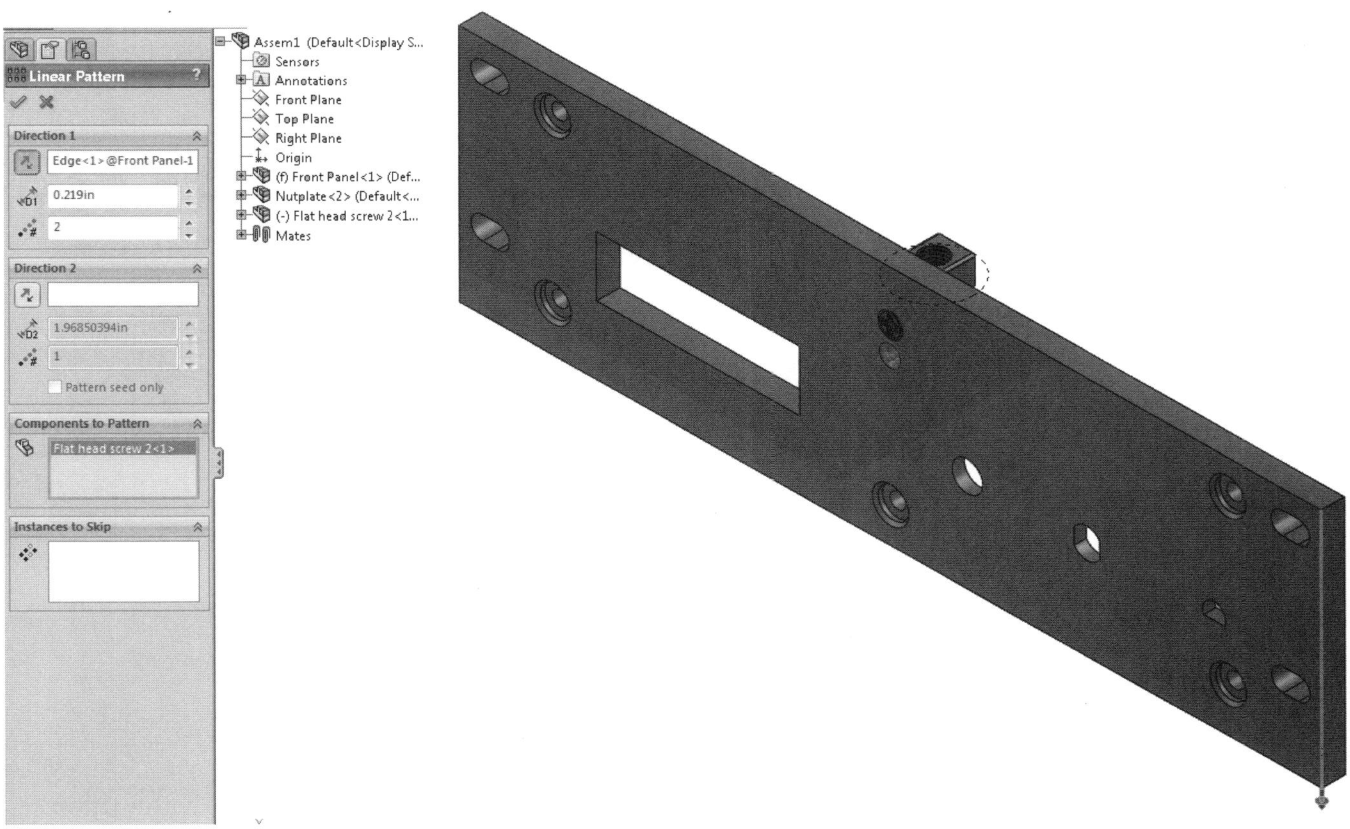

Figure 9-46
Copy the Screw Using Linear Pattern

The flat head screw is copied as shown in Figure 9-47, and the front panel assembly is now complete.

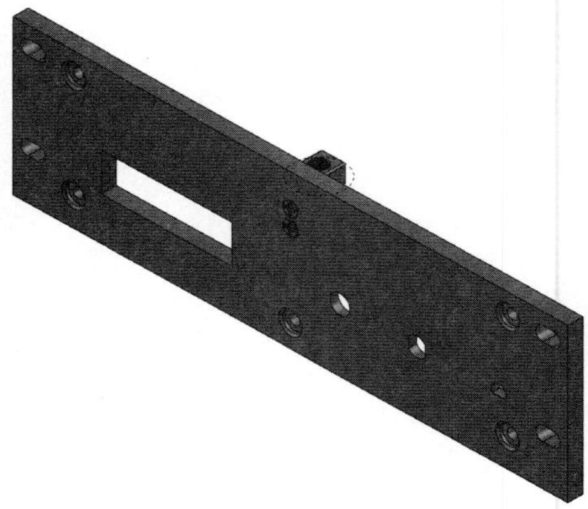

Figure 9-47
Screw Is Copied and Front Panel Assembly Is Complete

Step 15. Save your drawing with the file name **Project 9-3(your initials)** in at least two places.

Step 16. Print the drawing and turn it in.

Project 9-4: Make an Assembly from the Chassis Assembly and the Front Panel Assembly

Figure 9-48 shows the chassis assembly and front panel assembly as an assembly.

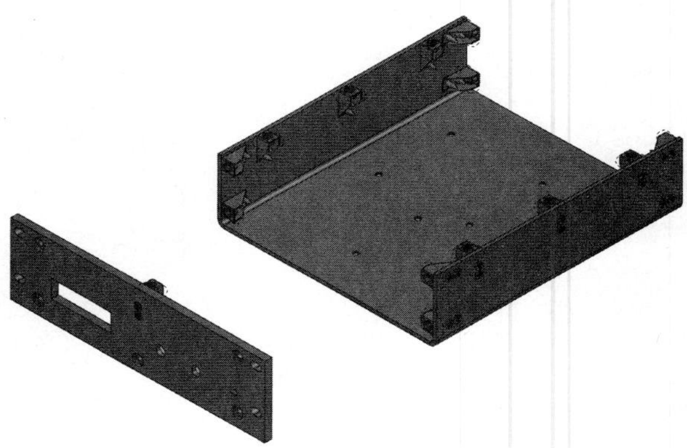

Figure 9-48
Front Panel and Chassis with Nutplates

Step 1. Start SolidWorks.

Step 2. Click **Open....**

The **Open SolidWorks Document** dialog box appears.

Step 3. Click **Front Panel Assembly** and click **Open.**

Step 4. Click **Make Assembly from Assembly.**

The **Begin Assembly Manager** appears.

Step 5. Click **OK** with the word **Front Panel Assembly** in the **Open documents:** box as shown in Figure 9-38.

Step 6. Click **Insert Components,** click the **Nutplate** as shown in Figure 9-39, and click **Open.**

Step 7. Click the **Nutplate,** and right-click and **Rotate** the **Nutplate** using **Move with Triad** as shown in Figure 9-40.

Step 8. Click **Mate,** then the **Concentric** icon as shown in the **Mate Manager.** Then, click the top small hole in the nutplate and the top hole at the right in the chassis as shown in Figure 9-41. Then, click **OK.** With the **Mate Manager** still active, repeat the **Concentric** mate for the bottom hole in the nutplate.

Step 9. Click **Mate,** then the **Coincident** icon as shown in the **Mate Manager.** Then, click the large vertical plane of the nutplate as shown in Figure 9-42. Then, click the inside vertical plane of the front panel as the second plane as shown in Figure 9-42. Then, click **OK.** Then, click **OK** again to close the **Mate Manager.**

Step 10. Click **Insert Components,** click the flat head screw as shown in Figure 9-43, and click **Open.**

Step 11. Click the flat head screw, and right-click and **Rotate** the flat head screw to the position shown in Figure 9-44.

Step 12. Click **Mate,** then the **Concentric** icon as shown in the **Mate Manager.** Then, click the cylindrical part of the screw and the bottom mounting hole for the nutplate as shown in Figure 9-44.

Step 13. Click **Mate,** then the **Coincident** icon as shown in the **Mate Manager.** Then, click the outside plane of the front panel as shown in Figure 9-45. Then, click the flat area on the screw as shown in Figure 9-46. Then, click **OK.** Then, click **OK** again to close the **Mate Manager.**

Step 14. Click **Linear Pattern.** Then, click a vertical edge of the front panel and make the settings of **.219** inch in **D1** text box and **2** in the **#** text box. Then, click the **Components to Pattern** area in the **Linear Pattern Manager** and click the **Flat head screw** in the **Feature Manager Design Tree** as shown in Figure 9-46. Then, click **OK.**

The flat head screw is copied as shown in Figure 9-47, and the front panel assembly is now complete.

Step 15. Save your drawing with the file name **Project 9-3(your initials)** in at least two places.

Step 16. Print the drawing and turn it in.

9 chapternine

Chapter Summary

This chapter contains the information necessary to make a complex model (a nutplate) to specific 3-place decimal dimensions. It also gives you the skills needed to make assemblies from parts, assemblies from assemblies, and drawings from complex parts. Now you have the skills and information necessary to make solid models, assemblies, and drawings.

Review Questions

Circle the most correct answer.

1. Where is the **Shell** command located?
   a. **Shell** toolbar
   b. **Sketch** toolbar
   c. **Features** toolbar
   d. Model menu
2. Which tab on the **Hole Wizard Manager** allows you to make dimensions to locate holes?
   a. **Type**
   b. **Position**
   c. **Sketch**
   d. **Model**
3. Where is the **Fillet** command for solids located?
   a. **Fillet** toolbar
   b. **Sketch** toolbar
   c. **Features** toolbar
   d. **Model** menu
4. Which of the following allows you to apply material to a part?
   a. **Edit Material**
   b. **Material**
   c. **Finish**
   d. **6061 Aluminum**
5. Which of the following settings is used to show threads in a tapped hole?
   a. **Cylindrical thread**
   b. **Edit thread**
   c. **Cosmetic thread**
   d. **Screw thread**
6. A part file must be open for it to appear in the **Open Documents** area when an assembly is begun.
   a. True
   b. False
7. To line up a circular surface with another circular surface, use the **Coincident** mate.
   a. True
   b. False
8. To add parts to an assembly, the **Insert Components** command must be used.
   a. True
   b. False
9. The **Linear Pattern** command cannot be used to copy holes.
   a. True
   b. False
10. A plane is not useful when you use the **Mirror Components** command.
    a. True
    b. False

9 chapternine

Matching

Place the correct letter on the line to the right of the numbered item.

1. A command that is used to copy items along an edge ______
2. A feature that allows you to mirror a part ______
3. A command that allows you to add parts to an assembly ______
4. The area where the **Cosmetic Thread** setting is located ______
5. A command that allows you to make models with thin walls ______

a. **Shell**
b. **Linear Pattern**
c. **Document Properties-Detailing**
d. **Plane**
e. **Insert Components**

General Questions

Review the following questions.

1. Why are assemblies needed?
2. How can I make an assembly from several subassemblies?
3. How can I use **Planes** to make models?
4. Why is it necessary to have consistent spacing between rows of dimensions?
5. What is the **Move with Triad** command used for and when should it be used instead of another command?

chapter ten

Assemble All of the Components for the Sheet Metal Assembly and Make an Exploded View

CHAPTER OBJECTIVES

After completing this chapter, you will be able to:

- Assemble parts onto the chassis.
- Assemble the front panel assembly, rear panel assembly, and cover onto the chassis assembly.
- Edit parts if needed.
- Make an exploded view.
- Correctly answer questions regarding SolidWorks commands used in this chapter.

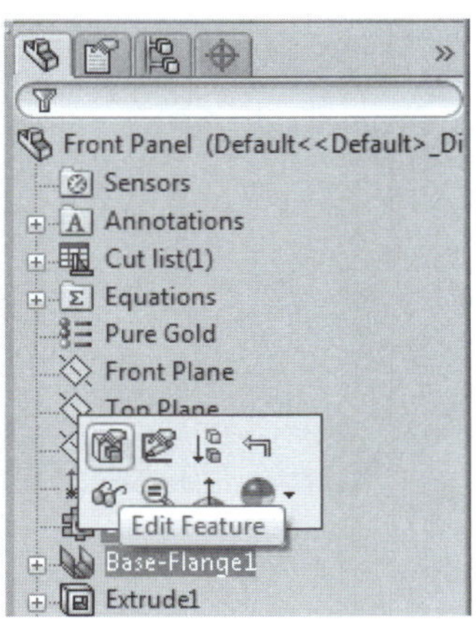

Introduction to the Real World of Sheet Metal Assembly

This chapter gives you the opportunity to see the difficulties that may occur when making a complex assembly. If holes do not align between parts in this assembly, check the chassis to make sure it is correct. Then, make any necessary corrections to the mating parts (the front panel, rear panel, or the chassis cover) by using **Edit Sketch** or the **Edit** feature. If you have not already used those commands to fix a feature, right-click on the item to be changed and click the icon that shows the command as shown in the margin art. Assembling several parts and subassemblies often involves making changes and corrections, so do not be surprised if you find yourself doing exactly that.

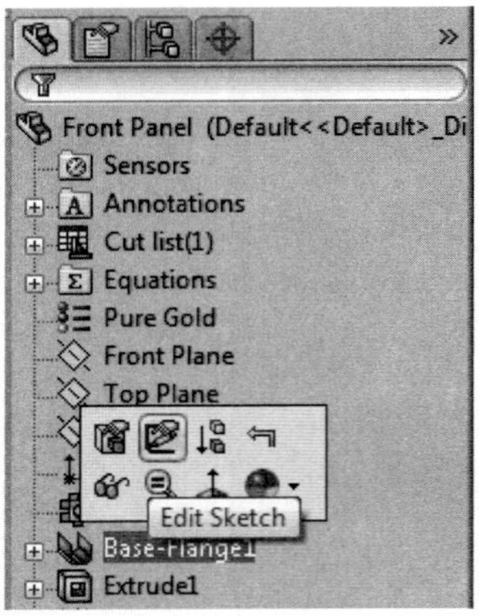

Using Copy with Mates

Copy with Mates can save you considerable time when it is appropriate. To use **Copy with Mates**, select the component or components to be copied, right-click on one of the components, and select **Copy with Mates.** The **Property Manager** shows the selected component to copy as shown in Figure 10-1. You can now add other components. Then, select the new mate or mates for where you will be copying the parts. In this chapter you may find it helpful to copy the screws and nut plates or other components to a different hole on the same face. To do that, check **Repeat** to repeat the **Coincident** mate because the new screw will be coincident to the same face as the original. For the **Concentric** mate reference, pick the face of the new hole that the screw will be inserted into. Right after you click the face, the new components appear. Click on **OK** (the green check) to accept them. This can be repeated as many times as you like. The **Parallel** mate works in the same manner.

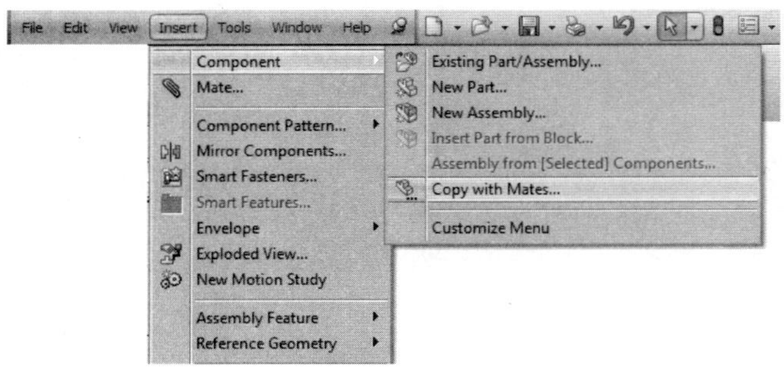

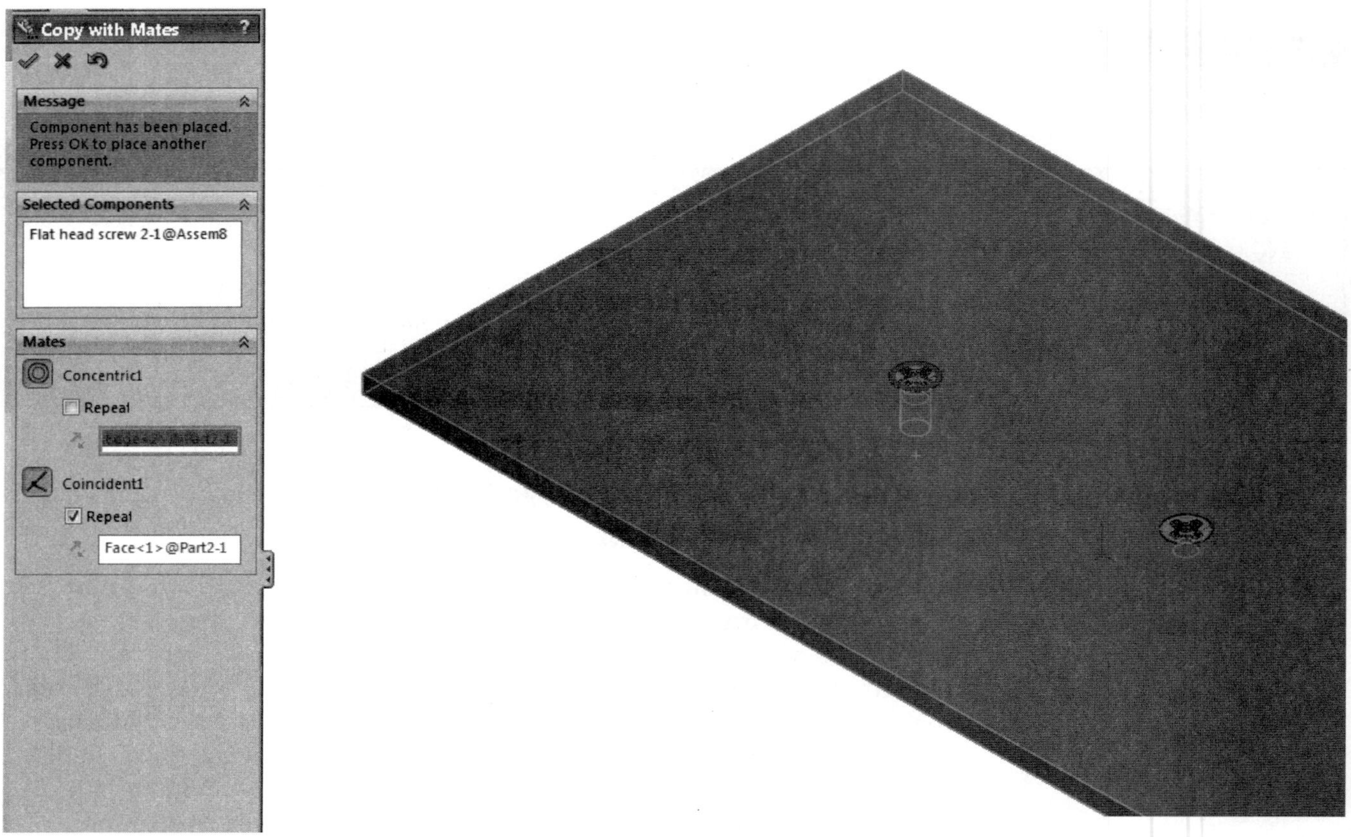

Figure 10-1
Copy with Mates

Inserting Components

Before you begin inserting components you can shorten the browsing time by opening all of the parts that are in the assembly. After you have all the parts, choose the largest or most important one and click **Make Assembly from Part.** If you have a subassembly, click **Make Assembly from Assembly.**

After you have clicked **Make Assembly from Part,** click **Insert Components** and click **OK** to insert the first part. After that, click **Insert Components** and click the next part in the **Insert Components Manager,** or click **Browse** and locate the next part.

Applying Mates

When you have all of the parts of a small assembly or some of the parts of a larger assembly, use mates to assemble the parts. The ones used in this chapter are:

Concentric. To use the **Concentric** mate, click **Mate,** then click the cylindrical surface of one part.

Next, click the cylindrical surface of the mating part and, if necessary, click the **Concentric** icon in the **Mate** manager.

Coincident. To use the **Coincident** mate, click **Mate,** then click the flat surface of one part.

Next, click the flat surface of the mating part and, if necessary, click the **Coincident** icon in the **Mate** manager.

Parallel. To use the **Parallel** mate, click **Mate,** then click an edge of one part. Next, click an edge of the mating part and, if necessary, click the **Parallel** icon in the **Mate** manager.

Project 10-1: Assemble Parts onto the Chassis and Make an Assembly (Figure 10-2)

If you have exited SolidWorks, open it again and start a new assembly using **Make Assembly from Part.**

Figure 10-2
Chassis Assembly GMC Complete

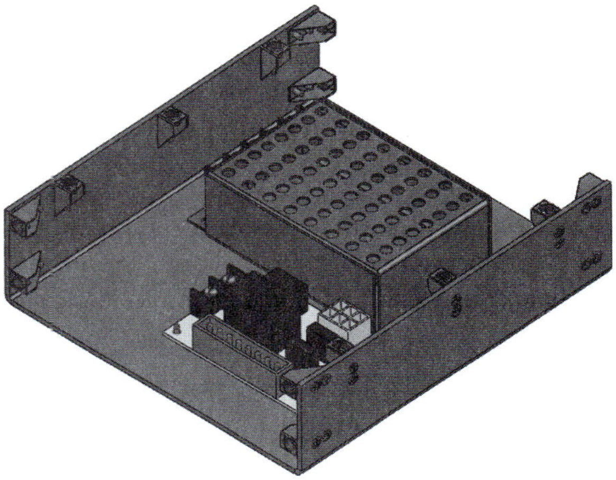

Step 1. Make sure **Units** are correctly set to inches.

Step 2. Open the drawings **Chassis, Nutplate,** and **Flat Head Screw #2.**

Step 3. Click **Insert Components, Insert Nutplates,** and **Flat Head Screws 2** as shown in Figure 10-3.

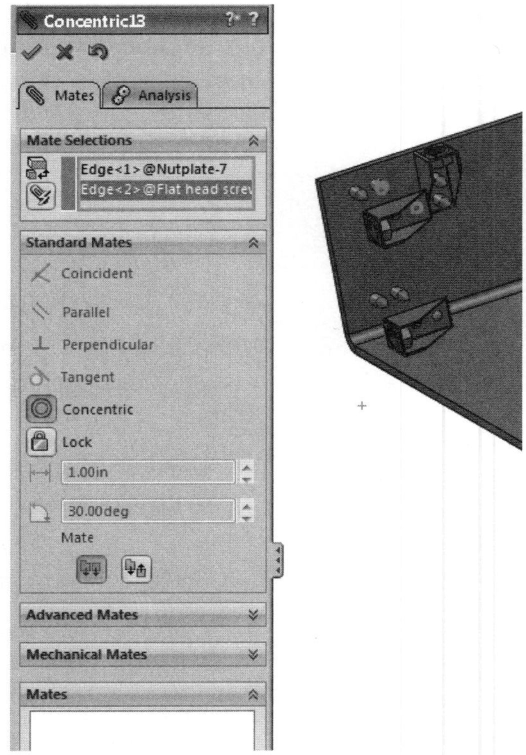

Figure 10-3
Insert Flat Head Screw 2 and Nutplates and Apply Concentric, Parallel and Coincident Mates

Step 4. Apply **Concentric, Parallel,** and **Coincident Mates.**

Step 5. Use **Insert Components,** applying mates and **Mirror** to assemble the remaining nutplates and screws, or use **Copy with Mates** to assemble them.

Step 6. Insert the **Power Control Board,** the **Power Supply,** and **Flat Head Screws 2** (insert the screws from the bottom) and apply **Concentric, Coincident,** and **Parallel** mates as shown in Figure 10-4.

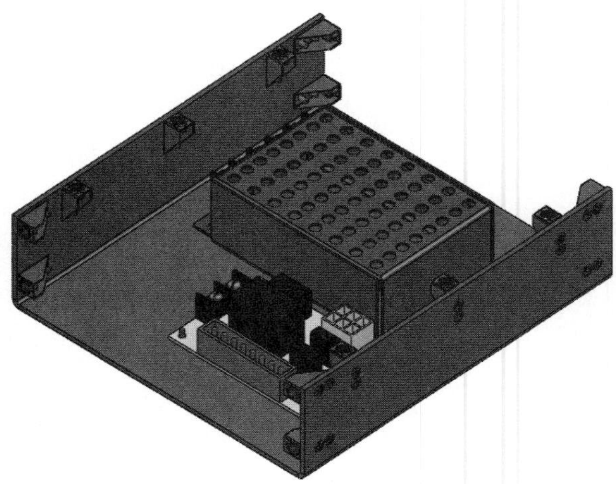

Figure 10-4
Insert Power Control Board, Power Supply, and Flat Head Screws 2 and Apply Concentric, Coincident, and Parallel Mates

Step 7. Select the **Isometric view** from the **Heads-Up View** toolbar.

Step 8. Put your name on the drawing.

Step 9. Save your model with the file name **Chassis Assembly GMC(your initials)** in at least two places.

Step 10. Print the drawing on an A-size sheet and turn it in.

The Chassis Assembly GMC is complete.

Project 10-2: Assemble Parts onto the Front Panel and Make an Assembly (Figure 10-5)

If you have exited SolidWorks, open it again and start a new assembly using **Make Assembly from Part.**

Step 1. Make sure **Units** are correctly set to inches.

Step 2. Open the drawings **Front Panel, Meter Assembly,** and **SPDT Switch.**

Step 3. Click **Insert Components** and insert **Front Panel, Meter Assembly,** and **SPDT Switch** as shown in Figure 10-6.

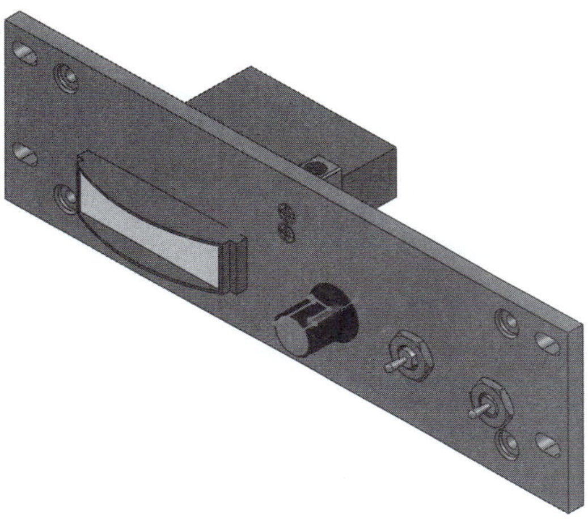

Figure 10-5
Front Panel Assembly Complete

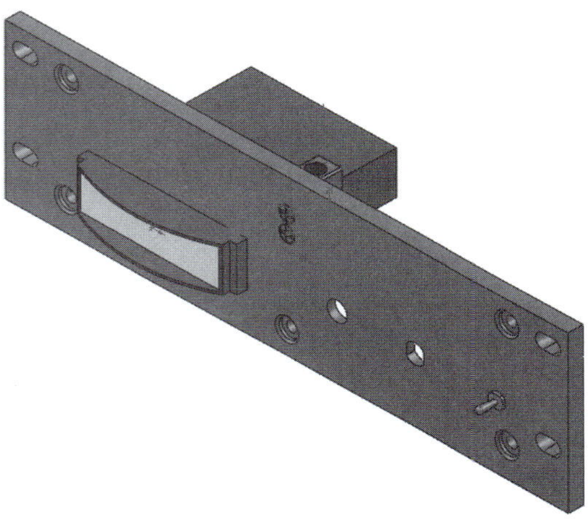

Figure 10-6
Insert Front Panel, Meter Assembly and SPDT Switch and Apply Concentric, Coincident, and Parallel Mates

Step 4. Apply **Concentric, Parallel,** and **Coincident** mates.

Step 5. Click **Insert Components** and insert **SPDT Switch, Potentiometer,** and **Knob** as shown in Figure 10-7.

Step 6. Apply **Concentric, Parallel,** and **Coincident** mates.

Step 7. Click **Insert Components** and insert **.25-20 Nut** as shown in Figure 10-8.

Step 8. Apply **Concentric, Parallel,** and **Coincident** mates.

Step 9. Use **Copy with Mates** to copy the nut as shown in Figure 10-8.

Step 10. Select the **Isometric** view from the **Heads-Up View** toolbar.

Step 11. Put your name on the drawing.

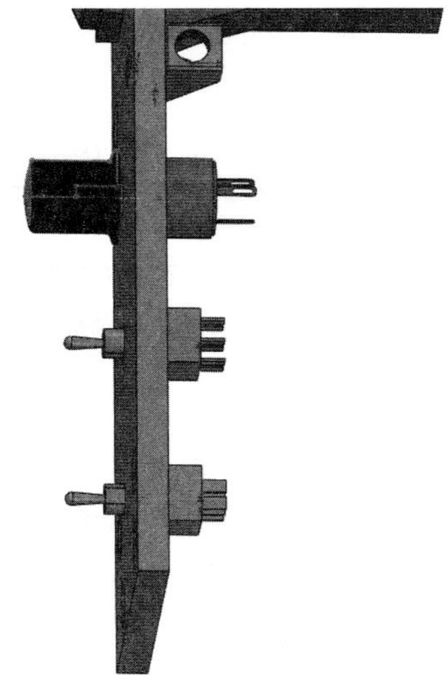

Figure 10-7
Insert Spot Switch, Potentiometer, and Knob and Apply Concentric, Coincident, and Parallel Mates

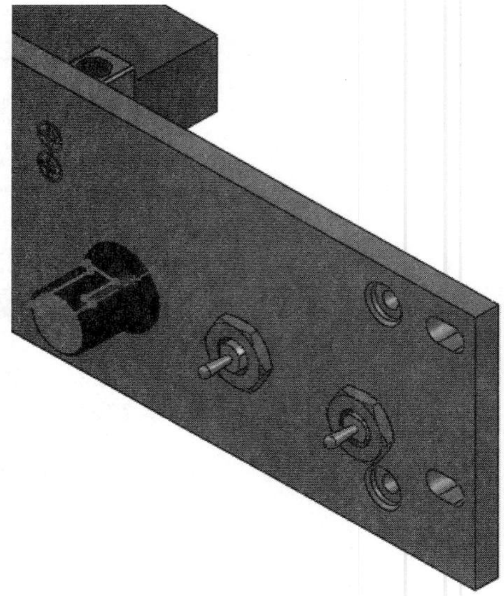

Figure 10-8
Insert .25-20 Nut and Apply Concentric, Coincident, and Parallel Mates and Use Copy with Mates to Copy the Nut

Step 12. Save your model with the file name **Front Panel Assembly (your initials)** in at least two places.

Step 13. Print the drawing on an A-size sheet and turn it in.

The Front Panel Assembly is complete.

Project 10-3: **Assemble Parts onto the Rear Panel and Make an Assembly (Figure 10-9)**

If you have exited SolidWorks, open it again and start a new assembly using **Make Assembly from Part.**

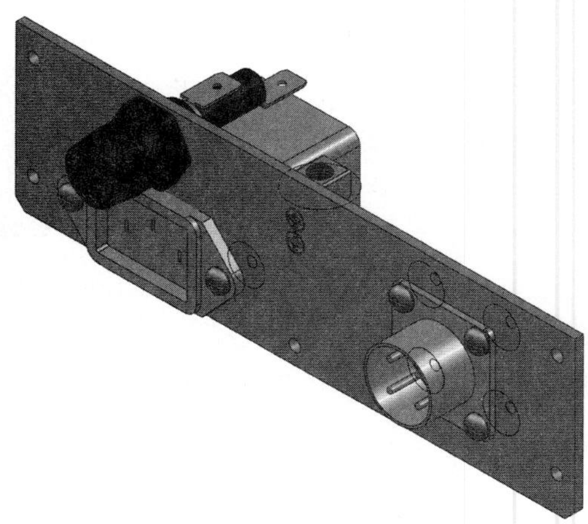

Figure 10-9
Rear Panel Assembly Complete

Step 1. Make sure the **Units** are correctly set to inches.

Step 2. Open the drawings **Rear Panel, Fuse, AC Connector, Connector 4 Pin, Nutplate, Flat Head Screw 2,** and **#6 × .375 Lg Round Head Machine Screw.**

Step 3. Click **Insert Components** and **Insert Rear Panel, Fuse, AC Connector, Connector 4 Pin, Nutplate, Flat Head Screw 2,** and **#6 × .375 Lg Round Head Machine Screw** as shown in Figure 10-10.

Step 4. Apply **Concentric, Parallel,** and **Coincident** mates.

Step 5. Click **Insert Components** and insert **#6 Nuts** as shown in Figure 10-11.

Figure 10-10
Insert Fuse, AC Connector, Connector 4 Pin, Nutplate, Flat Head Screw 2, and #6 × .375 LG Round Head Machine Screws and Apply Concentric, Coincident, and Parallel Mates

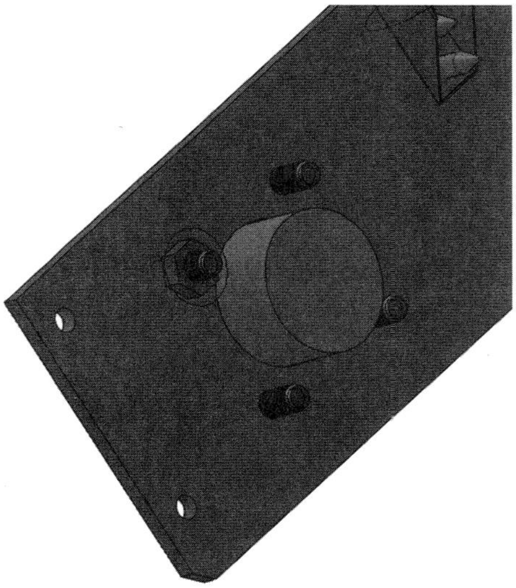

Figure 10-11
Insert #6 Nuts and Apply Concentric, Coincident, and Parallel Mates

Step 6. Apply **Concentric, Parallel,** and **Coincident** mates.

Step 7. Select the **Isometric** view from the **Heads-Up View** toolbar.

Step 8. Put your name on the drawing.

Step 9. Save your model with the file name **Rear Panel Assembly(your initials)** in at least two places.

Step 10. Print the drawing on an A-size sheet and turn it in.

The Rear Panel Assembly is complete.

Project 10-4: **Assemble Chassis Assembly, Front Panel Assembly, Rear Panel Assembly, and the Cover and Make the Top Assembly GCM (Figure 10-12)**

If you have exited SolidWorks, open it again and start a new assembly using **Make Assembly from Part.**

Step 1. Make sure the **Units** are correctly set to inches.

Step 2. Open the drawings **Front Panel Assembly, Rear Panel Assembly, Chassis Cover,** and **Chassis Assembly.**

Figure 10-12
Top Assembly GCM Complete

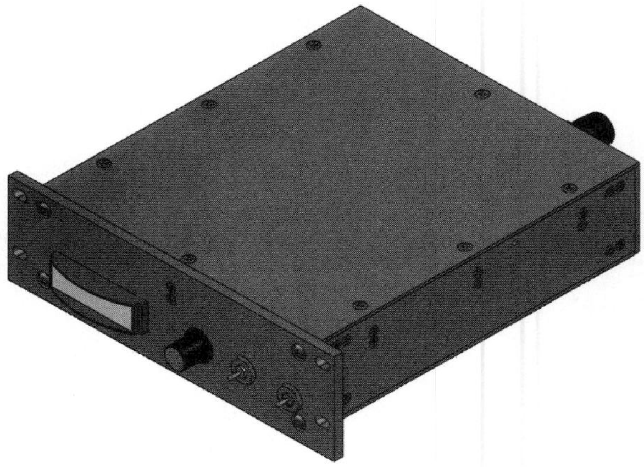

Step 3. Click **Insert Components** and insert **Chassis Assembly, Front Panel Assembly,** and **Rear Panel Assembly,** as shown in Figure 10-13.

Figure 10-13
Top Assembly GCM with Chassis Assembly, Front Panel Assembly, and Rear Panel Inserted and Mated

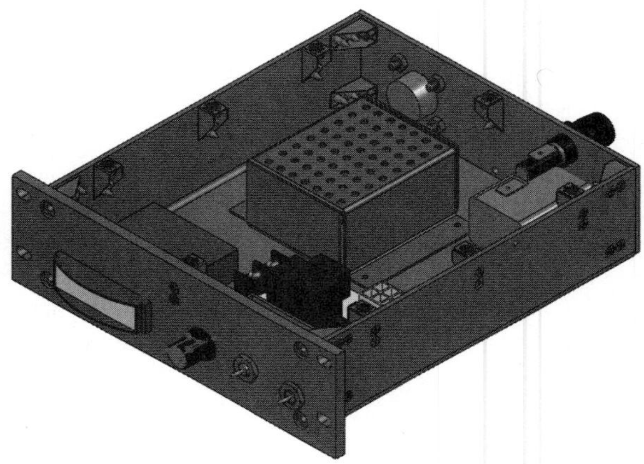

Step 4. Apply **Concentric, Parallel,** and **Coincident** mates.

Step 5. Click **Insert Components** and insert **Chassis Cover, Flat Head Screws 2,** and **#6 × .375 Lg Round Head Machine Screws** as shown in Figure 10-14.

Figure 10-14
Insert Chassis Cover, Flat Head Screws 2, and #6 × .375 Round Head Screws and Apply Concentric and Coincident Mates

Step 6. Apply **Concentric, Parallel,** and **Coincident** mates.

Step 7. Select the **Isometric** view from the **Heads-Up View** toolbar.

Step 8. Put your name on the drawing.

Step 9. Save your model with the file name **Top Assembly GCM(your initials)** in at least two places.

Step 10. Print the drawing on an A-size sheet and turn it in.

The Top Assembly GCM is complete.

Project 10-5: **Make An Exploded View (Figure 10-15)**

If you have exited SolidWorks, open it again.

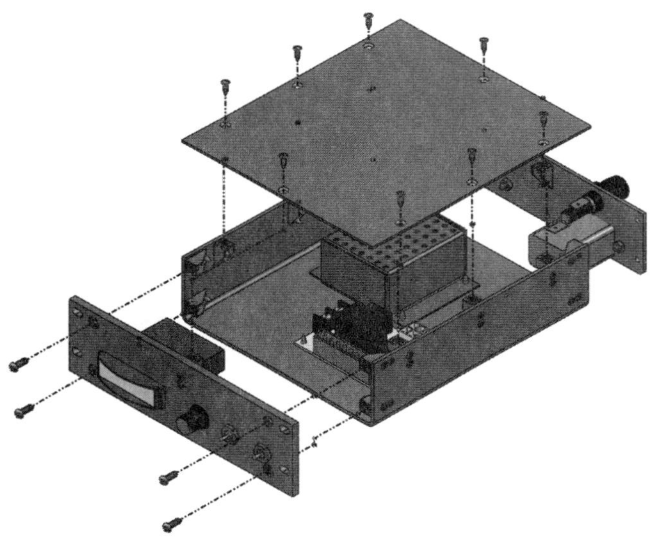

Figure 10-15
Top Assembly GCM Exploded View Complete

Step 1. Open the drawing **Top Assembly GCM.**

Step 2. Make sure **Units** are correctly set to inches.

Step 3. Click **Exploded View** from the **Command Manager.** Then click one of the **#6 × .375 Round Head Machine Screws** and drag it into position using a **Manipulator Handle** (one of the colored arrows that point either directly toward or directly away from the hole where the screw is assembled) as shown in Figure 10-16.

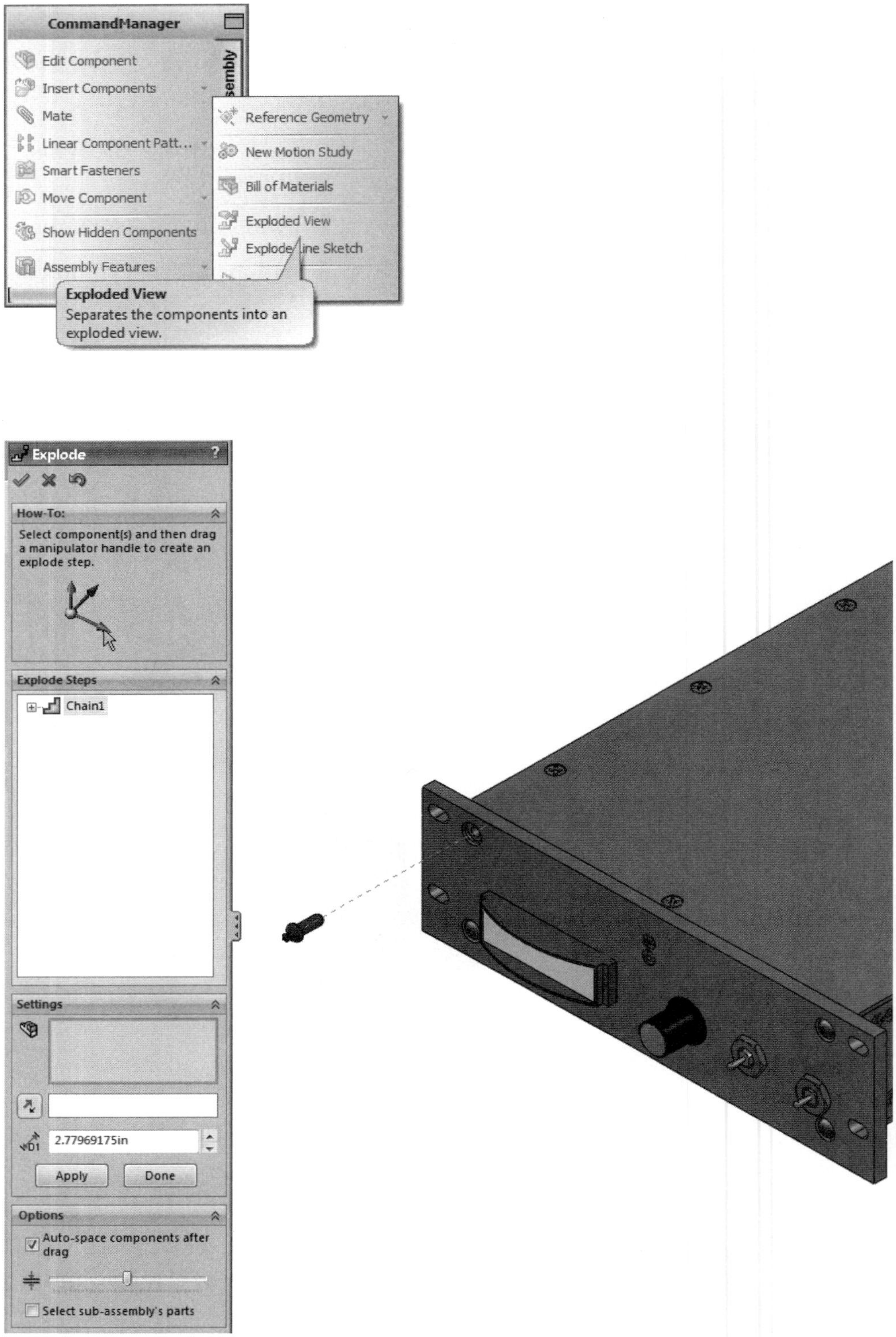

Figure 10-16
Select Exploded View from the Command Manager and Click One of the #6 × .375 Round Head Machine Screws and Drag It into Position Using a Manipulator Handle

Step 4. Click and drag the remaining machine screws (one at a time) on the front panel into position and drag the front panel assembly into position as shown in Figure 10-17.

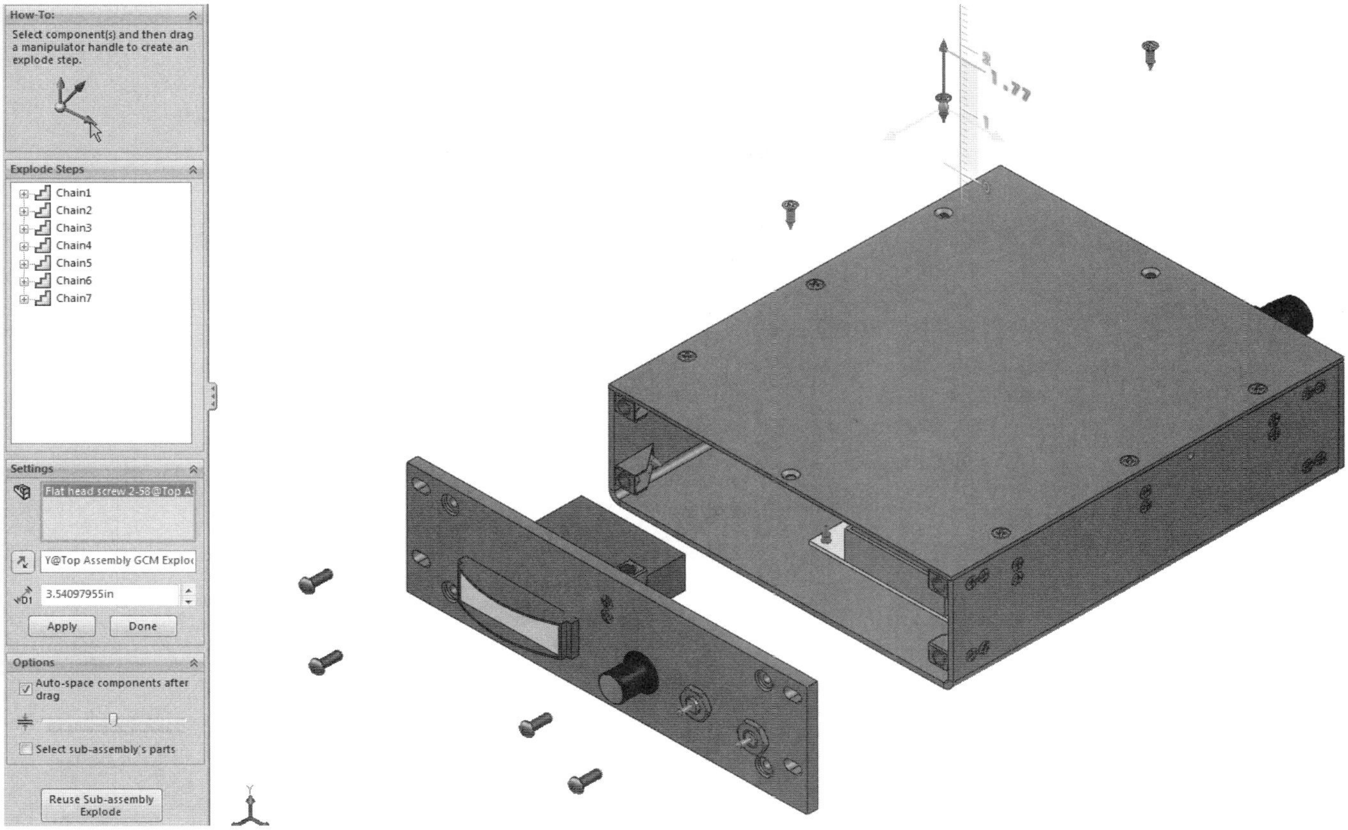

Figure 10-17
Drag the Remaining Machine Screws on the Front Panel into Position and Drag the Front Panel Assembly and Three Flat Head Screws 2 into Position

Step 5. Click and drag three **Flat Head Screws 2** into position as shown in Figure 10-17.

Step 6. Click and drag the remaining flat head screws into position as shown in Figure 10-18.

Step 7. Click and drag the rear panel assembly into position as shown in Figure 10-18.

Figure 10-18
Drag Remaining Flat Head Screws into Position and Drag the Rear Panel Assembly into Position

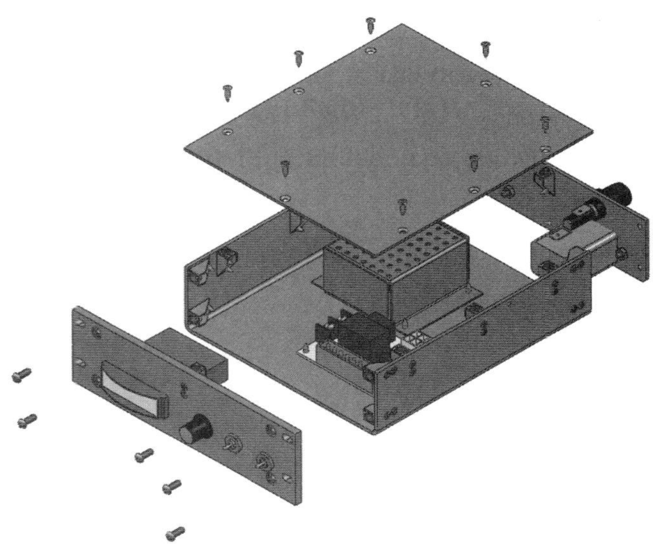

Step 8. Click **Exploded Line Sketch** from the **Command Manager** to show where parts are assembled by clicking on one feature of a part (the end of a screw) then clicking on the mating feature of another part (the hole in a nutplate) as shown in Figure 10-19.

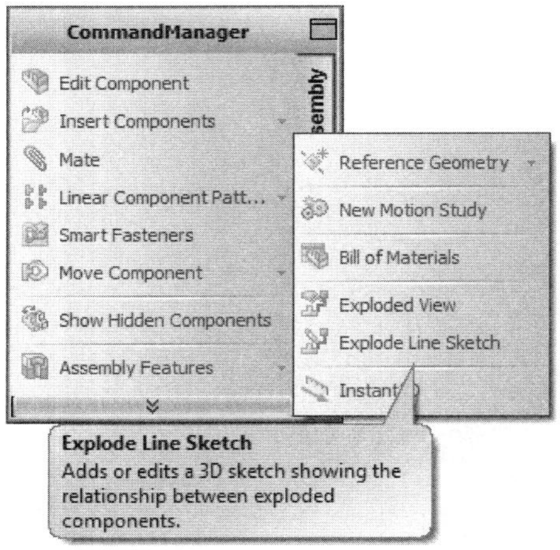

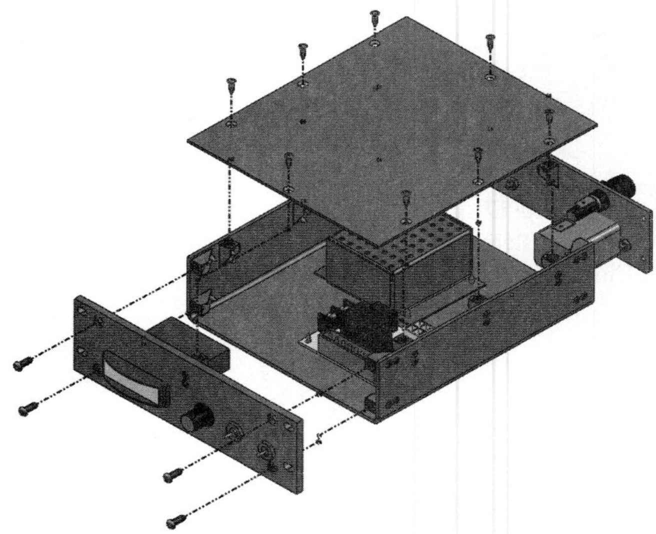

Figure 10-19
Use Exploded Line Sketch from the Command Manager to Show Where Parts Are Assembled by Clicking on One Feature of a Part (the End of a Screw) Then Clicking on the Mating Feature of Another Part (the Hole in a Nutplate)

Step 9. Select the **Isometric** view from the **Heads-Up View** toolbar.

Step 10. Put your name on the drawing.

Step 11. Save your model with the file name **Top Assembly GCM Exploded View(your initials)** in at least two places.

Step 12. Print the drawing on an A-size sheet and turn it in.

The Top Assembly GCM Exploded View is complete.

10 chapter ten

Chapter Summary

This chapter contains the information necessary to make subassemblies, assemblies, and exploded views. Now you have the skills and information necessary to produce subassemblies, assemblies, and exploded views.

Review Questions

Circle the most correct answer.

1. Which of the following is used to edit a sketch?
   a. **Edit** feature c. **Edit Drawing**
   b. **Edit Sketch** d. **Edit Sheet**
2. Which of the following is used to edit a feature?
   a. **Edit** feature c. **Edit Drawing**
   b. **Edit Sketch** d. **Edit Sheet**
3. On which menu is the **Copy with Mates** command found?
   a. **Edit** c. **Insert**
   b. **Tools** d. **View**
4. On which tab of the **Command Manager** is the **Exploded View** command found?
   a. **Assembly** c. **Sketch**
   b. **Features** d. **Layout**
5. Which of these commands shows the lines between parts in an exploded view?
   a. **Line** c. **Exploded View**
   b. **Explode Line Sketch** d. **Instant 3D**
6. **Copy with Mates** can copy only one part at a time.
   a. True b. False
7. The **Parallel** mate can line up parts vertically or horizontally.
   a. True b. False
8. You cannot insert parts into a subassembly.
   a. True b. False
9. A subassembly can be inserted into a top assembly.
   a. True b. False
10. Explosion sketch lines can be used to show where one part connects to another part.
    a. True b. False

chapter ten

Matching

Place the correct letter on the line to the right of the numbered item.

1. A command that is used to place one part into an assembly ______	a. **Copy with Mates**
2. A model containing only parts ______	b. **Insert Components**
3. A command that allows you to copy a part with a concentric mate ______	c. Subassembly
4. A command that can be used to place a part parallel to another part ______	d. Assembly
5. A model containing both parts and subassemblies ______	e. **Mates**

General Questions

Answer the following questions.

1. In what situations would you use an assembly?
2. When would you want to use an exploded view?
3. Can **Copy with Mates** really save time in assembling parts?
4. Why are explosion sketch lines necessary?
5. How can you show a part in an exploded view that is hidden in the isometric view?

chapter eleven

Make Three Subassembly Drawings and a Top Assembly Drawing with Bills of Materials

CHAPTER OBJECTIVES

After completing this chapter, you will be able to:

- Make assembly drawings.
- Make exploded views.
- Make parts lists.
- Correctly answer questions regarding SolidWorks commands used in this chapter.

Introduction

In this chapter, Project 11-1 contains very detailed instructions on making the subassembly drawings for the Chassis Subassembly GCM, the Front Panel Subassembly GCM, the Rear Panel Subassembly GCM, and the Top Assembly GCM, all of which contain Bills of Materials.

Project 11-1: Make the Assembly Drawing and Bill of Materials for the Chassis Subassembly GCM (Figure 11-1)

If you have exited SolidWorks, open it again.

Step 1. Open the drawing **Chassis Assembly GCM.**

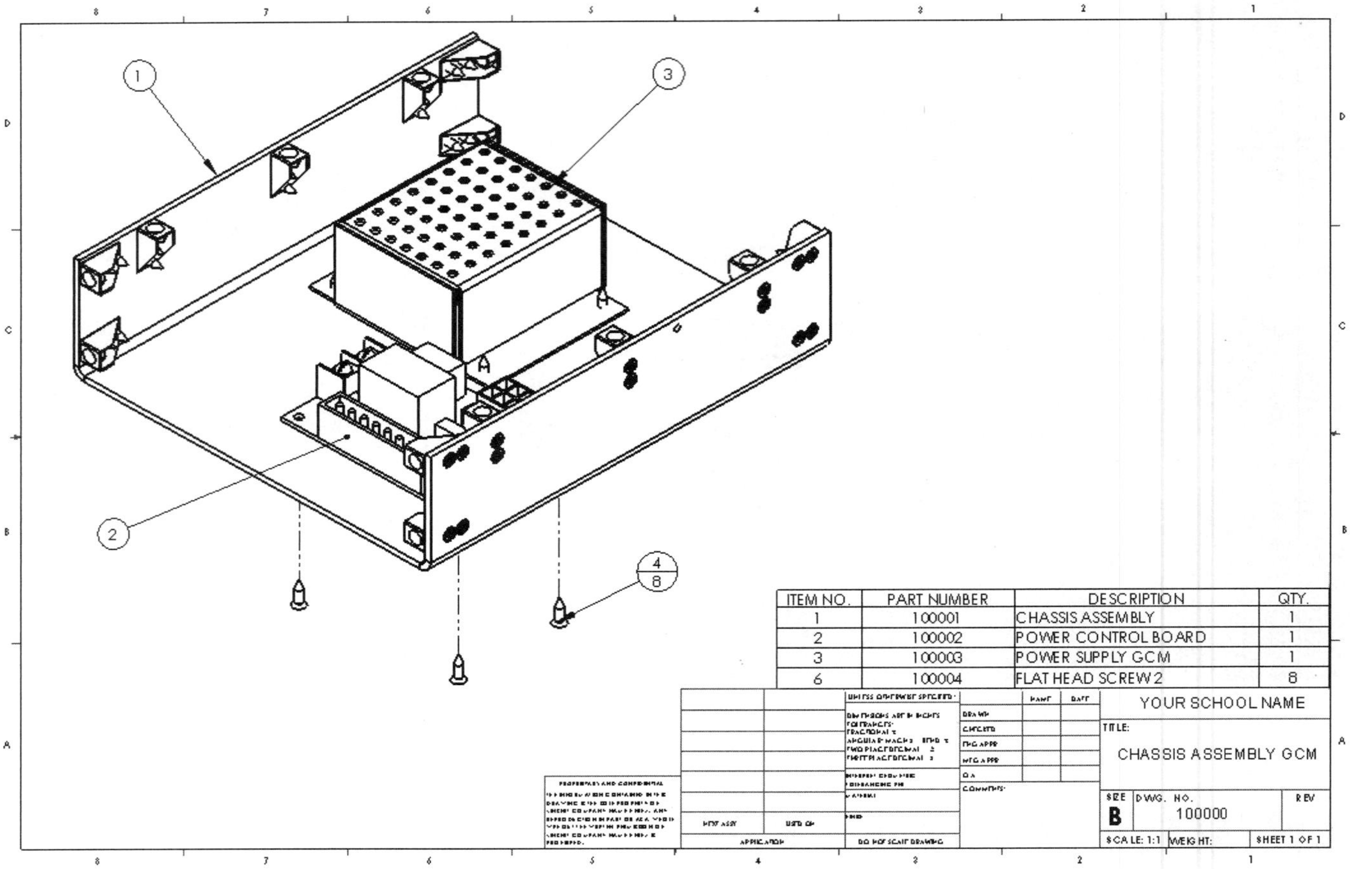

Figure 11-1

Chassis Subassembly GCM Drawing with Bill of Materials Complete

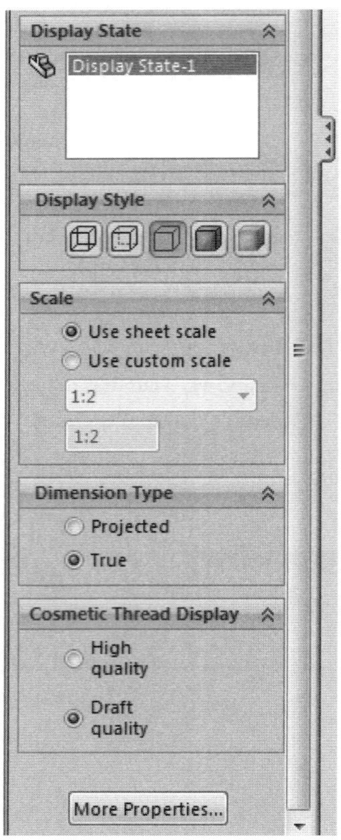

Step 2. Click **Make Drawing from Assembly.**

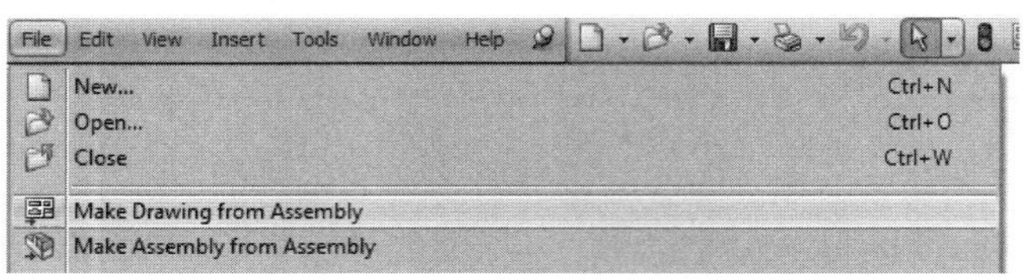

Step 3. Select the drawing format **b - landscape.slddrt** (or another one that suits your situation).

Step 4. Drag the isometric view from the **View** palette onto the format and set **Use custom scale** to **1:1** as shown in Figure 11-2.

Step 5. Click **More Properties...** at the bottom of the **Drawing View1 Manager** and check **Show in exploded state** in the **Drawing View Properties** dialog box as shown in Figure 11-3.

If you have not completed the exploded view as shown in Figure 11-1, you will need to do so now before you complete this project.

Step 6. Right-click in the drawing view and click **Bill of Materials....**

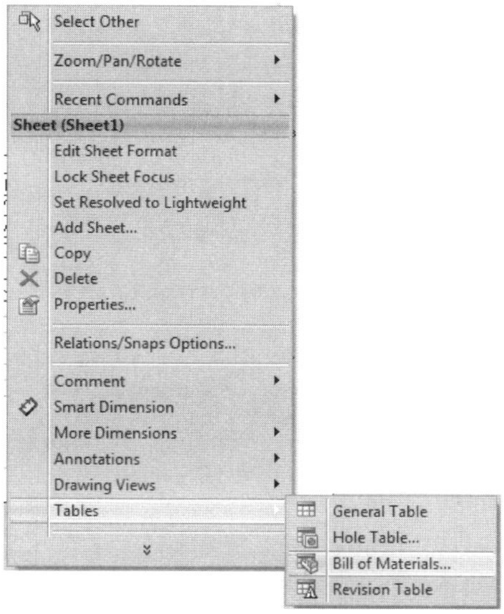

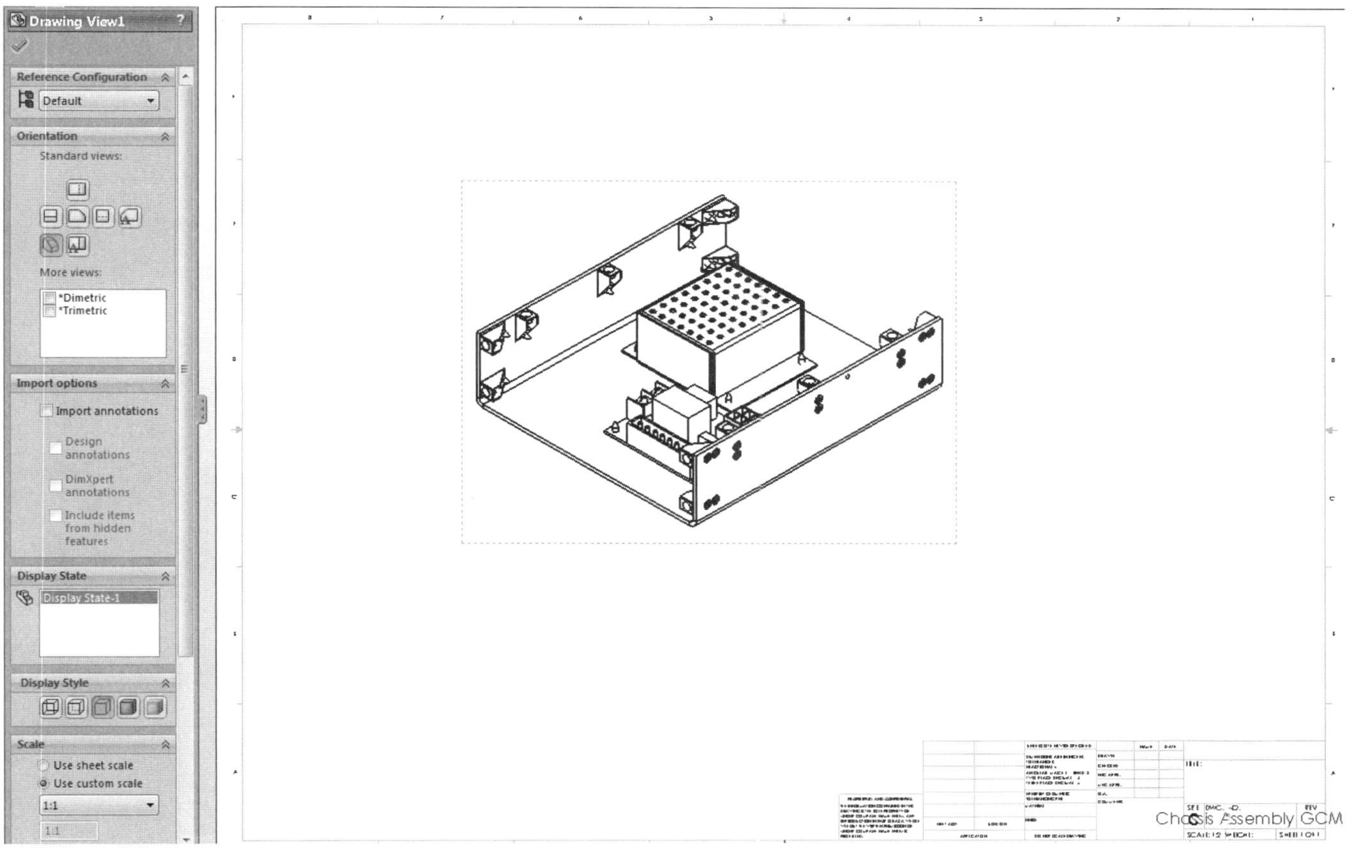

Figure 11-2
Drag Isometric View onto the Sheet and Set Use Custom Scale to 1:1

Step 7. Click **OK** and place the **Bill of Materials** in the drawing format just above the title block as shown in Figure 11-1.

Step 8. Add **Circular** and **Circular Split Line** balloons as shown in Figure 11-4.

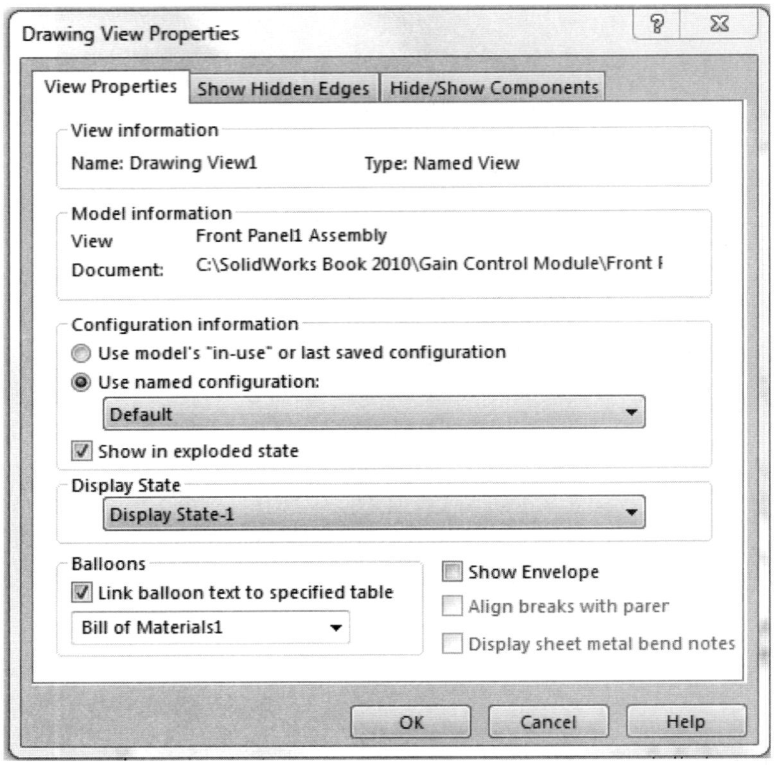

Figure 11-3
Show the Assembly in Exploded State (Check: Show in Exploded State)

Figure 11-4
Make the Bill of Materials, Add Circular and Circular Split Line Balloons, and Edit the Numbers and Descriptions

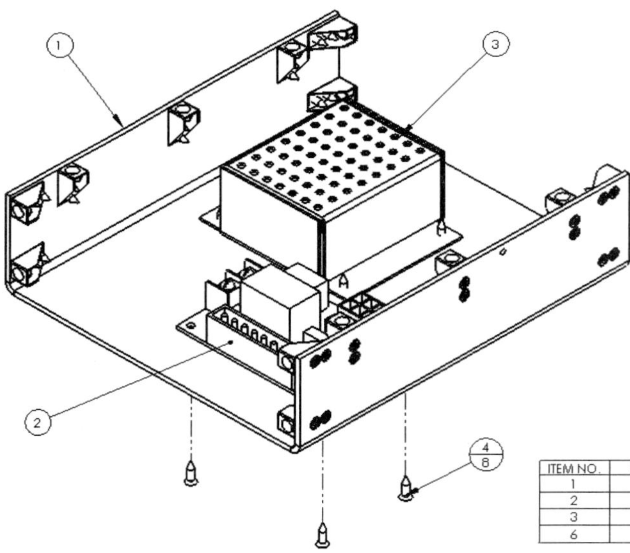

ITEM NO.	PART NUMBER	DESCRIPTION	QTY.
1	100001	CHASSIS ASSEMBLY	1
2	100002	POWER CONTROL BOARD	1
3	100003	POWER SUPPLY GCM	1
6	100004	FLAT HEAD SCREW 2	8

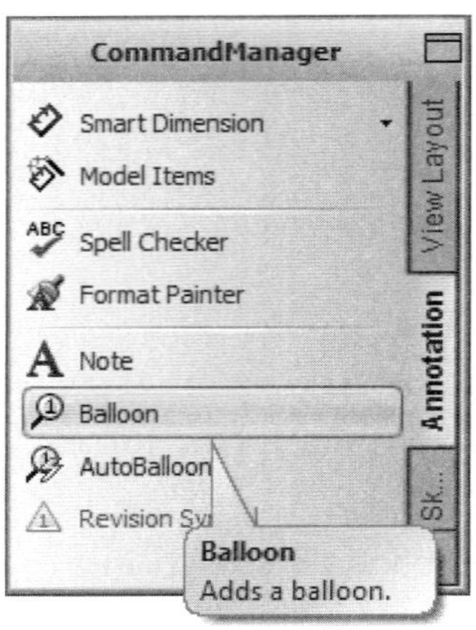

Step 9. Edit the numbers and descriptions in the **Bill of Materials** by double-clicking them.

Step 10. After the Bill of Materials has been edited, use the **Shift** key to select all rows in the **DESCRIPTION** column and click **Left Align** as shown in Figure 11-5.

Step 11. Use the **<Shift>** key to select all rows in the **PART NUMBER** column and click **Center Align** as shown in Figure 11-5.

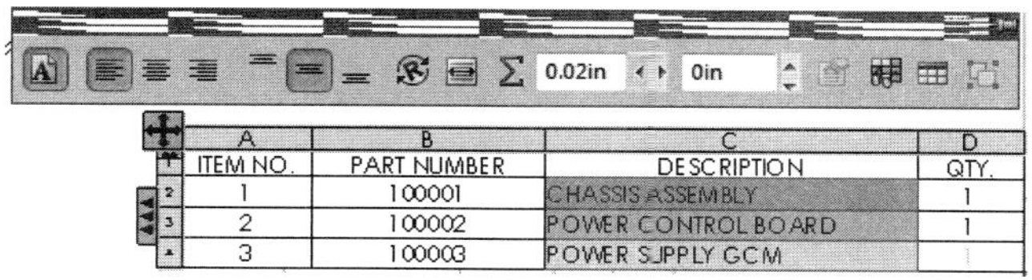

Figure 11-5
Use <Shift> to Select All Rows in the Description Column and Click: Left Align

Step 12. Right-click between the drawing view and the border and click **Edit Sheet Format.**

Step 13. Edit the format by double-clicking **SCALE:** and **DWG. NO.** and use **Note** to add **TITLE** and **YOUR SCHOOL NAME** as shown in Figure 11-6.

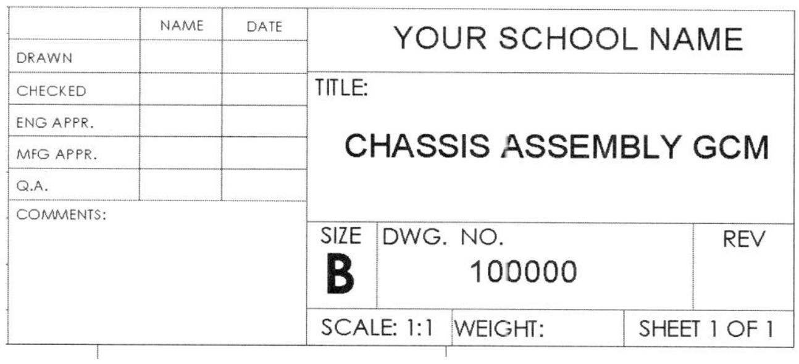

Figure 11-6
Edit Sheet Format by Double-Clicking SCALE: and DWG. NO. and Use Note to Add Title and Your School Name

Step 14. Right-click in the drawing format and click **Edit Sheet.**

Step 15. Save the drawing as **Chassis Assembly GCM with Bill of Materials(your initials).**

Step 16. Print the drawing.

Project 11-1 is complete.

Project 11-2: **Make the Assembly Drawing and Bill of Materials for the Front Panel Subassembly GCM (Figure 11-7)**

If you have exited SolidWorks, open it again.

Step 1. Open the drawing **Front Panel Assembly GCM.**

Step 2. Click **Make Drawing from Assembly.**

Step 3. Select the drawing format **b - landscape.slddrt** (or another one that suits your situation).

Step 4. Drag the isometric view from the **View** palette onto the format and set **Use custom scale** to **1:1** as shown in Figure 11-8.

Step 5. Click **More Properties...** at the bottom of the **Drawing View1 Manager** and check **Show in exploded state** in the **Drawing View Properties** dialog box as shown in Figure 11-7.

If you have not completed the exploded view as shown in Figure 11-7, you will need to do so now before you complete this project.

Step 6. Right-click in the drawing view and click **Bill of Materials....**

Step 7. Click **OK** and place the **Bill of Materials** in the drawing format just above the title block as shown in Figure 11-7.

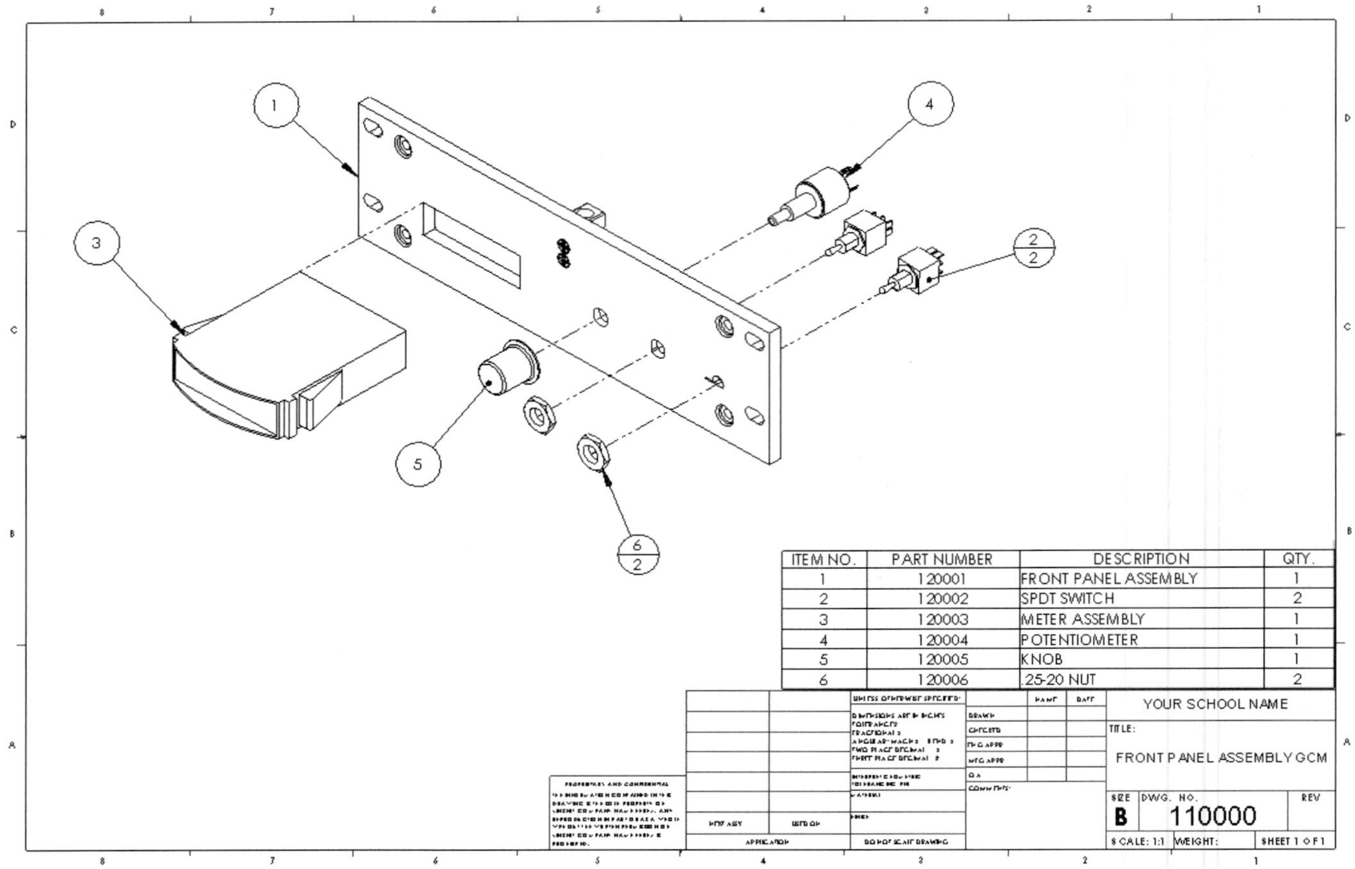

Figure 11-7
Front Panel Subassembly Drawing with Bill of Materials Complete

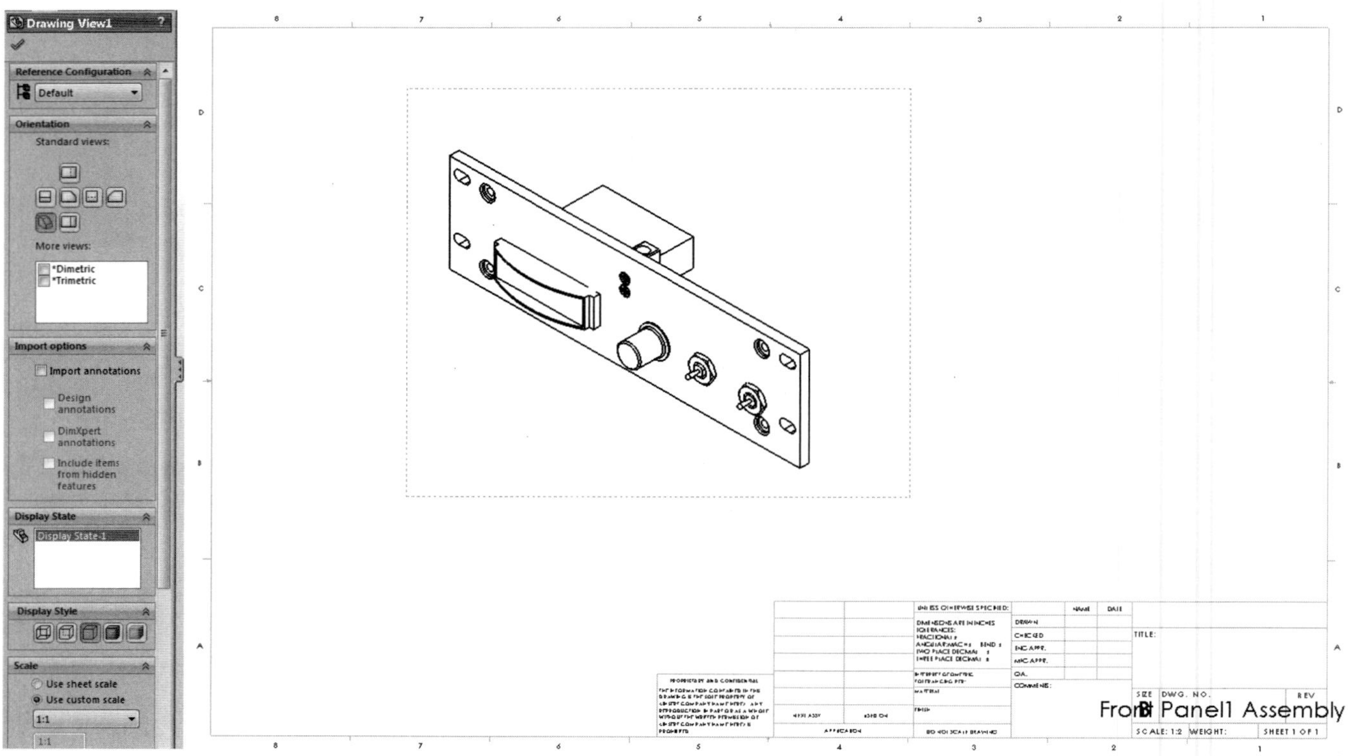

Figure 11-8
Drag Isometric View onto the Sheet and Set Use Custom Scale to 1:1

Step 8. Add **Circular** and **Circular Split Line** balloons as shown in Figure 11-9.

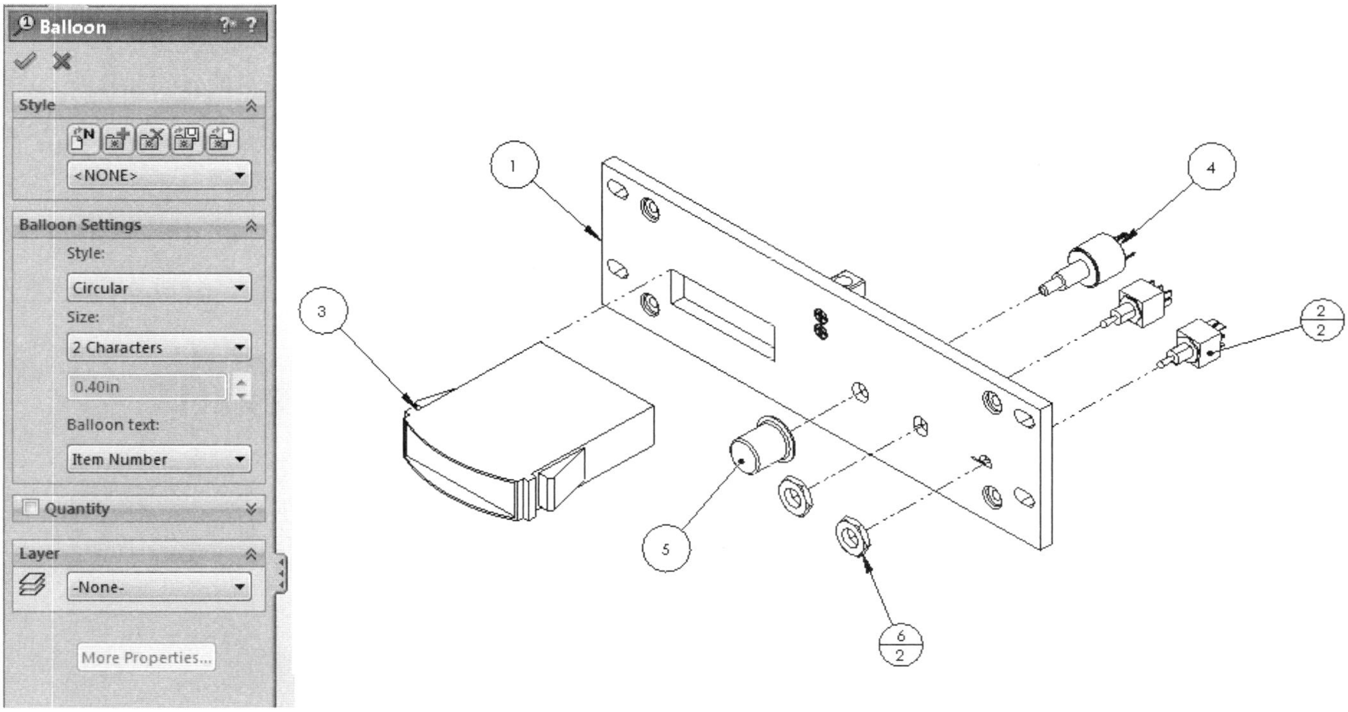

Figure 11-9
Add Circular and Circular Split Line Balloons

Step 9. Edit the numbers and descriptions in the **Bill of Materials** by double-clicking them.

Step 10. After the Bill of Materials has been edited, use the **<Shift>** key to select all rows in the **DESCRIPTION** column and click **Left Align.**

Step 11. Use the **<Shift>** key to select all rows, right-click, and click **Formatting - Row Height** and set the row height for all rows to **.25** as shown in Figure 11-10.

Step 12. Use the **<Shift>** key to select all rows in the **PART NUMBER** column and click **Center Align.**

Figure 11-10
Use the <Shift> Key to Select All Rows, Right-Click, and Set Row Height for All Rows to .25

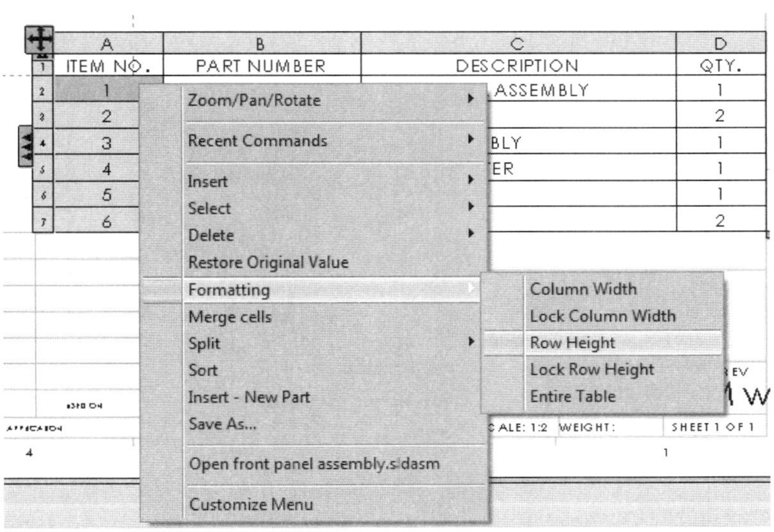

Step 13. Right-click between the drawing view and the border and click **Edit Sheet Format.**

Step 14. Edit the format by double-clicking **SCALE:** and **DWG. NO.** and use **Note** to add **TITLE** and **YOUR SCHOOL NAME.**

Step 15. Right-click in the drawing format and click **Edit Sheet.**

Step 16. Save the drawing as **Front Panel Assembly GCM with Bill of Materials(your initials).**

Step 17. Print the drawing.

Project 11-2 is complete.

Project 11-3: **Make the Assembly Drawing and Bill of Materials for the Rear Panel Subassembly GCM (Figure 11-11)**

If you have exited SolidWorks, open it again.

Step 1. Open the drawing **Rear Panel Assembly GCM.**

Step 2. Click **Make Drawing from Assembly.**

Step 3. Select the drawing format **b - landscape.slddrt** (or another one that suits your situation).

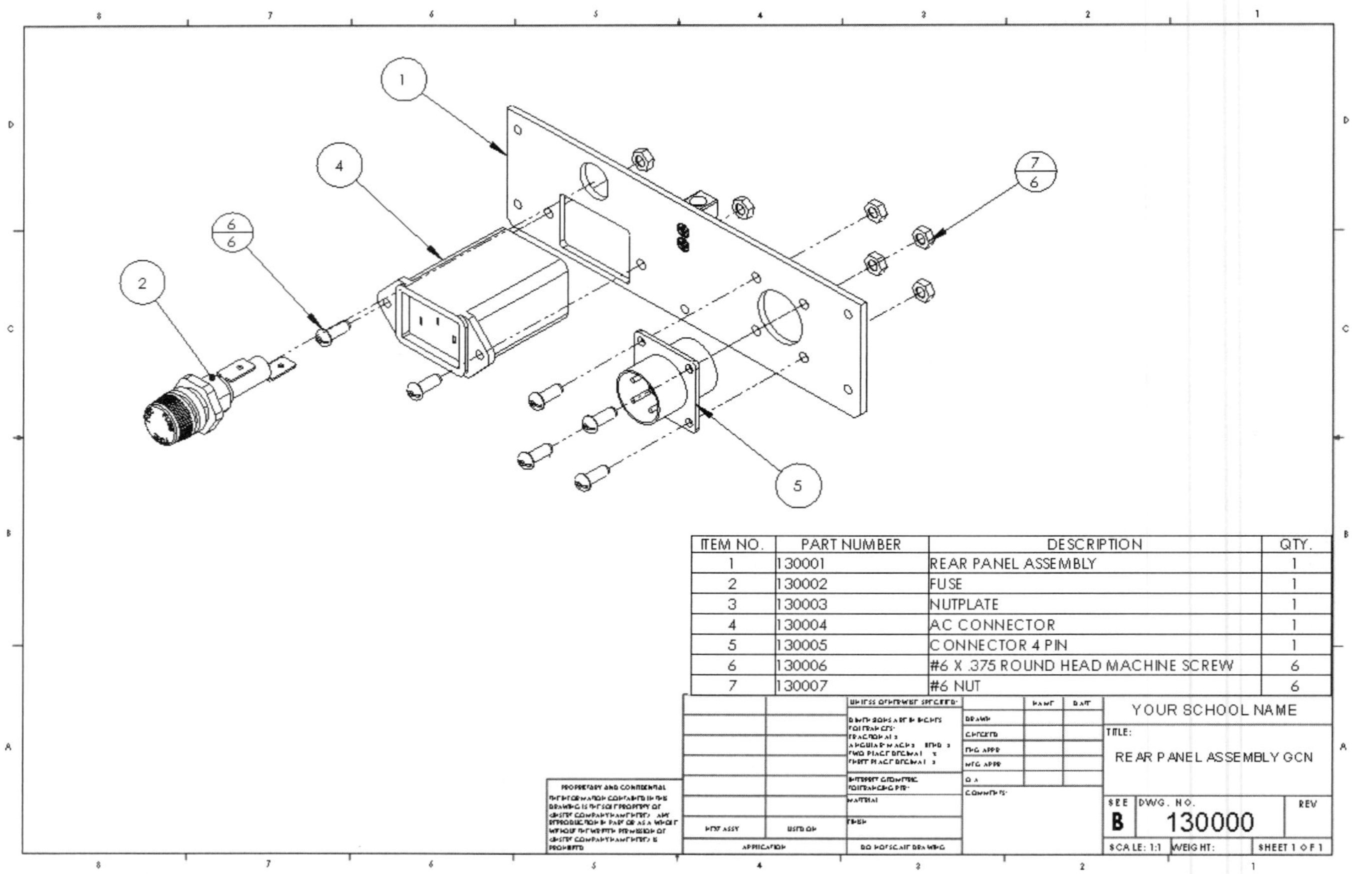

Figure 11-11
Rear Panel Subassembly GCM Drawing with Bill of Materials Complete

Step 4. Drag the isometric view from the **View** palette onto the format and set **Use custom scale** to **1:1.**

Step 5. Click **More Properties...** at the bottom of the **Drawing View1 Manager** and check **Show in exploded state** in the **Drawing View Properties** dialog box.

If you have not completed the exploded view as shown in Figure 11-11, you will need to do so now before you complete this project. While making the exploded view, you can select four parts at the same time when you are making the **Explode Sketch Lines** as shown in Figure 11-12.

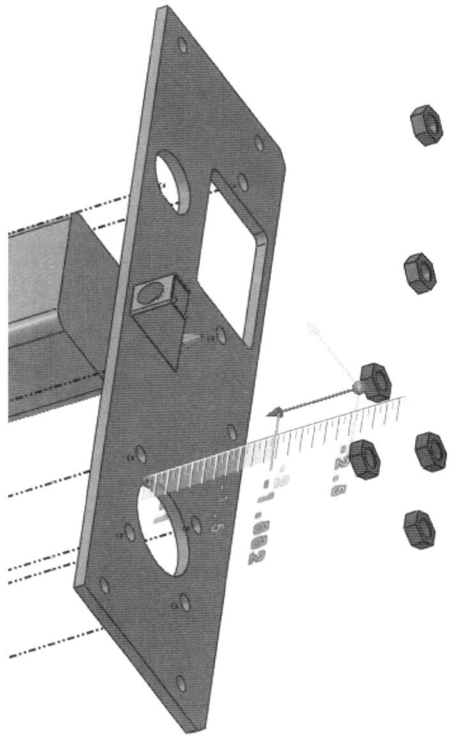

Figure 11-12
Selecting Four Parts for Explode Line Sketch Line

Step 6. Right-click in the drawing view and click **Bill of Materials....**

Step 7. Click **OK** and place the **Bill of Materials** in the drawing format just above the title block as shown in Figure 11-11.

Step 8. Add **Circular** and **Circular Split Line** balloons as shown in Figure 11-13.

Step 9. Edit the numbers and descriptions in the **Bill of Materials** by double-clicking them.

Step 10. After the Bill of Materials has been edited, use the **<Shift>** key to select all rows in the **DESCRIPTION** column and click **Left Align.**

Step 11. Click the **DESCRIPTION** column for item number 6 and click **Formatting - Column Width** and set column width for this column to **4.08.**

Step 12. Use the **<Shift>** key to select all rows, right-click, and click **Formatting - Row Height** and set the row height for all rows to **.25.**

Step 13. Use the **<Shift>** key to select all rows in the **PART NUMBER** column and click **Center Align.**

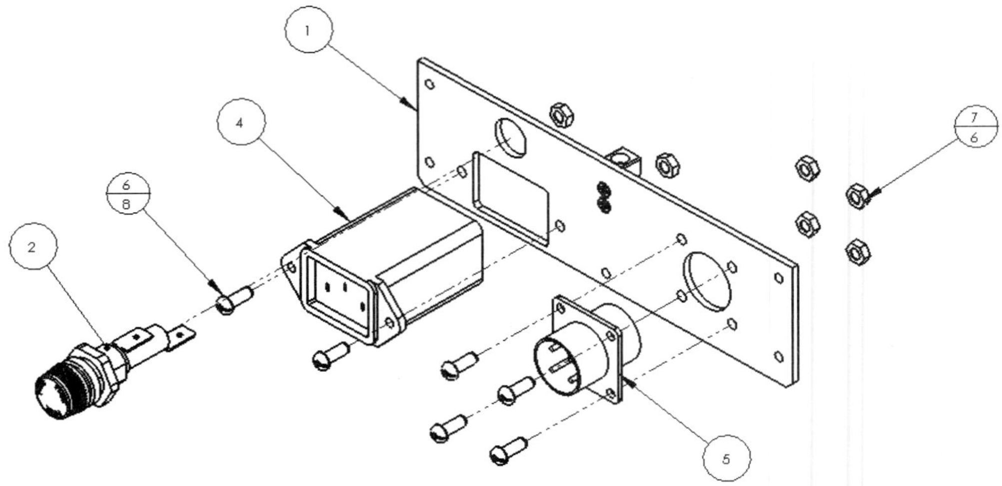

Figure 11-13
Add Circular and Circular Split Line Balloons

Step 14. Right-click between the drawing view and the border and click **Edit Sheet Format.**

Step 15. Edit the format by double-clicking **SCALE:** and **DWG. NO.** and use **Note** to add **TITLE** and **YOUR SCHOOL NAME.**

Step 16. Right-click in the drawing format and click **Edit Sheet.**

Step 17. Save the drawing as **Rear Panel Assembly GCM with Bill of Materials(your initials).**

Step 18. Print the drawing.

Project 11-3 is complete.

Project 11-4: Make the Top Assembly Drawing and Bill of Materials for the Top Assembly GCM (Figure 11-14)

If you have exited SolidWorks, open it again.

Step 1. Open the drawing **Top Assembly GCM.**

Step 2. Click **Make Drawing from Assembly.**

Step 3. Select the drawing format **b - landscape.slddrt** (or another one that suits your situation).

Step 4. Drag the isometric view from the **View** palette onto the format and set **Use custom scale** to **1:1.**

Step 5. Click **More Properties...** at the bottom of the **Drawing View1 Manager** and check **Show in exploded state** in the **Drawing View Properties** dialog box as shown in Figure 11-14.

If you have not completed the exploded view as shown in Figure 11-14, you will need to do so now before you complete this project.

Step 6. Right-click in the drawing view and click **Bill of Materials....**

Step 7. Click **OK** and place the **Bill of Materials** in the drawing format just above the title block as shown in Figure 11-14.

Step 8. Add **Circular** and **Circular Split Line** balloons.

Step 9. Edit the numbers and descriptions in the **Bill of Materials** by double-clicking them.

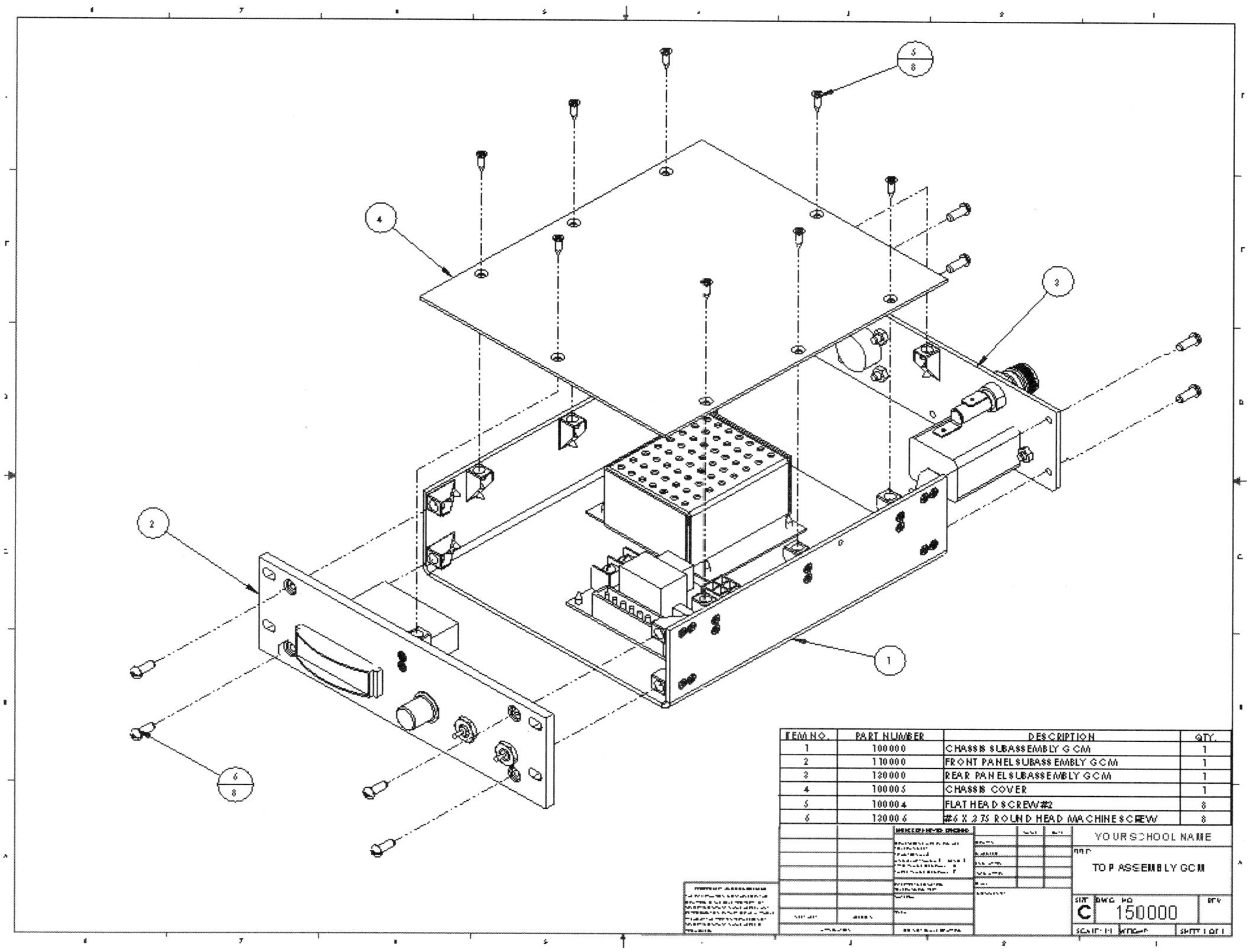

Figure 11-14
Top Assembly Drawing with Bill of Materials Complete

Step 10. After the Bill of Materials has been edited, use the **<Shift>** key and select all rows in the **DESCRIPTION** column and click **Left Align.**

Step 11. Use the **<Shift>** key to select all rows, right-click, click **Formatting - Row Height,** and set the row height for all rows to **.25.**

Step 12. Use the **<Shift>** key to select all rows in the **PART NUMBER** column and click **Center Align**.

Step 13. Right-click between the drawing view and the border and click **Edit Sheet Format.**

Step 14. Edit the format by double-clicking **SCALE:** and **DWG. NO.** and use **Note** to add **TITLE** and **YOUR SCHOOL NAME.**

Step 15. Right-click in the drawing format and click **Edit Sheet.**

Step 16. Save the drawing as **Top Assembly GCM with Bill of Materials (your initials).**

Step 17. Print the drawing.

Project 11-4 is complete.

11 chapter eleven

Chapter Summary

This chapter contains further information necessary to make assemblies. In addition, you have also made assembly drawings as exploded views with parts lists. Now you have the skills and information necessary to produce complex isometric assemblies, exploded views, and parts lists.

Review Questions

Circle the most correct answer.

1. Which of the following can be used to make an assembly drawing from an assembly?
   a. **Make Assembly from Assembly**
   b. **Make Drawing Format from Assembly**
   c. **Make Drawing from Assembly**
   d. **Make Drawing from Drawing**
2. Which of the following is used to show an assembly as an exploded view?
   a. **More Properties**
   b. **Display State**
   c. **Properties**
   d. **Display Style**
3. On which of the items with flyouts is the **Bill of Materials** located?
   a. **Add Sheet**
   b. **Drawing Views**
   c. **Annotations**
   d. **Tables**
4. Which tab on the **Command Manager** contains the **Balloon** command?
   a. **View Layout**
   b. **Annotation**
   c. **Sketch**
   d. **Feature**
5. The standard Bill of Materials contains how many columns?
   a. 3
   b. 4
   c. 5
   d. 6
6. A row cannot be added to the Bill of Materials.
   a. True
   b. False
7. The column width can be changed.
   a. True
   b. False
8. All row heights cannot be changed at the same time.
   a. True
   b. False
9. The scale of a drawing cannot be changed after it is inserted into the drawing format.
   a. True
   b. False
10. The **Explode Sketch Line** command can be used on several parts at the same time.
    a. True
    b. False

chapter eleven

Matching

Place the correct letter on the line to the right of the numbered item.

1. A dialog box that allows the drawing to be shown as an exploded view ______
2. A setting that is used to change the size of a drawing ______
3. A command that is used to show the item number and its quantity ______
4. An item in the drawing title block ______
5. An item that is clicked on the **Drawing View Manager** to show the **Drawing View Properties** dialog box ______

a. **Balloon Split Line**

b. **More Properties...**

c. **Use custom scale**

d. **Drawing View Properties**

e. **Edit Sheet Format**

General Questions

Answer the following questions.

1. In what situations would you use subassemblies?
2. When would you want to use an exploded view for an assembly?
3. Why is **Explode Sketch Lines** needed?
4. How does the **Edit Sheet Format** command work?
5. Why don't the column width and height change automatically?

chapter twelve

On-Your-Own Projects with Detailed Instructions for Making Assemblies, Exploded Views, Assembly Drawings, and Parts Lists

CHAPTER OBJECTIVES

After completing this chapter, you will be able to:

- Make assemblies.
- Make assembly drawings.
- Make exploded views.
- Make parts lists.
- Correctly answer questions regarding SolidWorks commands used in this chapter.

Introduction

In this chapter, Project 12-1 contains very detailed instructions on making the parts, assembly, assembly drawing, parts list, and exploded view. The remaining projects and exercises give overall instructions and refer you to parts of Project 12-1 and earlier chapters as needed.

Project 12-1: Make the Four Parts of the Tool Tray and Make an Assembly of It

There are four parts to the tool tray assembly (Figure 12-1). They are:

Body

Handle

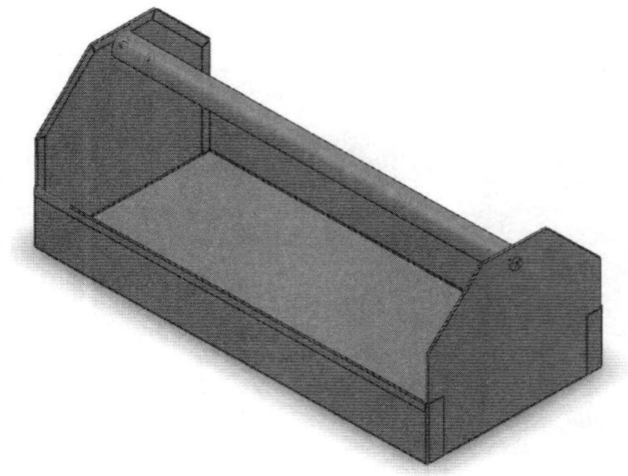

Figure 12-1
Project 12-1 Complete

Machine Screw

Lock Washer

The first part to be made in this project is the tool tray body.

Make the Tool Tray Body

If you have exited SolidWorks, open it again and start a new part.

Step 1. Click **Options** on the menu bar, select the **Document Properties** tab, and click **Units** in the list to the left.

Step 2. Make sure the **Units** are correctly set [notice that **IPS** (inch, pound, second) is checked], then click **OK.**

Step 3. Draw and dimension a rectangle on the top plane as shown in Figure 12-2.

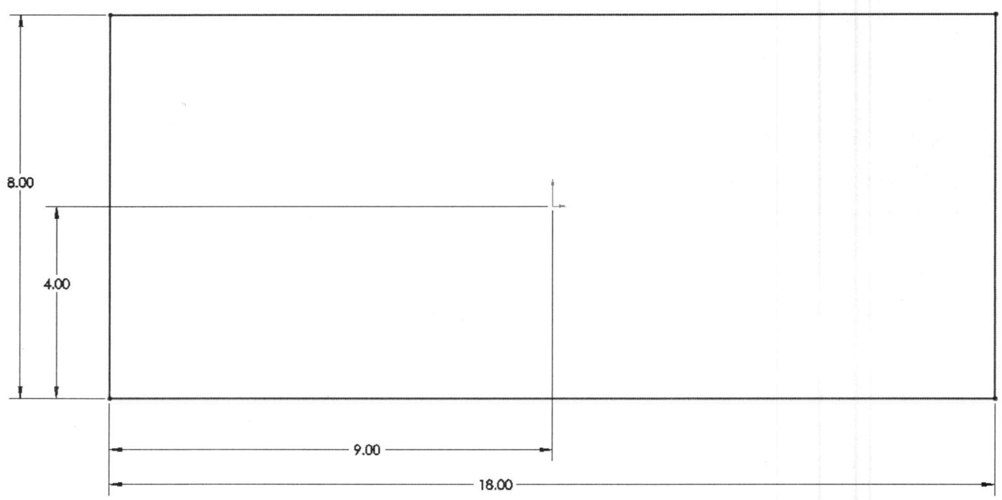

Figure 12-2
Draw and Dimension a Rectangle on the Top Plane

Step 4. Extrude the rectangle as gauge 20 sheet metal using the **Base-Flange** command as shown in Figure 12-3.

Step 5. Make the 2.50 inches edge flanges using the **Edge Flange** command as shown in Figure 12-4.

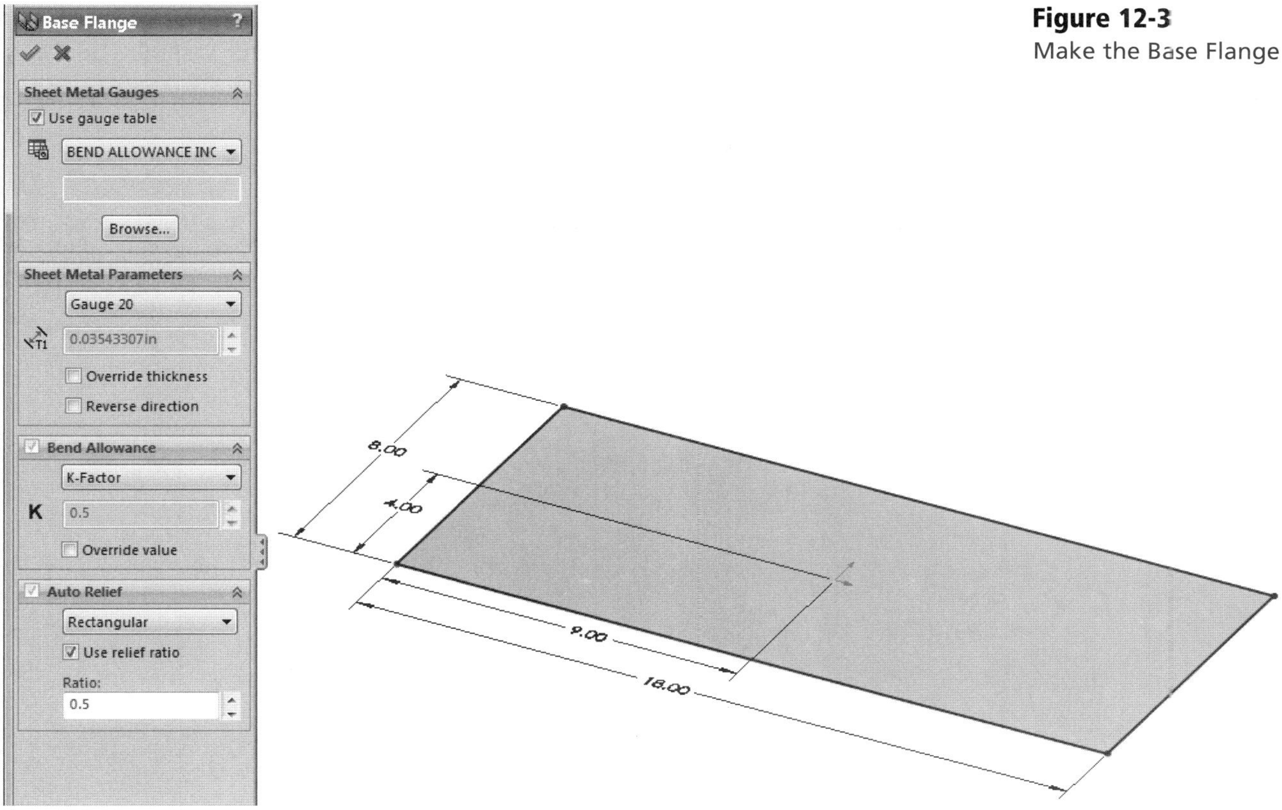

Figure 12-3
Make the Base Flange

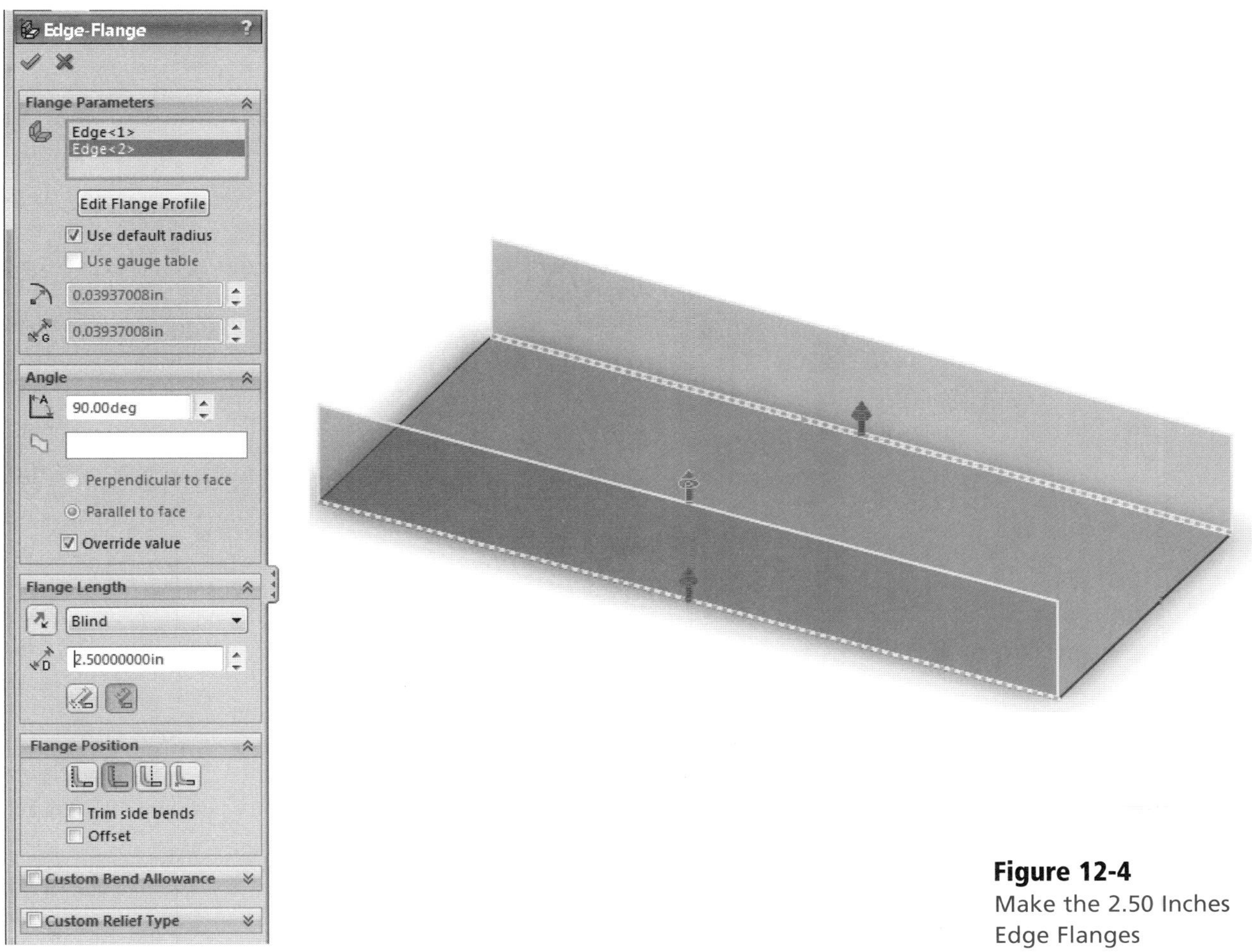

Figure 12-4
Make the 2.50 Inches Edge Flanges

Step 6. Make the 7.00 inches edge flange using the **Edge Flange** command and make two 2.50 inches chamfers using the **Sketch Chamfer** command on it as shown in Figure 12-5.

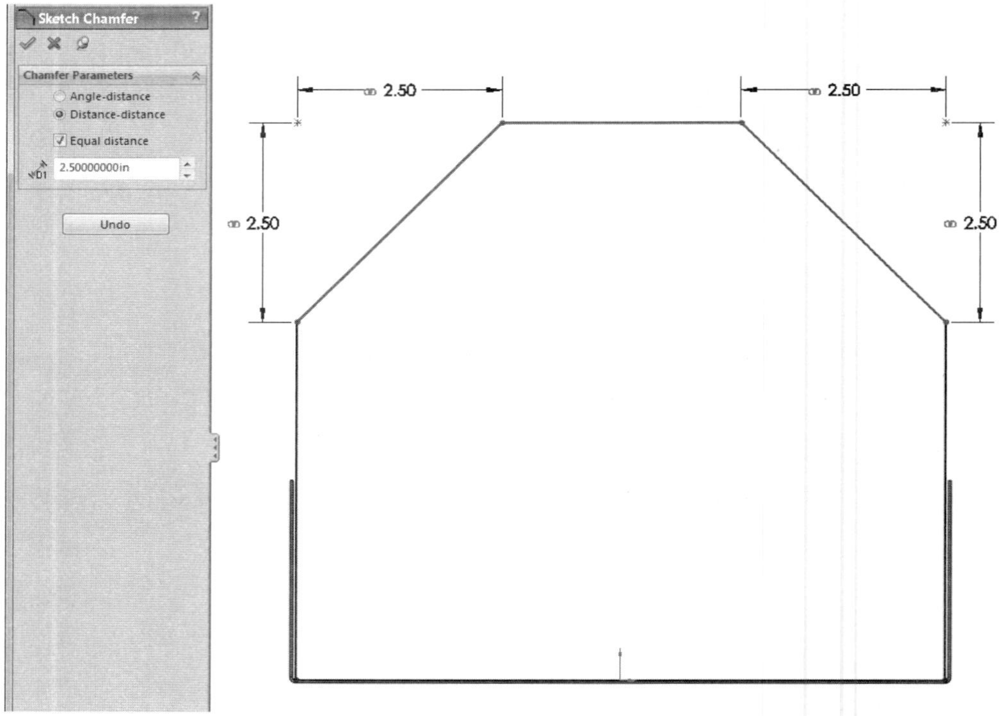

Figure 12-5
Make the 7.00 Inches Flange and Edit It

Step 7. Make a sketch for the miter flange from the lower left corner using the **Line** command and dimension it as shown in Figure 12-6. Make sure the line is perpendicular to the long edge of the body.

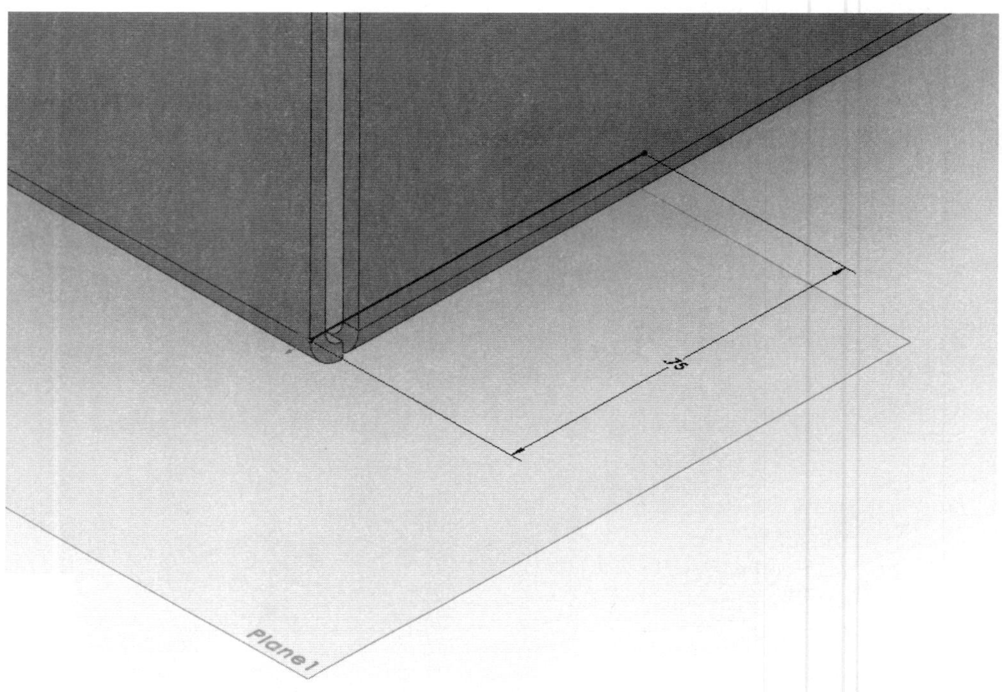

Figure 12-6
Make a Sketch for the Miter Flange and Dimension It

Step 8. Make a miter flange around the lower left corner of the 7 inches edge flange using the **Miter Flange** command as shown in Figure 12-7. Then, make another miter flange around the lower right corner.

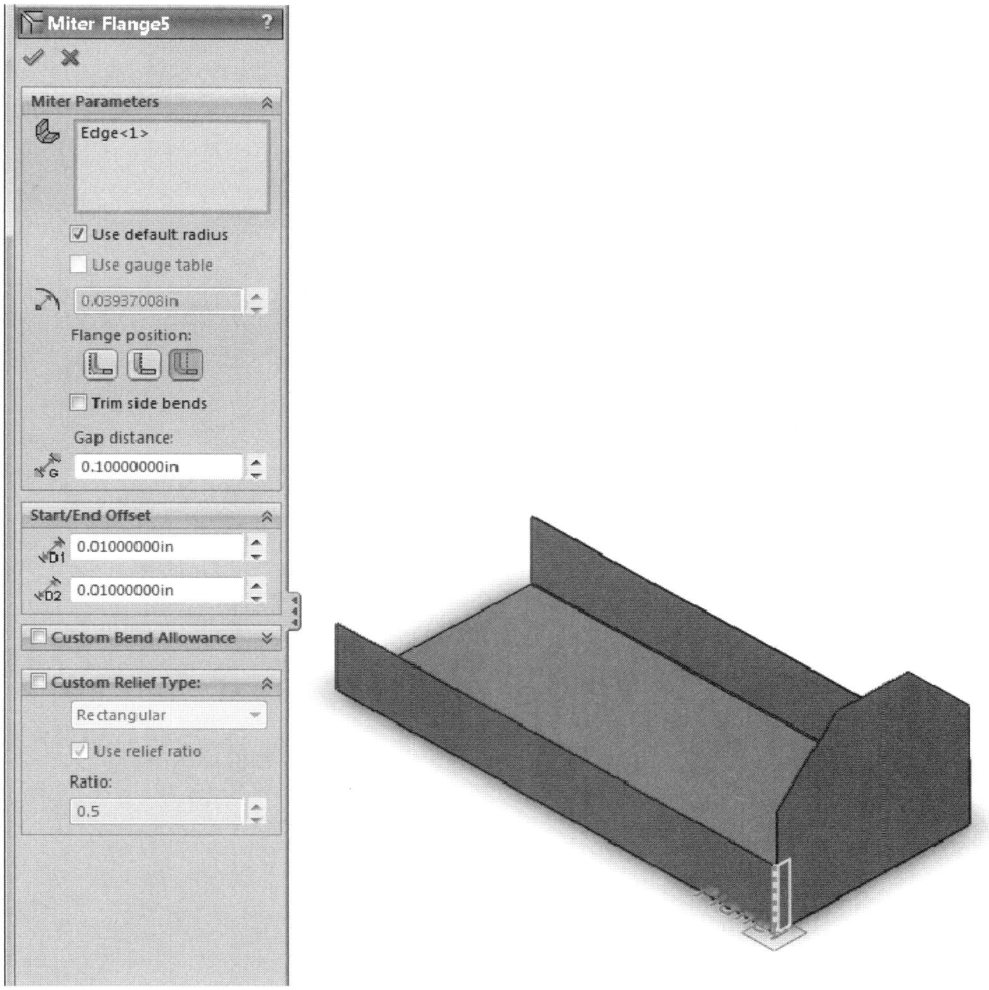

Figure 12-7
Make a Miter Flange around the 7 inches Edge Flange

Step 9. Mirror the 7 inches edge flange and the two miter flanges as shown in Figure 12-8. Make sure the **Edge-Flange** and the two **Miter Flanges** show in the **Features to Mirror** box and the **Right Plane** shows in the **Mirror Face/Plane** box.

Step 10. Make .375 inch hems on 5 edges of both 7 inches edge flanges using the **Hem** command as shown in Figure 12-9.

Step 11. Make .375 inch hems on the top edges of the 2.50 inches edge flanges as shown in Figure 12-10.

Step 12. Draw and dimension a .25 inch diameter circle on one of the 7 inches edge flanges and dimension it as shown in Figure 12-11. Then make an extruded cut using that circle **Through All** so you have a hole in both 7 inches edge flanges.

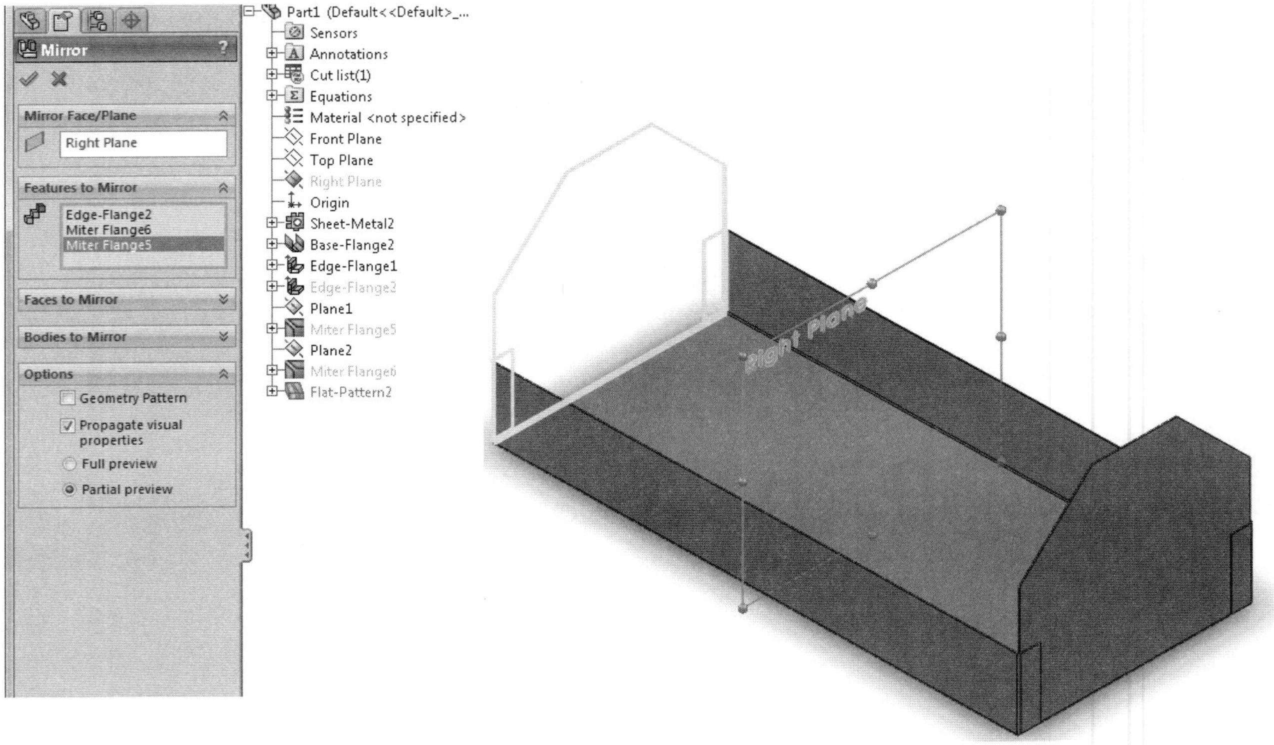

Figure 12-8
Mirror the 7 inches Edge Flange and the Two Miter Flanges

Figure 12-9
Make Hems on 5 Edges of the Edge Flanges

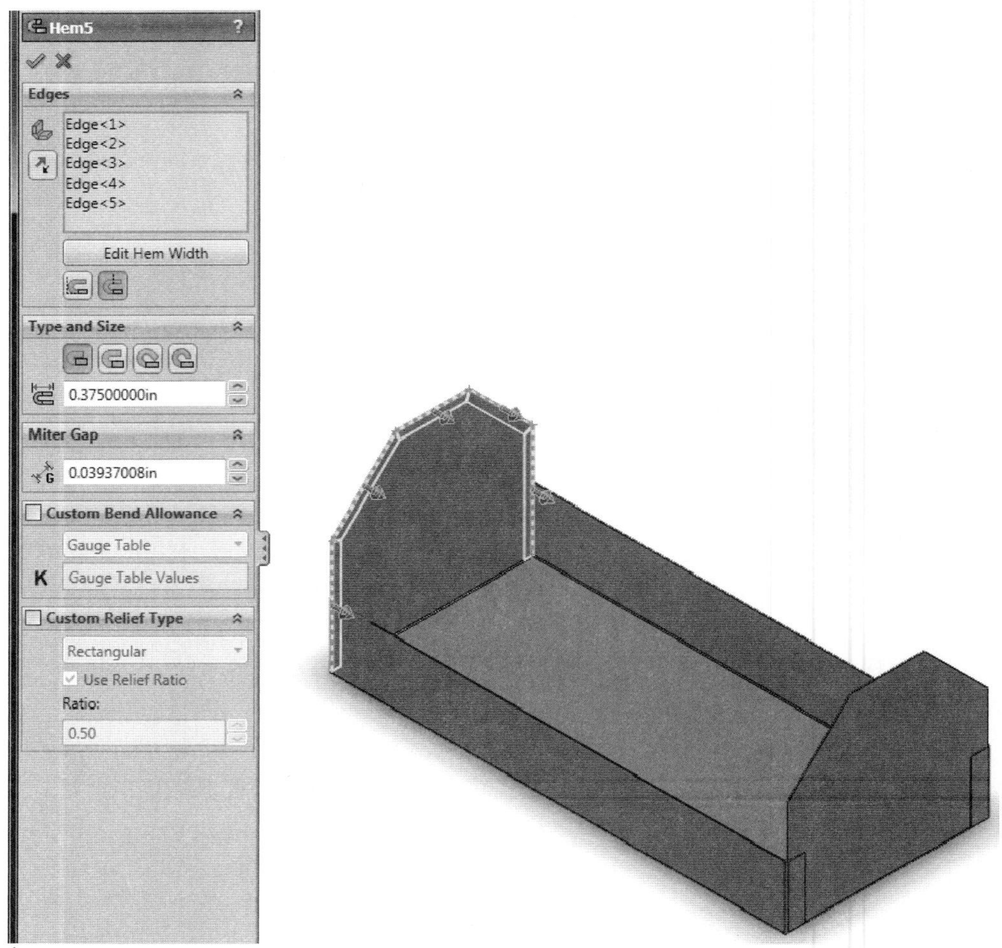

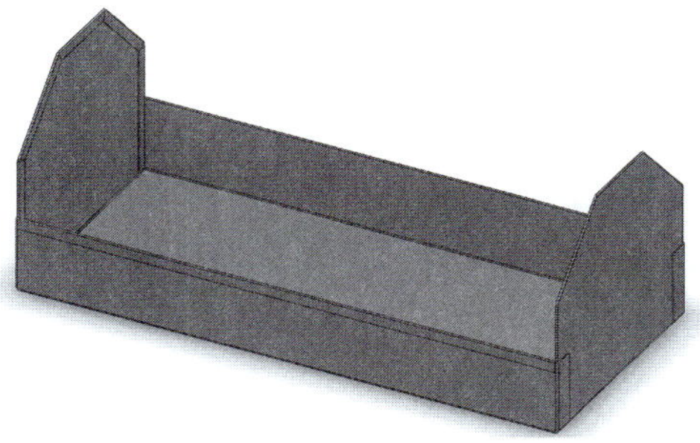

Figure 12-10
Make .375 inch Hems on the Top Edges of the 2.50 inches Edge Flanges

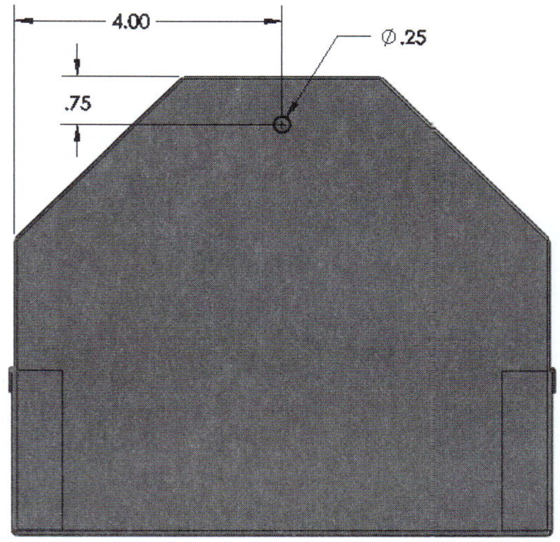

Figure 12-11
Draw and Dimension a .25 Inch Diameter Circle and Make an Extrude Cut Through All

Step 13. Save the model as **Tool Tray Body.**

The tool tray body is complete as shown in Figure 12-12.

Figure 12-12
Tool Tray Body Complete

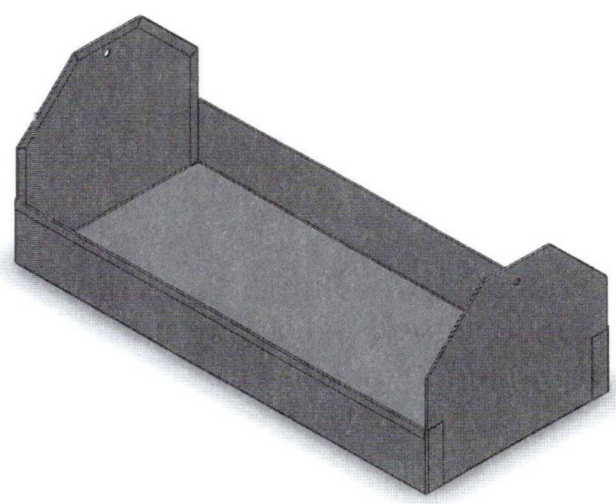

Make the Tool Tray Handle

If you have exited SolidWorks, open it again and start a new part.

Step 1. Click **Options** on the menu bar, select the **Document Properties** tab, and click **Units** in the list to the left.

Step 2. Make sure the **Units** are correctly set [notice that **IPS** (inch, pound, second) is checked], then click **OK.**

Step 3. Draw and dimension a circle on the right plane as shown in Figure 12-13.

Step 4. Extrude the circle 17.875 inches as shown in Figure 12-13.

Step 5. Make a 1/4-20 inches tapped hole 1.50 inches deep in the end of the handle using **Hole Wizard** as shown in Figure 12-14.

Figure 12-13
Draw a 1 Inch Circle on the Right Plane and Extrude It 17.875 Inches from the Mid Plane

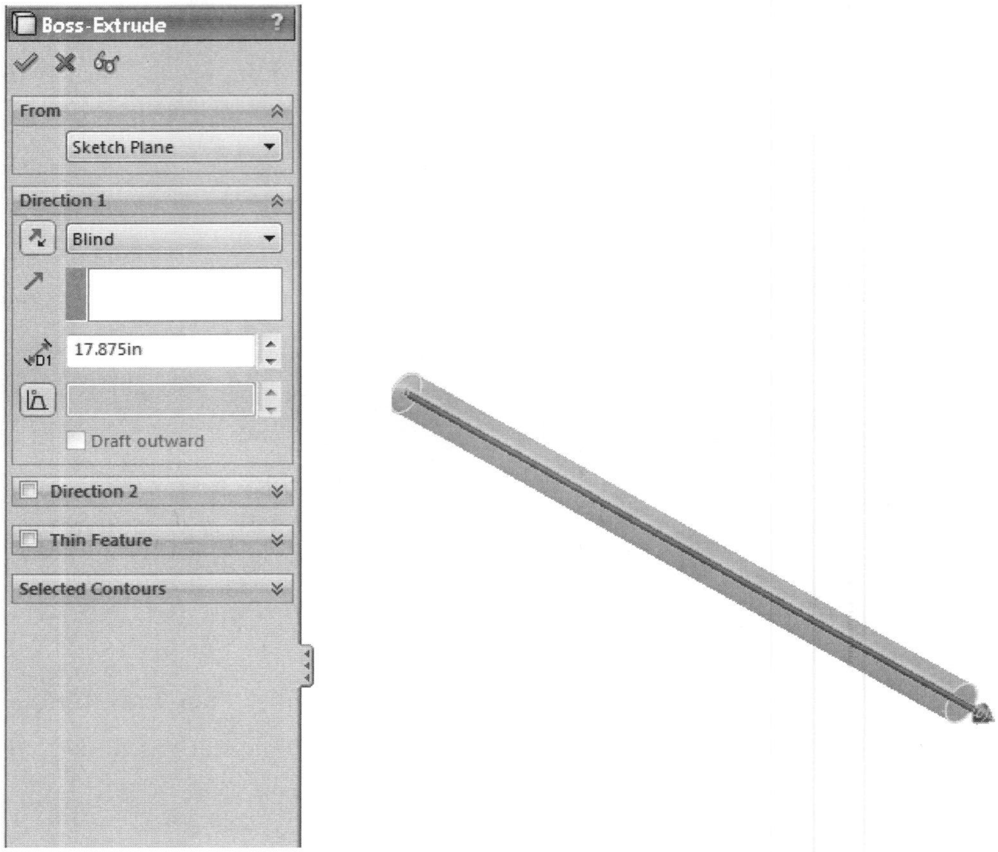

Figure 12-14
Make a 1/4-20 Inches Tapped Hole in the End of the Handle

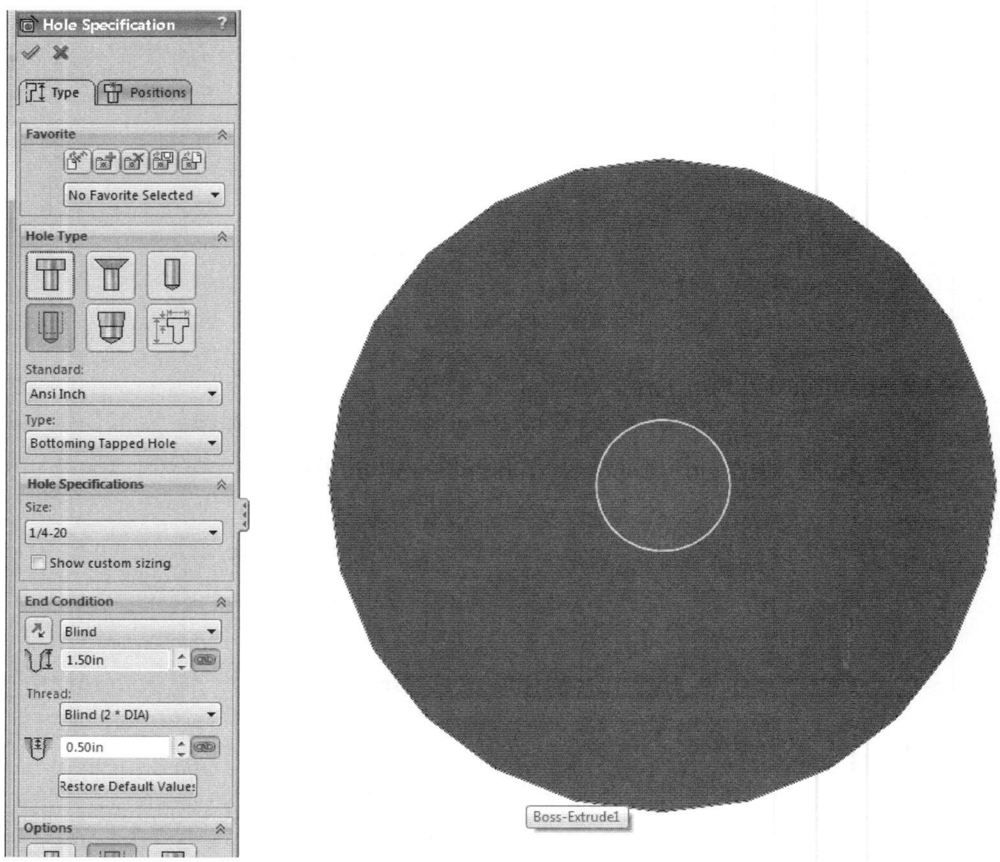

Step 6. Mirror the 1/4-20 inches tapped hole using the right plane as shown in Figure 12-15.

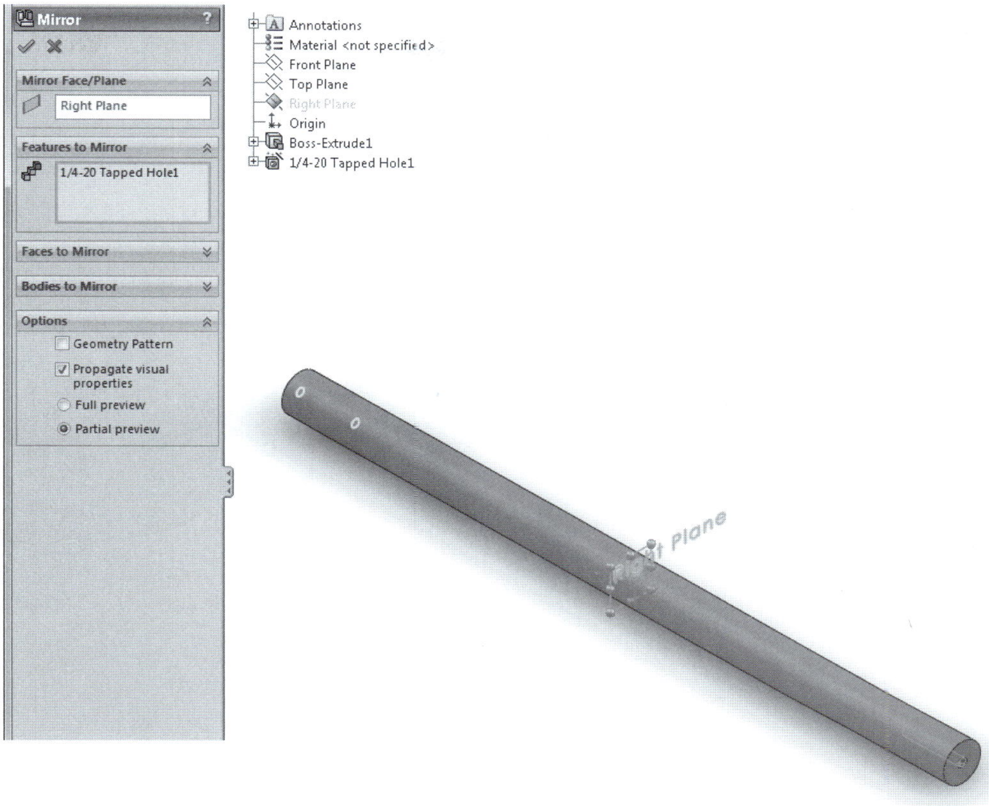

Figure 12-15
Mirror the 1/4-20 Inches Tapped Hole

Step 7. Save the model as **Tool Tray Handle.**

The tool tray handle is complete.

Make the 1/4-20 Inches Machine Screw

If you have exited SolidWorks, open it again and start a new part.

Step 1. Click **Options** on the menu bar, select the **Document Properties** tab, and click **Units** in the list to the left.

Step 2. Make sure the **Units** are correctly set [notice that **IPS** (inch, pound, second) is checked], then click **OK.**

Step 3. Make and dimension the sketch on the front plane as shown in Figure 12-16.

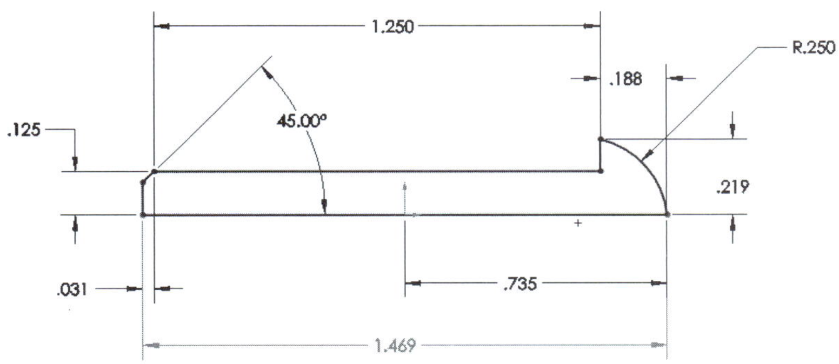

Figure 12-16
Draw and Dimension the 1/4-20 Inches Machine Screw on the Front Plane

Step 4. Revolve the sketch as shown in Figure 12-17.

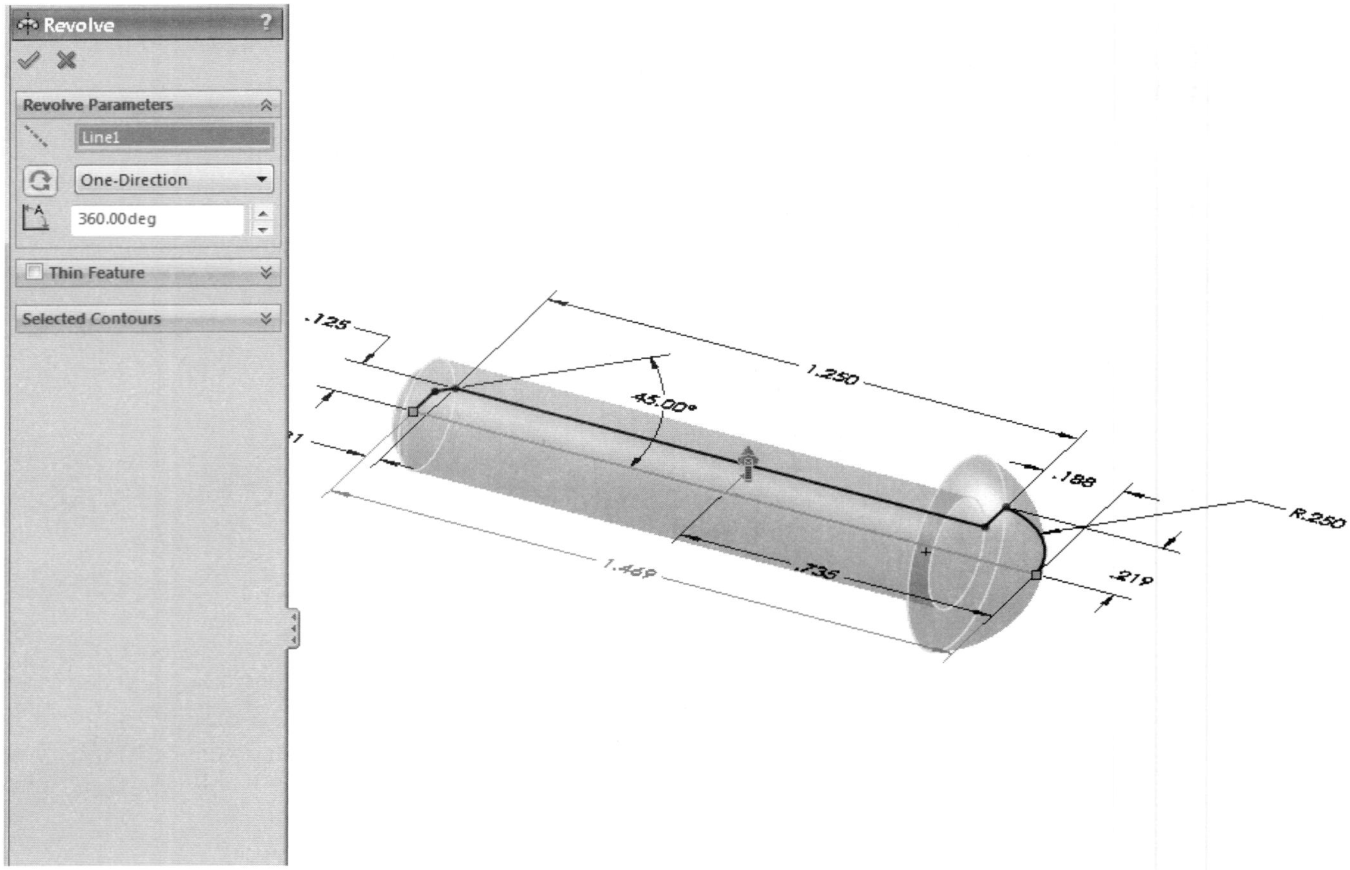

Figure 12-17
Revolve the Sketch

Step 5. Make a new plane 0.735 inch from the right plane as shown in Figure 12-18.

Figure 12-18
Make a New Plane 0.735 Inch from the Right Plane

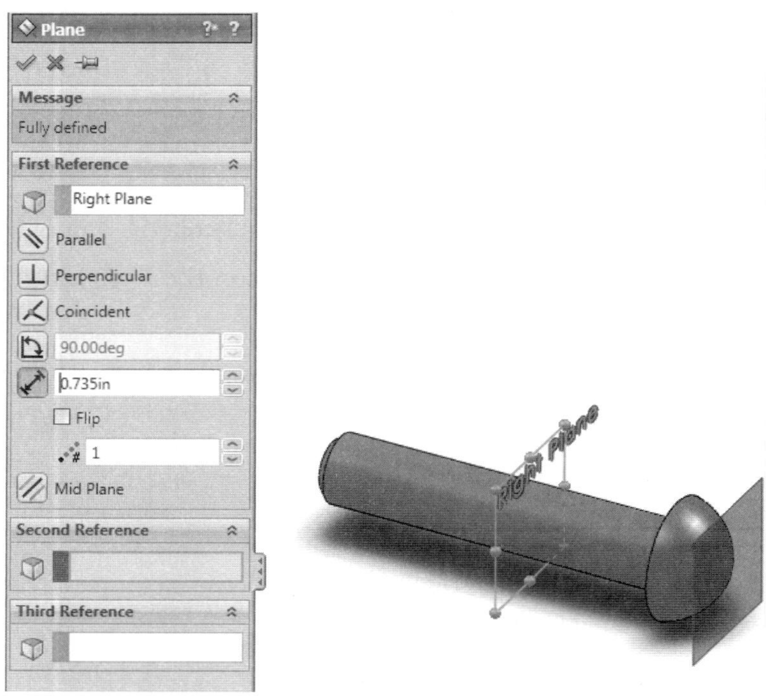

Step 6. Draw and dimension a rectangle on **Plane1** as shown in Figure 12-19.

The length of the rectangle is not important. Just make sure it is long enough to extend past the screw head.

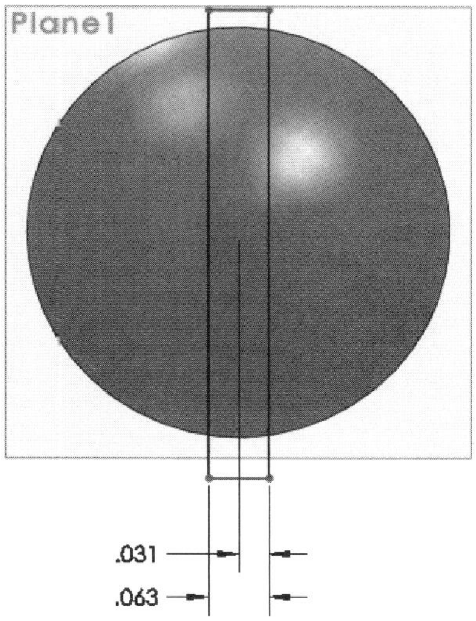

Figure 12-19
Draw and Dimension a Rectangle for the Screw Head Slot on Plane1

Step 7. **Extrude Cut** the rectangle .125 inch as shown in Figure 12-20.

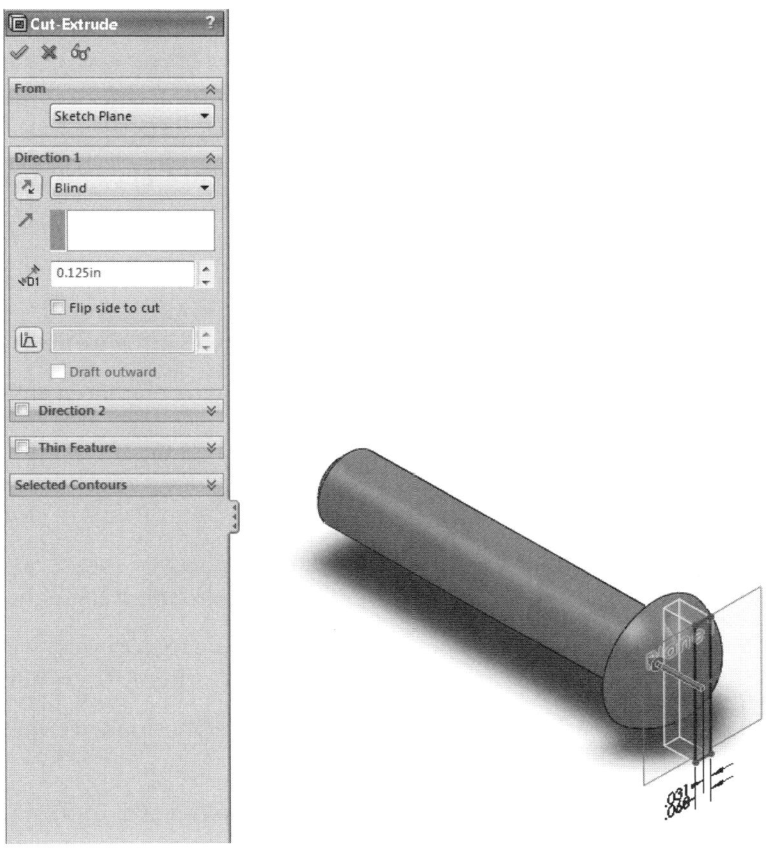

Figure 12-20
Extrude Cut the Rectangle 0.125 Inch

Step 8. Apply cosmetic threads as shown in Figure 12-21.

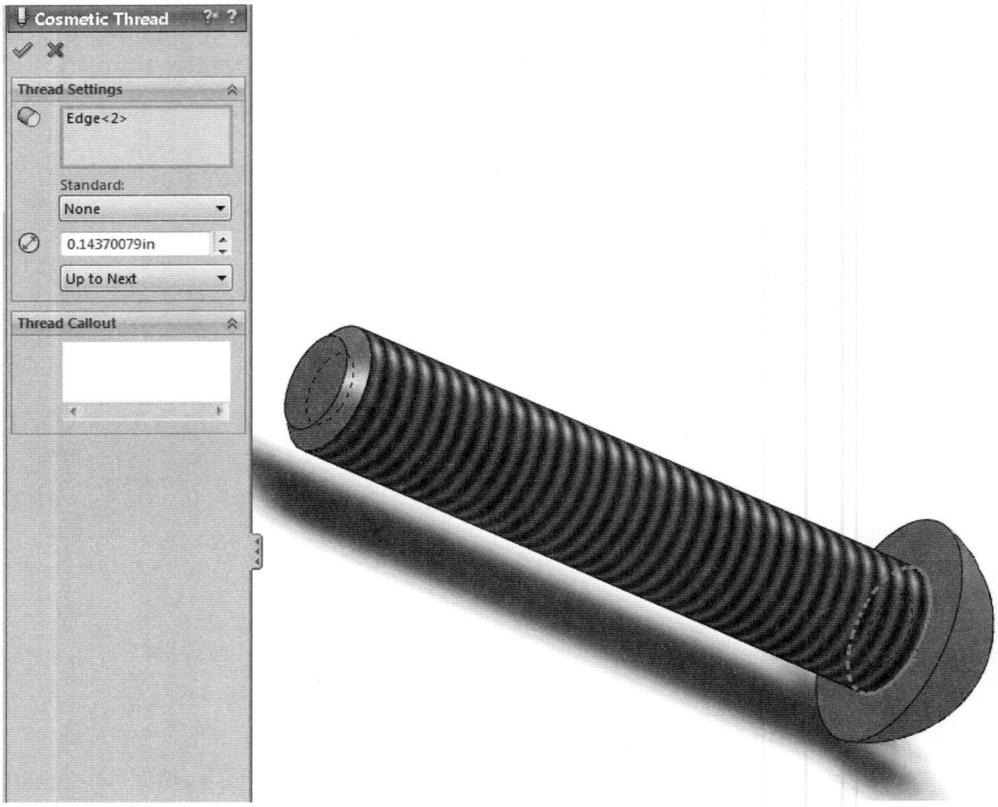

Figure 12-21
Apply Cosmetic Threads

1. Check: **Shaded cosmetic threads** in the **Detailing** area of the **Document Properties** tab on the **Options** dialog box (at the top of your screen).
2. Click **Cosmetic Thread...** on **Annotations** from the **Insert** menu on the menu bar.
3. Click the lower 1/4 inch edge of the screw and **Up to Next** as shown in Figure 12-21.

Step 9. Save the model as **1/4-20 Machine Screw.**

The 1/4-20 inches machine screw is complete.

Make the 1/4 Inch Lock Washer

If you have exited SolidWorks, open it again and start a new part.

Step 1. Click **Options** on the menu bar, select the **Document Properties** tab, and click **Units** in the list to the left.

Step 2. Make sure the **Units** are correctly set [notice that **IPS** (inch, pound, second) is checked], then click **OK**.

Step 3. Draw and dimension the lock washer on the top plane as shown in Figure 12-22.

Step 4. Extrude the lock washer .05 inch as shown in Figure 12-23.

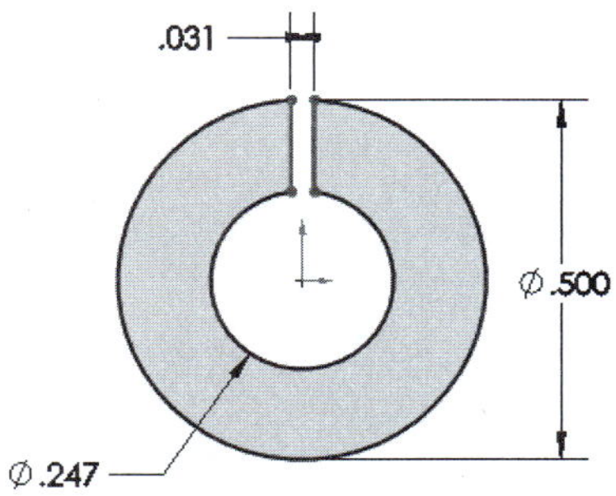

Figure 12-22
Draw and Dimension the Lock Washer on the Top Plane

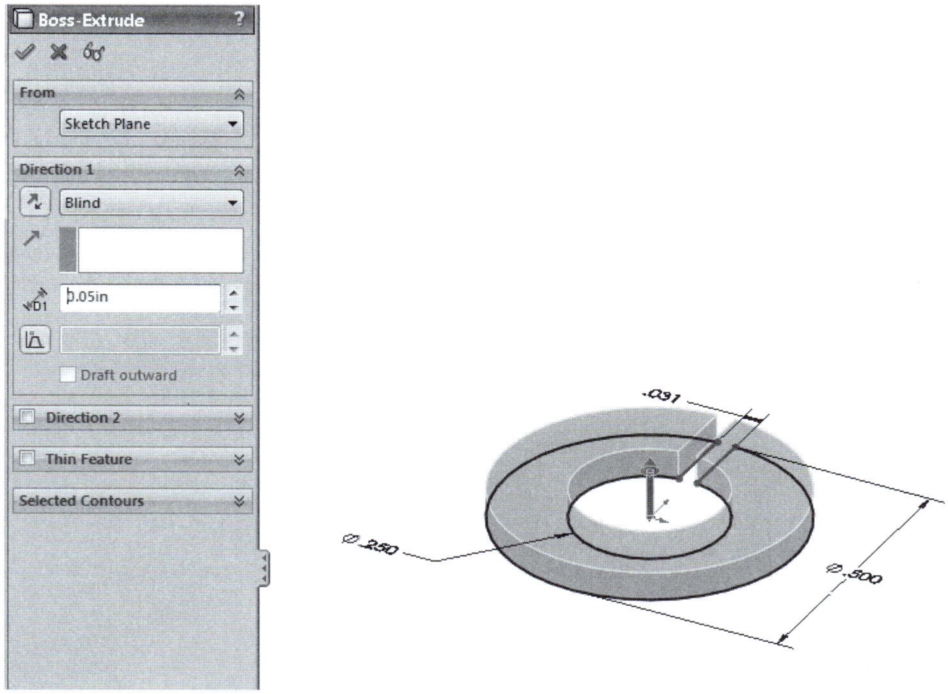

Figure 12-23
Extrude the Lock Washer 0.05 Inch

Step 5. Save the model as **Lock Washer.**

The lock washer is complete.

Make the Assembly

If you have exited SolidWorks, open it again and open the tool tray body.

Step 1. Click **Options** on the menu bar, select the **Document Properties** tab, and click **Units** in the list to the left.

Step 2. Make sure the **Units** are correctly set [notice that **IPS** (inch, pound, second) is checked], then click **OK.**

Step 3. Insert the tool tray handle using **Insert Components.**

Step 4. Apply the **Concentric** mate as shown in Figure 12-24. Be sure you click the tapped hole or another cylindrical shape on the handle and the hole in the body.

Step 5. Apply the **Coincident** mate as shown in Figure 12-25. Be sure you click the flat end on the handle and the 7 inches flange on the body.

Step 6. Insert the lock washer and the machine screw and assemble them using the **Concentric** and **Coincident** mates as shown in Figure 12-26.

Step 7. Mirror the lock washer and the machine screw using the **Right Plane** as the **Mirror** plane as shown in Figure 12-27.

Figure 12-24
Apply the Concentric Mate

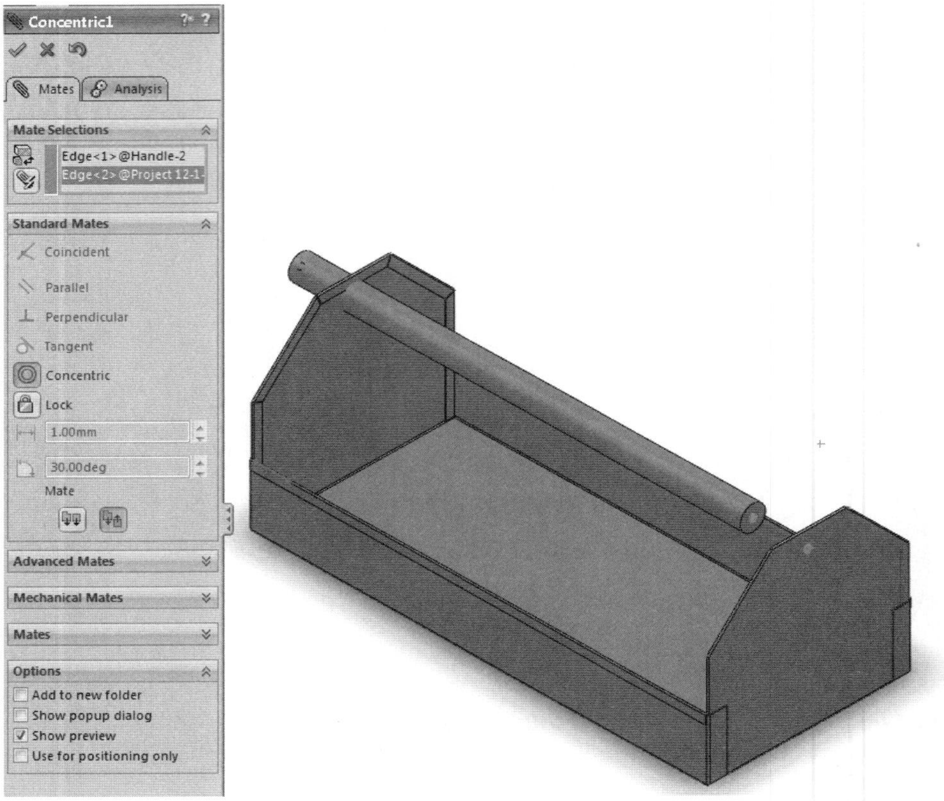

Figure 12-25
Apply the Coincident Mate

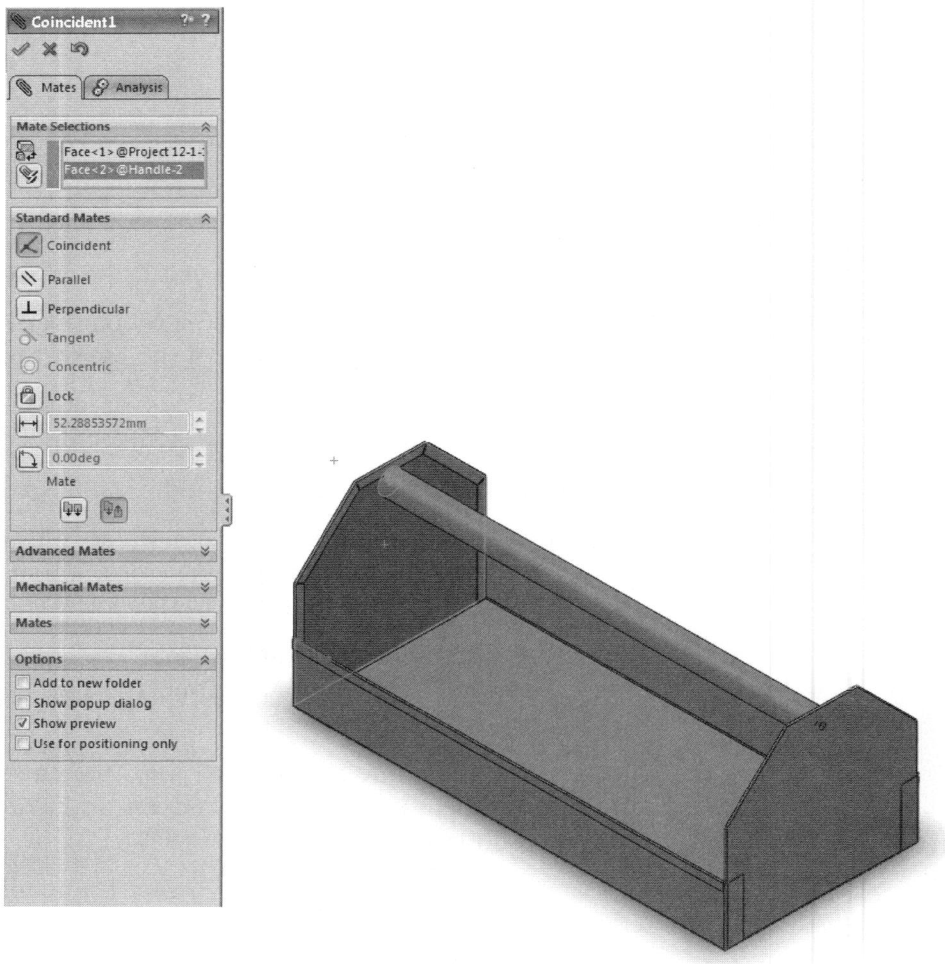

Figure 12-26
Assemble the Lock Washer and Machine Screw

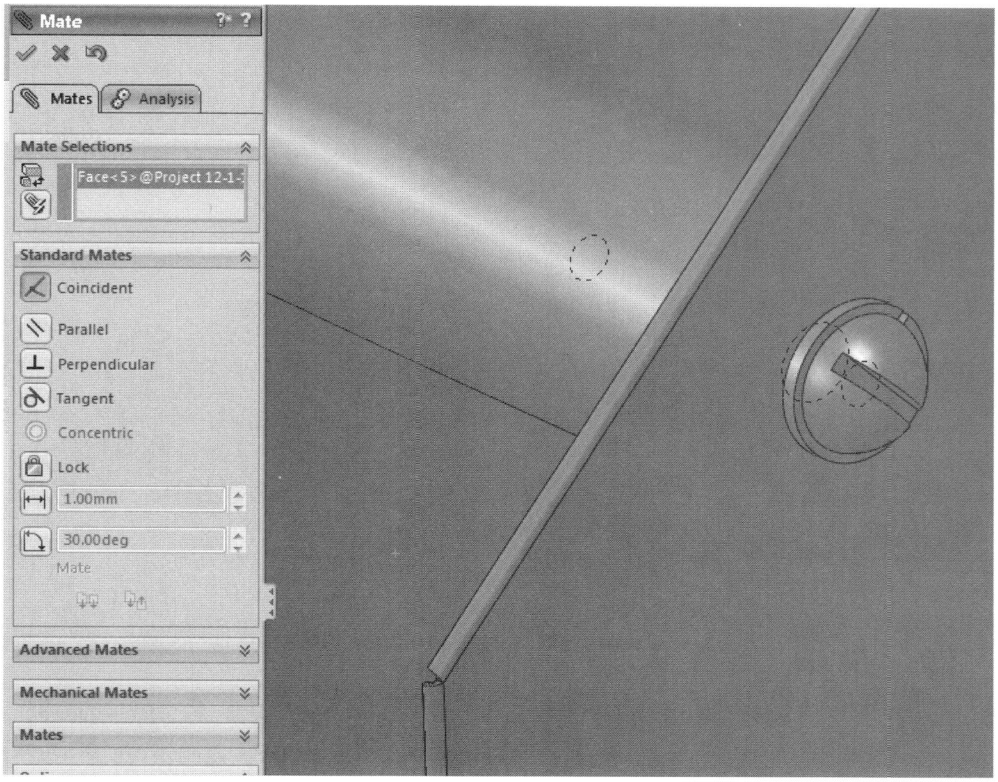

Figure 12-27
Mirror the Lock Washer and Machine Screw

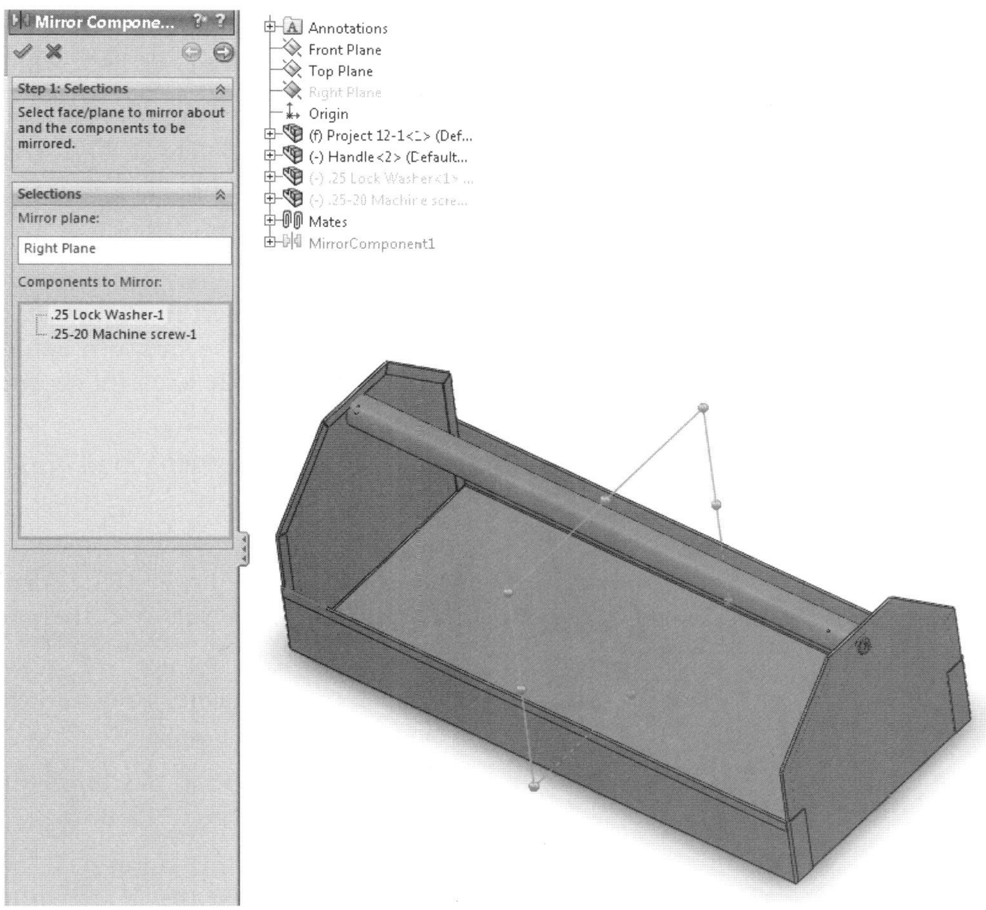

Step 8. Save the assembly as **Tool Tray Assembly.**

Project 12-1 is complete as shown in Figure 12-28.

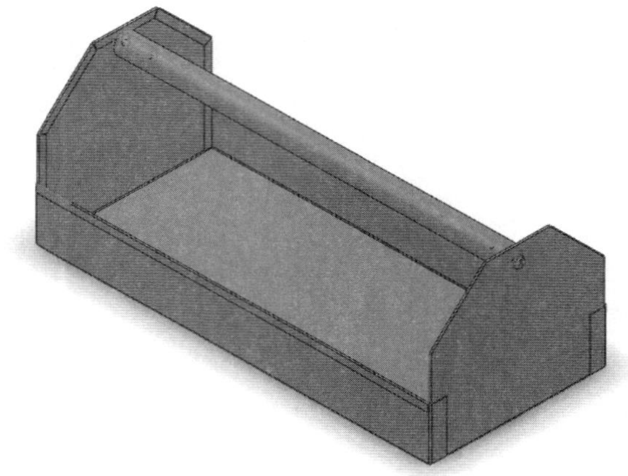

Figure 12-28
Project 12-1 Complete

Project 12-2: Make an Assembly Drawing, an Exploded View, and a Parts List

If you have exited SolidWorks, open it again and open the tool tray assembly.

Step 1. Click **Options** on the menu bar, select the **Document Properties** tab, and click **Units** in the list to the left.

Step 2. Make sure the **Units** are correctly set [notice that **IPS** (inch, pound, second) is checked], then click **OK.**

Step 3. Click **Make Drawing from Assembly,** click the **B(ANSI) Landscape** format, and click **OK.**

Step 4. Drag the isometric view onto the sheet and use a custom scale of 1:3 as shown in Figure 12-29. Notice that you will have to click **Use custom scale** and **User Defined** to type **1:3** in the **Scale** area as shown in the lower part of the **Drawing Manager.**

Step 5. Right-click in the drawing area outside of the drawing view and click **Bill of Materials...** as shown in Figure 12-30.

Step 6. Click the isometric drawing view, click **OK.** place the **Bill of Materials** in the lower right corner of the format and edit the bill of materials and the sheet format as shown in Figure 12-31.

Step 7. Right-click in the drawing area and click **Open tool tray assembly** as shown in Figure 12-32.

Step 8. Use the steps described in Chapter 10 to make an exploded view as shown in Figure 12-33.

Step 9. Return to the drawing, right-click, click **Properties,** and check **Show in exploded state** as shown in Figure 12-34.

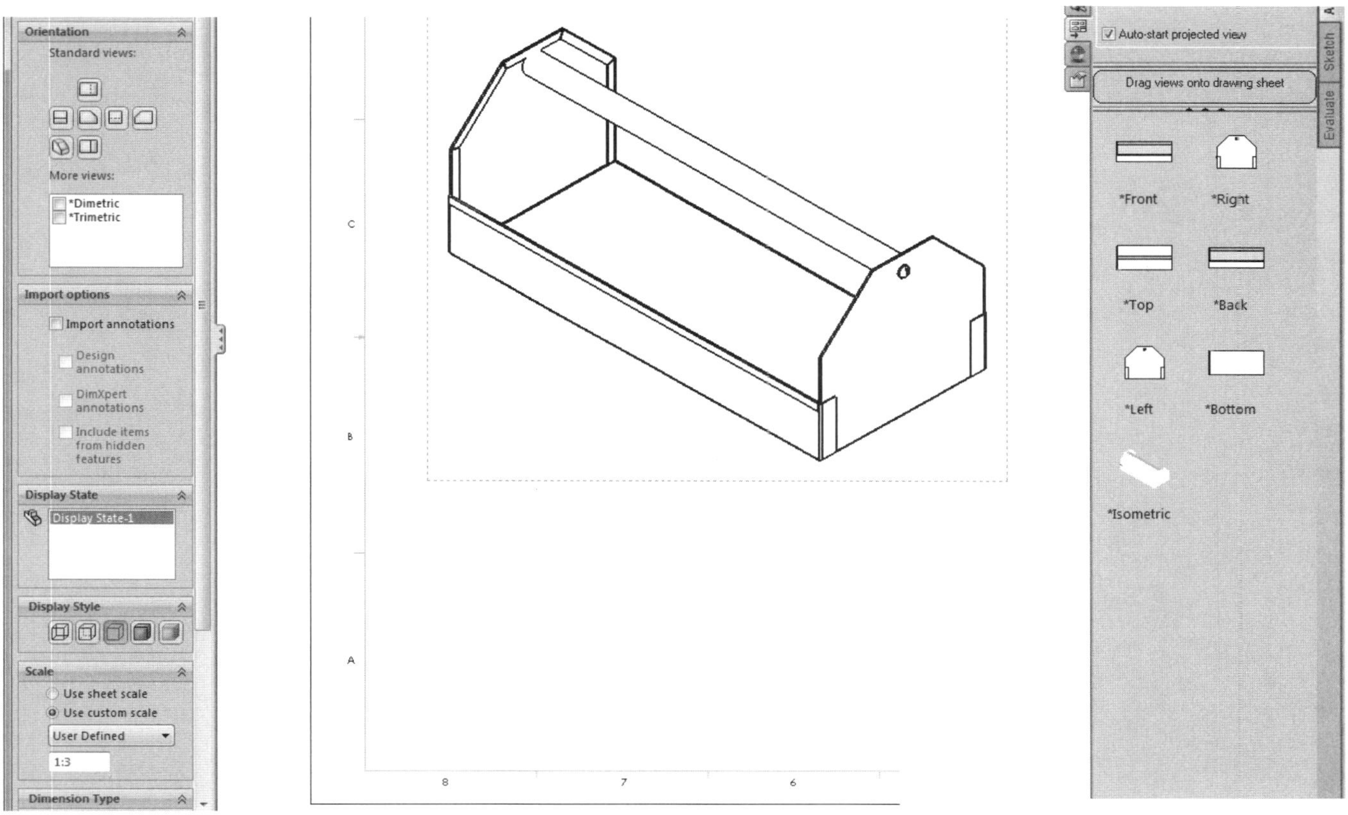

Figure 12-29
Drag the Isometric View onto the B-Size Format and Select a Custom Scale of 1:3

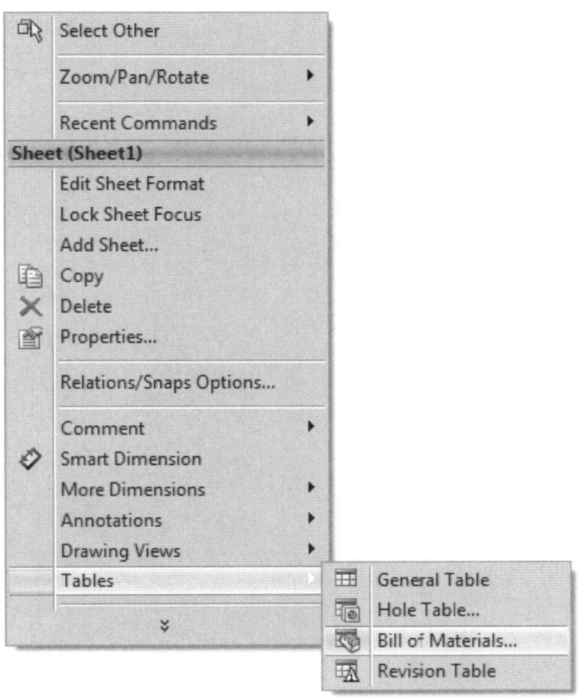

Figure 12-30
Right-Click and Click Bill of Materials...

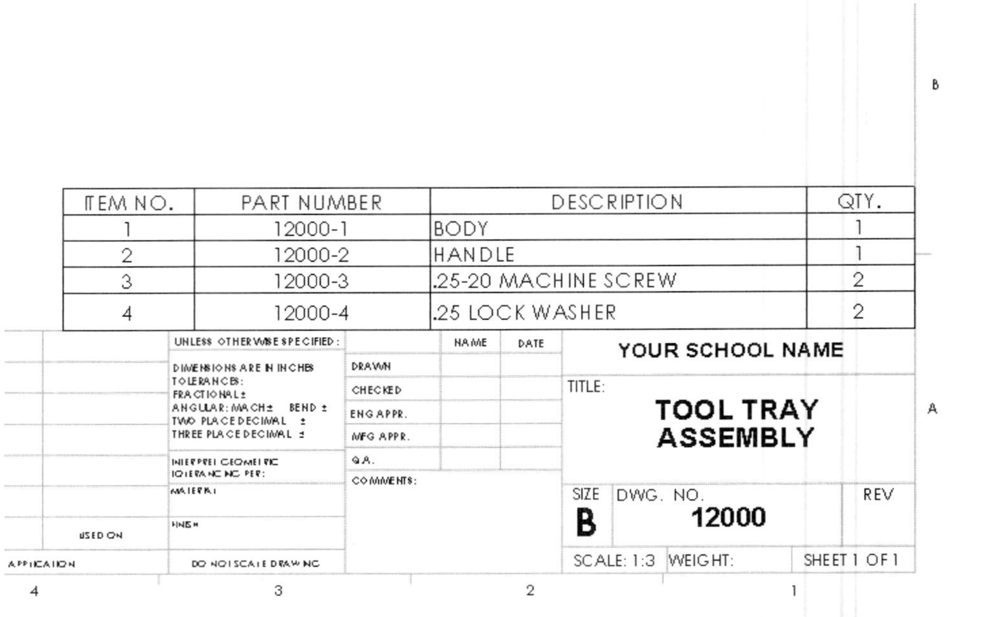

Figure 12-31
Make a Bill of Material and Edit Sheet Format

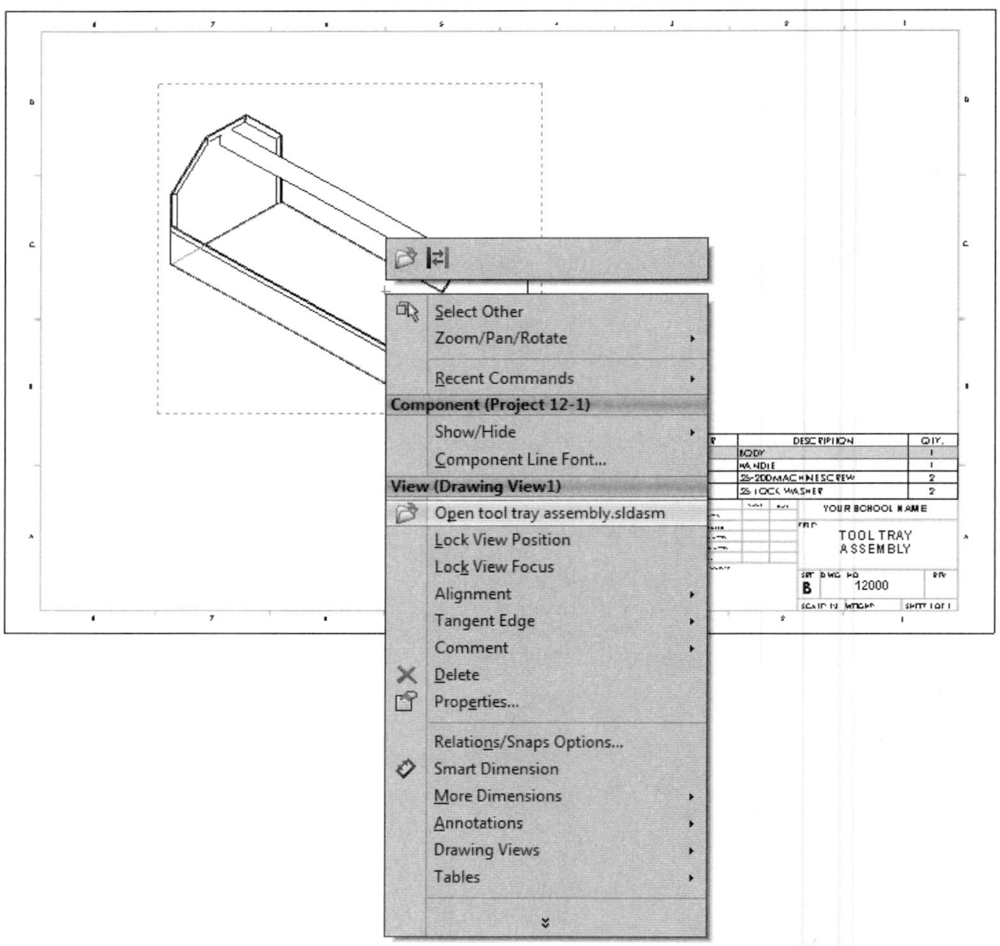

Figure 12-32
Click: Open Tool Tray Assembly

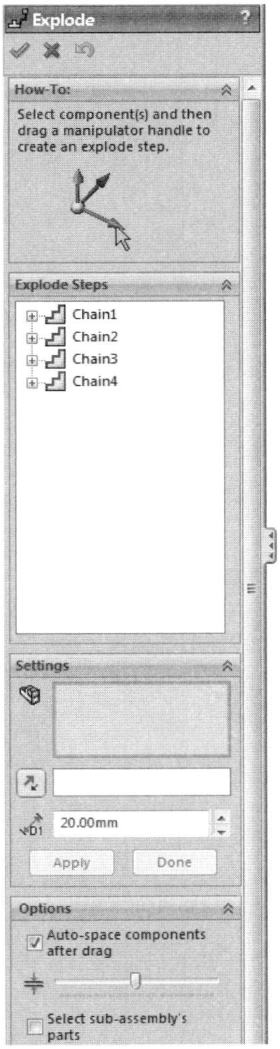

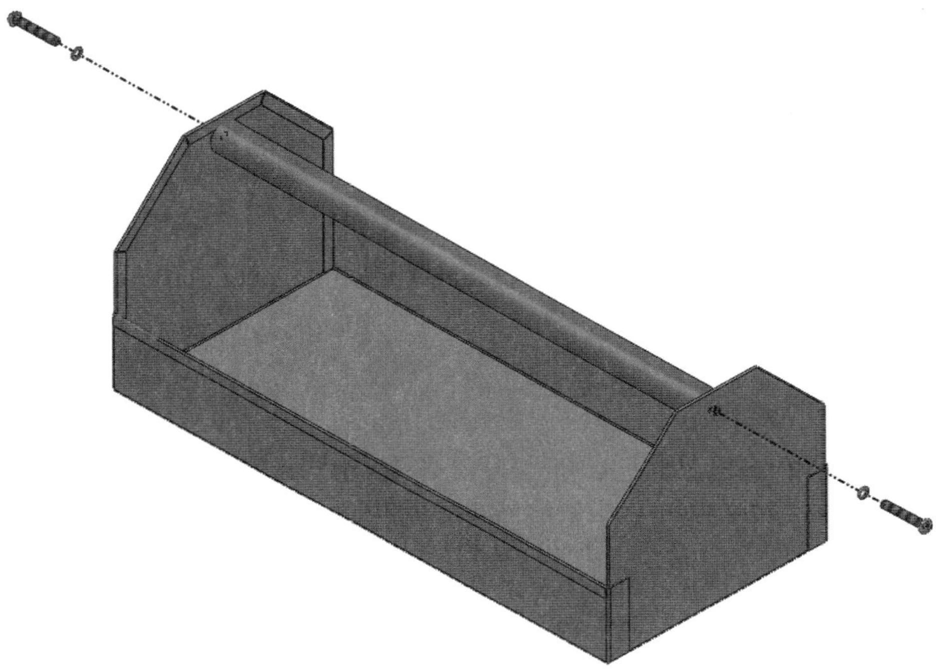

Figure 12-33
Explode the Screws and Lock Washers and Add Explode Line Sketch Lines

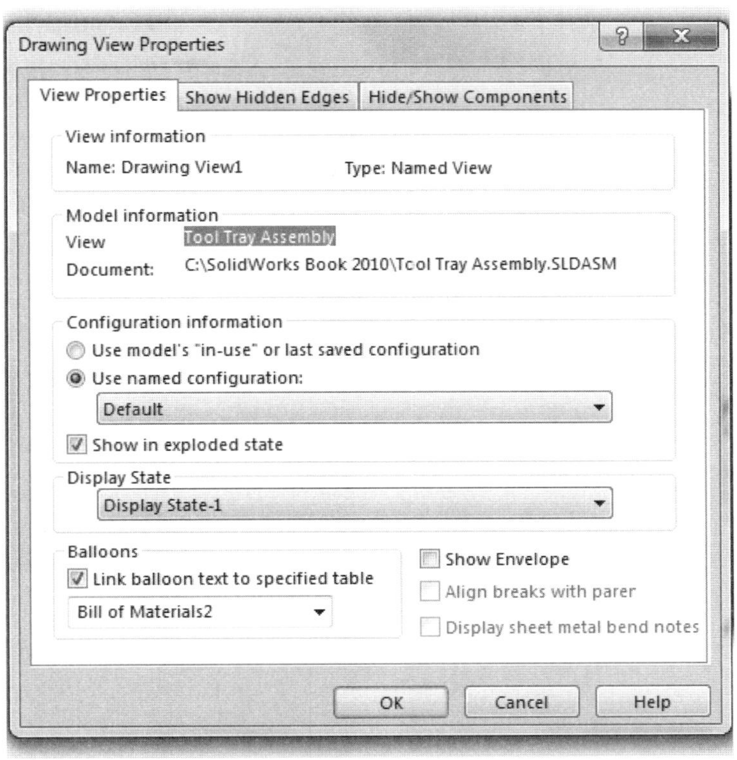

Figure 12-34
Check Show in Exploded State

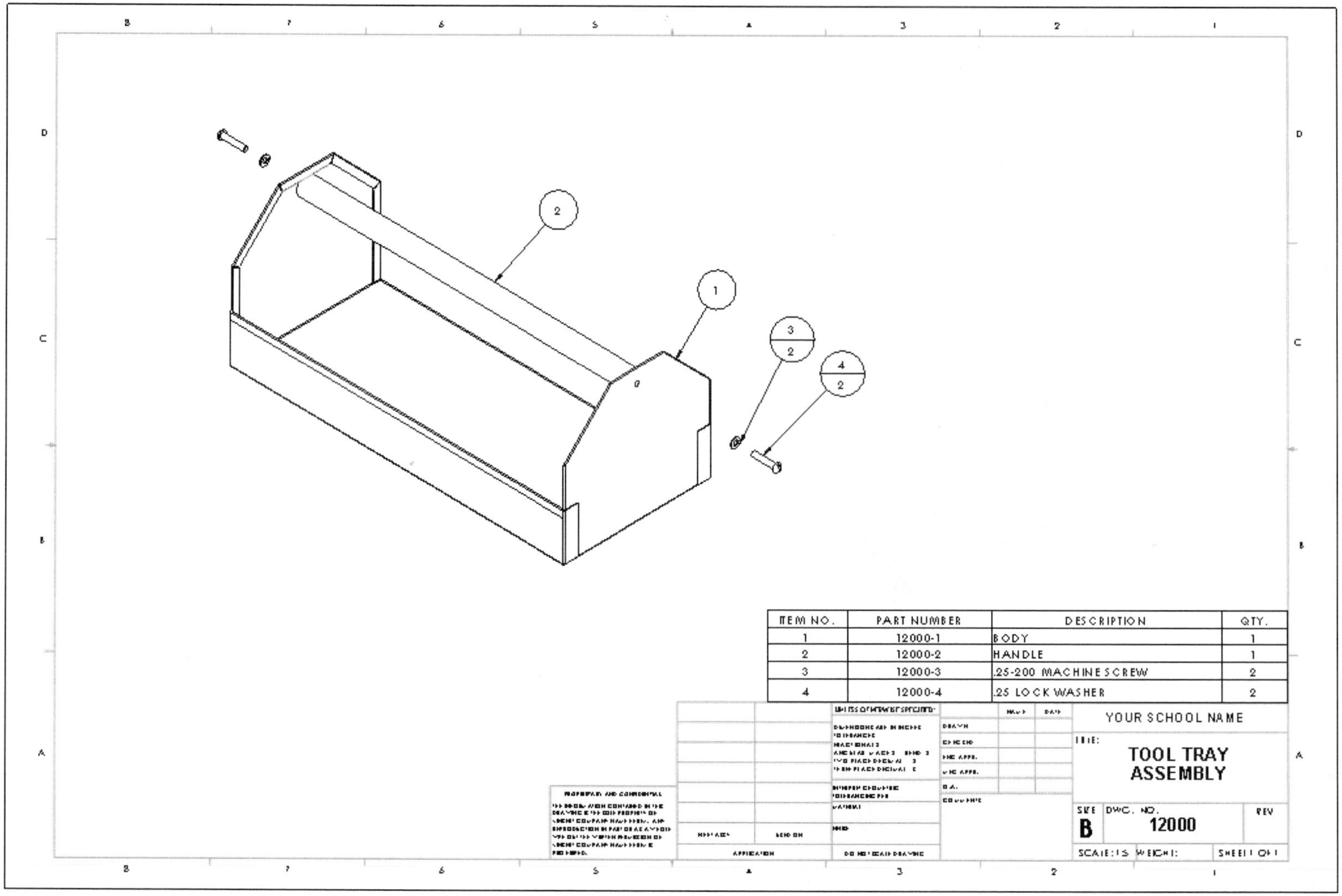

Figure 12-35
Project 12-2 Complete

Step 10. Right-click, click **Balloon,** and complete the drawing (using the information from Chapters 10 and 11 if needed) as shown in Figure 12-35. Be sure to use two split balloons.

Step 11. Save the drawing as **Tool Tray Top Assembly.**

Project 12-2 is complete.

Project 12-3: Make the Subassemblies of the Stroller and Make an Assembly of It

Figure 12-36 shows the completed assembly of Project 12-3, the stroller.

On this project, instructions that are needed are included in the figure title of each illustration. You may need to refer to earlier chapters and Project 12-1 for examples of commands that are unfamiliar to you for constructing and assembling.

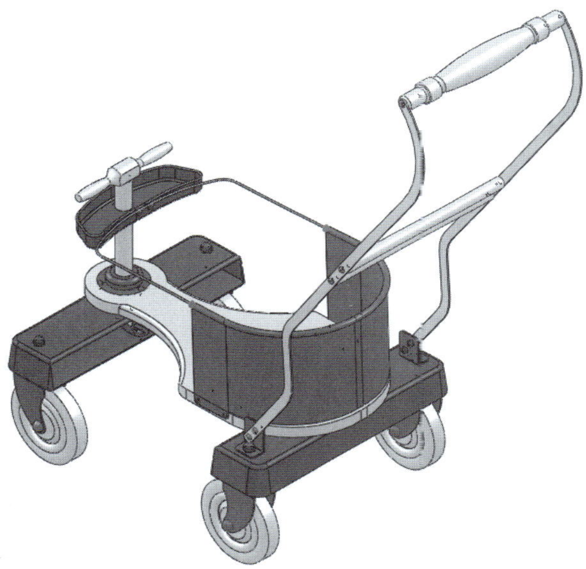

Figure 12-36
Stroller Top Assembly (Project 12-3) Complete

After you have constructed all of the models (Figures 12-37 through 12-111) you can begin assembling them using the following:

1. Make the following subassemblies using **Make Assembly from Part:**

A. Handle Subassembly
The Handle Subassembly consists of:

1	Handle	1 ea
2	Brace	2 ea
3	Spacers	2 ea
4	.25-20 × .75 Screw	2 ea
5	.188-20 × 1 Machine Screw	4 ea
6	.188 Lock Washer	4 ea

B. Wheel Subassembly

The Wheel Subassembly consists of:

1	Wheel Bracket	2 ea
2	Wheel	2 ea
3	Wheel Mounting Screw	2 ea
4	Wheel Mounting Bracket	1 ea
5	Wheel Mounting Pin	2 ea

2 Open the **Wheel Subassembly** and make the **Top Stroller Assembly** using **Make Assembly from Assembly.** The **Top Stroller Assembly** consists of the following:

1	Handle Subassembly	1 ea
2	Handle Assembly Bracket	2 ea
3	Wheel Subassembly	2 ea
4	Seat	1 ea
5	Seat Enclosure Subassembly	1 ea
6	Tray	1 ea
7	Tray Bracket	2 ea
8	Post	1 ea
9	Post Support	1 ea
10	Post Bracket	1 ea
11	Front Handle	1 ea
12	Wheel Mounting Nut	4 ea
13	.188-20 × .5 Machine Screw	7 ea
14	.188 Lock Washer	7 ea
15	.188 Nut	7 ea
16	.188 × .5 Long Screw	3 ea
17	.25-20 × .75 Screw	4 ea

Figure 12-36 shows the completed drawing of Project 12-3.

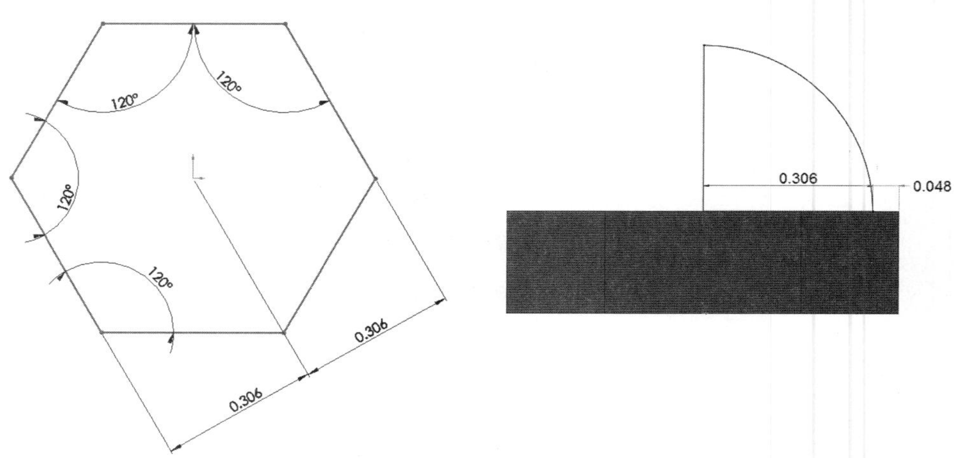

Figure 12-37
Draw a Hexagon on the Top Plane and Extrude It .25 Inch. Draw a Sketch for a Revolve on the Front Plane and Revolve It

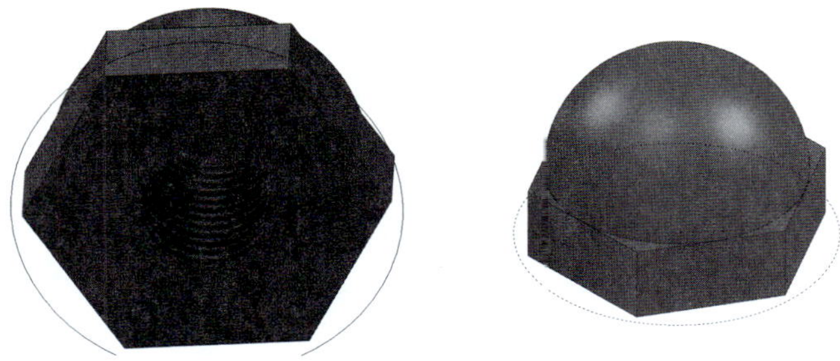

Figure 12-38
Draw a .250 Circle on the Bottom of the Part, Do an Extrude Cut .4125 Inch Deep, and Apply Cosmetic Threads Cap Nut Complete

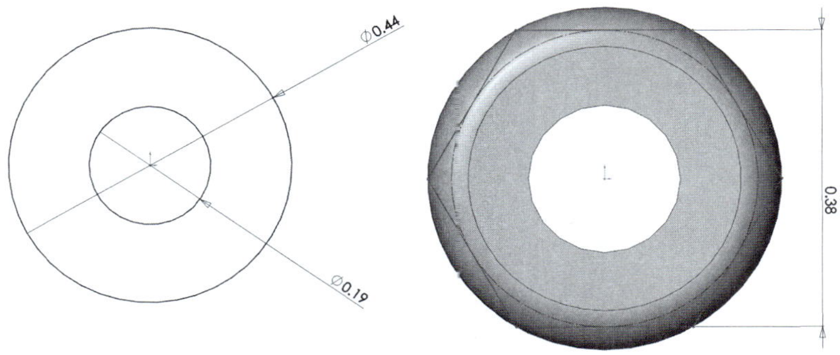

Figure 12-39
Draw 2 Circles for an Extrude on the Top Plane and Extrude It .188 Inch. Draw an Inscribed .38 Inch Hexagon on Top of the Extrusion

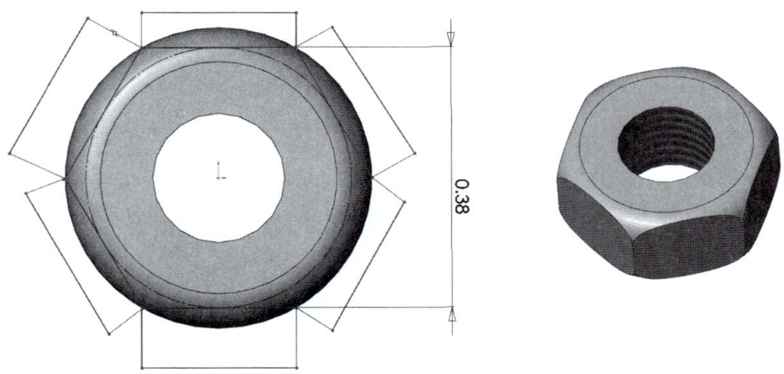

Figure 12-40
Draw Lines from the Polygon for an Extrude Cut and Do an Extrude Cut .188. Nut Complete

> **TIP**
>
> You can also make this part by drawing a circle on top of the extruded hexagon and using the circle to make an extude cut with a 60° draft angle.

Figure 12-41
Draw a Sketch for the .25-20 × .75 Inch Lg Screw on the Front Plane and Revolve It. Draw a Sketch for an Extruded Cut on Top of the Screw and Extrude It .06251 Inch. Apply Cosmetic Threads .25-20 × .75 Inch Lg Screw Complete

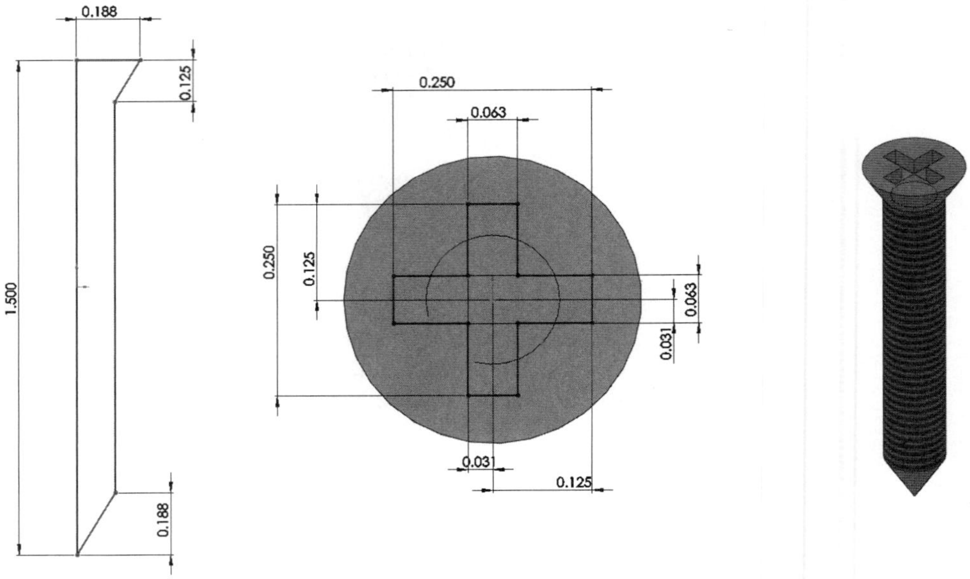

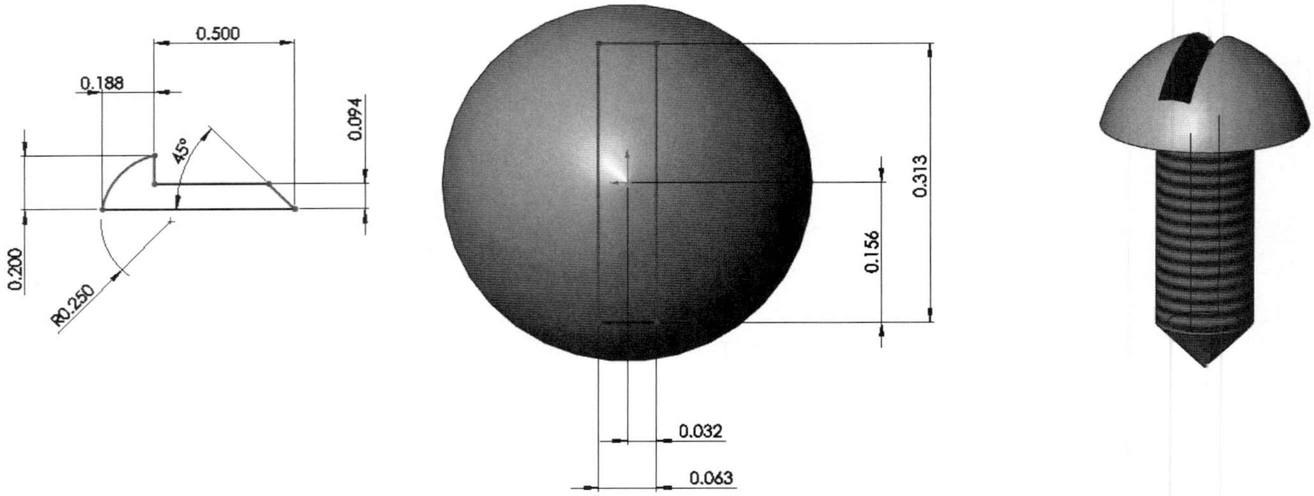

Figure 12-42
Draw a Sketch for the .188 × .5 Inch Lg Screw on the Front Plane and Revolve it. Draw a Sketch for an Extruded Cut on Top of the Screw and Extrude It .156 Inch. Apply Cosmetic Threads .188 × .5 Inch Lg Screw Complete

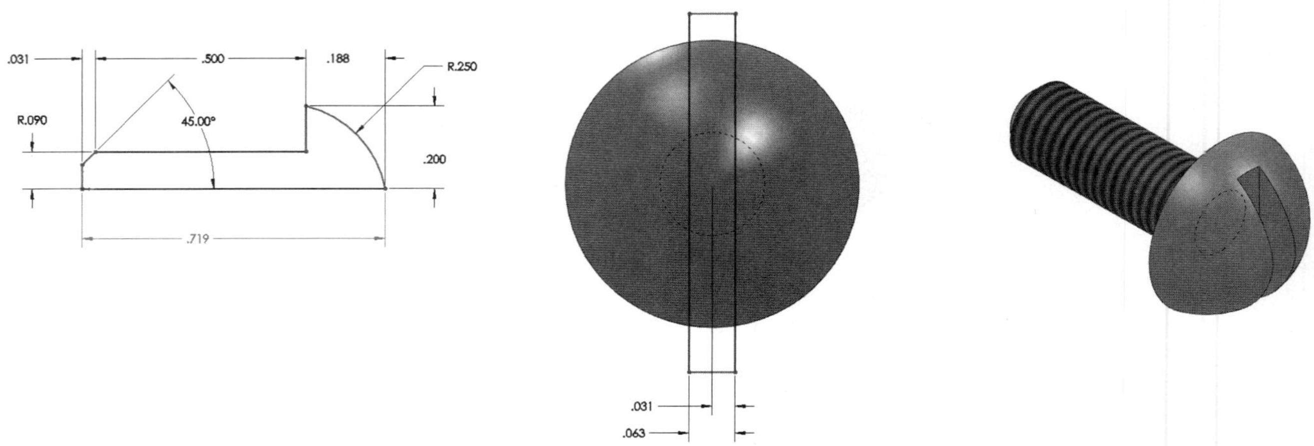

Figure 12-43
Draw a Sketch for the .188-20 × .5 Inch Lg Machine Screw on the Front Plane and Revolve It. Draw a Sketch for an Extruded Cut on a Plane .735 Inch from the Right Plane and Do an Extruded Cut .125 Inch Deep. Apply Cosmetic Threads. .188-20 × 5 Inches Lg Machine Screw Complete

Figure 12-44
Draw a Sketch for a Brace on the Front Plane and Extrude It .125 Inch

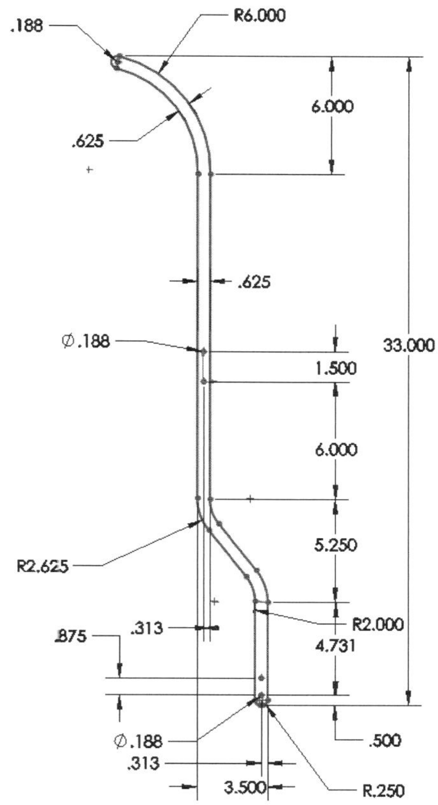

Figure 12-45
Side Braces Complete

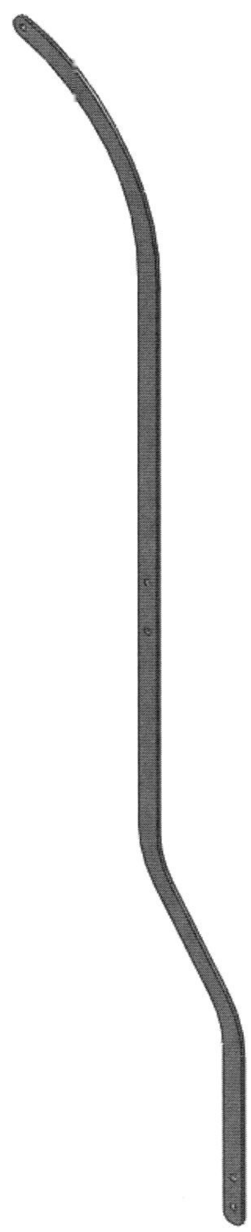

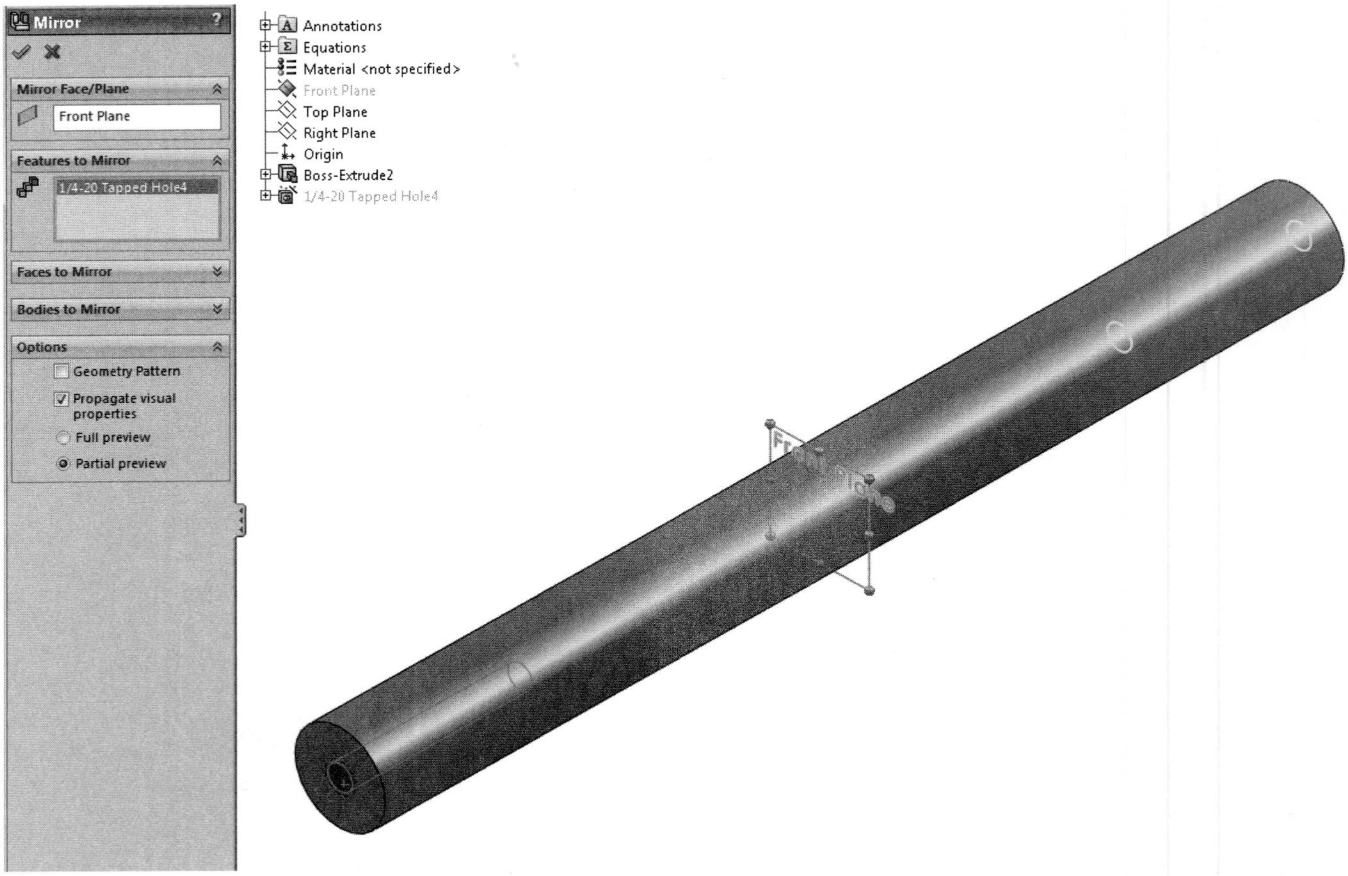

Figure 12-46
Draw a .75 Diameter on the Front Plane, Extrude It 12.750 Inches, Make a 1/4-20 × 1.50 Inches Long Tapped Hole and Mirror It to the Other End

Figure 12-47
The Spacer Complete

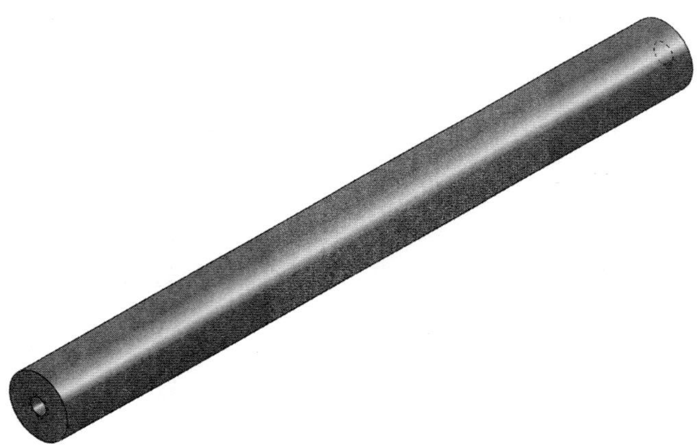

Figure 12-48
Draw a Sketch for the Grip on the Front Plane and Revolve It

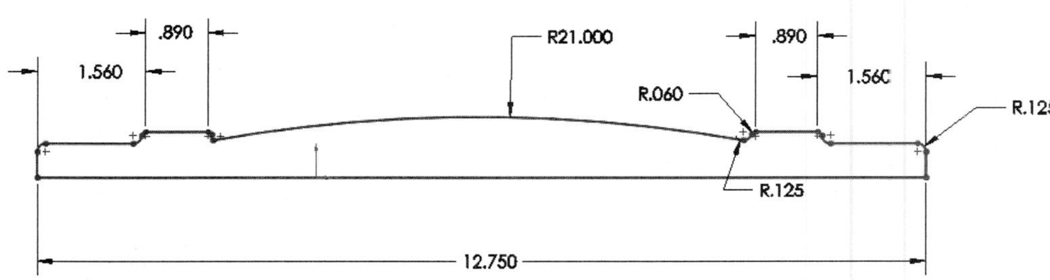

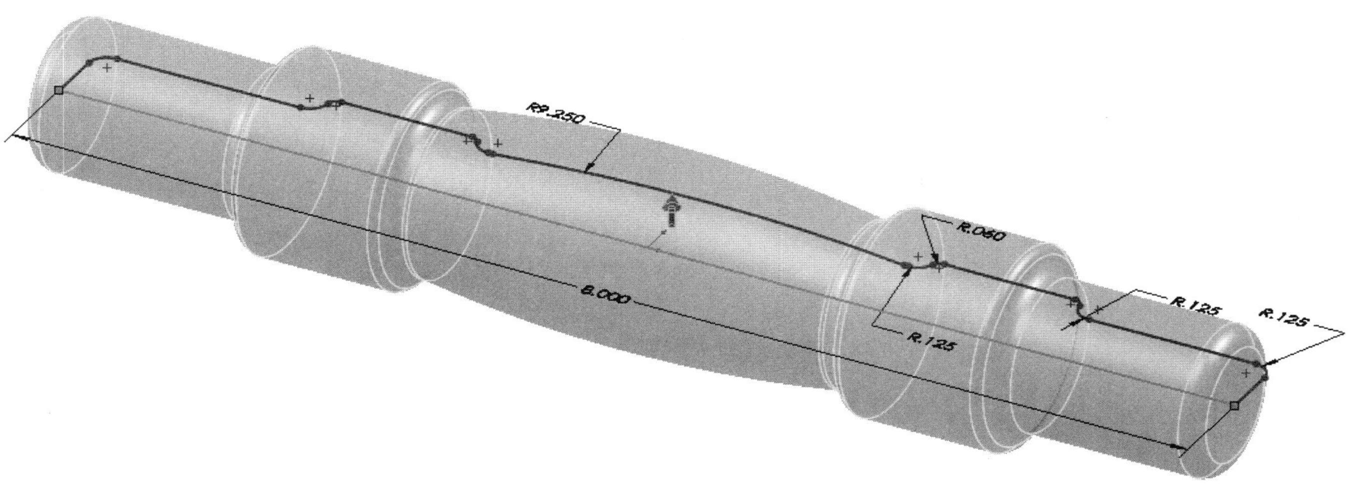

Figure 12-49
Handle Grip Revolved

Figure 12-50
Make a 1/4-20 × .75 Inch Deep Tapped Hole in One End and Mirror It

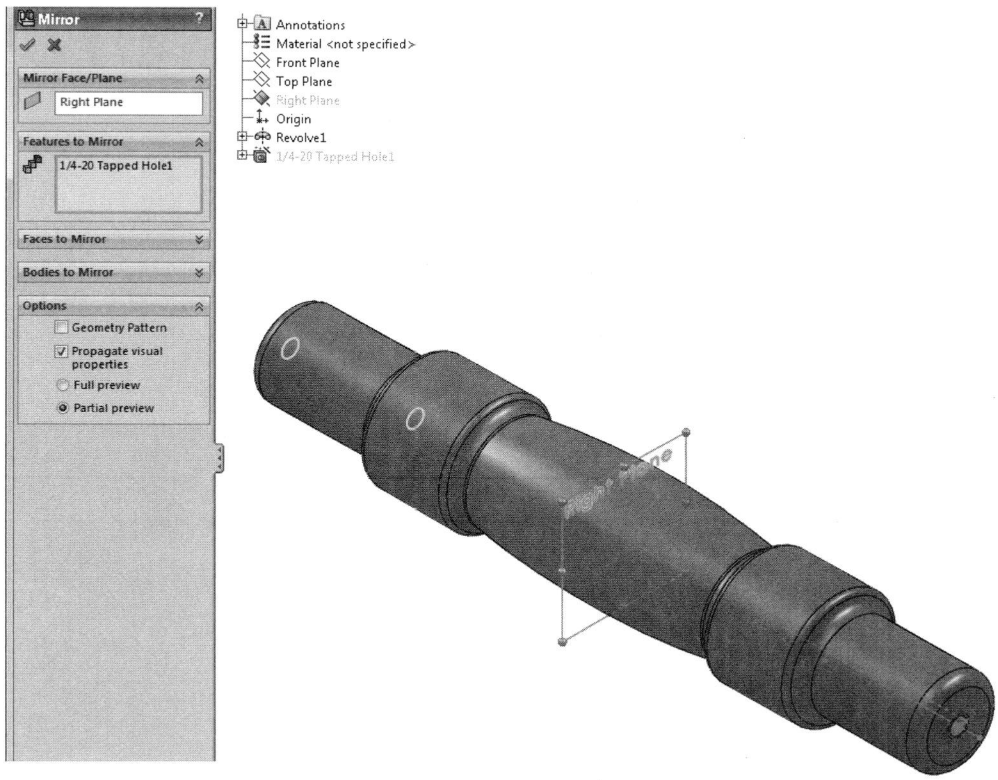

Figure 12-51
Handle Grip Complete with Teak Wood Material

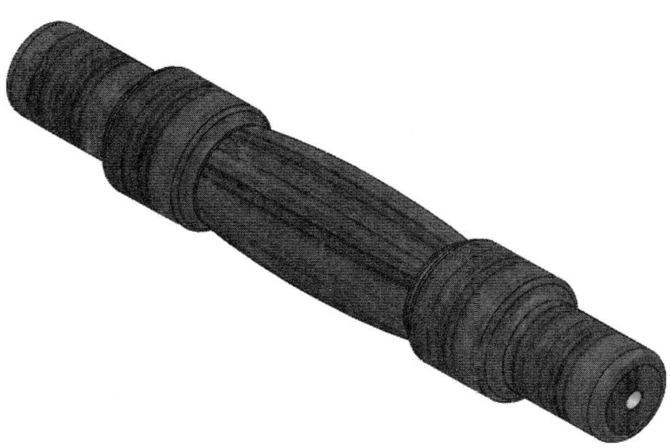

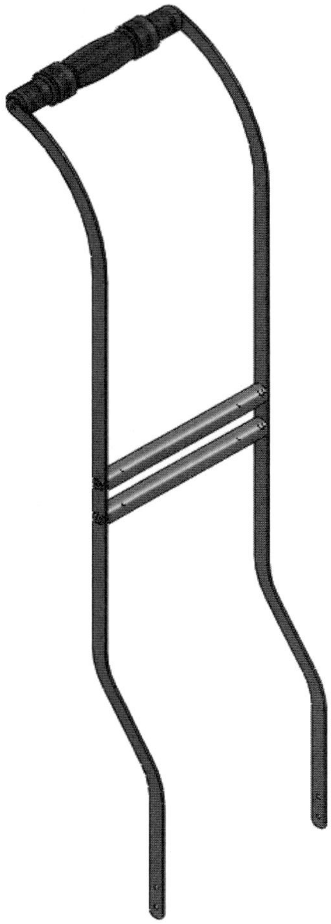

Figure 12-52
Handle Subassembly Complete Using Concentric and Coincident Mates. Assembly Consists of: Two Braces, Two Spacers, a Handle, Two .188 × .75 Long Screws, Four .188 Lock Washers and Four .188-20 × .50 Machine Screws

Figure 12-53
Draw a Sketch for the Seat Enclosure on the Top Plane

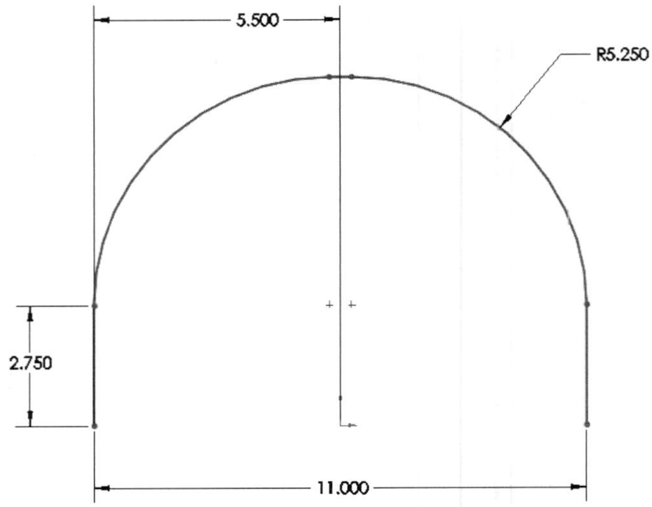

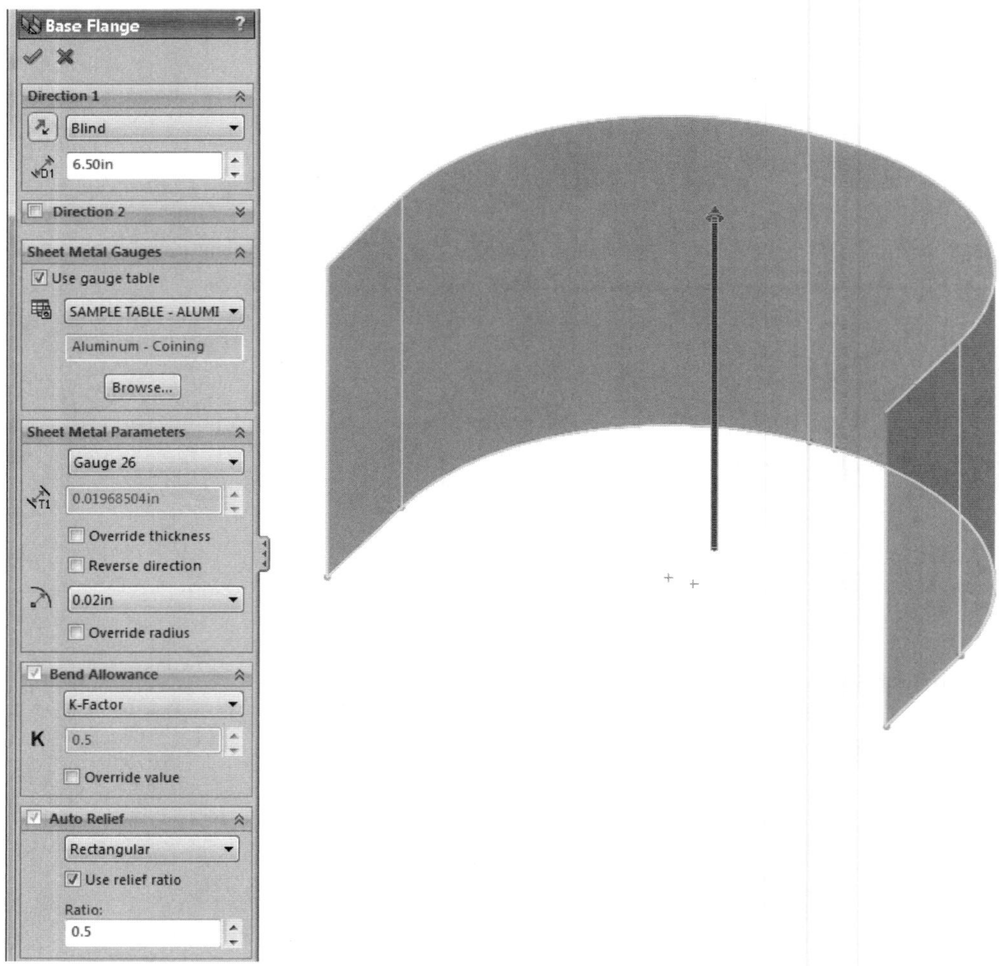

Figure 12-54
Extrude the Seat Enclosure Using Base Flange 6.50 Inches

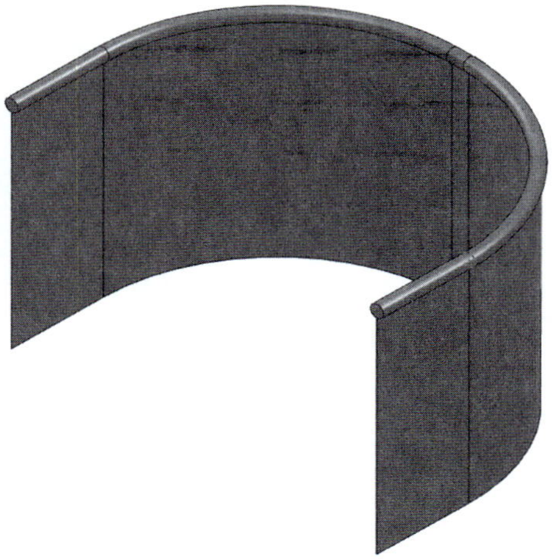

Figure 12-55
Draw a Path on Top of the Extrusion for a Swept Boss/Base the Same Dimensions as the Sketch for the Extrusion. Then, Draw a .375 Diameter Circle at 90° to the Path and Make the Sweep. Seat Enclosure with Swept Boss/Base Complete

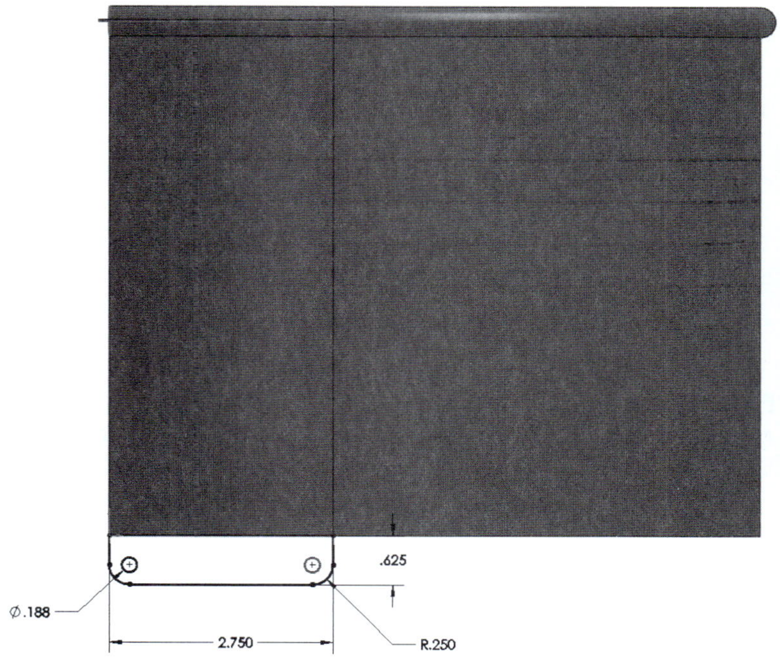

Figure 12-56
Make a Sketch for the Base Flange Tab and Extrude It to a Gauge 26

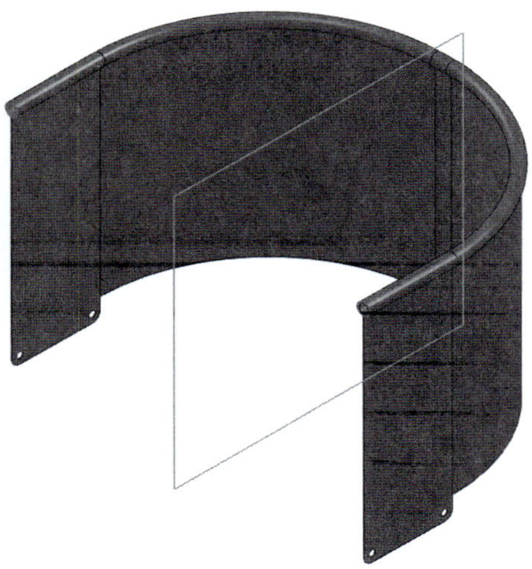

Figure 12-57
Mirror the Base Flange/Tab Using the Right Plane and Make an Extrude Cut .25 Inch Diameter × 1.25 Inches Deep in Each Side of the Swept Boss/Base

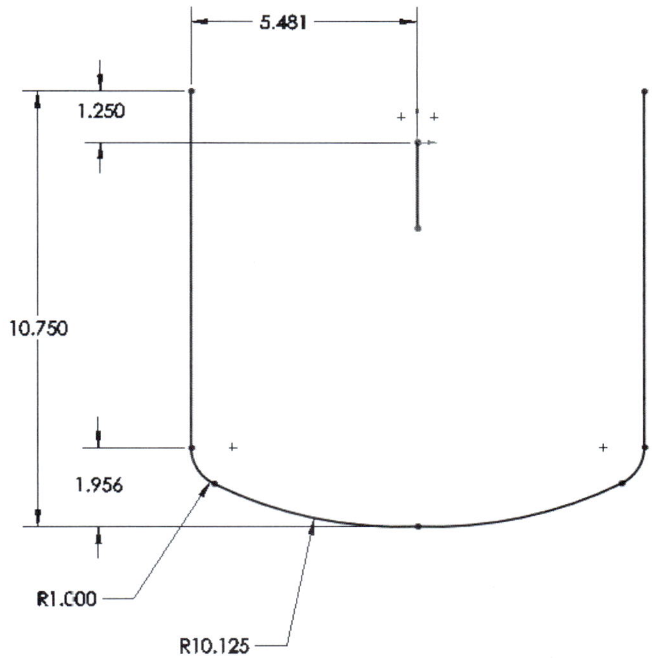

Figure 12-58
Draw a Sketch for the Path on the Top Plane (for the Swept Boss/Base) for the Seat Enclosure Front

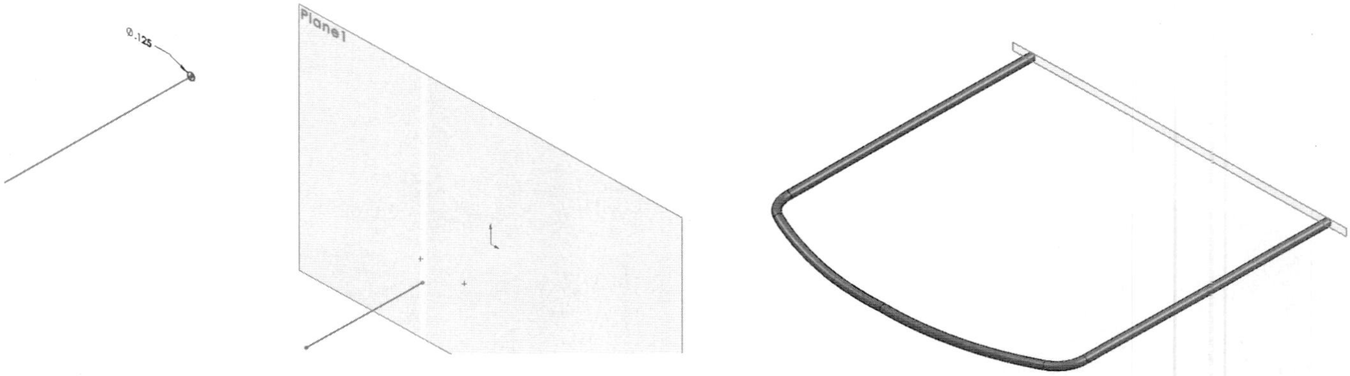

Figure 12-59
Make a New Plane Offset 1.25 Inches from the Front Plane and Draw a .125 Diameter Circle for the Profile of the Seat Enclosure Front. Using Swept Boss/Base, Sweep the Profile. Seat Enclosure Front Complete

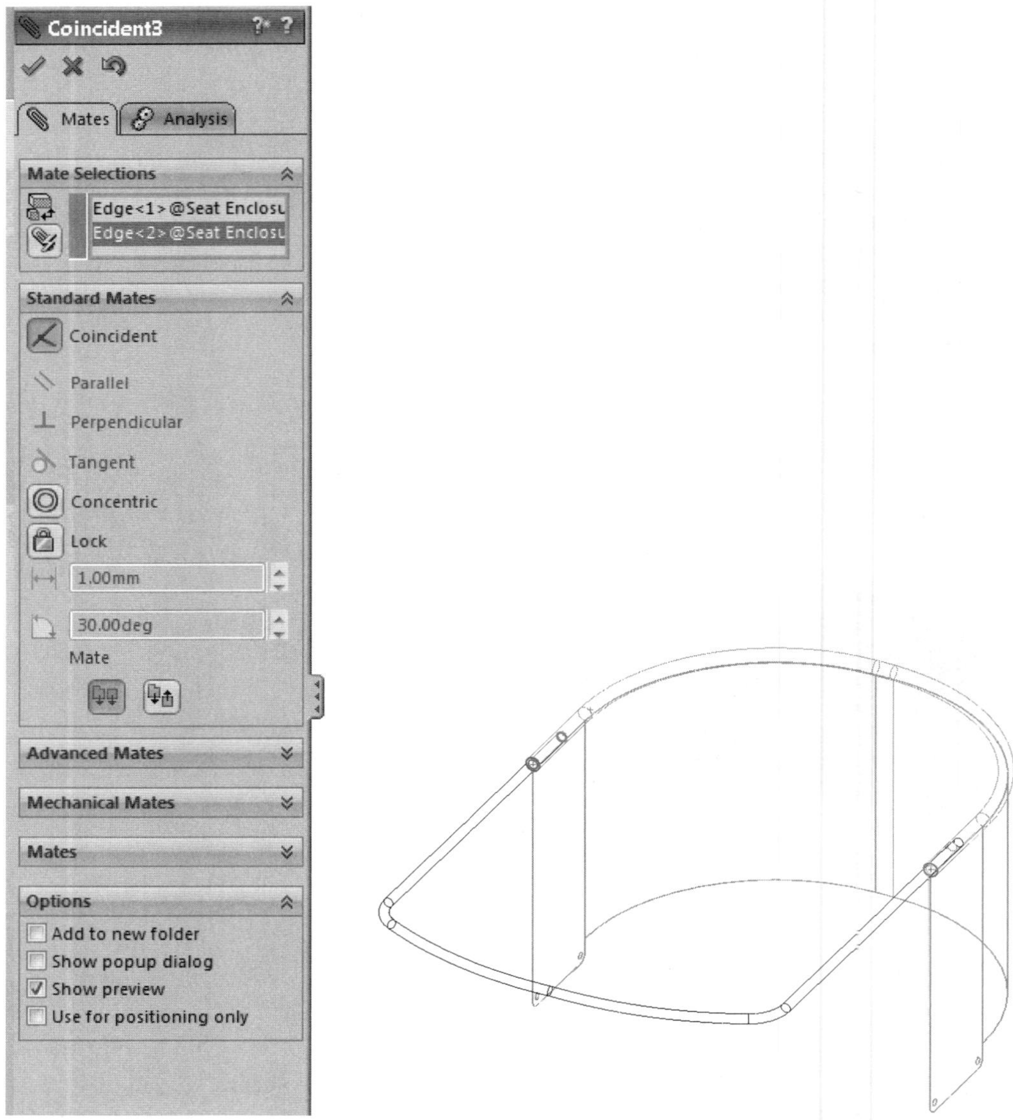

Figure 12-60
Open Seat Enclosure, Click: Make Assembly from Part, Click: OK, Insert the Seat Enclosure Front Using Insert Components and Apply Concentric and Coincident Mates

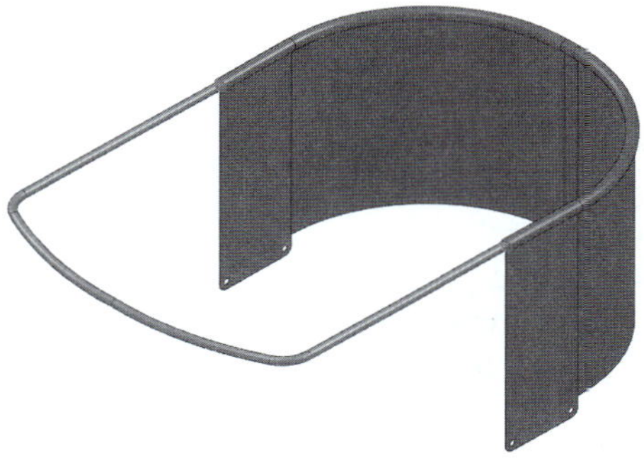

Figure 12-61
Seat Enclosure Assembly Complete

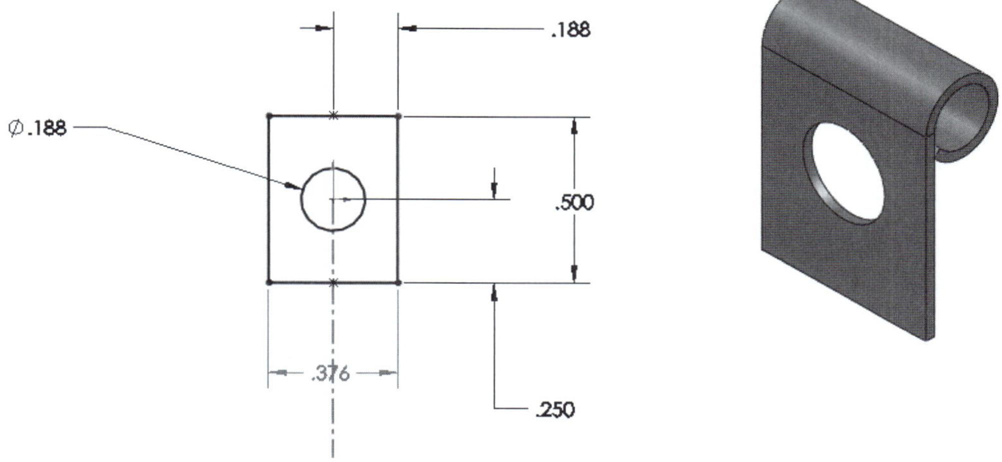

Figure 12-62
Make a Sketch for the Tray Bracket, Extrude It to 26 Gauge Using Base Flange and Put a .125 Diameter Open Hem with 322 Degree Bend Angle on It. Tray Bracket Complete

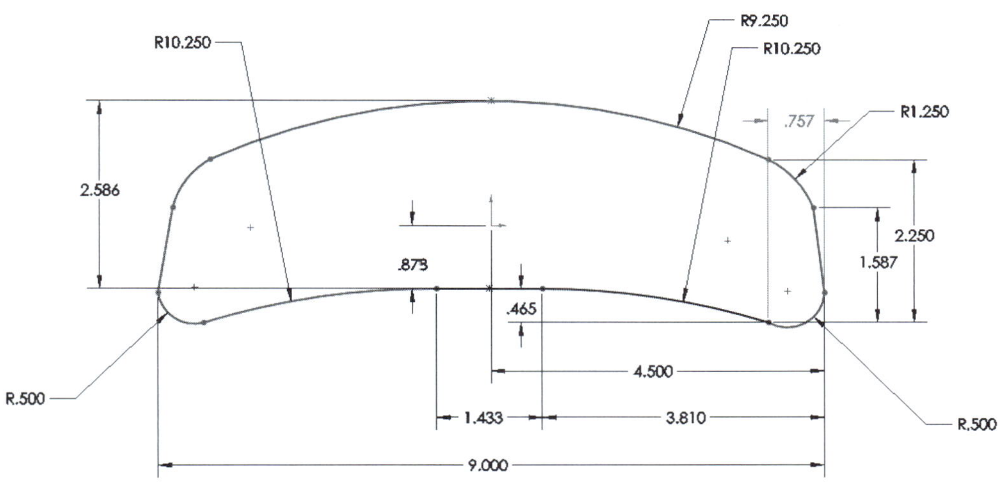

Figure 12-63
Draw a Sketch for the Tray Base on the Top Plane and Extrude It .035 inch Using Base Flange. Make 1 inch Edge Flanges on All of the Edges. Draw Another Sketch Using the Same Dimensions on Top of the Edge Flanges as a Path for a Sweep. Draw a .125 Diameter Circle at 90 degrees to the Path for the Profile. Sweep the Profile Along the Path Using Swept Boss/Base

Figure 12-64
Make a Sketch for the Extrude Cut and Make the Extrude Cut

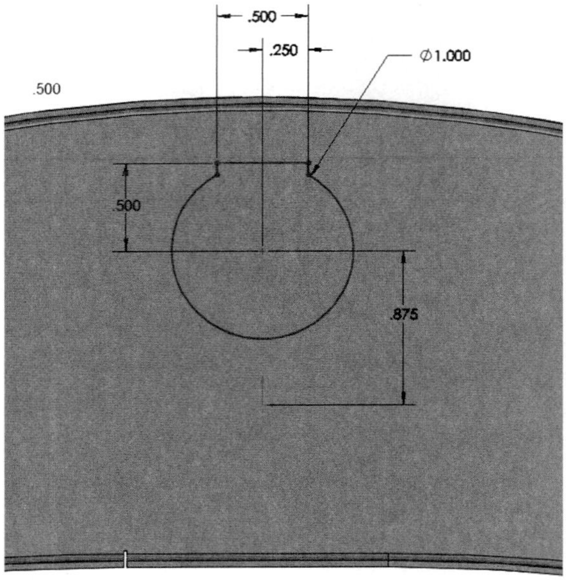

Figure 12-65
Make a .500 Edge Flange, Two .25 Fillets, and a Sketch for a .188 Diameter Hole, and Make the Extrude Cut

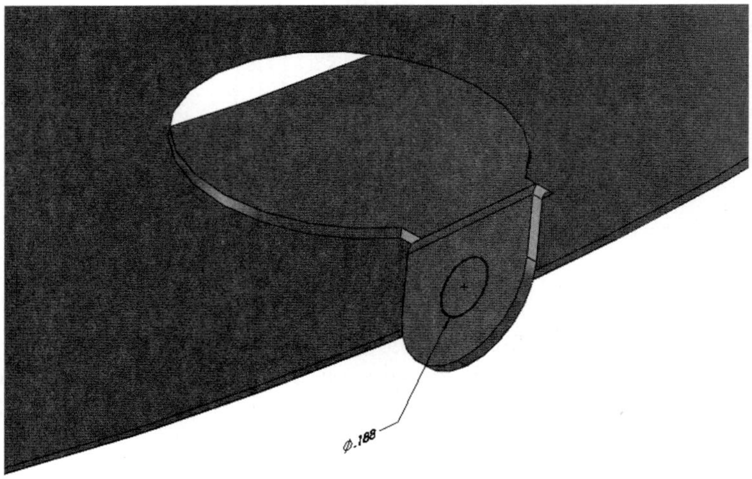

Figure 12-66
Make a Sketch for Two .188 Diameter and Make the Extrude Cut Up to Next

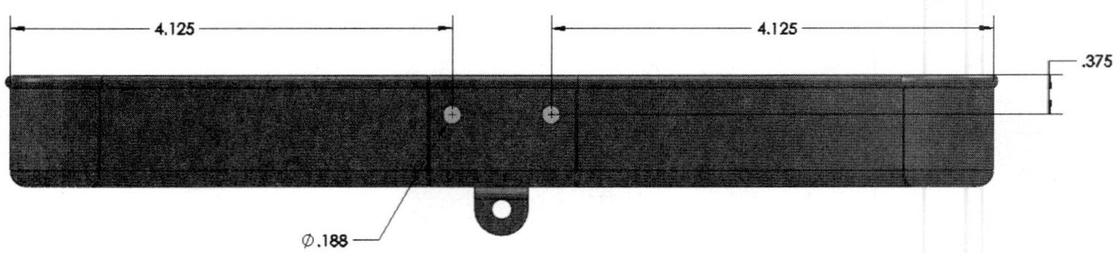

Figure 12-67
Tray Complete

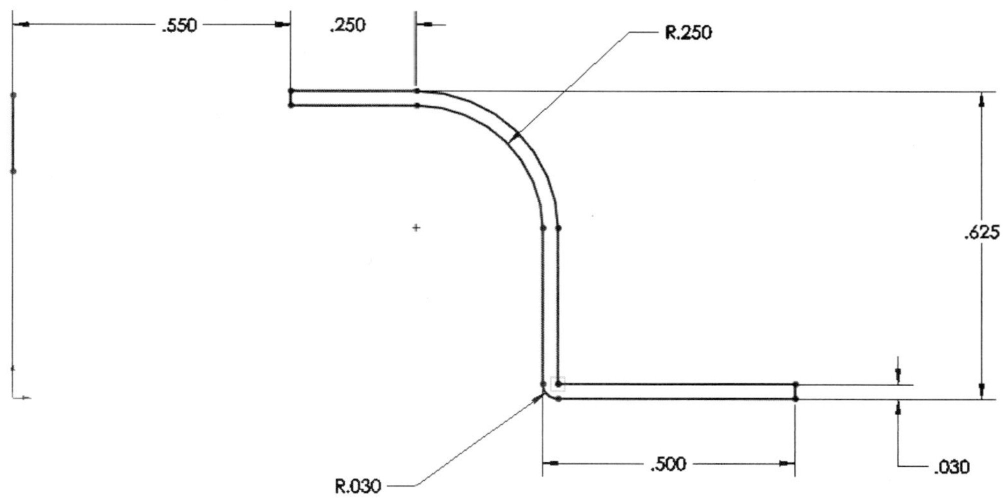

Figure 12-68
Draw a Sketch for the Post Support on the Front Plane and Revolve It

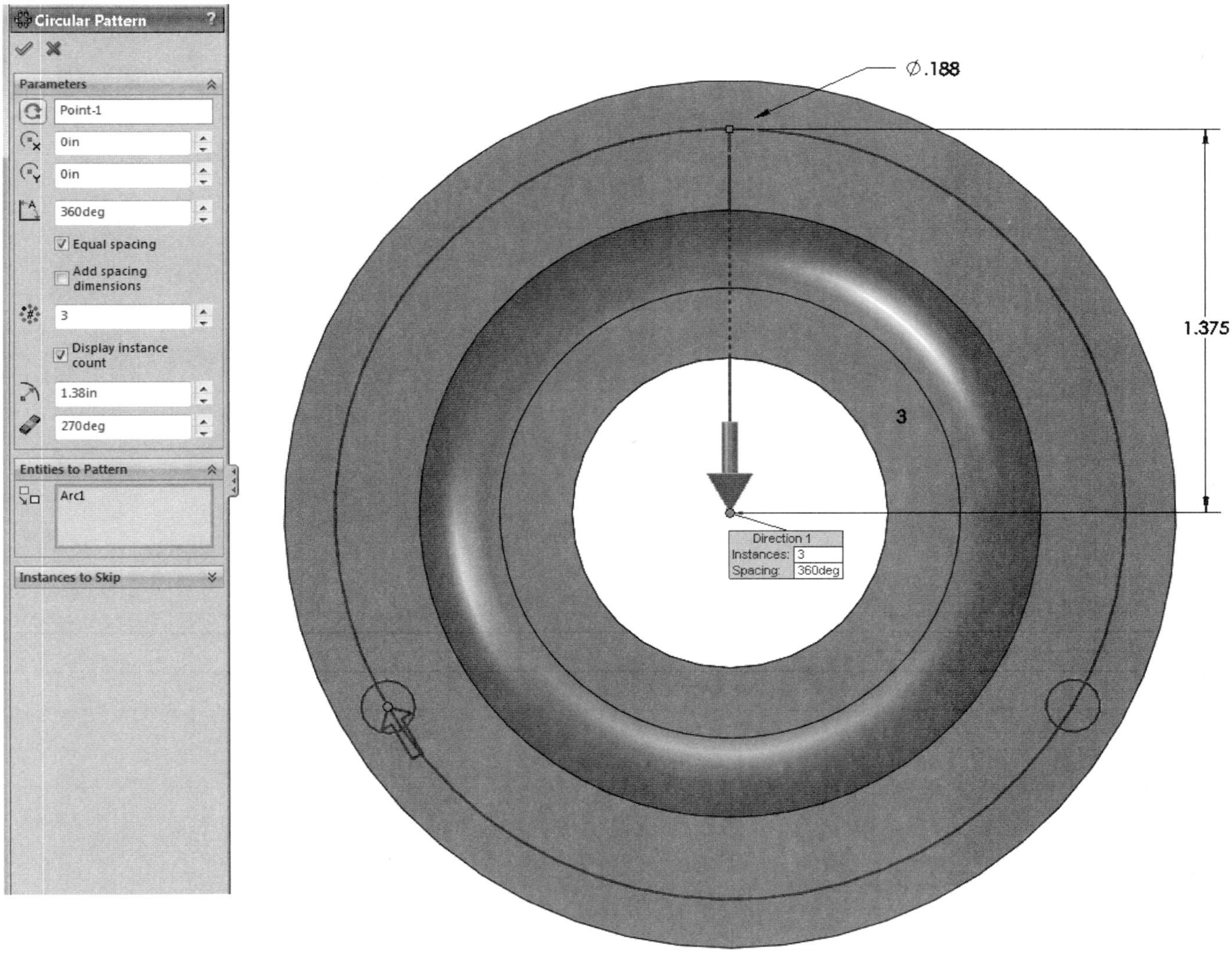

Figure 12-69
Make a Circular Pattern for Three .188 Holes and Make an Extrude Cut from Them

Figure 12-70
Post Support Complete

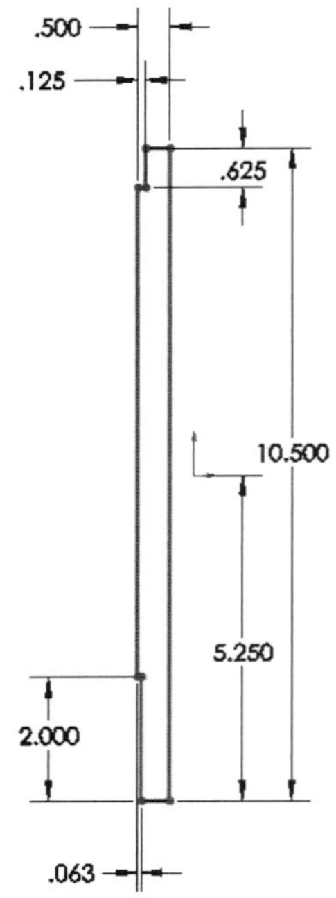

Figure 12-71
Make a Sketch for the Post on the Front Plane and Revolve It

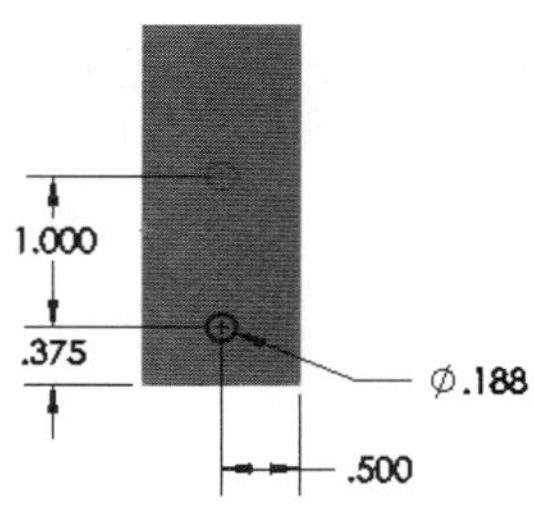

Figure 12-72
Make a Sketch for Two .188 Diameter Holes at the Bottom of the Post and Make an Extrude Cut Through All

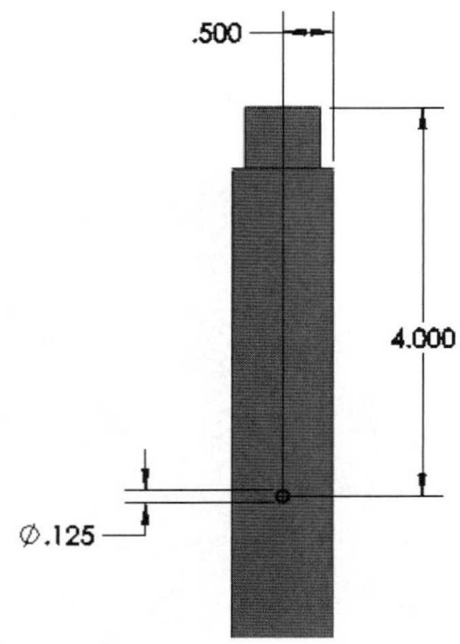

Figure 12-73
Make a Sketch for a .125 Diameter Hole and Make an Extrude Cut Through All

Figure 12-74
Post Complete

Figure 12-75
Make a Sketch for a Revolved Front Handle on the Front Plane. Mirror the Sketch and Revolve It

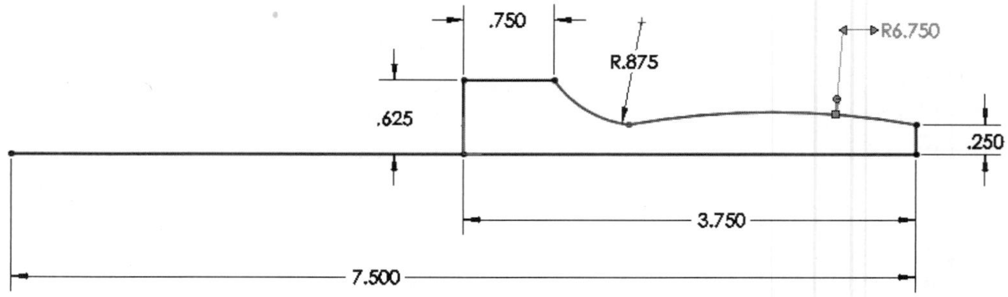

Figure 12-76
Draw Rectangles on Two Sides and the Bottom and Dimension Them from the Origin for Extruded Cuts and Make the Through All Cuts

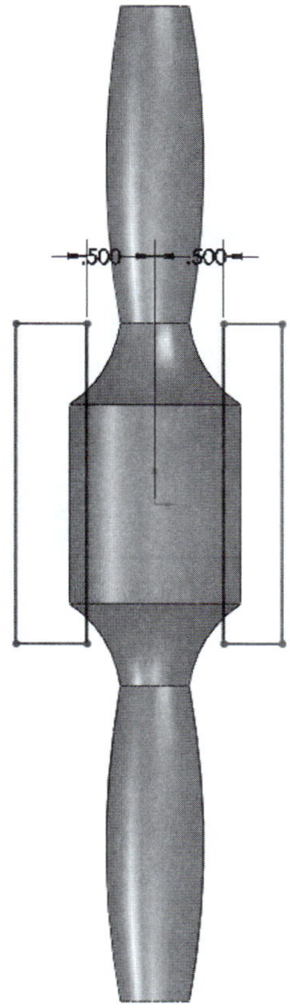

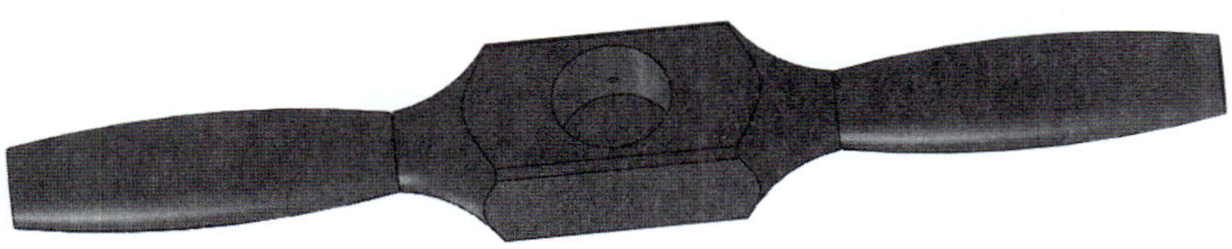

Figure 12-77
Draw a .750 Diameter Circle on the Bottom and Use Extrude Cut to Make a Hole .625 Inch Deep. Front Handle Complete

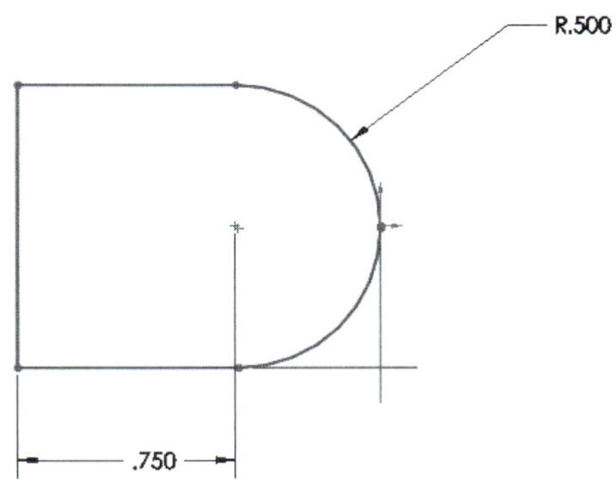

Figure 12-78
Make a Sketch for the Post Bracket Base on the Top Plane and Extrude It to 26 Gauge Using Base Flange

Figure 12-79
Make a Sketch for an Extrusion and Extrude It to 26 Gauge Using Base Flange

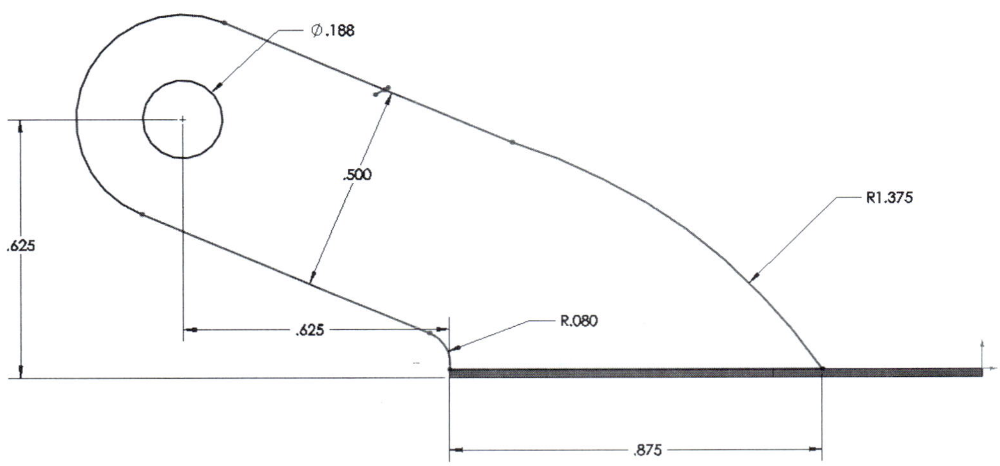

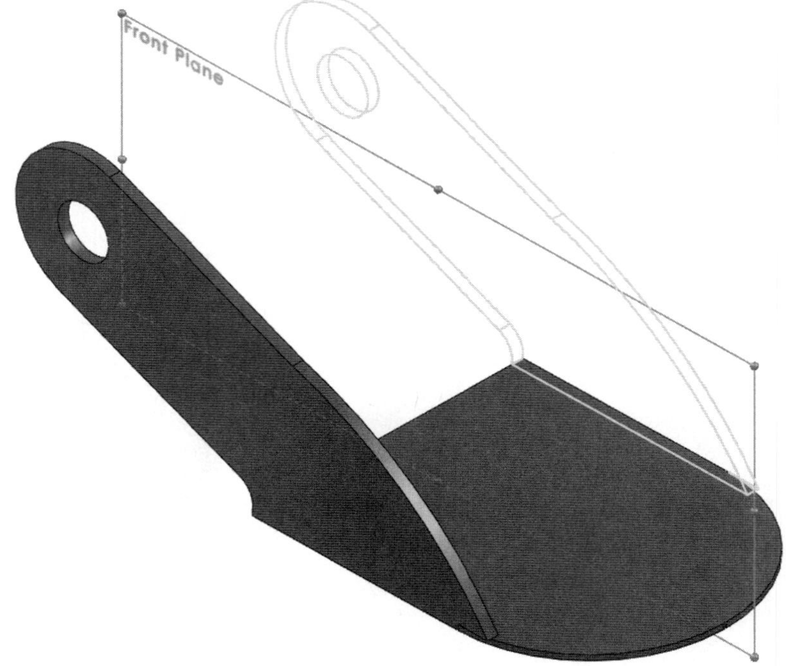

Figure 12-80
Mirror the Extrusion Using the Front Plane as the Object to Mirror About. You May Need to Select Bodies in the Mirror Manager

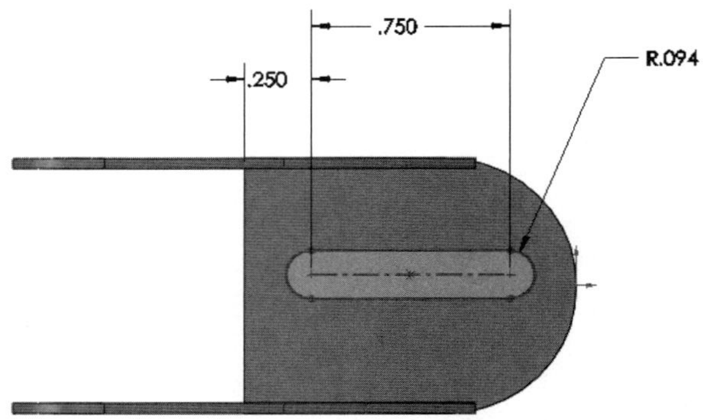

Figure 12-81
Make a Sketch for a Through All Extrude Cut Using Straight Slot and Make the Extrude Cut

Figure 12-82
Post Bracket Complete

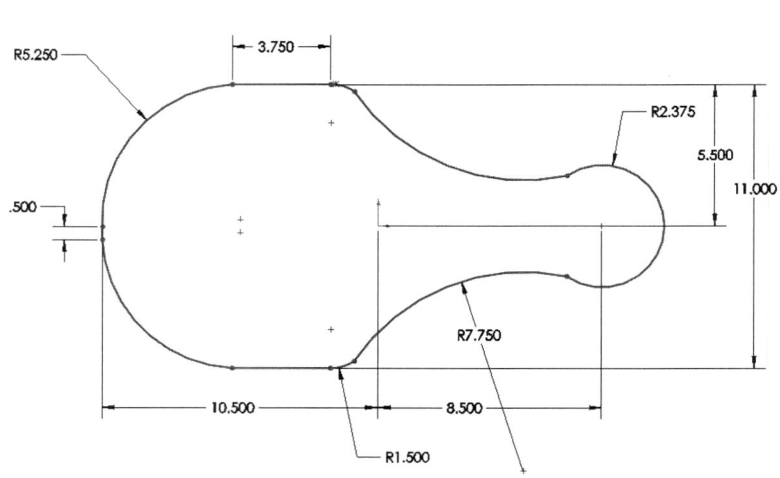

Figure 12-83
Make a Sketch for the Seat on the Top Plane and Extrude It

Figure 12-84
Make a Sketch for the Through Holes and Make the Extrude Cut

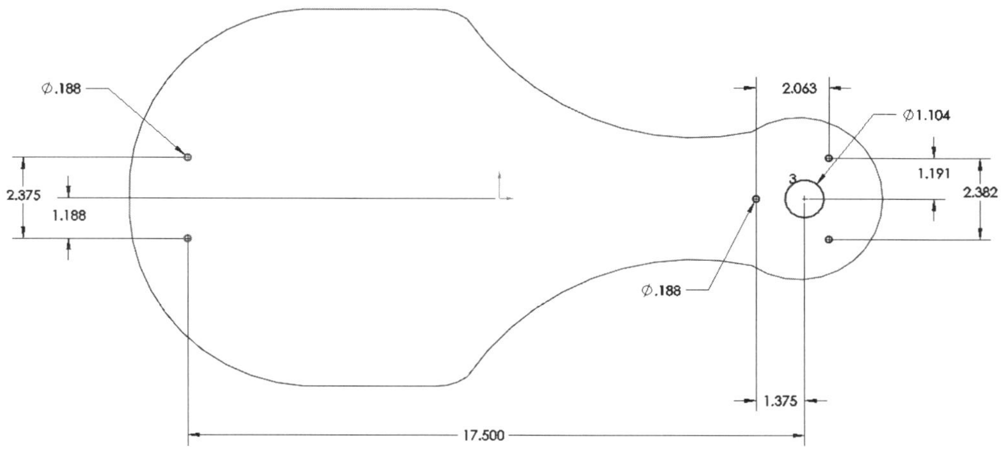

Chapter 12

Figure 12-85
Make a Sketch for the .125 Diameter Hole 1/2 Inch Deep on the Bottom and Make the Extrude Cut

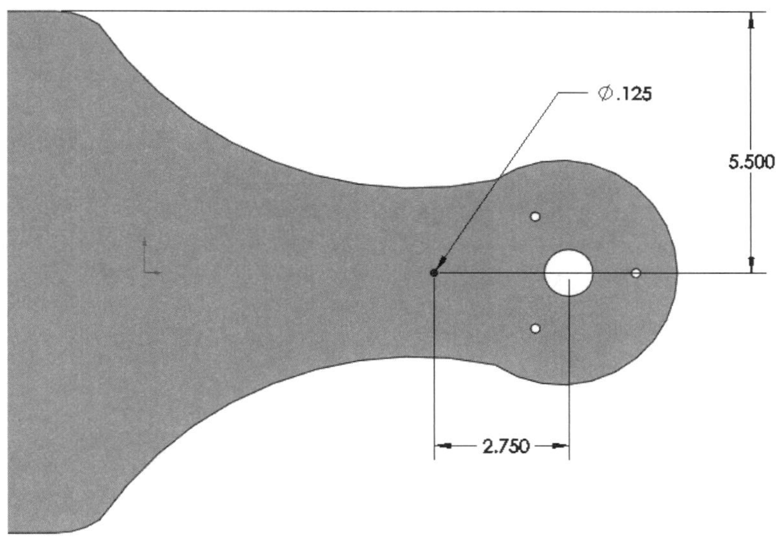

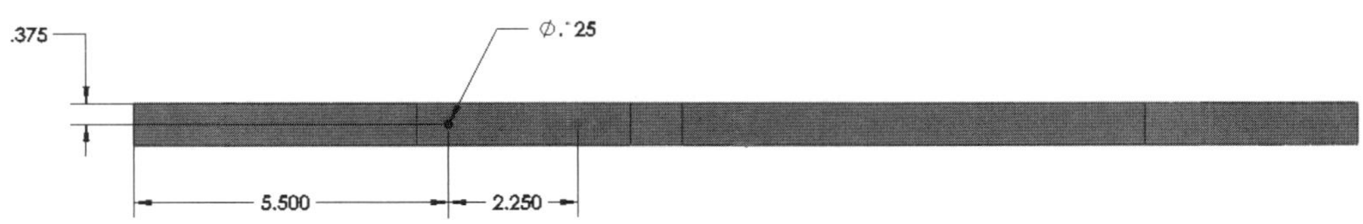

Figure 12-86
Make a Sketch for Four .125 Diameter Holes, 1/2 Inch Deep, Two on the Left Side and Two on the Right Side, and Make the Extrude Cuts

Figure 12-87
Fillet Two Edges with a .25 Radius. Seat Complete

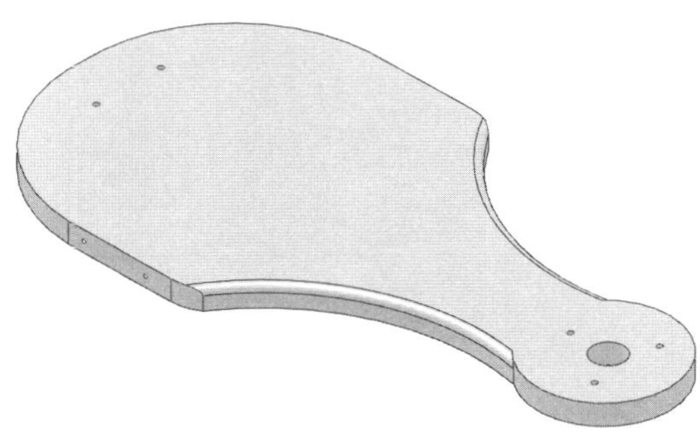

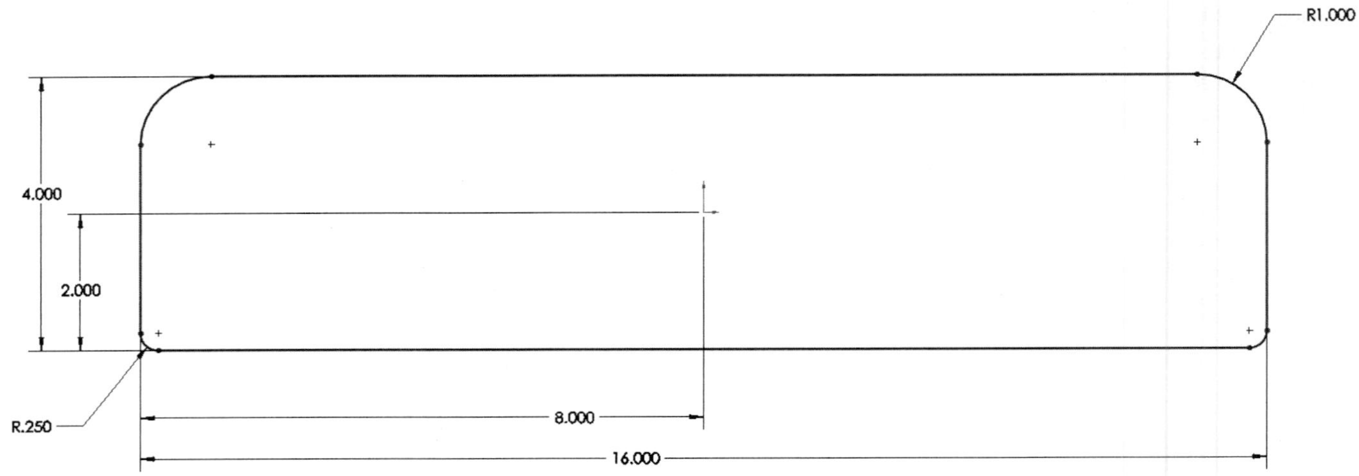

Figure 12-88
Make a Sketch for the 2 Inches Extrusion on the Top Plane with a 7 Degree Draft Angle and Extrude It

Figure 12-89
Make a .25 Fillet on the Top of the Extrusion. Use the Shell Command to Shell the Extrusion with .08 Thickness and the Back Face Removed. Make a Sketch for the .313 Diameter Through All Holes and Make the Extrude Holes

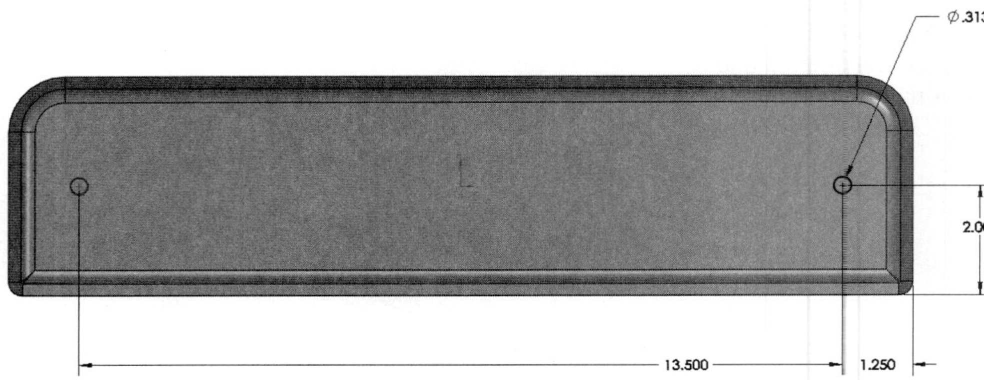

Figure 12-90
Make a Sketch for the Three .188 Diameter and the 1.040 Diameter Through All Holes and Make the Extrude Cut

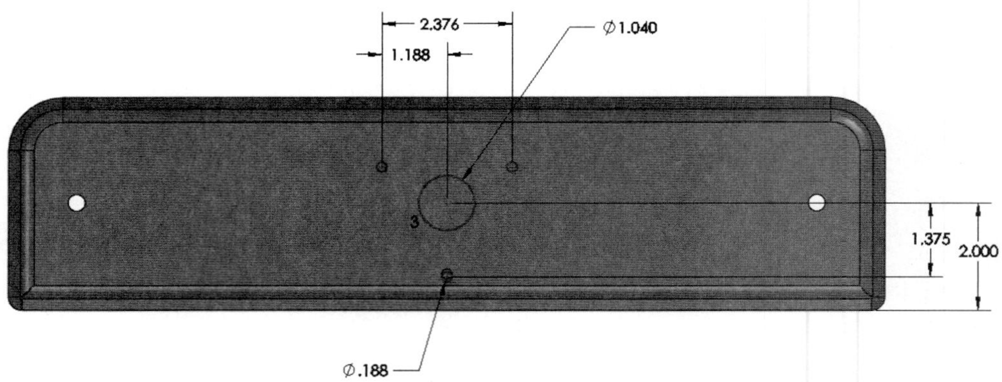

Figure 12-91
Wheel Mounting Bracket Complete

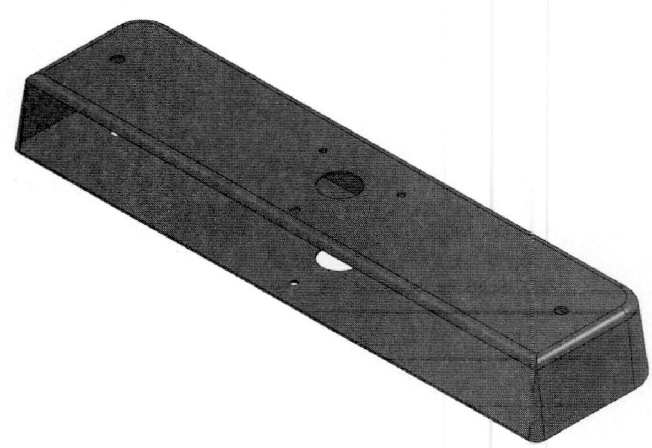

Figure 12-92
Make a Sketch for the Left Side of the Wheel Bracket on the Front Plane, Mirror It, Offset It .100 Inch and Draw a Line Across Both Ends and Extrude It .250 Inch

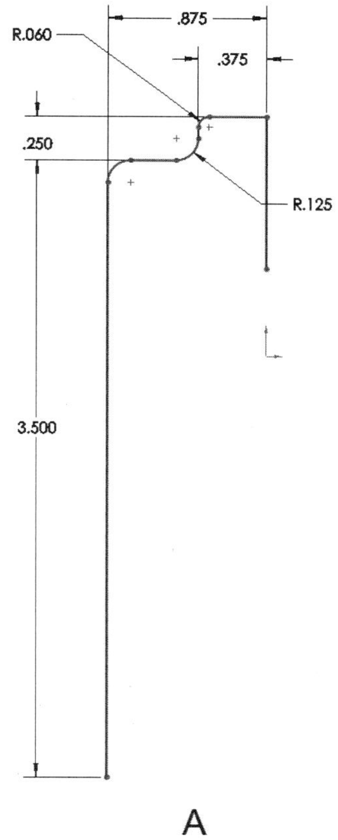

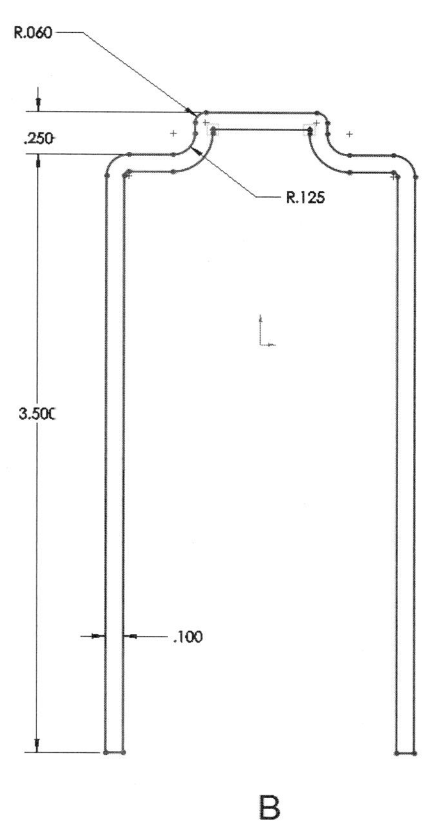

Figure 12-93
Draw a Sketch for the Extrude Cut and Make the Extrude Cut

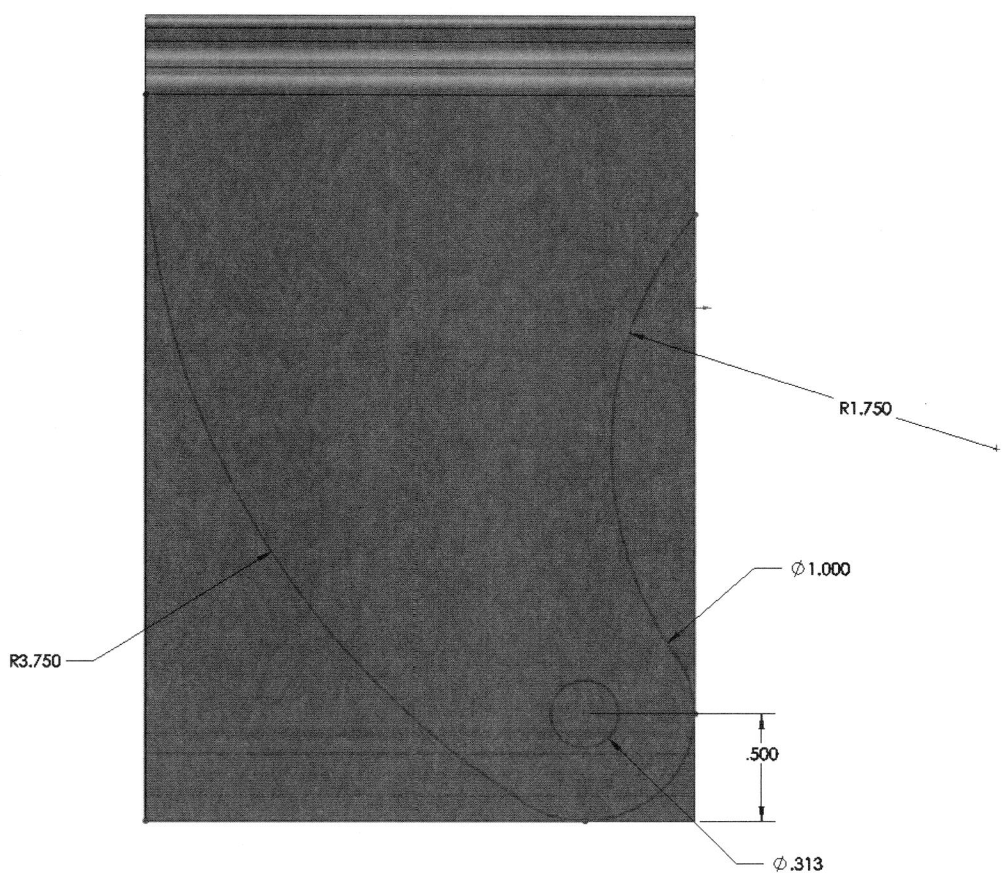

Figure 12-94
Make a .313 Diameter Circle Centered in the Top of the Wheel Mounting Bracket and Make an Extrude Cut Through All. Wheel Mounting Bracket Complete

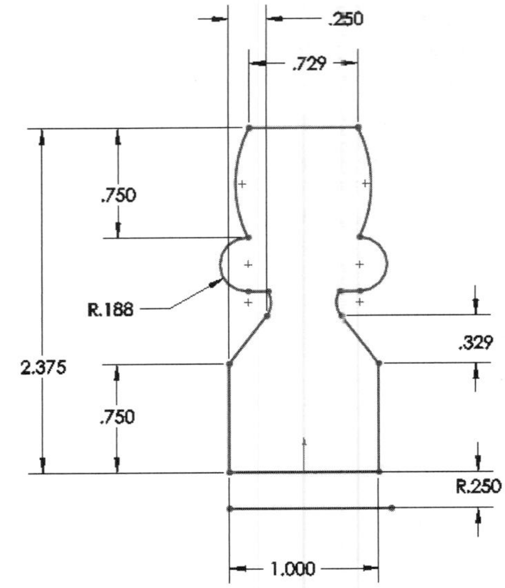

Figure 12-95
Make a Sketch for the Wheel on the Front Plane and Revolve It

Figure 12-96
Wheel Complete

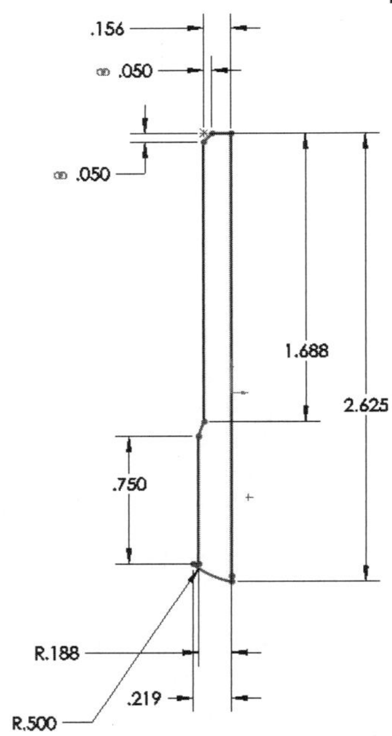

Figure 12-97
Make a Sketch for the Wheel Mounting Screw on the Front Plane and Revolve It

Figure 12-98
Add Cosmetic Threads. Wheel Mounting Screw Complete

Figure 12-99
Make a Sketch for the Wheel Mounting Pin on the Front Plane and Revolve It

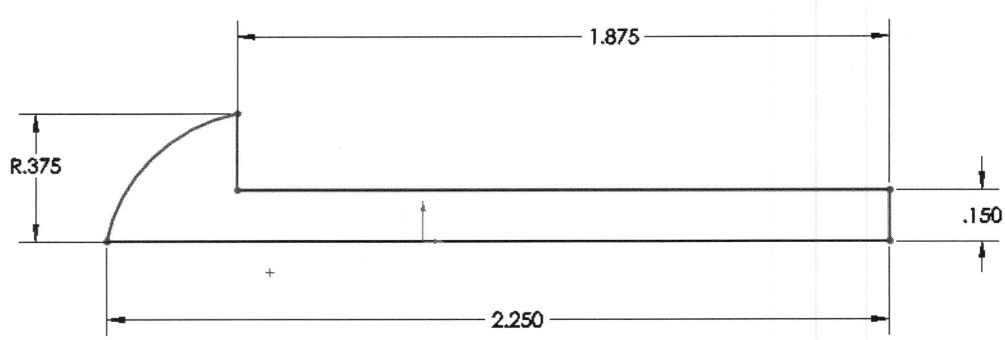

Figure 12-100
Wheel Mounting Pin Complete

Figure 12-101
Use Make Assembly from Part to Make the Wheel Subassembly Using Insert Component and Concentric and Coincident Mates

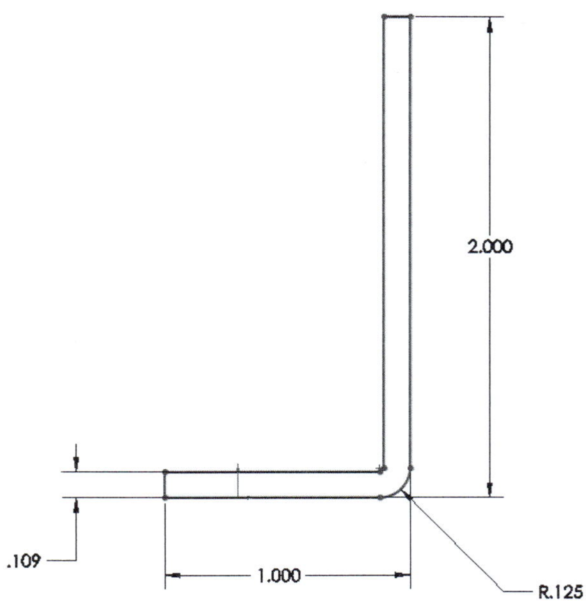

Figure 12-102
Make a Sketch for the Right Handle Assembly Bracket on the Front Plane and Extrude It 1.50 Inches

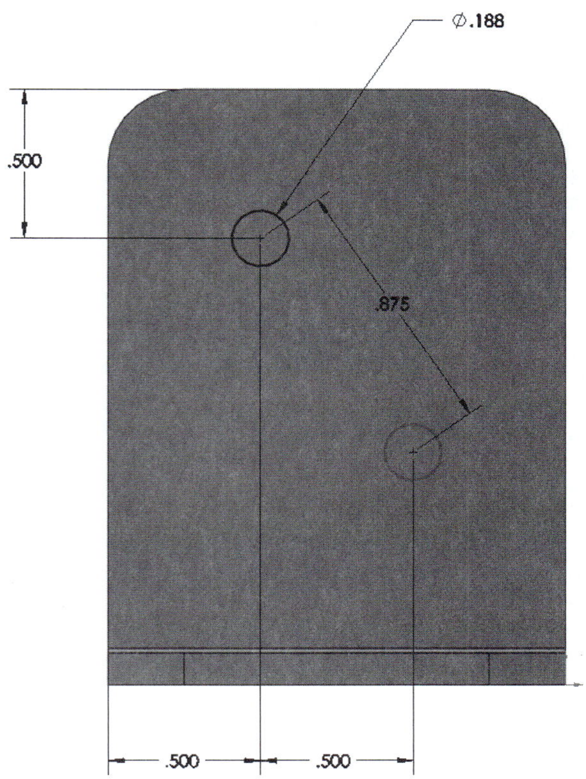

Figure 12-103
Make a Sketch for Two .188 Diameter Holes and Do an Extrude Cut Through All

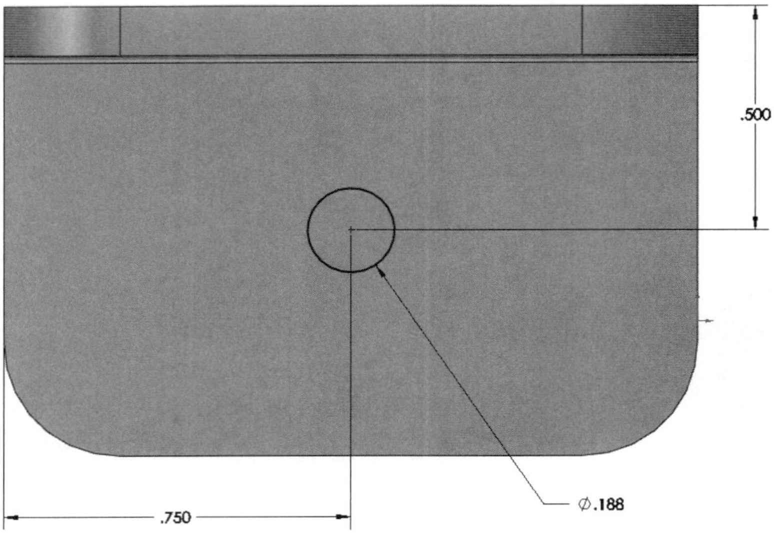

Figure 12-104
Make a Sketch for a .188 Diameter Hole and Make the Cut. Make Four .25 Radius Fillets on Four Edges

Figure 12-105
Handle Assembly Bracket Right Complete

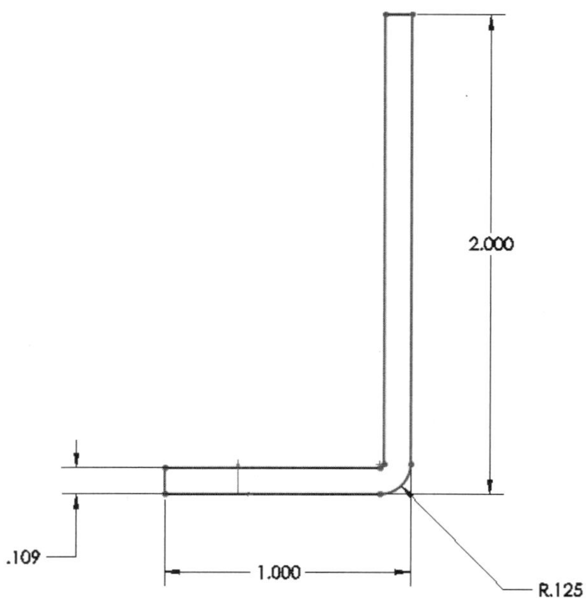

Figure 12-106
Make a Sketch for the Left Handle Assembly Bracket on the Front Plane and Extrude It 1.50 Inches

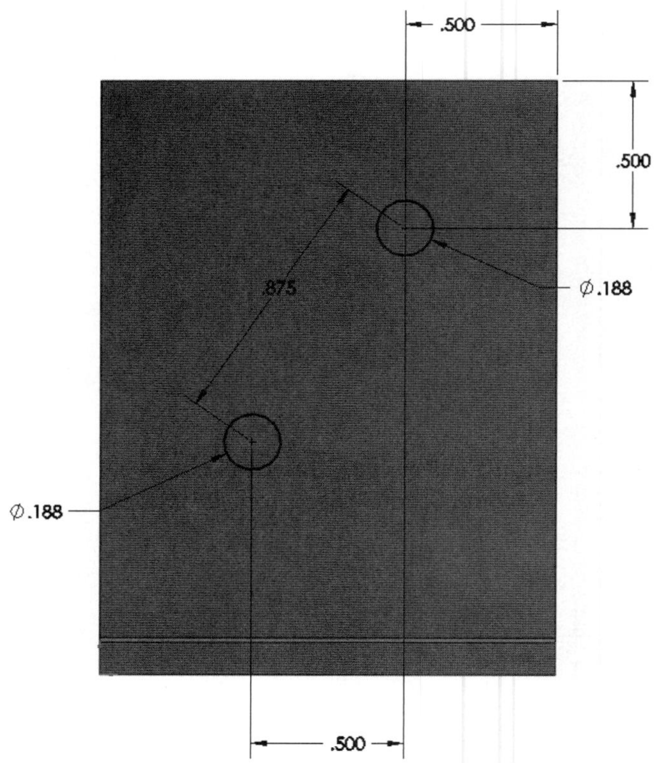

Figure 12-107
Make a Sketch for Two .188 Diameter Holes and Do an Extruded Cut Through All

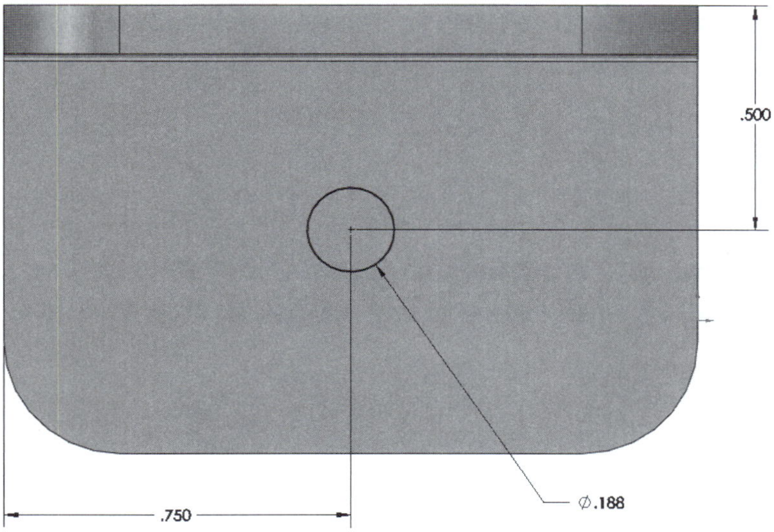

Figure 12-108
Make a Sketch for a .188 Diameter Hole and Make the Cut. Make Four .25 Radius Fillets on Four Edges

Figure 12-109
Handle Assembly Bracket Left Complete

Figure 12-110
Make a Sketch for a Revolve Boss/Base on the Front Plane and Revolve It. Wheel Mounting Pin Cap Complete

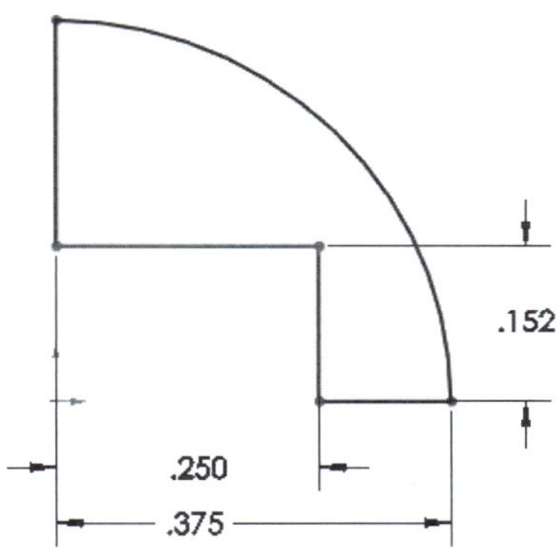

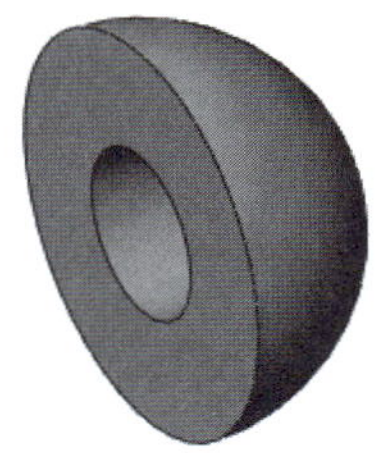

Figure 12-111
Stroller Top Assembly (Project 12-3) Complete

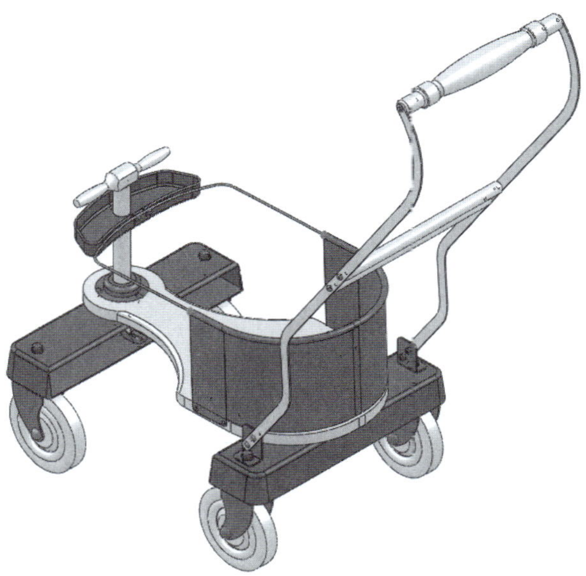

Project 12-4: **Make an Isometric Assembly Drawing of the Handle Assembly With a Parts List**

Figure 12-112 shows the completed drawing of Project 12-4, the handle assembly.

Use the instructions in Chapter 11 and Project 12-2 to make this drawing.

Project 12-5: **Make an Isometric Assembly Drawing of the Seat Enclosure Assembly with a Parts List**

Figure 12-113 shows the completed drawing of Project 12-5, the seat enclosure assembly.

Use the instructions in Chapter 11 and Project 12-2 to make this drawing.

Project 12-6: **Make An Isometric Assembly Drawing of the Wheel Assembly with a Parts List**

Figure 12-114 shows the completed drawing of Project 12-6, the wheel assembly.

Use the instructions in Chapter 11 and Project 12-2 to make this drawing.

Project 12-7: **Make an Isometric Assembly Drawing of the Stroller Assembly with a Parts List**

Figure 12-115 shows the completed drawing of Project 12-7, the stroller assembly.

Use the instructions in Chapter 11 and Project 12-2 to make this drawing.

Project 12-8: **Make the 13 Parts of the Churn and Make an Assembly of It**

There are 13 parts to the churn assembly (Figure 12-116). They are:

Churn Gear Support

Gear

Gear Pin

Handle

Handle Grip

Lid

.188-20 × .625 inch Machine Screw

.188 Lock Washer

.188-20 Nut

Shaft

Paddle Holder

Paddle

Glass Jar

Make the 13 parts and assemble them using the instructions from previous chapters and Project 12-2 if necessary.

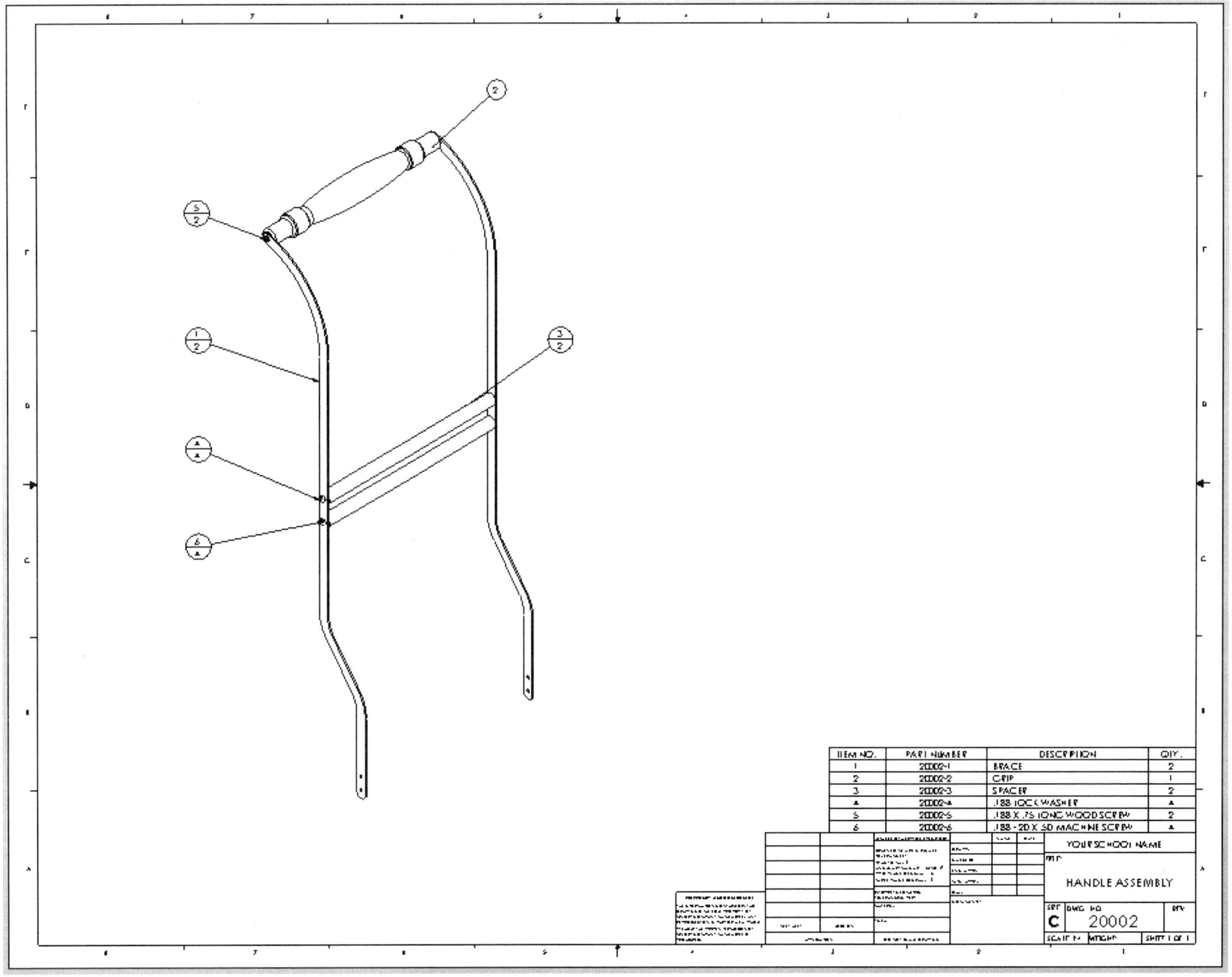

Figure 12-112
Handle Assembly with Parts List. Project 12-4 Complete

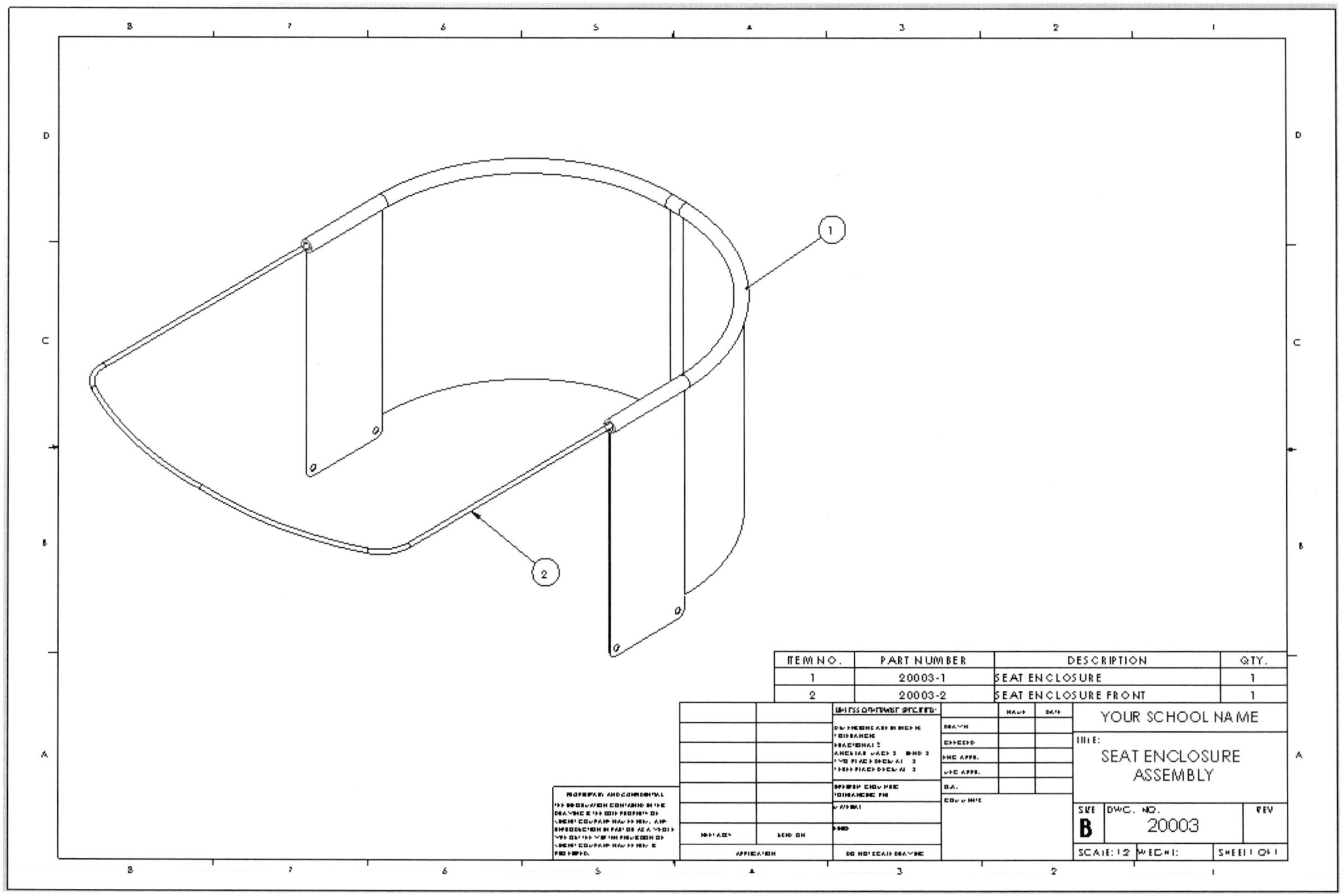

Figure 12-113

Seat Enclosure Assembly Drawing with Parts List. Project 12-5 Complete

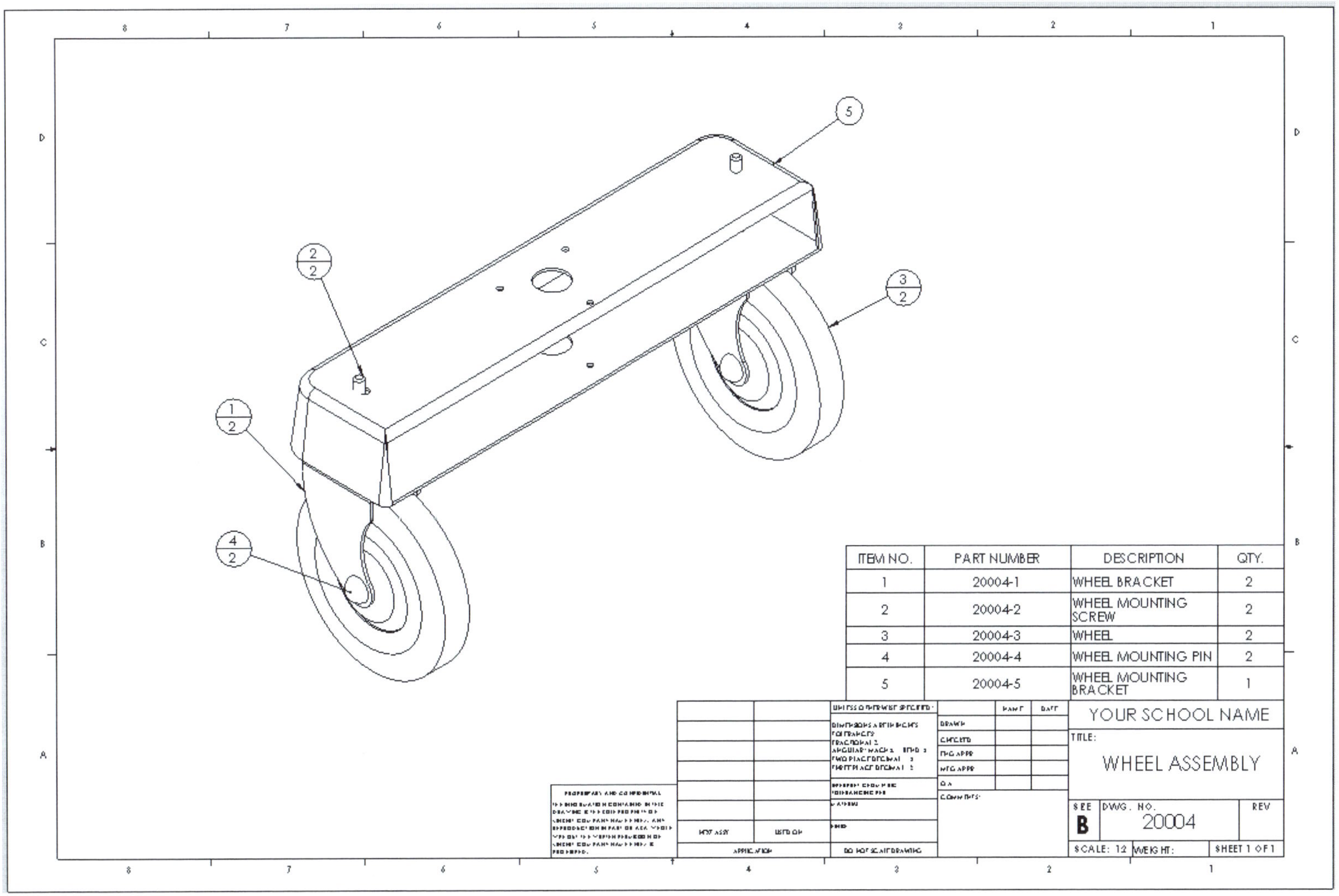

ITEM NO.	PART NUMBER	DESCRIPTION	QTY.
1	20004-1	WHEEL BRACKET	2
2	20004-2	WHEEL MOUNTING SCREW	2
3	20004-3	WHEEL	2
4	20004-4	WHEEL MOUNTING PIN	2
5	20004-5	WHEEL MOUNTING BRACKET	1

Figure 12-114
Wheel Assembly Drawing with Parts List. Project 12-6 Complete

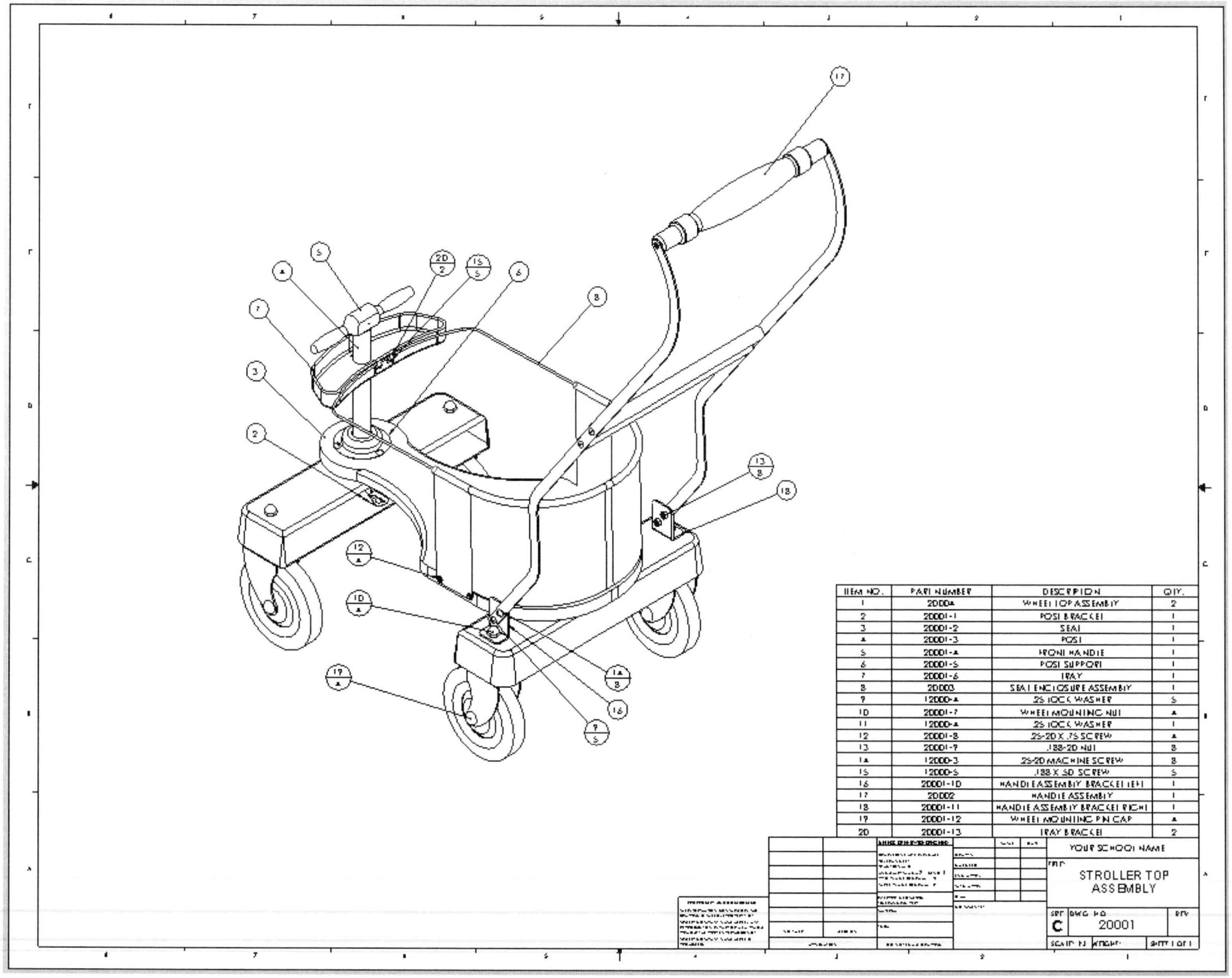

ITEM NO.	PART NUMBER	DESCRIPTION	QTY.
1	20004	WHEEL TOP ASSEMBLY	2
2	20001-1	POST BRACKET	1
3	20001-2	SEAT	1
4	20001-3	POST	1
5	20001-4	FRONT HANDLE	1
6	20001-5	POST SUPPORT	1
7	20001-6	TRAY	1
8	20003	SEAT ENCLOSURE ASSEMBLY	1
9	12000-4	.25 LOCK WASHER	5
10	20001-7	WHEEL MOUNTING NUT	4
11	12000-4	.25 LOCK WASHER	1
12	20001-8	.25-20 X .75 SCREW	4
13	20001-9	.188-20 NUT	8
14	12000-3	.25-20 MACHINE SCREW	8
15	12000-5	.188 X .50 SCREW	5
16	20001-10	HANDLE ASSEMBLY BRACKET LEFT	1
17	20002	HANDLE ASSEMBLY	1
18	20001-11	HANDLE ASSEMBLY BRACKET RIGHT	1
19	20001-12	WHEEL MOUNTING PIN CAP	4
20	20001-13	TRAY BRACKET	2

Figure 12-115
Stroller Top Assembly Drawing with Parts List. Project 12-7 Complete

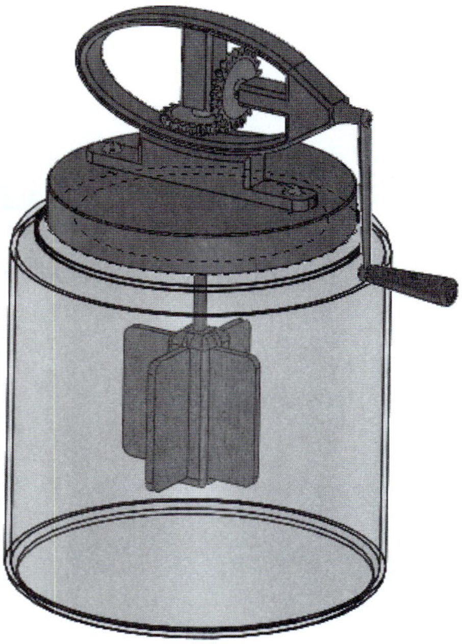

Figure 12-116
Project 12-8 Complete

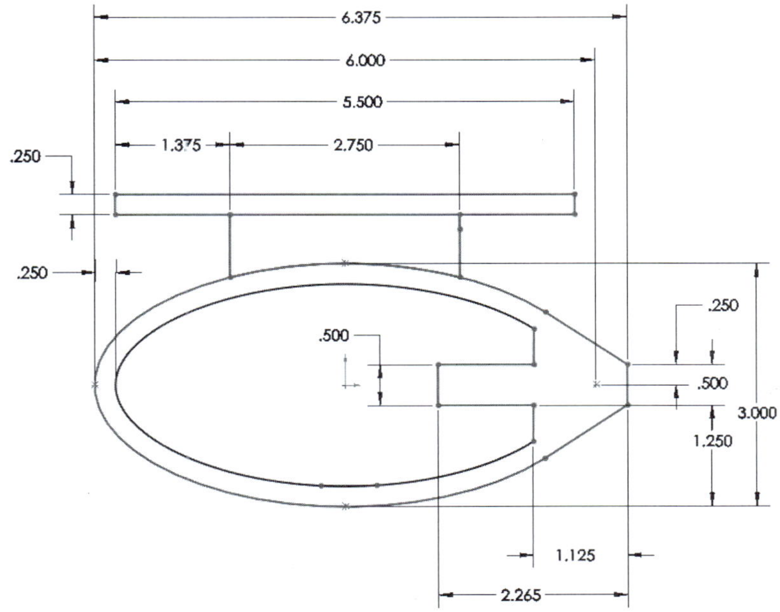

Figure 12-117
Make a Sketch for the Churn Gear Support on the Front Plane and Extrude It .75 Inch

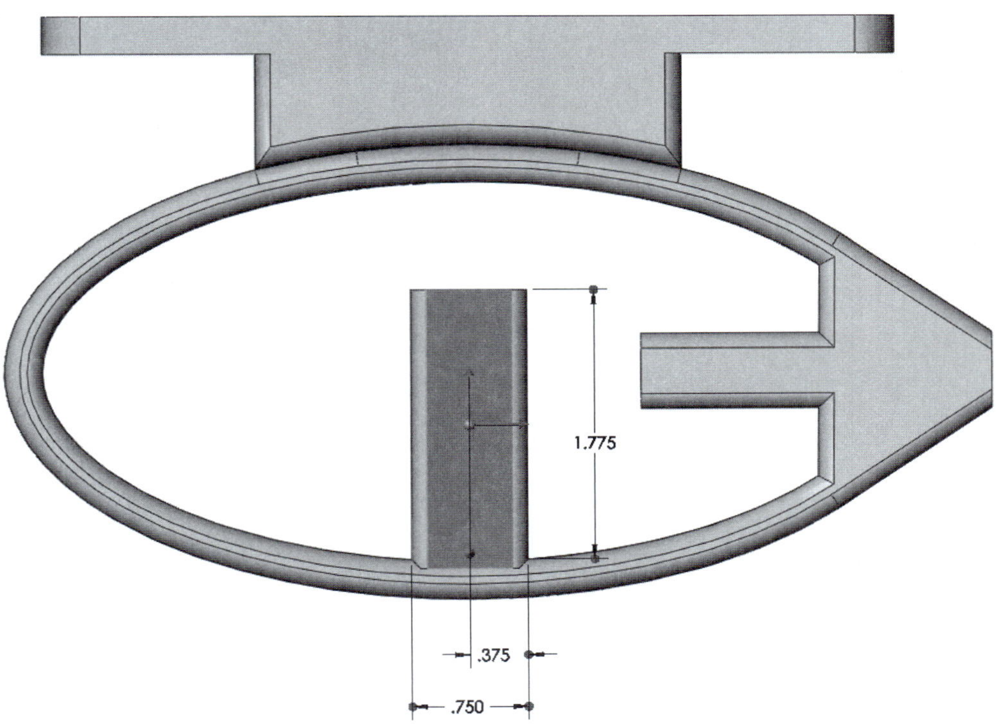

Figure 12-118
Draw a Sketch for the Center Post for the Churn Gear Support and Extrude It .75 Inch. Make .100 in Fillets on All Edges as Shown

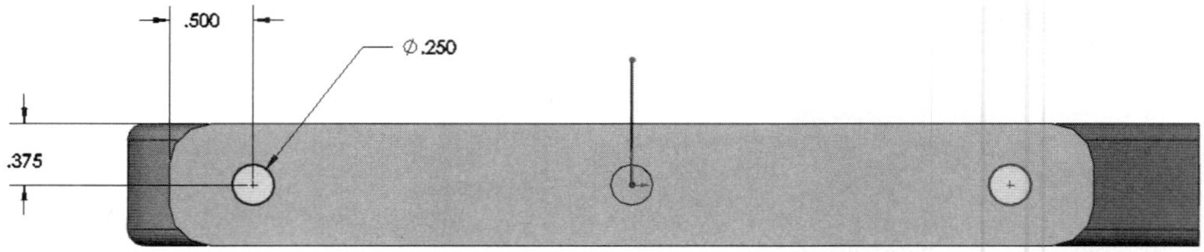

Figure 12-119
Draw a Sketch for Two .25 Diameter Holes and Do an Extrude Cut Through the Support Flange

Figure 12-120
Draw a Sketch for a .250 inch Diameter Hole × 3.000 Inches Deep in the Center of the Flange and Do the Extruded Cut. Make a .250 Inch Diameter Circle in the Right Horizontal Post Centered. Do an Extruded Cut Up to Next. Churn Gear Support Complete

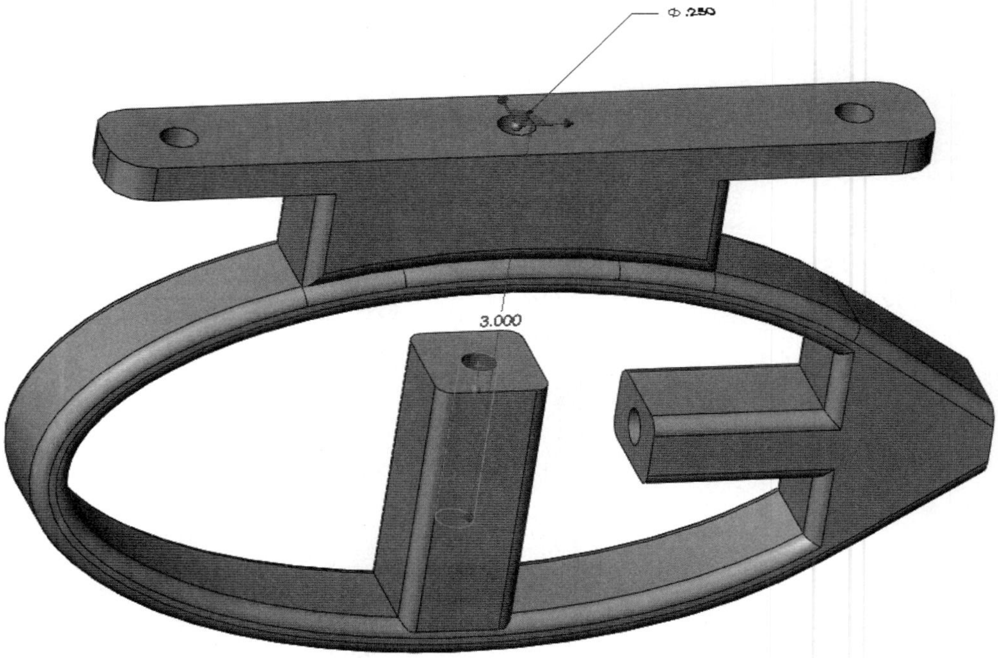

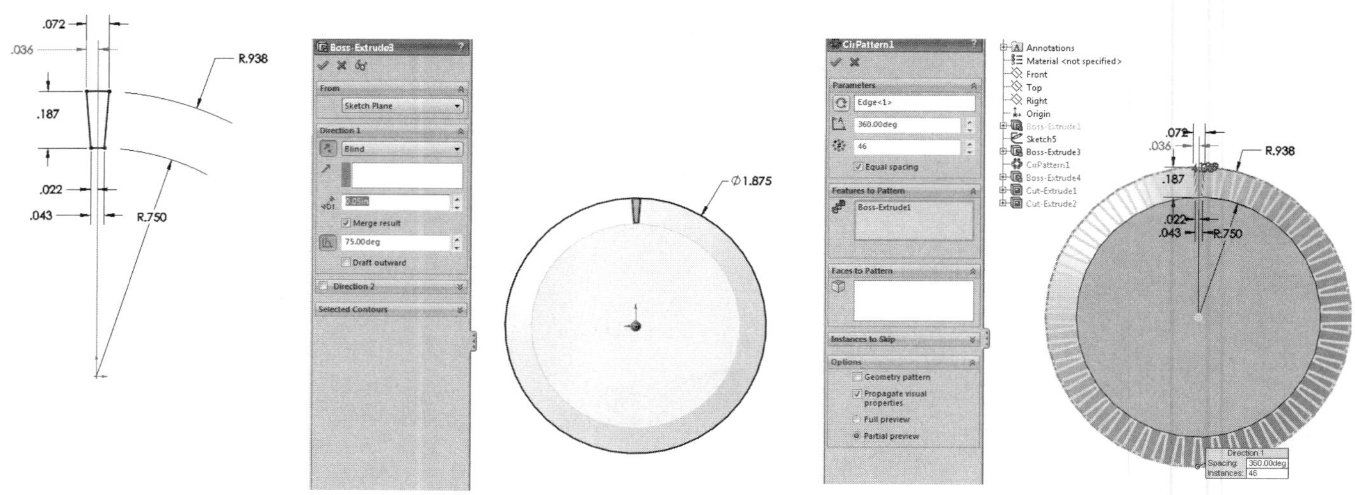

Figure 12-121
Draw a Sketch for the Gear Teeth on the Front Plane. Use a .938 Radius Outside and a .750 Radius Inside. Draw the Tooth as Shown; Trim All Lines and Arcs and Extrude the Tooth .125 Inch with a 4 degree Draft Angle. Draw a Circle on the Front Plane and Extrude It .05 Inch with a 75 Degree Draft Angle. Make a Circular Pattern of 46 Teeth Using the Outside Edge of the Extruded Circle as the Pattern Axis

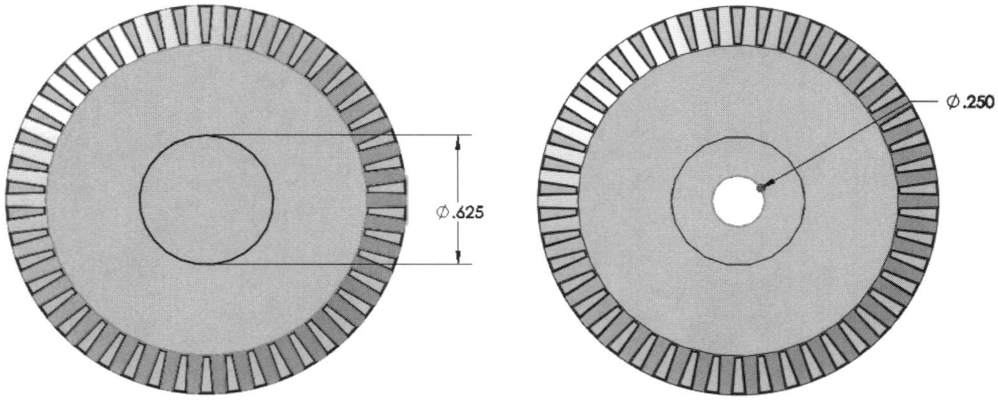

Figure 12-122
Draw a .625 Diameter on the Face of the Gear and Extrude It .25 Inch. Draw a Circle on the Top of the Face Just Extruded for a .750 Diameter Extrude Cut and Extrude It Through All

Figure 12-123
Draw a .090 Diameter Circle for an Extrude Cut on the Right Plane and Make the Cut 1.00 Inch from the Mid Plane Gear Complete

Figure 12-124
Make a Sketch for the Gear Pin on the Front Plane and Revolve It. Gear Pin Complete

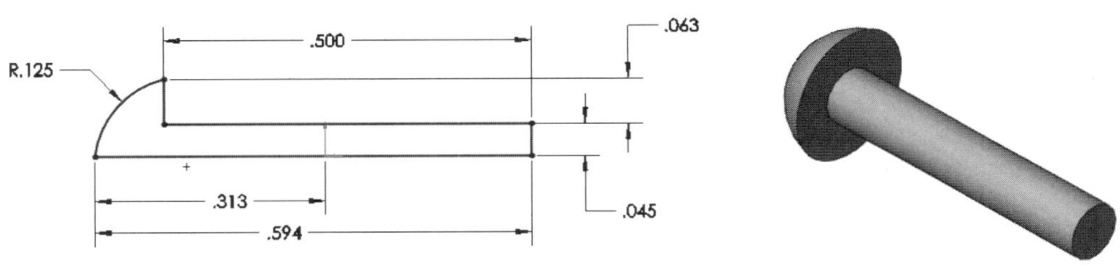

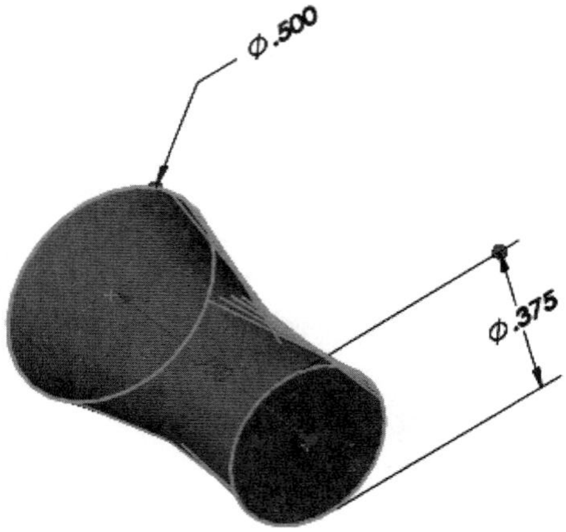

Figure 12-125
Draw Two Circles for a Lofted Boss/Base. Draw the .500 Diameter Circle on the Right Plane and the .375 Diameter Circle on a Plane .750 Inch from the Right Plane and Make the Lofted Boss/Base

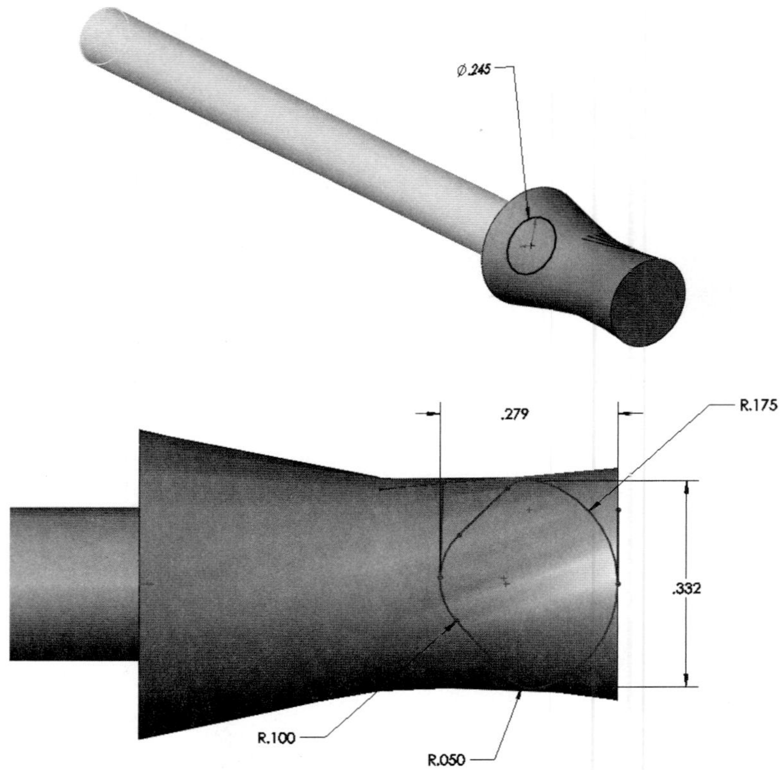

Figure 12-126
Make a .245 Diameter Circle on the Right Plane and Extrude It 2.75 Inches. Draw a Sketch on the Top Plane Using the Dimensions Shown

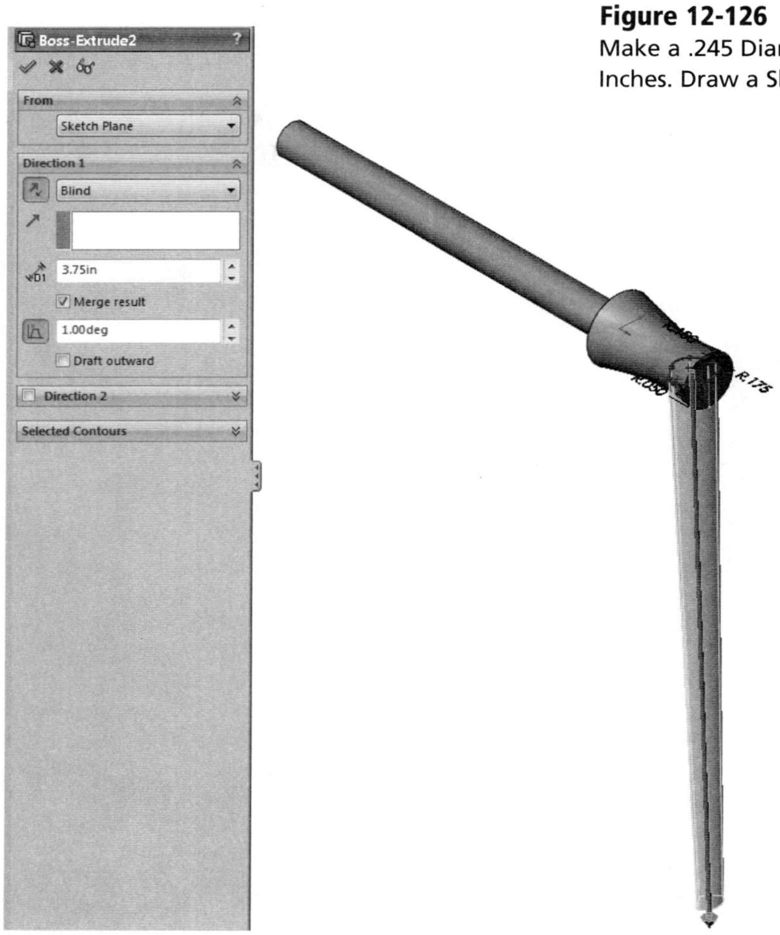

Figure 12-127
Extrude the Sketch 3.75 Inches Using a 1.00 Degree Draft Angle

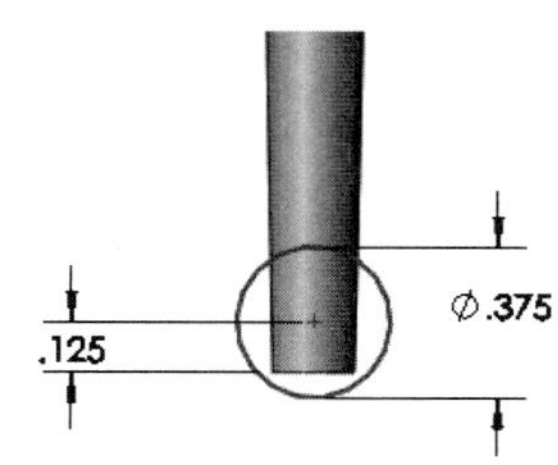

Figure 12-128
Draw a .375 Diameter Circle Using the Dimensions Shown for an Extrusion on the Plane where the .375 Diameter Circle for the Lofted Boss/Base Was Drawn and Extrude It .25 Inch

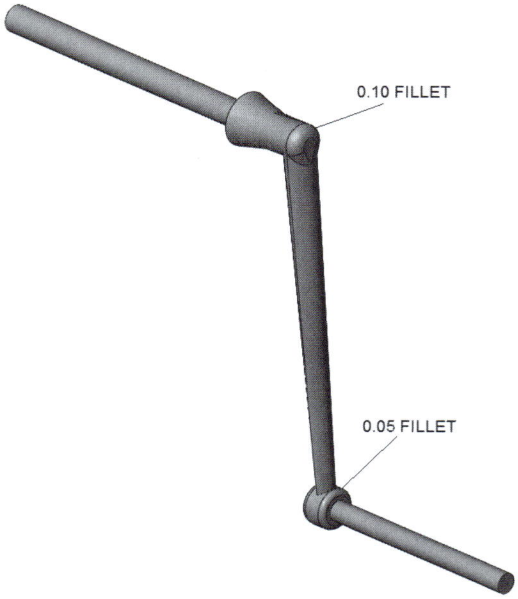

Figure 12-129

Draw 0.10 Fillets on the Lofted Boss/Base. Draw 0.05 Fillets on the Lower Extrusion as Shown. Draw a .188 Diameter Circle on the Lower Extrusion and Extrude It 2.25 Inches. Handle Complete

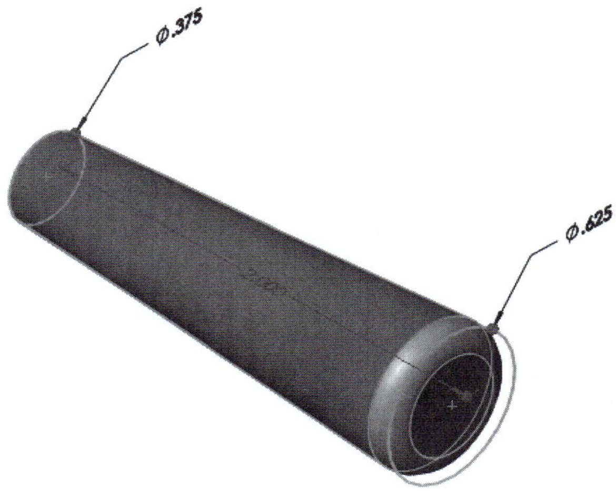

Figure 12-130

Draw a Sketch for a Boundary Boss/Base. Draw the .375 Diameter Circle on the Right Plane and the .625 Circle on a Plane 2.000 Inches. From the Right Plane and Make the Boundary Boss Base. Draw a 0.10 Inch Fillet On the .625 Diameter End of the Boundary Boss/Base

Figure 12-131

Draw a .188 Diameter Circle on the Handle Grip as Shown and Make an Extrude Cut Through All. Assign the Teak Wood Material Using Edit Material. Handle Grip Complete

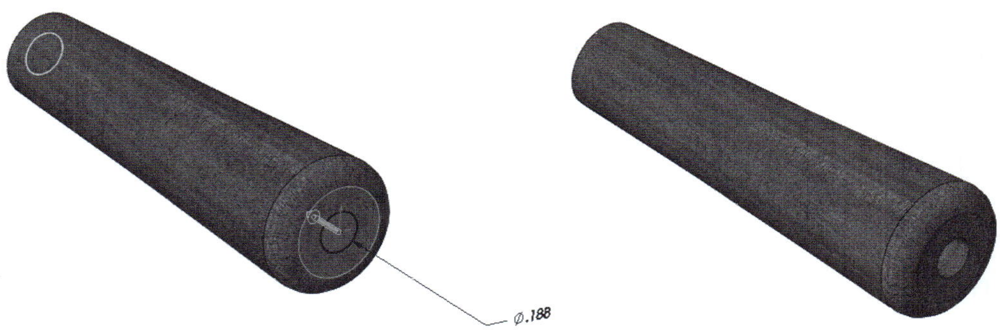

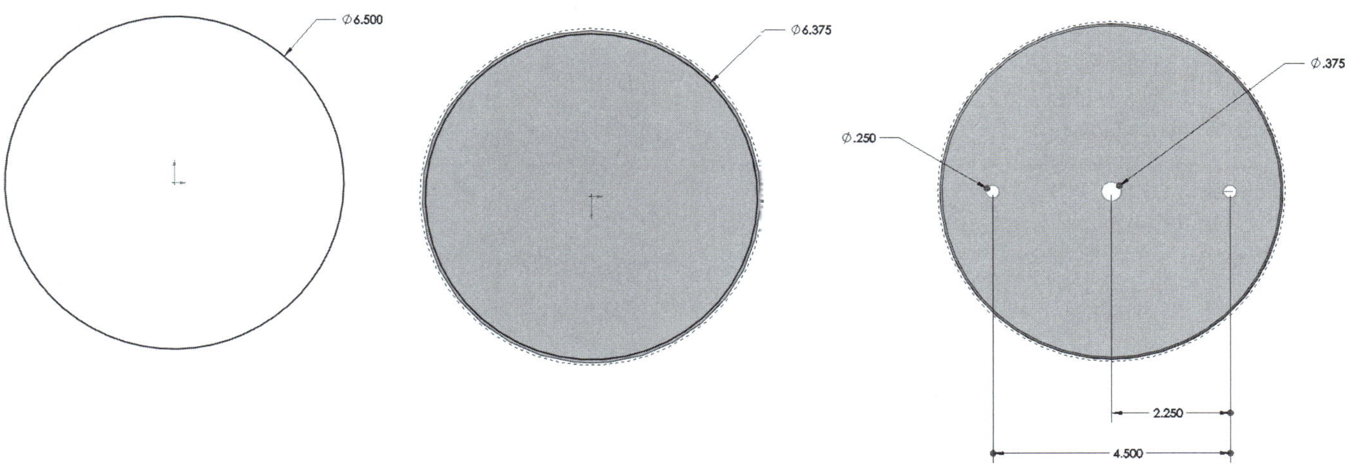

Figure 12-132

Draw a 6.500 Diameter Circle on the Top Plane and Extrude It .875 Inch. Draw a 6.375 Diameter Circle on the Bottom of the Extrusion for an Extrude Cut and Make the Cut 13/16 Inch Deep. Draw Two .250 Diameter and one .375 Diameter Circles on Top of the Lid for an Extrude Cut and Make the Cut Through All. Use Dimensions Shown

Figure 12-133
Add Cosmetic Threads as Shown. Lid Complete

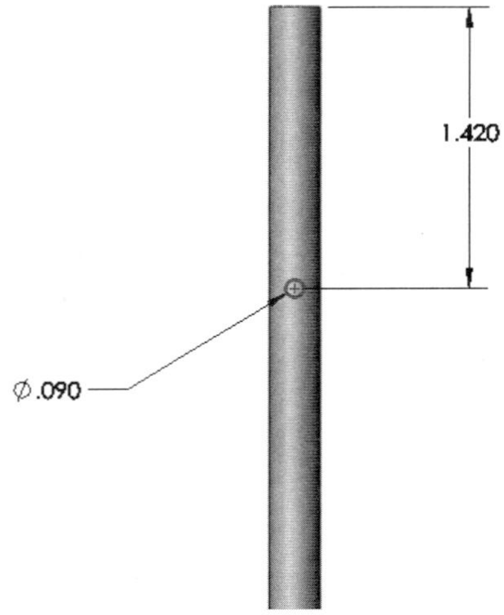

Figure 12-134
Draw a .250 Circle on the Top Plane and Extrude It 7.00 Inches. Draw a 2.090 Circle for an Extrude Cut on the Front Plane Using the Dimensions Shown and Make the Cut. Shaft Complete

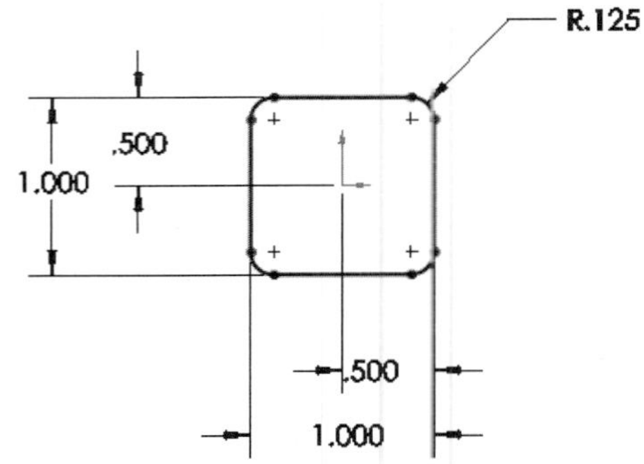

Figure 12-135
Draw a Rectangle with Fillets as Shown on the Top Plane and Extrude It 3.50 Inches

Figure 12-136
Draw a .250 Circle, a .188 × .188 Rectangle (Make 4 of Them Using Circular Pattern) on the Top of the Paddle Holder as Shown for an Extrude Cut and Make the Cut 3.25 Inches Down. Assign the Oak Wood Material Using Edit Material. Paddle Holder Complete

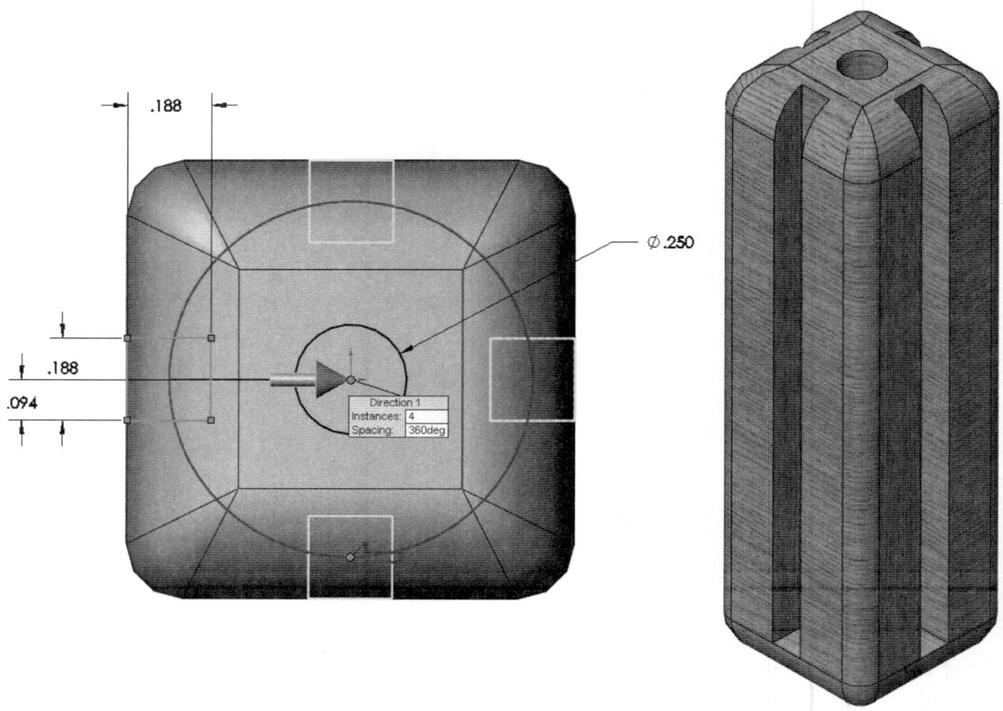

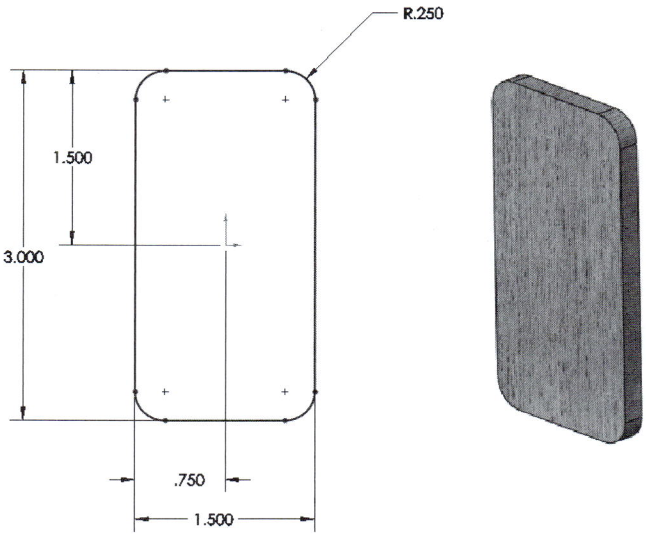

Figure 12-137
Draw a Rectangle with Fillets on the Right Plane as Shown and Extrude It .188 Inch. Assign Oak Material Using Edit Material. Paddle Complete.

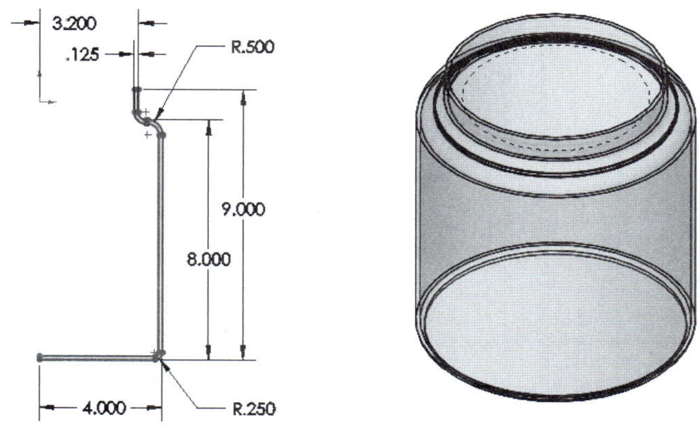

Figure 12-138
Draw a Sketch for a Revolved Boss/Base on the Front Plane and Revolve It. Assign Glass Material Using Edit Material. Glass Jar Complete

Figure 12-139
Make an Assembly Using Make Part from Assembly and Concentric, Coincident, and Parallel Mates

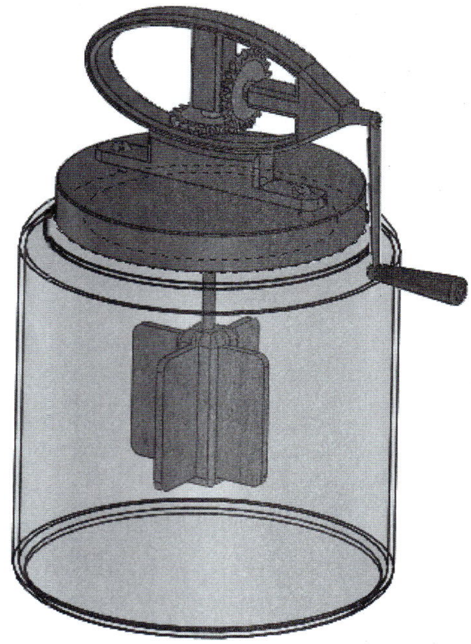

Project 12-9: Make an Exploded View Assembly Drawing of the Churn Assembly with a Parts List

Figure 12-140 shows the completed drawing of Project 12-9, the exploded view of the churn assembly with a parts list.

Use the instructions in Chapters 10 and 11 and Project 12-2 to make this drawing.

To make all the lines of the format appear black choose Inactive Entities from Color and then Click the Edit button and change the color to black.

To make all the dimensions appear black choose Dimensions, Non Imported (Driven) from Color and then Click the Edit button and change the color to black as shown in Figure 12-141.

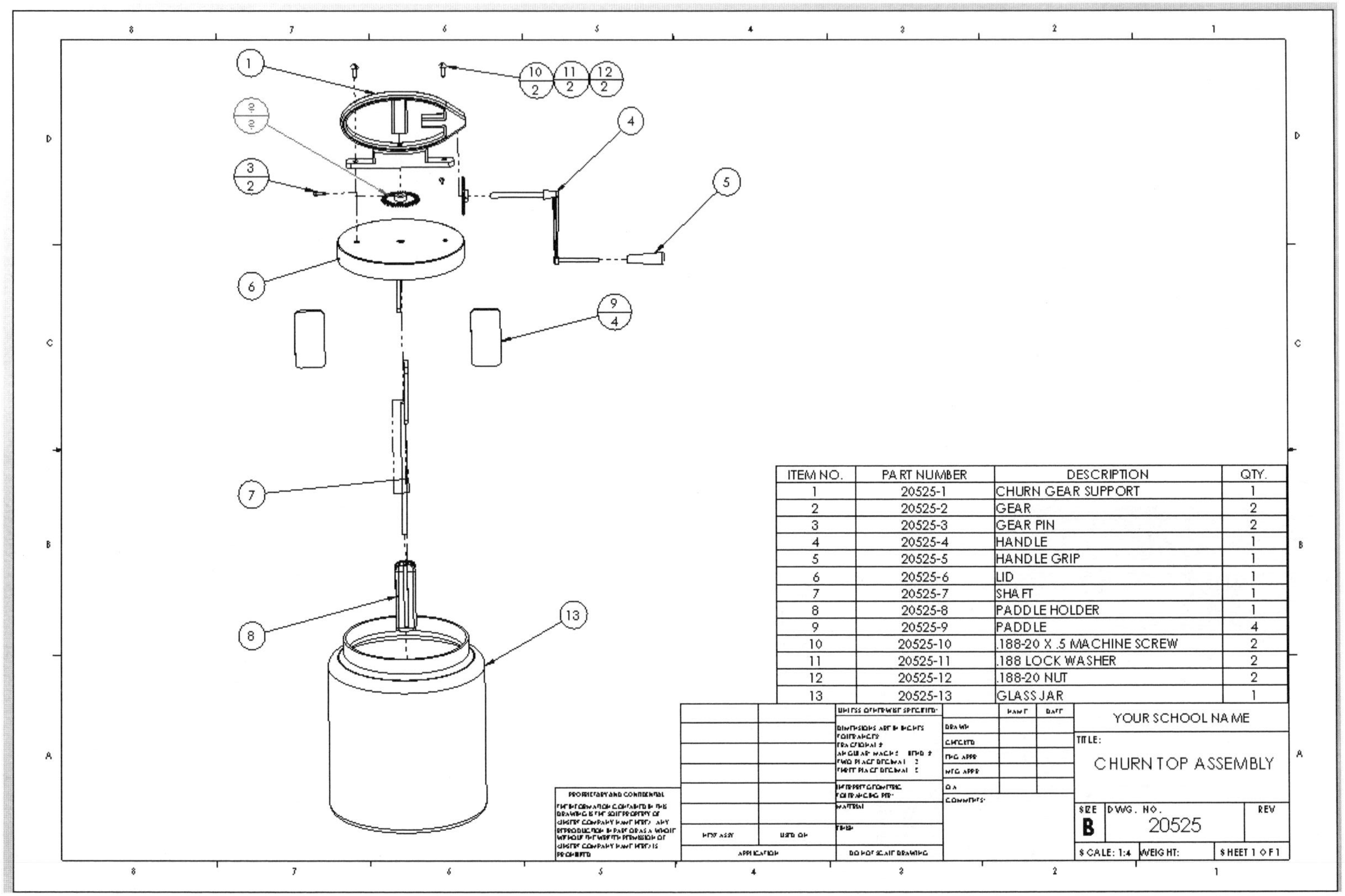

Figure 12-140
Churn Top Assembly with Parts List. Project 12-9 Complete

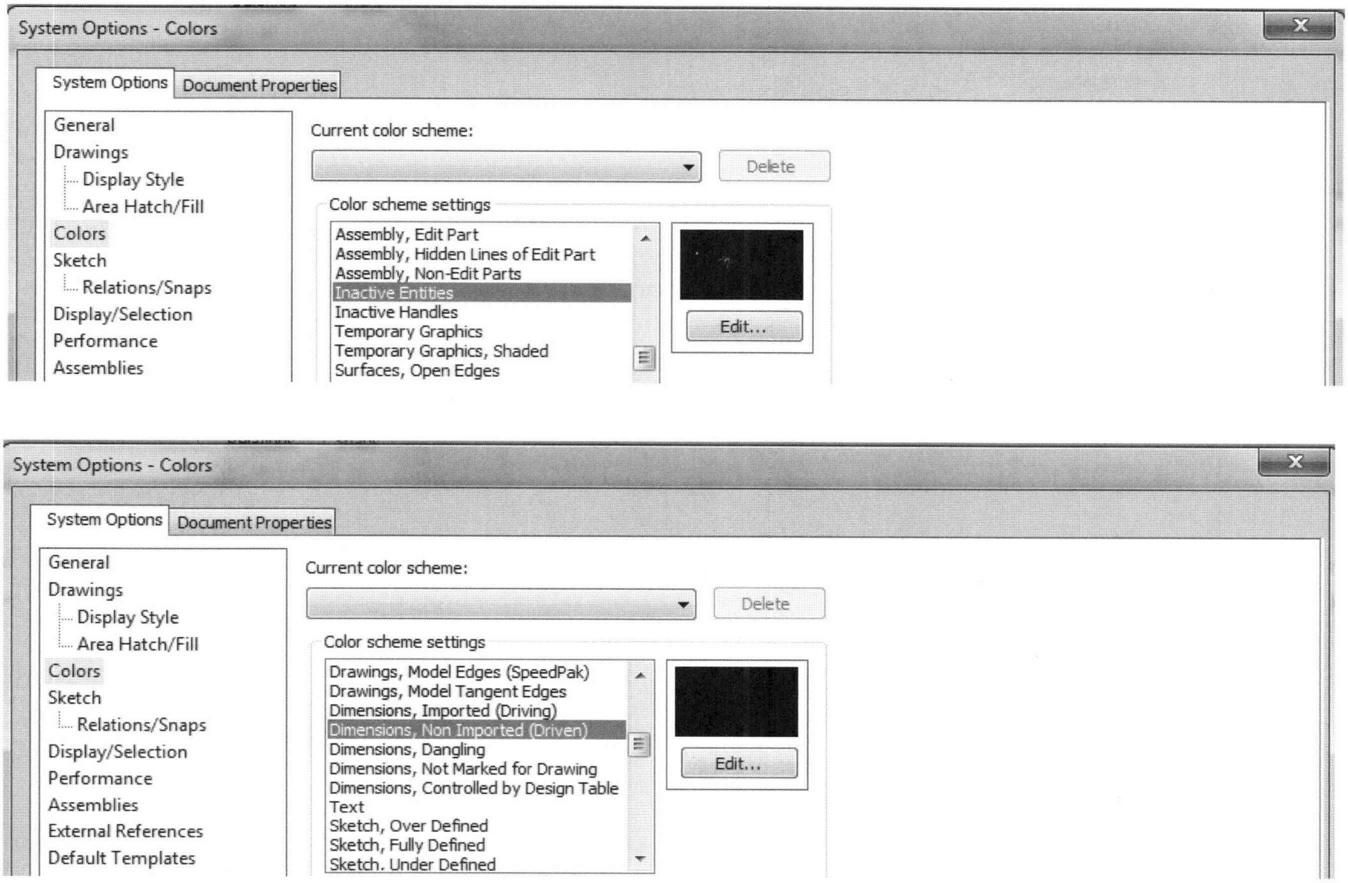

Figure 12-141

Option Settings for Making Dimensions and Lines of the Format Black

Chapter Summary

This chapter contains further information necessary to make assemblies. In addition, you have also made assembly drawings with parts lists; some of these drawings are isometric assemblies, and two are exploded views. Now you have the skills and information necessary to produce complex isometric assemblies and exploded views with parts lists.

Review Questions

Circle the most correct answer.

1. Which of the following can be used as the mirror/face plane to mirror an edge flange?
   a. **Top** c. **Left**
   b. **Front** d. All of these
2. Which of the following is NOT one of the **Mates** that can be used to assemble parts?
   a. **Concentric** c. **Coincident**
   b. **Isometric** d. **Parallel**
3. Where is the **Shaded cosmetic threads** setting located in the **Options** dialog box?
   a. **Detailing** area of the **Document Properties** tab
   b. **Annotations** area of the **System Properties** tab
   c. **Drawings** area of the **System Properties** tab
   d. **Annotations** area of the **Document Properties** tab
4. Which of these must be chosen to divide a balloon to show the item number on top and the quantity on bottom?
   a. **Circular** c. **Underline**
   b. **Circular split line** d. **None**
5. Which of these commands can be used to make the rounded edges appear on a hex nut as shown in Figure 12-40?
   a. **Chamfer** c. **Extrude Cut**
   b. **Scale** d. **Edge Flange**
6. A single line can be used to make a sketch for a miter flange.
   a. True b. False
7. Hems can be made easily on a circular base flange.
   a. True b. False
8. Balloons can be automatically added by clicking **AutoBalloon.**
   a. True b. False
9. Cosmetic threads can be added to square extrusions.
   a. True b. False
10. The **Reference Triad** can be used to rotate components.
    a. True b. False

Matching

Place the correct letter on the line to the right of the numbered item.

1. A setting that can be used to show the entire fillet on an edge ______
2. The box you must check so metal gauges can be used to show metal thickness ______
3. A command that shows the item number of an object you click ______
4. A sketch that can be used to make a swept boss/base ______
5. A place on the screen where the object you select is shown ______

a. Balloon
b. **Full Preview**
c. **Use gauge table**
d. **Entities to Pattern**
e. Path

General Questions

Answer the following questions.

1. In what situations would you use **AutoBalloon**?
2. When would you want to use the **Parallel** mate?
3. Why are some assemblies better shown as exploded views?
4. How does the SolidWorks **Bill of Materials** command work?
5. When is it best to use an orthographic view for an assembly drawing?

Index

E

F

G

H

T

U

V

W

Z